A Guide to

LATEX

A Guide to LaTeX

Document Preparation for Beginners and Advanced Users

Third edition

Helmut Kopka

Patrick W. Daly

Addison-Wesley

An imprint of **Pearson Education**

Harlow, England · London · New York · Reading, Massachusetts · San Francisco
Toronto · Don Mills, Ontario · Sydney · Tokyo · Singapore · Hong Kong · Seoul
Taipei · Cape Town · Madrid · Mexico City · Amsterdam · Munich · Paris · Milan

PEARSON EDUCATION LIMITED

Head Office:
Edinburgh Gate
Harlow CM20 2JE
Tel: +44 (0)1279 623623
Fax: +44 (0)1279 431059

London Office:
128 Long Acre
London WC2E 9AN
Tel: +44 (0)20 7447 2000
Fax: +44 (0)20 7240 5771

Web site: *www.awl.com/cseng/*

First published in Great Britain 1993
Second Edition published 1995
Third edition published 1999

© Pearson Education Ltd 1999

The rights of Helmut Kopka and Patrick W. Daly to be identified as authors of this
by them in accordance with the Copyright, Designs and Patents Act 1988.

ISBN 0-201-39825-7

British Library Cataloguing in Publication Data
A catalogue record for this book is available from the British Library.

Library of Congress Cataloging in Publication Data
Kopka, Helmut.
    A guide to LaTeX : document preparation for beginners and
advanced users / Helmut Kopka, Patrick W. Daly -- 3rd ed.
        p. cm.
    Includes bibliographical references and index.
    ISBN 0-201-39825-7
    1. LaTeX (Computer file) 2. Computerized typesetting. I. Daly,
Patrick W. II. Title.
    Z253.4.L38 K66 1998
    686.2'2544--dc21   98-47054
                                                                CIP

The programs in this book have been included for their instructional value. The pul
any warranties or representations in respect of their fitness for a particular purpose
accept any liability for any loss or damage arising from their use. A list of trademar
appears on page v.

Many of the designations used by manufacturers and sellers to distinguish their pro
trademarks. Pearson Education has made every attempt to supply trademark inforr
manufacturers and their products mentioned in this book.

10 9 8 7 6 5 4 3

Cover designed by Designers & Partners, Oxford.
Typeset by the authors with the LaTeX Documentation System
Printed and bound in Great Britain by Henry Ling Ltd, at the Dorset Press, Dorches

The publishers' policy is to use paper manufactured from sustainable forests.

Trademark notices

METAFONT™ is a trademark of Addison-Wesley Publishing Company.

TEX™, $\mathcal{A}_{\mathcal{M}}$S-TEX™, and $\mathcal{A}_{\mathcal{M}}$S-LATEX™ are trademarks of the American Mathematical Society.

Lucida™ is a trademark of Bigelow & Holmes.

Microsoft®, MS-DOS®, Windows®, Internet Explorer® are registered trademarks of Microsoft Corporation.

PostScript®, Acrobat Reader®, Acrobat logo® are registered trademarks and PDF™ a trademark of Adobe Systems Incorporated.

UNIX® is a registered trademark in the United States and other countries, licensed exclusively through X/Open Company, Limited.

VAX™ and VMS™ are trademarks of Digital Equipment Corporation.

IBM® is a registered trademark and **techexplorer** Hypermedia Browser™ a trademark of International Business Machines Corporation.

Netscape™ and Netscape Navigator™ are trademarks of Netscape Communications Corporation.

TrueType™ is a trademark and Apple® and Macintosh® are registered trademarks of Apple Computer Inc.

Preface

Four years after the release of LATEX 2_ε and almost as long since the appearance of the second edition of *A Guide to LATEX*, the time is ripe to consider a third edition. How has LATEX changed in this interval? What has to be altered in the book?

First of all, LATEX 2_ε is now well established as the official version of LATEX; for this reason the title of this book reverts to the original form used for the first edition. (The second edition was titled *A Guide to LATEX 2_ε* to emphasize that it covered the new LATEX.) Nevertheless, we continue to point out those features that are exclusive to LATEX 2_ε and which were not available under version 2.09.

LATEX is upgraded every six months. The first few updates to LATEX 2_ε saw a number of important changes, but now it has become very stable, at least for standard features at the user level. Improvements and changes occur mostly at deeper levels, or in supporting packages. For example, the number of input encoding tables and graphics drivers has steadily increased. The 256-character DC fonts have now been replaced by their EC equivalents. However, the major change since 1994 is the prevalence of the Internet and World Wide Web; new programs are now available to enable LATEX documents to be 'put online'. These do not reflect changes to LATEX itself but rather to the entire LATEX environment and its applications. This is now dealt with in Section D.4.

A new edition provides an opportunity to reorganize much material, to change emphasis, and to correct mistakes. In this light, we have decided that the importation of graphics files is no longer an extension for advanced users, but a basic part of LATEX application. The usage has become standardized; many problems have been identified and solved. Thus a very detailed explanation of the `graphics` and `color` packages is now given in Chapter 6 and the emphasis on the LATEX `picture` environment has been reduced.

The use of PostScript fonts has also become more relevant, to such an extent that Computer Modern fonts are no longer the hallmark of a LATEX document. Appendix F (*TEX Fonts*) has been revised to reflect this.

Several example packages in Appendix C (*LATEX programming*) have been removed, in particular those dealing with language adaptation and author–year citations. These examples contained far too much TEX code to be appropriate as demonstrations, and their usefulness as packages is questionable considering the widespread availability of the `babel` and `natbib` packages. As compensation, a new package is offered for redefining the sectioning commands.

It has always been our intention only to describe the standard LaTeX features, and not to elaborate on many of the excellent contributed packages available. This is not because we consider them to be inferior; on the contrary, a large number of them are indispensable and should be part of any standard installation. It is simply that we must limit the material in this book somehow, and these packages are dealt with elsewhere, for example in the *LaTeX Companion* (Goossens *et al.*, 1994) and *LaTeX Graphics Companion* (Goossens *et al.*, 1997). We have decided to make two exceptions. Many of the 'tools' packages mentioned in Section D.3.3 are now described in the main text where their application would be most appropriate. Packages like `multicol`, `array`, `longtable` should be used in everyday situations, and are by no means exotic.

The `amsmath` and `amsfonts` packages are the other exception. An overview to these important mathematical tools is now provided in Appendix E and tables of the extra $\mathcal{A}_{\mathcal{M}}\mathcal{S}$ symbol fonts are given on pages 552–554. For mathematical typesetting, these additional commands must also be considered indispensable.

We feel the changes will make this book even more relevant and applicable to the effective production of high-class documents with LaTeX.

Helmut Kopka and Patrick W. Daly
September 1998

Preface to the second edition

The great success of Helmut Kopka's book *LaTeX, eine Einführung* in Germany has led to many requests for an English translation, especially from non-German-speaking visitors to our Institute. The standard manual on LaTeX by Leslie Lamport, *LaTeX, A Documentation Preparation System*, provides a reasonable introduction to this text formatting program, but leaves many open questions about advanced topics and user-defined extensions. Kopka's book includes many tips and practical examples for solving common problems that are not to be found in the Lamport manual.

A straightforward translation of the German book would not be particularly useful, since it contains considerable information on the application to German texts. For this reason, Patrick W. Daly has undertaken to produce an English version of Kopka's work, adding the sections on international and multilingual LaTeX, and including some applications from his own experience.

At the time of the first edition, late 1992, LaTeX was undergoing major revisions. Some test versions of the new extensions were described in that book, but were in fact out of date before it appeared. Now the revised and consolidated new LaTeX has been released, mid-1994, with many improvements and additions.

This second edition of the *Guide* attempts to present both the classical version (designated 2.09) and the new official standard, named LaTeX 2_ε. Of course, the emphasis is on the newer version.

The entire book has been revised in the light of LaTeX 2_ε, although the basic material in Chapters 2–7 is much the same. The Advanced Features of Chapter 8 now include the New Font Selection Scheme, a list of standard packages, and the `slides` class (formerly SliTeX, which had an appendix to itself previously). The error messages of Chapter 9 have been extended. Appendix C is totally new, describing the advanced programming features of LaTeX, with useful examples. Finally, Appendix D, on Extensions, has been expanded; it also contains a description of the CTAN network, the electronic source for TeX and its accessories.

As for the first edition, it is hoped that this book will be particularly useful to those readers who wish to go beyond the basic features of the LaTeX system but who do not want to venture into the murky depths of the TeX world on which it is all based. The Command Summary and Summary Tables and Figures in Appendix G should become standard reference tools for any LaTeX user of any degree of experience.

Helmut Kopka and Patrick W. Daly
March 1995

Contents

Appendices

List of Tables

List of Figures

Introduction

1.1 Text formatting

The electronic processing of text, one of the most common functions of computers today, consists essentially of four stages:

1. the text is *entered* into the computer for storage and later corrections, extensions, deletions;

2. the input text is *formatted* into lines of equal length and pages of a certain size;

3. the output text is *displayed* on the computer monitor screen;

4. the final output is sent to a *printer*.

Many word processing systems handle all four functions together in one package so that the user is not aware of this division of labor. Furthermore, stages 3 and 4 are actually one and the same: sending the results of the formatting to an output device, which is in one case a monitor, in the other, a printer.

A *text formatting program*, such as TEX, is only concerned with the second stage. Any text *editor* program may be used to enter and modify the input text. If you have an *editor* with which you are familiar and comfortable, then you may use it. A *word processor* program is not always appropriate for this purpose because it normally adds many unseen *control characters*. It may very well be that 'what you see is what you get' with a *word processor*, but what you see is not necessarily what you've got.

The text that is produced with the *editor* as input for the *formatting program* must also contain special commands or instructions, but these are written with normal characters that are not hidden. To some extent, the instruction set for a *formatter* is much like a *markup language*, indicating where paragraphs, sections, chapters, etc. begin without explicitly formatting them in the text itself. How these instructions are to be interpreted during formatting depends on the layout that is chosen. The same text may be formatted quite differently if another layout is selected.

A formatter can be much more. In fact TₑX is also a powerful *programming language* that allows knowledgeable users to write code for all sorts of additional features. LATₑX itself is just such a set of complex macros. Furthermore, anyone can program extra *extensions* to LATₑX, or make use of public ones that have been contributed by other programmers. The capabilities of TₑX and LATₑX are by no means limited to those of the original package.

The last stage of text processing with a formatter is to send the results to an output device, either a printer or the computer monitor or a file. The programs that accomplish this are called *drivers*; they translate the encoded output of the formatter program into the specific instructions for the particular printers that the user has at hand. This means that for each type of printer, one has to have the corresponding driver program.

1.2   TₑX and its offspring

The most powerful formatting program for producing book quality text of scientific and technical works is that of Donald E. Knuth (Knuth, 1984, 1986a, 1986b, 1986c, 1986d). The program is called TₑX, which is a rendering in capitals of the Greek letters $\tau\epsilon\chi$. For this reason the last letter is pronounced not as an *x*, but as the *ch* in Scottish *loch* or German *ach*, or as the Spanish *j* or Russian *kh*. The name is meant to emphasize that the printing of mathematical texts is an integral part of the program and not a cumbersome add-on. In addition to TₑX, the same author has developed a further program called METAFONT for the production of character fonts. The standard TₑX program package contains 75 fonts in various design sizes, each of which is also available in up to eight magnification steps. All these fonts were produced with the program METAFONT. With additional applications, further character fonts have been created, such as for Cyrillic and even Japanese, with which texts in these alphabets can be printed in book quality.

1.2.1   The TₑX program

The basic TₑX program only understands a set of very primitive commands that are adequate for the simplest of typesetting operations and programming functions. However, it does allow more complex, higher-level commands to be defined in terms of the primitive ones. In this way, a more user-friendly environment can be constructed out of the low-level building blocks.

During a processing run, the program first reads in a so-called *format file* which contains the definitions of the higher-level commands in terms of the primitive ones, and which also contains the hyphenation patterns for word division. Only then does it read in the *source file* containing the actual text to be processed, including formatting commands that are predefined in the format file.

Creating new formats is something that should be left to very knowledgeable programmers. The definitions are written to a source file which is then processed with a special version of the TEX program called initex. It stores the new format file in a compact manner so that it can be read in quickly by the regular TEX program.

Although the normal user will almost never write such a format, he or she may be presented with a new format source file that will need to be installed with initex. For example, this is just what must be done to upgrade to the LATEX format every six months, as described in Section D.5.2.

1.2.2 Plain TEX

Knuth has provided a basic format named *Plain TEX* to interact with TEX at its simplest level. This is such a fundamental part of word processing with TEX that one tends to forget the distinction between the actual processing program TEX and this particular format. Most people who claim to 'work only with TEX' really mean that they only work with Plain TEX.

Plain TEX is also the basis of every other format, something that only reinforces the impression that TEX and Plain TEX are one and the same.

1.2.3 LATEX

The emphasis of Plain TEX is still very much at the typesetter's level, rather than the author's. Furthermore, the exploitation of all its potential demands considerable experience with programming techniques. Its application thus remains the exclusive domain of typographic and programming professionals.

For this reason, the American computer scientist Leslie Lamport has developed the LATEX format (Lamport, 1985), which provides a set of higher-level commands for the production of complex documents. With it, even the user with no knowledge of typesetting or programming is in a position to take extensive advantage of the possibilities offered by TEX, and to be able to produce a variety of text outputs in book quality within a few days, if not hours. This is especially true for the production of complex tables and mathematical formulas.

LATEX is very much more a *markup language* than the original Plain TEX, on which it is based. It contains provisions for automatic running heads, sectioning, tables of contents, cross-referencing, equation numbering, citations, floating tables and figures, without the author having to know just how these are to be formatted. The layout information is stored in additional *class files* which are referred to but not included in the input text. The layouts may be accepted as they are.

Since its introduction in the mid-1980s, LaTeX has been periodically updated and revised, like all software products. For many years the version number was fixed at 2.09 and the revisions were only identified by their dates. The last major update occurred on December 1, 1991, with some minor corrections up to March 25, 1992, at which point LaTeX 2.09 became frozen.

1.2.4   LaTeX 2ε

The enormous popularity of LaTeX and its expansion into fields for which it was not originally intended, together with improvements in computer technology, especially dealing with cheap but powerful laser printers, have created a diversity of formats bearing the LaTeX label. In an effort to re-establish a genuine, improved standard, the LaTeX3 Project was set up in 1989 by Leslie Lamport, Frank Mittelbach, Chris Rowley, and Rainer Schöpf. Their goal is to construct an optimized and efficient set of basic commands complemented by various *packages* to add specific functionality as needed.

As the name of the project implies, its aim is to achieve a version 3 for LaTeX. However, since that is the long-term goal, a first step towards it has been the release of LaTeX 2ε in mid-1994 together with the publication of the second edition of Lamport's basic manual (Lamport, 1994) and of an additional book (Goossens *et al.*, 1994) describing many of the extension packages available and LaTeX programming in the new system. A more recent third book (Goossens *et al.*, 1997) deals with the inclusion of graphics and color, two weak points in the early days. LaTeX 2ε is now the standard version, pending the appearance of the promised LaTeX3 at some future date.

1.2.5   TeX on various computers

TeX and LaTeX were originally designed to run on central mainframe computers, in the days before there were such things as graphics monitors, mice, and joysticks. The underlying TeX formatting program is meant to be called from a command line, not by mouse clicks from some graphics interface. Today this may be considered old-fashioned, but it has led to TeX being one of the most stable and portable programs in the computer world, since it is not rigidly bound to any particular computer or operating system.

Modern *workstations* and powerful *personal computers*, or *PCs*, are the standard tools for computing now, far exceeding the performance of those early giants. Many TEX installations are available that succeed in integrating file editing, processing, previewing, and printing in a near WYSIWYG ('what-you-see-is-what-you-get') manner. Some of these are listed in Section D.5.1 on page 397.

One extremely useful output driver available with workstations and PCs is a screen previewer, which displays the output on the graphics monitor. Such a previewer was once a luxury but today is an indispensable tool for preparing LATEX documents. Especially for complicated tables or mathematical formulas, the ability to check the results quickly after a trial change is essential. Fast, convenient previewers are vital in order to compete with WYSIWYG text processors.

On a mainframe computer, there is an operator responsible for the installation and maintenance of the TEX program and its accessories. However, the user of a workstation or PC must take on the role of computer manager as well, meaning that he or she must learn something more about the setting up of both TEX and LATEX than was previously necessary. Or, as is more often the case, somebody must be found who will do this for them. This person is normally referred to as the local TEX *guru*.

On the other hand, there are many excellent ready-to-run TEX installations for most computer types (Section D.5.1).

Everything in this book about writing LATEX text files applies equally well to all computer types. It is this portability that makes LATEX such a useful tool for scientific publishing.

1.3 What's different in LATEX 2$_\varepsilon$?

This section is intended primarily for readers who are familiar with the older LATEX 2.09. It gives a brief overview of the differences between it and the current standard, LATEX 2$_\varepsilon$.

1.3.1 Classes and packages

The most fundamental difference between LATEX 2.09 and LATEX 2$_\varepsilon$ is the very first command which declares the overall layout: this difference actually makes compatibility between the two versions possible.

In LATEX 2.09, one declared the *main style* together with any options that go with it, as for example:

```
\documentstyle[ifthen,12pt,titlepage]{article}
```

Here the main style is *article*, stored in a file named `article.sty`, with 12 pt as the basic font size, and with the title on a separate page. The additional option `ifthen` is not understood by the main style; instead a file named `ifthen.sty` is read in.

This method of treating undefined options was the principal way in which LATEX extensions, or supplemental coding, were read into a document. The hundreds of such additional files available were therefore referred to as *style options* or *sub-styles*.

Under LATEX 2_ε, a clear distinction is made between such supplements and true internal options to the main layout. The main style is renamed the *class* and any extension files are called *packages*. The above initial declaration now becomes:

```
\documentclass[12pt,titlepage]{article}
\usepackage{ifthen}
```

The layout information is contained in the file `article.cls`, which is designed to handle the options `12pt` and `titlepage`. The file `ifthen.sty` is read in with an extra command; however, in contrast to the situation under LATEX 2.09, it may also have its own internal options. Not only that, but those options listed in the `\documentclass` command are considered global and are therefore valid for all further packages (see Section 3.1.2).

The initial declaration `\documentstyle` may still be used in LATEX 2_ε, in which case it switches to a compatibility mode to reproduce the behavior of LATEX 2.09.

Many features have been added to assist LATEX programmers, those intrepid individuals who write class and package files. The handling of options has been improved, and, as mentioned above, it is also possible to add options to packages. Some safety devices have been included to ensure that version numbers match. Better tests exist for reading in other files so that alternative action may be taken if they are not found. These programming elements are described in Appendix C.

1.3.2   Font management

Under LATEX 2.09, TEX's Computer Modern fonts were firmly hardwired into the format. This was acceptable in the days when these were the only fonts that one was likely to use, but today when so many other fonts are available, especially with PostScript printers, a more flexible system is required. The New Font Selection Scheme (NFSS) meets this demand and has been fully incorporated in LATEX 2_ε. Changing the basic fonts from Computer Modern to something else is only a matter of a few simple redefinitions (see Section 8.5).

NFSS also changes the way in which fonts are addressed within the document. LATEX 2.09 inherited the Plain TEX font commands like \bf (bold) and \it (italic) which rigidly select a particular font. Only the size of the font remains unchanged by these commands. Under NFSS, fonts are described by certain *attributes* that may be selected independently of one another. Thus it is possible to select first bold and then italic to obtain a bold, italic font, something that was not possible in LATEX 2.09.

LaTeX 2ε encourages the use of *font commands* as opposed to *declarations*. For example, to emphasize some text, the command \emph{word} is preferred to the declaration {\em word}. Such commands are more logical to the beginner, although the experienced LaTeX 2.09 user who is thoroughly used to the latter form might object.

The use of text fonts in math mode is achieved by special *math alphabet commands* rather than with the older font declarations. That is, the declarations \rm, \bf, \cal and so on are not allowed in math mode, and have been replaced by the commands \mathrm, \mathbf, \mathcal which operate on arguments.

1.3.3 Float placement

One annoying problem in LaTeX 2.09 is getting floats (figures and tables) to end up where one would optimally prefer to see them. There are very complicated rules for placing floats and the human does not always see through them. LaTeX 2ε provides two mechanisms to intervene in this process, one to discourage and the other to encourage the placement.

The current page may be declared to be free of floats by issuing the command \suppressfloats. Optionally, one may specify an argument t or b to suppress only the floats at the top or bottom of the current page.

On the other hand, an added float location specifier ! suspends all restrictions for that one float on the amount of text and number of floats that may appear on one page. This overcomes a very common grievance which is normally assuaged only by redefining \textfraction or some other float placement parameter, and that with much trial and error. The float location specifier ! is given just like the others: for example,

```
\begin{figure}[!]
```

It is possible to define horizontal rules to be drawn above or below floats, in order to set them off better from the text. All these features are described in Section 6.4.

1.3.4 Extended syntaxes

Several commands in LaTeX 2ε have an extended syntax compared to that in version 2.09. The older, limited syntax still functions as before, of course.

- The commands \newcommand, \renewcommand, \newenvironment, and \renewenvironment may be used to define new commands and environments that contain one optional argument in addition to several mandatory ones (Sections 7.3 and 7.4).

- The box commands \makebox, \framebox, and \savebox can refer to their natural dimensions while specifying their actual size. That is, one can make them to be, say, twice their natural width. See Section 4.7.

- The \parbox command and minipage environment can specify a vertical size in addition to the horizontal width. An internal placement parameter determines whether the text in the box is to be shoved to the top, centered, pushed to the bottom, or even stretched to fill the whole box (Section 4.7.5).

- Under LATEX 2.09 it was possible to measure the width of some text with the command \settowidth; LATEX 2ε complements this with \settoheight and \settodepth (Section 7.2).

1.3.5 Compatibility with LATEX 2.09

Every effort has been made to make LATEX 2ε as compatible as possible with version 2.09. This means that documents written for the older standard should produce the same results under the newer one. It also means that most style options (or sub-styles) should function just as before as a package, without any revisions.

That at least is the theory. In practice, incompatibilities must exist. How likely is it that one will encounter them?

- If only upper-level commands are used, those not containing any @ signs in their names, then the compatibility should be 100%.

- If internal commands have been manipulated, there could be problems, especially if they involve boxes or output routines.

- If internal font control commands have been used, especially those from the LATEX 2.09 lfonts.tex file, problems are likely to appear. However, it should be pointed out that even here there are fewer incompatibilities than were present in the first and second releases of NFSS.

From our own experience so far, we have found very few packages that do not work under LATEX 2ε, even when boxes and output routines are involved. Our worst case was when we explicitly loaded a support file from LATEX 2.09 that is not supposed to exist under LATEX 2ε. It was loaded anyway, causing serious error messages to be issued.

1.3.6 'Which version do I have?'

If you do not know whether or not LATEX 2ε is installed on your system, there are a number of ways to find out. At the start of any LATEX job, the name and date of the format is printed, both on the monitor screen and in the transcript file, for example

```
LaTeX Version 2.09 <25 March 1992>    for LATEX 2.09, or
LaTeX2e <1998/06/01>    for LATEX 2ε.
```

Your dates may differ from these.

Alternatively, try processing a LATEX file with \documentclass on the first line. If you receive the error message

```
! Undefined command sequence
        \documentclass
```

then you do not have LATEX 2ε.

1.3.7 Upgrades to LATEX

Since LATEX 2ε was first released on June 1, 1994, upgrades have been issued every June and December, consisting not only of improved internal coding, but also of additional features. This means that it is not sufficient to state that a document has been written for LATEX 2ε, but rather one must specify the necessary version it requires as well as the date when it was written.

In this book, any commands that were not present in the first official release are so indicated, along with their release date, both in the main text and in the command summary of Appendix G.

It is possible to declare in a document file the earliest possible version date needed to be able to process all the features it employs. This is done by placing an identification command near the beginning (Section C.2.1). For the first release, this would be

```
\NeedsTeXFormat{LaTeX2e}[1994/06/01]
```

The date must be numerical, in exactly the format shown above, as year/month/day, including slashes and zeros. If a file with this command is processed by an earlier version of LATEX, a warning message is printed.

1.4 How to use this book

This *Guide* is meant to be a mixture of textbook and reference manual. It explains all the basic elements of LATEX, especially those of the current standard LATEX 2ε, restricting itself as much as possible to what is in the standard distribution. Compared to Lamport (1985, 1994), it goes into more detail at times, offers more examples and exercises, and describes many 'tricks' based on the authors' experiences. Unlike *The LATEX Companion* (Goossens *et al.*, 1994), it does not describe the many additional packages available outside the standard distribution, since these may not always be easily obtainable for all readers. Exceptions are the tools collection of packages, graphics inclusion, use of color, and $\mathcal{A}_{\mathcal{M}}\mathcal{S}$-LATEX.

This book is designed for LATEX users who have little or no experience with computers. It contains no information about computers or system-dependent activities such as *logging on, calling the editor,* or *using the editor.*

There is considerable repetition in the text, especially within the first half, so that the reader, having learned a brief definition of some expression, is not expected to have fully mastered it when it reappears a few pages later. Nevertheless, the user should become familiar right from the beginning with the basic elements described in Sections 2.1–2.4.

In the description of command syntax, typewriter type is used to indicate those parts that must be entered exactly as given, while *italic* is reserved for those parts that are variable or for the text itself. For example, the command to produce tables is presented as follows:

\begin{tabular}{*col_form*}   *lines*   \end{tabular}

The parts in typewriter type are obligatory, while *col_form* stands for the definition of the column format that must be inserted here. The allowed values and their combinations are given in the detailed descriptions of the commands. In the above example, *lines* stands for the line entries in the table and are thus part of the text itself.

!

Sections of text that are printed in a smaller typeface together with the boxed exclamation mark at the left are meant as an extension to the basic description. They may be skipped over on a first reading. This information presents deeper insight into the workings of LATEX than is necessary for everyday usage, but which is invaluable for creating more refined control over the output.

1.4.1   Distinguishing LATEX 2ε from version 2.09

Since LATEX 2ε is now the current standard, all the descriptions of syntax and functions given here will conform to it. However, we are aware that some users are still working with version 2.09, either because they lack the possibility of upgrading, or because they do not have the inclination to change. This book must also serve these clients.

For this reason, those commands that are specific to, or have been extended under, LATEX 2ε are marked with the symbol 2ε. Similarly, any sections of text that apply strictly to LATEX 2.09 are marked with 2.09.

Since LATEX 2ε is downward compatible with LATEX 2.09, it should not be necessary to mark any command as belonging only to the older version. Nevertheless, some commands have been retained strictly for compatibility reasons (such as \bf and \rm) and are discouraged in LATEX 2ε, while in LATEX 2.09 they are indispensable.

1.5   Basics of a LaTeX file

1.5.1   Text and commands

A LaTeX file contains the *input text* that is to be processed to produce the printed output. Splitting the text up into lines of equal width, formatting it into *paragraphs*, and breaking it into *pages* with page numbers and running heads are all functions of the processing program and not of the input text itself.

For example, words in the input text are strings of letters terminated by some non-letter, such as *punctuation*, *blanks*, or *carriage returns*; whereas punctuation marks will be transferred to the output, blanks and carriage returns merely indicate a gap between words. Multiple blanks in the input, or blanks at the beginning of a line, have no effect on the interword spacing in the output.

Similarly, a new paragraph is indicated in the input text by an empty line; multiple empty lines have the same effect as a single one. In the output, the paragraph may be formatted either by indentation of the first line, or by extra interline spacing, but this is not affected in any way by the number of blank lines or extra spaces in the input.

The input file contains more than just text, however; it is also interspersed with commands that control the formatting. If is therefore necessary for the author to be able to recognize what is text and what is a command. Commands consist either of certain single characters that cannot be used as text characters, or of words preceded immediately by a special character, the backslash \.

1.5.2   Structure of a LaTeX document

In every LaTeX file there must be a *preamble* and a *body*.

The preamble is a collection of commands that specify the global processing parameters for the following text, such as the paper format, the height and width of the text, the form of the output page with its pagination and automatic page heads and footlines. As a minimum, the preamble must contain the command \documentclass to specify the document's overall processing type. This is normally the first command in the preamble.

If there are no other commands in the preamble, LaTeX selects standard values for the line width, margins, paragraph spacing, page height and width, and much more. In the original version, these specifications are tailored to the American norms. For European applications, built-in options exist to alter the text height and width to the A4 standard. Furthermore, there are language-specific packages to translate certain headings such as 'Chapter' and 'Abstract'.

The preamble ends with \begin{document}. Everything that follows this command is interpreted as *body*. It consists of the actual text mixed with additional commands. In contrast to those in the preamble, these commands have only a local effect, meaning they apply only to a part of the text, such as *indentation*, *equations*, temporary change of *font*, and so on. The body ends with the command \end{document}. This is normally the end of the file as well.

The general syntax of a LaTeX file is as follows:

⎡2ε⎤ \documentclass[*options*]{*class*}
Further global commands and specifications
\begin{document}
Text mixed with additional commands of local effect
\end{document}

The possible *options* and *classes* that may appear in the \documentclass command are presented in Section 3.1.1.

For LaTeX 2.09, or for compatibility with it under LaTeX 2ε, the initial command \documentclass must be replaced by \documentstyle. Thus the general syntax would appear as follows in version 2.09:

⎡2.09⎤ \documentstyle[*options*]{*class*}

        . . . .

\begin{document}

        . . . .

\end{document}

At a computing center, information about the available LaTeX classes and packages is described in a so-called *Local Guide*. This should also include how the LaTeX program is to be run, as well as what output devices are installed, such as printer, microfilm, plotters, etc., and how they are to be implemented. The device drivers may also recognize options to produce the output in various discrete magnifications.

1.5.3   LaTeX processing modes

⎡!⎤ (This section is marked with a boxed exclamation mark and is printed in a smaller typeface. For the meaning of this, see the end of Section 1.4 on page 10.)

During processing, LaTeX is always in one of three modes:

1. paragraph mode,
2. math mode,
3. LR, or left-to-right mode.

The *paragraph mode* is the normal processing mode in which LaTeX treats the input text as a sequence of words and sentences to be broken up (automatically) into lines, paragraphs, and pages.

LATEX switches to *math mode* when it encounters certain commands that tell it that the following text represents a *formula*. In math mode, blanks are ignored. The two texts is and i s are both interpreted as the product of i and s and appear as *is*. LATEX switches back to paragraph mode when it encounters the corresponding command that the formula is completed.

LR mode is similar to paragraph mode: LATEX treats the text input from left to right as a string of words *that cannot be split up by a line break*. LATEX will be in this mode when, for example, normal text is to appear embedded within a formula, or with a special command \mbox{*short text*} that forces *short text* to be printed on one line.

Understanding and recognizing the processing modes is important since some commands are allowed only in certain modes or their effects are different depending on the mode.

In the following we will refer to paragraph and LR modes as *text mode* to the extent that their properties are identical. This is to distinguish them from *math mode*, which is normally quite different.

1.5.4　Producing a LATEX document

The creation of a LATEX document from text input to printer output is a three-stage process. First a text file is created (or corrected) using the computer editor. This text file consists of the actual text mixed with LATEX commands.

The full name of the text file consists of a *root* name plus the *extension* .tex (for example, sample.tex). The operating system of the computer usually puts additional limitations on the choice of file name, such as the maximum number of characters in the name and extension, or the exclusion of special characters. For example, if the root name is limited to eight characters, finances.tex is an allowed name, but financial.tex is not.

The text file is then processed by LATEX. As explained in Section 1.2.1, this means running the TEX program with the LATEX format. Most installations have a special shorthand command, called latex, to accomplish this, followed by the name of the text file, with or without the extension .tex.

For example, if the name of the text file is sample.tex, one would start the LATEX processing with the call

```
latex sample
```

During the processing, the terminal monitor displays the page numbers together with possible warnings and error messages. Chapter 9 deals with the interpretation of these error messages and their consequences. After LATEX has finished processing, it creates a new file with the selected root name and the extension .dvi. In the above example, this would be sample.dvi.

This .dvi (*device-independent*) file contains the *formatted* text along with information about the required character fonts, but in a form that is independent of the characteristics of the printer to be used. Such a device-independent file is called a *metafile*.

Finally the information in the .dvi metafile has to be converted to a form that can be output on the chosen printer by means of a so-called *printer driver* that is specific for that printer type. The call to the driver contains only the root name of the file, just like the LATEX call, that is, without the ending .dvi. After processing with the printer driver, another file is created with the selected root name and an extension specifying the printer type.

For example, if the output is to go to a LaserJet printer, the driver might be called as

```
dvilj sample
```

which either produces a file sample.lj or sends the output directly to the printer.

Other drivers exist for letter-quality dot-matrix printers as well as for PostScript. However, it is not the purpose of this book to describe the DVI drivers, since there are a great many available, with different names and extra features, such as printing only certain pages or printing them in reverse order. A computing center will have the important drivers installed for the output devices available. A summary of these calls and further information should be available in a TEX or LATEX *Local Guide* that is accessible to all system users. The best source of information, however, is often other experienced users, the *gurus* mentioned previously.

2 Commands and Environments

The text that is to be the input to a LaTeX processing run must be prepared with a *text editor* (any one will do) in a file with the extension .tex. The contents of this file are printable characters only—there are no invisible characters or commands hiding behind the screen. However, commands are still necessary to issue instructions to LaTeX. These commands are visible in the input text but will not appear in their literal form in the output. It is therefore vital to know how commands are distinguished from text that is to be printed, and, of course, how they function.

2.1 Command names and arguments

A *command* is an instruction to LaTeX to do something special, like print some symbol or text not available to the restricted character set used in the input file, or to change the current typeface or other formatting properties. There are three types of command names:

- the single characters # $ & ~ _ ^ % { } all have special meanings that are explained later in this chapter;

- the backslash character \ plus a single non-letter character; for example \\$ to print the $ sign; all the special characters listed above have a corresponding two-character command to print them literally;

- the backslash character \ plus a sequence of letters, ending with the first non-letter; for example, \large to switch to a larger typeface.

Many commands operate on some short piece of text, which then appears as an *argument* in curly braces following the command name. For example, \emph{stress} is given to print the word stress in an emphasized typeface (here italic) as *stress*. Such arguments are said to be *mandatory* because they must always be given.

Other commands take *optional* arguments, which are normally employed to modify the effects of the command somehow. The optional arguments appear in square braces.

In this book we present the general syntax of commands as

$$\backslash commandname[optional\_argument]\{mandatory\_argument\}$$

where typewriter characters must be typed exactly as illustrated and italic text indicates something that must be substituted for. Optional arguments are put into square brackets [ ] and the mandatory ones into curly braces { }. A command may have several optional arguments, each one in its set of brackets in the specified sequence. If none of the optional arguments is used, the square brackets may be omitted. Any number of blanks may appear between the command name and the arguments.

Some commands have several mandatory arguments. Each one must be put into a { } pair and their sequence must be maintained as given in the command description. For example,

$$\backslash rule[lift]\{width\}\{height\}$$

produces a black rectangle of size *width* and *height*, raised by an amount *lift* above the current baseline. A rectangle of width 10 mm and height 3 mm sitting on the baseline is made with \rule{10mm}{3mm}. The optional argument *lift* is not given here. The arguments must appear in the order specified by the syntax and may not be interchanged.

Some commands have a so-called *-form in addition to their normal appearance. A * is added to their name to modify their functionality slightly. For example, the \section command has a *-form \section* which, unlike the regular form, does not print an automatic section number. For each such command, the difference between the normal and *-form will be explained in the description of the individual commands.

Command names end before the first character that is not a letter. If there are optional or mandatory arguments following the command name, then it ends before the [or { bracket, since these characters are not letters. Many commands, however, possess no arguments and are composed of only a name, such as the command \LaTeX which produces the LaTeX logo. If such a command is followed by a punctuation mark, such as comma or period, it is obvious where the command ends. If it is followed by a normal word, the blank between the command name and the next word is interpreted as the command terminator: The \LaTeX logo results in 'The LaTeXlogo', that is, the blank was seen only as the end of the command and not as spacing between two words. This is a result of the special rules for blanks, described in Section 2.5.1.

In order to insert a space after a command that consists only of a name, either an empty structure {} or a space command (\ and blank) must be placed after the command. The proper way to produce 'The LaTeX logo' is to type either The \LaTeX{} logo or The \LaTeX\ logo. Alternatively, the command itself may be put into curly braces, as The {\TeX} logo, which also produces the correct output with the inserted blank: 'The T_EX logo'. Incidentally, the L^AT_EX 2_ε logo is produced with \LaTeXe. Can the reader see why this logo command cannot be named \LaTeX2e?

2.2   Environments

An *environment* is initiated with the command \begin{*environment*} and is terminated by \end{*environment*}.

An environment has the effect that the text within it is treated differently according to the environment parameters. It is possible to alter (temporarily) certain processing features, such as indentation, line width, typeface, and much more. The changes apply only within the environment. For example, with the quote environment,

> *previous text*
> \begin{quote}
> *text1* \small *text2* \bfseries *text3*
> \end{quote}
> *following text*

the left and right margins are increased relative to those of the previous and following texts. In the example, this applies to the three texts *text1*, *text2*, and *text3*. After *text1* comes the command \small, which has the effect of setting the next text in a smaller typeface. After *text2*, there is an additional command \bfseries to switch to bold face type. Both these commands only remain effective up to the \end{quote}.

> The three texts within the quote environment are indented on both sides relative to the previous and following texts. The *text1* appears in the normal typeface, the same one as outside the environment. The *text2* and *text3* appear in a smaller typeface, and *text3* **furthermore appears in bold face.**

After the end of the quote environment, the subsequent text appears in the same typeface that was in effect beforehand.

Most command names may also be used as environment names. In this case the command name is used *without* the preceding \. For example, the command \em switches to an emphatic typeface, usually *italic*, and the corresponding environment \begin{em} will set all the text in *italic* until \end{em} is reached.

An environment without a name can be simulated by bracketing with a {...} pair. The effect of any command within it ends with the closing curly brace.

The user can even create his or her own environments, as described in Section 7.4.

2.3   Declarations

A *declaration* is a command that changes the values or meanings of certain parameters or commands without immediately printing any text. The effect of the declaration begins immediately and ends when another declaration of the same type is encountered. However, if the declaration occurs within an environment or a {...} pair, its scope extends only to the corresponding \end command, or to the closing brace }. The commands \em, \bfseries, and \small mentioned in the previous section are examples of such non-printing declarations that alter the current typeface.

Some declarations have associated arguments, such as the command \setlength which assigns a value to a *length parameter* (see Sections 2.4 and 7.2).

Examples:

{\bfseries This text appears in bold face} The \bfseries declaration changes the typeface: **This text appears in bold face**. The effect of this declaration ends with the closing brace }.

\setlength{\parindent}{0.5cm} The paragraph indentation is set to 0.5 cm. The effect of this declaration ends with the next encounter of the command \setlength{\parindent}, or at the latest with the \end command that terminates the current environment.

\pagenumbering{roman} The page numbering is to be printed in Roman numerals.

Some declarations, such as the last example, are global, that is, their effects are not limited to the current environment. The following declarations are of this nature, the meanings of which are given later:

\newcounter	\pagenumbering	\newlength
\setcounter	\thispagestyle	\newsavebox
\addtocounter		

Declarations made with these commands are effective right away and remain so until they are overridden by a new declaration of the same type. In the last example above, page numbering will be done in Roman numerals until countermanded by a new \pagenumbering{arabic} command.

2.4   Lengths

2.4.1   Fixed lengths

Lengths consist of a decimal number with a possible sign in front (+ or -) followed by a mandatory dimensional unit. Permissible units and their abbreviated names are:

cm   centimeter,
mm   millimeter,
in   inch (1 in = 2.54 cm),
pt   point (1 in = 72.27 pt),
bp   big point (1 in = 72 bp),
pc   pica (1 pc = 12 pt),
dd   didôt point (1157 dd = 1238 pt),
cc   cicero (1 cc = 12 dd),
em   a font-specific size, the width of the capital M,
ex   another font-related size, the height of the letter x.

Decimal numbers in TeX and LaTeX may be written in either the English or European manner, with a *period* or a *comma*: both 12.5cm and 12,5cm are permitted.

Note that 0 is not a legitimate length since the unit specification is missing. To give a zero length it is necessary to write 0pt or 0cm.

Values are assigned to a length parameter by means of the LaTeX command \setlength, which is described in Section 7.2 along with other commands for dealing with lengths. Its syntax is:

\setlength{*length_command*}{*length_spec*}

For example, the width of a line of text is specified by the parameter \textwidth, which is normally set to a default value depending on the class, paper type, and font size. To change the line width to be 12.5 cm, one would give:

\setlength{\textwidth}{12.5cm}

2.4.2   Rubber lengths

Some parameters expect a *rubber* length. These are lengths that can be stretched or shrunk by a certain amount. The syntax for a rubber length is:

nominal_value plus *stretch_value* minus *shrink_value*

where the *nominal_value*, *stretch_value*, and *shrink_value* are each a length. For example,

\setlength{\parskip}{1ex plus0.5ex minus0.2ex}

means: the extra line spacing between paragraphs, called \parskip, is to be the height of the x in the current font, but it may be increased to 1.5 or reduced to 0.8 times that size.

One special *rubber* length is \fill. This has the natural length of *zero* but can be stretched to any size.

2.5   Special characters

2.5.1   Spaces and carriage returns

The *space* or *blank* character has some properties different from those of normal characters, some of which have already been mentioned in Section 2.1. During processing, blanks in the input text are replaced by rubber lengths (Section 2.4.2) in order to allow the line to fill up to the full line width. As a result, some peculiar effects can occur if one is not aware of the following rules:

- one blank is the same as a thousand, only the first one counts;

- blanks at the beginning of an input line are ignored;

- blanks terminating a command name are removed;

- *carriage returns* (a new line) are treated as blanks.

Some of the consequences of these rules are that there may be as many blanks as desired between words or at the beginning of a line (to make the input text more legible) and that a word may come right at the end of a line without the spacing between it and the next word disappearing. To force a space to appear where it would otherwise be ignored, one must give the command \␣ (a \ followed by a space character, made visible here by the symbol ␣).

To ensure that certain words remain together on the same line, a *protected space* is inserted between them with the ˜ character (Section 3.5.1, page 45). Multiple protected spaces are all printed out, in contrast to normal spaces.

Sometimes it is necessary to suppress the space that appears because of the new line. In this case, the line must be ended with the *comment* character % (Section 4.11) before the carriage return.

Two carriage returns in a row produce a blank line, the signal for a new paragraph. As for blank characters, one blank line is the same as a thousand.

2.5.2   Quotation marks

The *quotation marks* found on the typewriter " are not used in book printing. Instead different characters are used at the beginning and end, such as 'single quotes' and "double quotes". Single quotes are produced with ' and ', while double quotes are made by typing the respective characters twice: '' for " and '' for ". Furthermore the typewriter character " will also generate the double closing quote ".

2.5.3 Hyphens and dashes

In book printing, the character that appears on the typewriter as - comes in various lengths: -, –, —. The smallest of these, the *hyphen*, is used for compound words such as *father-in-law* and for word division at the end of a line; the middle sized one, the *en dash*, is used in ranges of numbers, for example, pages 33–36; and the largest, the *em dash*, is used as punctuation—what is normally called the *dash*. These are generated by typing the hyphen character one, two, or three times, so that - yields -, while -- makes –, and --- produces —. A fourth type of dash is the minus sign −, which is entered in math mode as $-$ (Chapter 5).

2.5.4 Printing command characters

As mentioned in Section 2.1, the characters # $ ~ _ ^ % { } are interpreted as commands. To print them as text, one must give a command consisting of \ plus that character.

$= \$ & = \&   % = \%   # = \#   _ = _   { = \{   } = \}

2.5.5 The special characters §, †, ‡, ¶, © and £

These special characters do not exist on the computer keyboard. They can however be generated by special commands as follows:

§ = \S   † = \dag   ‡ = \ddag   ¶ = \P   © = \copyright   £ = \pounds

The production of Greek letters and other mathematical symbols is described in Chapter 5.

The euro symbol € is not yet part of basic LaTeX, but it can be produced with the help of some standard packages. See page 452.

2.5.6 Foreign letters

Special letters that exist in European languages other than English can also be generated with TeX. These are:

œ={\oe}   Œ={\OE}   æ={\ae}   Æ={\AE}   å={\aa}   Å ={\AA}   ¡=!'
ø ={\o}   Ø ={\O}   ł ={\l}   Ł ={\L}   ß={\ss}   SS={\SS}   ¿=?'

Ångstrøm may be written as {\AA}ngstr{\o}m while *Karlstraße* can be input as Karlstra{\ss}e. The upper case version \SS is new to LaTeX 2ε.

2.5.7 Accents

In non-English languages, there is a multiplicity of *diacritic marks* or *accents*, most of which can be printed with TeX:

$$\grave{o} =\backslash`\{o\} \quad \acute{o}=\backslash'\{o\} \quad \hat{o}=\backslash\char94\{o\} \quad \ddot{o}=\backslash"\{o\} \quad \tilde{o}=\backslash\char126\{o\}$$
$$\bar{o} =\backslash=\{o\} \quad \dot{o}=\backslash.\{o\} \quad \breve{o}=\backslash u\{o\} \quad \check{o}=\backslash v\{o\} \quad \H{o}=\backslash H\{o\}$$
$$\widehat{oo}=\backslash t\{oo\} \quad o\!\!\cc=\backslash c\{o\} \quad o\!\!\ud=\backslash d\{o\} \quad o\!\!\ub=\backslash b\{o\} \quad \r{o}=\backslash r\{o\}$$

(The last command, \r, is new to LATEX 2$_\varepsilon$.) The *o* above is given merely as an example: any letter may be used. With *i* and *j* it should be pointed out that the dot must first be removed. This is carried out by prefixing these letters with \. The commands \i and \j yield ı and ȷ. In this way ĭ and J̃ are formed by typing \u{\i} and \H{\j}.

The accent commands consisting of a non-letter may also be given without the curly braces:

$$\grave{o}=\backslash`o \quad \acute{o}=\backslash'o \quad \hat{o}=\backslash\char94 o \quad \ddot{o}=\backslash"o \quad \tilde{o}=\backslash\char126 o \quad \bar{o}=\backslash=o \quad \dot{o}=\backslash.o$$

The letter accent commands should always be used with the curly braces.

!

The above foreign letters and accent commands are necessary for keyboards that do not otherwise provide these characters directly, but for typing large amounts of text in a language that makes frequent use of such symbols, this can become very tedious. LATEX 2$_\varepsilon$ provides the package inputenc which permits the proper interpretation of such special characters, according to the computer system being used. Thus a French writer with a non-English keyboard can produce *le château enchanté* with le château enchanté. See Section 8.6 on page 237.

2.5.8 Ligatures

In book printing, certain combinations of letters are not printed as individuals but as a single symbol, a so-called *ligature*. TEX processes the letter combinations ff, fi, fl, ffi, and ffl not as

ff, fi, fl, ffi, ffl   but rather as   ff, fi, fl, ffi, ffl

Section 3.5.1 describes how one may force TEX to print the letters of such combinations separately. This is sometimes desired for such words as *shelfful*, which looks rather strange when printed with the normal *ff* ligature, *shelfful*.

2.5.9 The date

The current date can be placed at any point in the text with the command \today. The standard form for the date is the American style of month, day, year (for example, November 15, 1998). The British form (15th November 1998) or the date in other languages can be generated with the help of the TEX commands \day, \month, and \year, which return the current values of these parameters as numbers. Examples of how such a new \today command may be made are shown on page 352 in Section C.4.2.

2.6   Fragile commands

!

The influence of certain commands is not limited to the place where they appear but may also affect other parts of the document. For example, the sectioning commands, such as \chapter{*heading*}, generate the headings not only at the point where they are issued but also possibly at the top of each of the following pages in a different typeface, and perhaps once again, in yet another typeface, in the table of contents. Such an argument that appears at several places in the document is called a *moving argument.*

Moving arguments can create problems if they contain commands that are prematurely interpreted before they are finally applied. Such commands are said to be *fragile*, while others that can withstand moving are called *robust*. A fragile command can still be safely used within a moving argument if it is preceded by \protect.

The issue of fragile commands was more important for LaTeX 2.09, but with LaTeX 2ε it has become of academic interest: all commands that are likely to be used in moving arguments are now robust. However, a problem might still arise with user-defined commands. One should therefore be aware of the concept of fragility and of its remedies: adding the \protect prefix or creating one's own robust commands (Section C.2.5).

It is only necessary to add \protect to secure fragile commands when they appear in moving arguments, that is, those that are written elsewhere such as in the table of contents, bibliography, or running heads. It does not hurt to precede most commands with \protect in a moving argument if their fragility is in doubt. Exceptions to this rule are *length commands* (Section 7.2) and *counter commands* (Section 7.1.4) such as \arabic and \value. These should never have a \protect before them.

2.7   Exercises

As part of a successful teach-yourself course of study, practical exercises are absolutely essential. In addition to the plentiful examples in the text, which the reader should try out, a number of further exercises will also be given in the course of this book, as suggested homework that should definitely be carried out.

All examples and exercises are given for LaTeX 2ε; if you are still using LaTeX 2.09, you must use the initial command \documentstyle in place of \documentclass.

Exercise 2.1: *This exercise tests the basic operations of running the LaTeX program with a short piece of text. A few simple commands are also included. Use a text editor to produce the following text and store it in a file named* exer.tex.

```
\documentclass{article}
\begin{document}
Today (\today) the rate of exchange between the British
pound and American dollar is \pounds 1 = \$1.63, an
```

```
increase of 1\% over yesterday.
\end{document}
```

Running the LATEX program is dependent on the computer system, but assuming that it is called with the command `latex`, the processing is run with

```
latex exer
```

Note: although the file name is `exer.tex`, only the root name `exer` needs to be given in the call to LATEX.

If the processing occurs without any error messages, the `.dvi` file `exer.dvi` has been successfully created and may be further converted for the printer with the help of the printer driver. How this is called also depends on the computer system. The final printed result should look as follows except that your current date will appear:

Today (November 15, 1998) the rate of exchange between the British pound and American dollar is £1 = $1.63, an increase of 1% over yesterday.

Note the following points about the commands used:

- no blank is necessary after \today because the) suffices to terminate it;
- the blank after \pounds is optional and it is not printed in the output;
- the commands \$ and \% do not require blanks to terminate them; if blanks are given, they will be printed.

Exercise 2.2: Take some text of about 3/4 of a page long out of a book or journal article and type it into a LATEX file. Pay attention that the paragraphs are separated by blank lines. Use the same set of commands as in Exercise 2.1, that is, put the text between the commands \begin{document}...\end{document} and repeat the procedures for obtaining the printed output.

Note: the selected text should not contain any special structures such as block indentation, different typefaces, centered text, enumerations, mathematical formulas, tables, etc. How to form such structures is explained in the next chapters.

3 Document Layout and Organization

3.1 Document class

The first command in the preamble of a LaTeX file normally determines the global processing format for the entire document. The syntax for this command is:

$\boxed{2_\varepsilon}$ \documentclass[*options*]{*class*}

or

$\boxed{2.09}$ \documentstyle[*options*]{*class*}

where the latter is the only possibility under LaTeX 2.09. With LaTeX 2_ε, the command \documentstyle is provided only for compatibility with older document files *and should not be used for new ones!* In fact, a very insistent warning is now issued when \documentstyle is encountered.

The possible values of *class*, of which one and only one may be given, are: book, report, article, or letter. (The properties of the letter class are explained in Appendix A.) The basic differences between these classes lie not only in the page layouts, but also in the organization. An article may contain *parts, sections, subsections*, and so on, while a report can also have *chapters*. A book also has chapters, but treats even and odd pages differently; also, it prints running heads on each page with the chapter and section titles.

3.1.1 Standard class options

The *options* available allow various modifications to be made to the formatting. They can be grouped as follows.

Selecting font size

The basic font size is selected with one of the options

$\boxed{2\varepsilon}$ 10pt   11pt   12pt

This is the size of the font in which the normal text in the document will be set. The default is 10pt, which means that this is the value assumed if no size option is specified. All other font sizes are relative to this standard size, so that the section titles, footnotes, and so on will all change size automatically if a different basic font size is selected. (Note that in LaTeX 2.09 there is no option 10pt; to select 10 pt as the basic size one specifies no size option.)

Specifying paper size

LaTeX calculates the text line width and lines per page according to the selected font size and paper mode. It also sets the margins so that the text is centered both horizontally and vertically. In order to do this, it needs to know which paper format is being used. This is specified by one of the following options, all of which exist only for LaTeX 2_ε:

$\boxed{2\varepsilon}$ letterpaper (11×8.5 in) $\boxed{2\varepsilon}$ a4paper (29.7×21 cm)

$\boxed{2\varepsilon}$ legalpaper (14×8.5 in) $\boxed{2\varepsilon}$ a5paper (21×14.8 cm)

$\boxed{2\varepsilon}$ executivepaper (10.5×7.25 in) $\boxed{2\varepsilon}$ b5paper (25×17.6 cm)

The default is letterpaper, American letter size paper, 11×8.5 in.

Normally, the paper format is such that the longer dimension is the vertical one, the so-called *portrait* mode. With the option

$\boxed{2\varepsilon}$ landscape

the shorter dimension becomes the vertical one, the *landscape* mode.

Page formats

The text on the page may be formatted into one or two columns with the options

$\boxed{2\varepsilon}$ onecolumn   twocolumn

The default is onecolumn. In the case of the twocolumn option, the separation between the columns as well as the width of any rule between them may be specified by \columnsep and \columnseprule, described below.

The even- and odd-numbered pages may be printed differently according to the options

$\boxed{2\varepsilon}$ oneside   twoside

With oneside, all pages are printed the same; however, with twoside, the running heads are such that the page number appears on the right on odd pages and on the left on even pages. *It does not force the printer to output double-sided.* The idea is that when these are later printed back-to-back, the page numbers are always on the outside where they are more easily noticed. This is the default for the book class. For article and report, the default is oneside.

With the book class, chapters normally start on a right-hand, odd-numbered page. The options

$\boxed{2\varepsilon}$ openright $\qquad$ $\boxed{2\varepsilon}$ openany

control this feature: with openany a chapter always starts on the next page, but with openright, the default, a blank page may be inserted if necessary.

Normally the title of a book or report will go on a separate page, while for an article, it is placed on the same page as the first text. With the options

$\boxed{2\varepsilon}$ notitlepage $\qquad$ titlepage

this standard behavior may be overruled. See Sections 3.3.1 and 3.3.2.

Further options

The remaining standard options are:

leqno   Equation numbers in displayed formulas will appear on the left instead of the normal right side (Section 5.1).

fleqn   Displayed formulas will be set flush left instead of centered (Section 5.1). The amount of indentation may be set with the parameter \mathindent described below.

openbib
    The format of bibliographies may be changed so that segments are set on new lines. By default, the text for each entry is run together.

draft   If the LaTeX line-breaking mechanism does not function properly and text must stick out into the right margin, then this is marked with a thick black bar, to make it noticeable.

$\boxed{2\varepsilon}$ final   The opposite of draft, and the default. Lines of text that are too wide are not marked in any way.

If multiple options are to be given, they are separated by commas, as for example, \documentclass[11pt,twoside,fleqn]{article}. The order of the options is unimportant. If two conflicting options are specified, say oneside and twoside, it is not obvious which one will be effective. That depends entirely on the definitions in the class file itself, so it would be best to avoid such situations.

Parameters associated with some options

Some options make use of parameters that have been given certain default values:

\mathindent
>   specifies the indentation from the left margin for the equation numbers when fleqn is selected (Section 5.1);

\columnsep
>   specifies the space between the two columns for the twocolumn option (see Figure G.2 on page 556);

\columnseprule
>   determines the width of the vertical line between the two columns for the twocolumn option. The default is zero width, that is, no vertical rule (see Figure G.2).

The standard values of these parameters may be changed with the LaTeX command \setlength. For example, to change \mathindent to 2.5 cm, give

>   \setlength{\mathindent}{2.5cm}

These parameters may be assigned values either in the preamble or at any place in the document. Parameters in the preamble apply to the entire document, whereas those within the text are in effect until the next change or until the end of the environment in which they were made (Section 2.3). In the latter case, the previous values become effective once more.

Exercise 3.1: Take your text file from Exercise 2.2 and change the initial command \documentclass{article} first to \documentclass[11pt]{article} and then to \documentclass[12pt]{article} and print the results of each LaTeX processing. Compare the line breaking of these outputs with that of Exercise 2.2.
Note: if there are some improper word divisions, you call tell LaTeX where the correct division should occur with the command \-, for example, man\-u\-script. (This is one of the few words that the TeX English word divider does not handle properly.) Additional means of modifying word division are given in Section 3.6. If there are warnings of the sort Overfull \hbox ... during the LaTeX processing, TeX was not able to break the lines cleanly. In the output, these lines will

extend beyond the right margin. The usual cause is that T$_E$X was not able to divide some word, either because it is indivisible or because T$_E$X's word division routines were not adequate. Here again a suggested hyphenation in the text can solve the problem. Other solutions will be given shortly.

Exercise 3.2: *Now employ* `\documentclass[twocolumn]{article}` *in your text file. If you now receive a number of warnings with* `Underfull \hbox ...`*, then these lines will indeed be left and right justified but will have too much empty space between the words. Check the output yourself to see whether the word spacing is acceptable. If not, try giving some hyphenation suggestions in the first words of the next line.*

Note: if you use the classes book *or* report *instead of* article *in the preceding exercises, you will notice no difference in the outputs. These classes affect the subsequent structural elements of the document. Basically, you should use* article *for short articles (say 10-20 pages) and* report *for longer reports that are to be organized into chapters. The chapters always begin on a new page. The class* book *is available for producing books.*

3.1.2   Adding features with packages $\boxed{2\varepsilon}$

Although the *class* of the document determines its overall general properties, such as layout and sectioning, it is possible to invoke more specific *packages* that might change the way certain commands behave, or even define totally new commands to add extra features that are not part of the standard LAT$_E$X. A number of such packages are provided in the LAT$_E$X distribution, or you may obtain some of the hundreds that have been made available by the LAT$_E$X user community (Appendix D), or you may even write your own (Appendix C).

A package is nothing more than a set of LAT$_E$X (or T$_E$X) commands stored in a file with the extension `.sty`, although there some special commands that may only appear within them. To invoke a package, simply call

$\boxed{2\varepsilon}$ `\usepackage{`*package*`}`

in the preamble, where *package* is the root name of the file. More than one package may be loaded with one call to `\usepackage`. For example, two packages provided with standard LAT$_E$X are stored in files `makeidx.sty` (Section 8.4) and `ifthen` (Section 7.3.5). They may be read in together with

`\usepackage{makeidx,ifthen}`

A package may have options associated with it, which may be selected in the same way as for document classes: by including the option names within square braces. The general syntax is thus:

`\usepackage[`*opt1,opt2 ...* `]{`*package1,package2, ...* `}`

where all the listed options will be applied to all the selected packages. If any of the packages does not understand one of the options, a warning message is output to the monitor.

Just what options are available depends on the package itself. One must read the accompanying description and/or documentation to discover not only its options, but also what the package does and what new commands it defines.

3.1.3   Style options `2.09`

The \usepackage command did not exist in LaTeX 2.09; in fact, it does not function at all if \documentstyle is issued in place of \documentclass. In LaTeX 2.09, packages were treated in the same way as *style options*: the file names were added to the list of options in the \documentstyle command. For this reason, they were called *style option files*, or simply *style files*, which explains the extension .sty that these files bear. This extension has been maintained in LaTeX 2ε in order that the older style files may still be employed as packages.

There is a fundamental difference between a true option, which is programmed within the class or package file, and a package, which is read in as a file of the same name. In LaTeX 2.09, this distinction was never clear and often led to confusion.

To load the package makeidx with the options 12pt and titlepage into the class (or 'main style') article in the older LaTeX version, one must write

`2.09` \documentstyle[makeidx,12pt,titlepage]{article}

Since packages are handled here as options, it is clear that they themselves cannot have any further options.

3.1.4   Global and local options `2ε`

> **!**

One interesting feature about options specified with the \documentclass command is that they also apply to any packages that follow. This means that if several packages all take the same option, it is only necessary to declare it once in \documentclass. For example, one might design a package to modify article for generating a local house style that might do different things for single or double column text; this package could make use of the class options onecolumn and twocolumn to achieve this. Or it could elaborate on the draft option to produce double line spacing, as for a manuscript. Alternatively, several packages might have language-dependent features that could be activated with options like french or german; it is sufficient to list such options only in \documentclass to apply it to all packages. Such options are called *global*, for they are passed on to all subsequent packages automatically.

Global options need not be limited to the standard class options listed in Section 3.1.1. A warning message is printed only if neither the class nor any of the packages understand one or more of them. By contrast, any options specified with \usepackage will be applied only to those packages listed in that one command; and it is applied to all of them. A warning is printed if one or more of those packages does not recognize any one of these *local* options.

3.2 Page style

The basic page format is determined by the *page style*. With one exception, this command is normally given in the preamble. Its form is:

> \pagestyle{*style*}

The mandatory argument *style* takes on one of the following values:

plain   The page head is empty, the foot contains the centered page number. This is the default for the article and report classes when no \pagestyle is given in the preamble.

empty   Both head and foot lines are empty; no page numbers are printed.

headings
>   The head contains the page number as well as title information (chapter and section headings); the foot is empty. This is the default for book class.

myheadings
>   The same as headings except that the page titles in the head are not chosen automatically but rather are given explicitly by the commands \markright or \markboth (see below).

The command

> \thispagestyle{*style*}

functions exactly as \pagestyle except that it affects only the current page. For example, the page numbering may be suppressed for just the current page with the command \thispagestyle{empty}. It is only the *printing* of the page number that is suppressed; the next page will be numbered just as though the command had never been given.

3.2.1 Heading declarations

For the page styles headings and myheadings, the information appearing in the headline may be given with the declarations

\markright{*right_head*}
\markboth{*left_head*}{*right_head*}

The declaration \markboth is used with the document class option twoside, with even-numbered pages considered to be on the *left* and odd-numbered pages on the *right*. Furthermore, the page number is printed on the left side of the head for a left page and on the right side for a right page.

For one-sided output, all pages are considered to be right-handed. In this case, the declaration \markright is appropriate. It may also be used with two-sided output to overwrite the *right_head* given in \markboth.

With the page style headings, the standard titles in the page headline are the chapter, section, or subsection headings, depending on the document and page style, according to the following scheme:

Style		Left Page	Right Page
book, report	one-sided	—	*Chapter*
	two-sided	*Chapter*	*Section*
article	one-sided	—	*Section*
	two-sided	*Section*	*Subsection*

If there are more than one \section or \subsection on a page, it is the heading of the last one that appears in the page head.

An example of two-sided output with headings is this book itself.

3.2.2   Page numbering

The declaration that specifies the style of the page numbering has the form

\pagenumbering{*num_style*}

The allowed values of *num_style* are:

arabic   for normal (Arabic) numerals,
roman   for lower case Roman numerals,
Roman   for upper case Roman numerals,
alph   for lower case letters,
Alph   for upper case letters.

The standard value is arabic. This declaration resets the page counter to 1. In order to paginate the foreword of a document with Roman numerals and the rest with Arabic numbers beginning with page 1 for chapter 1, one must declare \pagenumbering{roman} at the start of the foreword and then reset the page numbering with \pagenumbering{arabic} immediately after the first \chapter command. (See Section 3.3.5 for an alternative method.)

Pages may be numbered starting with a value different from 1 by giving the command

> `\setcounter{page}{`*page_num*`}`

where *page_num* is the number to appear on the current page.

Exercise 3.3: Expand your exercise text file so that it fills more than one page of output and include the following preamble:

> *\documentclass{article}*
> *\pagestyle{myheadings} \markright{Exercises}*
> *\pagenumbering{Roman}*
> *\begin{document}*

3.2.3   Paragraph formatting

The following parameters affect the appearance of a paragraph and may be given new values with `\setlength` as explained in Section 7.2:

`\parskip`
> The distance between paragraphs, expressed in units of `ex` so that it will automatically change with character font size. This should be a *rubber* length.

`\parindent`
> The amount of indentation for the first line of a paragraph.

`\baselinestretch`
> This is a number that magnifies the normal distance between *baselines*, the line on which the letters sit, that is, neglecting the hanging parts of letters such as *g* and *y*. This number is initially 1, for standard line spacing. It may be changed to another number with
>
> > `\renewcommand{\baselinestretch}{`*factor*`}`
>
> where *factor* is any decimal number, such as 1.5 for a 50% increase. This then applies to all font sizes. If this command is given outside the preamble, it does not come into effect until another font size has been selected (Section 4.1.2).

These parameters may be set either in the preamble or anywhere in the text of the document. In the latter case, the changes remain in effect until the next change or until the end of the environment in which they were made (Section 2.3).

The true interline spacing is contained in the `\baselineskip` length parameter, which is set automatically whenever a font size declaration is called. For each

!

font size, there is a normal value, which is then multiplied by \baselinestretch to produce \baselineskip. See Section 4.1.2 for more details.

If \baselineskip is changed within a paragraph, the new value determines the interline spacing for the entire paragraph. More precisely, it is the value at the end of the paragraph that applies to the whole.

Exercise 3.4: *Add the following to the preamble of your exercise file:*

```
\setlength{\parindent}{0em}
\setlength{\parskip}{1.5ex plus0.5ex minus 0.5ex}
\renewcommand{\baselinestretch}{1.2}
```

After processing this exercise, repeat it with another value for the parameter \baselinestretch, *say* 1.5, *in order to get a feeling for how it works. Remove these lines from the exercise file afterwards.*

3.2.4   Page format

Each page consists of a *head*, the *body* containing the actual text, and a *foot*. The selection of the page style determines what information is to be found in the head and footlines.

LATEX uses default values for the distances between the head, body, and foot, for the upper and left margins, and for the text line width and heights of the head, body, and foot. These formatting lengths are illustrated in Figure 3.1 on the opposite page. They may be changed by declaring new values for them, preferably in the preamble, with the command \setlength (Section 7.2). For example, give

```
\setlength{\textwidth}{12.5cm}
```

to make the text line width to be 12.5 cm.

More detailed diagrams of the page formats for one- and two-column outputs are shown at the end of Appendix G in Figures G.1 and G.2.

In order to calculate the page layout precisely, one must realize that LATEX measures all distances from a point one inch from the top of the paper and one inch from the left edge. Thus the total left margin is \oddsidemargin plus one inch. The LATEX 2$_\varepsilon$ parameters \paperwidth and \paperheight, which include this extra inch, are given their values by the paper size option in the \documentclass command; they are used internally to calculate the margins so that the text is centered. The user may also take advantage of them.

With document class book or with the option twoside, the bottom edge of the body will always appear at exactly the same position on every page. In the other classes or options, it will vary slightly. In the first two cases, the constant bottom edge is produced by the internal command \flushbottom, whereas the varying bottom is produced by the command \raggedbottom. The user may apply these declarations to change the behavior of the bottom edge at any time, independent of document class and options.

Exercise 3.5: *You can change the page format of your text by altering the above parameters. Add the following to the preamble of your text:*

\oddsidemargin
  left margin for odd pages,
\evensidemargin
  left margin for even pages,
\topmargin
  upper margin to top of head,
\headheight
  height of head,
\headsep
  distance from the bottom of headline to top of body,
\topskip
  distance from top of body to baseline of first line of text,
\textheight, \textwidth
  height and width of main text,
\footskip
  distance from bottom of body to bottom of foot,
\paperwidth, \paperheight
  total width and height of paper as given by paper size option, including all margins.

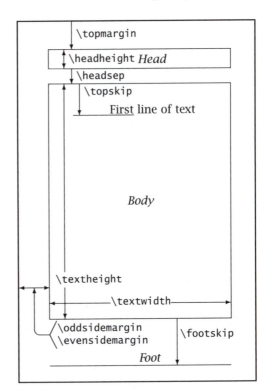

Figure 3.1: Page layout parameters

\setlength{\textwidth}{13cm}   \setlength{\textheight}{20.5cm}

The upper and left margins of your output will now seem too small. Select new values for \oddsidemargin and \topmargin to correct this.

Note: do not forget the 1 inch margin at the left and top from which additional margins are measured. You must take this into account when you select \oddsidemargin and \topmargin.

Exercise 3.6: *Expand your text so that the output requires more than two full pages with the reduced page format. Add \flushbottom to the preamble and check that the last line of all pages is at exactly the same location.*

Exercise 3.7: *Remove the command \flushbottom and select the document class \documentclass[twoside]{article}. Now the last lines are at the same location without the \flushbottom command. On the other hand, the left margin of the odd pages probably does not agree with the right margin of the even pages. Use the declaration \evensidemargin to correct this.*

3.2.5   Single and double column pages

The document class option `twocolumn` sets the entire document in two columns per page. The default is one column per page. Individual pages may be output in one or two columns with the declarations:

`\twocolumn[`*text*`]`
> Terminates the current page, starting a new one with *two columns* per page. The optional *text* is written at the top of the page in one column with the width of the whole page.

`\onecolumn`
> Terminates the current two-column page and continues with one column per page.

! The option `twocolumn` automatically changes certain page style parameters, such as indentation, compared with the one-column format. This does not occur with the command `\twocolumn`. These additional changes must be made with the corresponding `\setlength` declarations if they are desired. If the bulk of the document is in two-column format, the class option is to be preferred.

! An additional page style parameter is `\columnwidth`, the width of one column of text. For single column text, this is the same as `\textwidth`, but when `twocolumn` has been selected, LATEX calculates it from the values of `\textwidth` and `\columnsep`. The author should never change this parameter, but he or she may make use of it, for example to draw a rule the width of a column of text.

! An alternative method of producing multi-column output, without page breaks, and with any number of columns, is provided by the package `multicol`, described in Section D.3.3 on page 387.

3.3   Parts of the document

Every document is subdivided into chapters, sections, subsections, and so on. There can be an appendix at the end and at the beginning a title page, table of contents, an abstract, etc. LATEX has a number of commands available to free the user from such formatting considerations. In addition, sequential numbering and sub-numbering of headings can take place. Even a table of contents may be generated and printed automatically.

The effects of some sectioning commands depend on the selected document class and not all commands are available in every class.

3.3.1   Title page

A title page can be produced either unformatted with the environment

> `\begin{titlepage}` *Title page text* `\end{titlepage}`

or with the commands

```
\title{
    How to Write DVI Drivers}

\author{
    Helmut Kopka\thanks{Tel.
        05556--401--451 FRG}\\
    Max--Planck--Institut\\
    f\"ur Aeronomie
\and
    Phillip G. Hardy
    \thanks{Tel.
        319--824--7134 USA}\\
    University\\of Iowa}

\maketitle
```

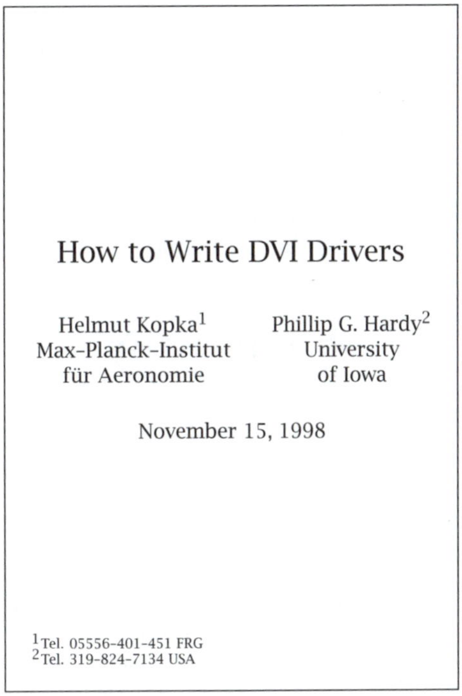

How to Write DVI Drivers

Helmut Kopka[1]   Phillip G. Hardy[2]
Max-Planck-Institut   University
für Aeronomie   of Iowa

November 15, 1998

[1] Tel. 05556–401–451 FRG
[2] Tel. 319–824–7134 USA

Figure 3.2: Sample title page and the text that produced it

> \title{*Title text*}
> \author{*Author names and addresses*}
> \date{*Date text*}

in the prescribed LaTeX format.

In the standard LaTeX format for the title page, all entries are centered on the lines in which they appear. If the title is too long, it will be broken up automatically. The author may select the break points himself with the \\ command, that is, by giving \title{...\\...\\...}.

If there are several authors, their names may be separated with \and from one another, such as \author{G. Smith \and J. Jones}. These names will be printed next to each other in one line. The sequence

> \author{*Author1**Institute1**Address1*
> \and *Author2**Institute2**Address2*}

separately centers the entries, one per line, in each of the sets *Author1, Institute1, Address1* and *Author2, Institute2, Address2* and places the two blocks of centered entries beside each other on the title page.

Instead of printing the author names next to each other, one may position them on top of one another by replacing \and with the \\

command. In this case, the vertical spacing may be adjusted with an optional length specification [*space*] following the \\.

If the command \date is omitted, the current date is printed automatically below the author entries on the title page. On the other hand, the command \date{*Date text*} puts the text *Date text* in place of the current date. Any desired text may be inserted here, including line break commands \\ for more than one line of centered text.

The command

    \thanks{*Footnote text*}

may be given at any point in the *title, author,* or *date text.* This puts a marker at that point where the command appears and writes *footnote text* as a footnote on the title page.

The title page is created using the entries in \title, \author, \date, and \thanks when the command

    \maketitle

is issued. The title page itself does not possess a page number and the first page of the following document is number 1. (For book, the page numbering is controlled by the special commands in Section 3.3.5.) A separate title page is only produced for document classes book and report. For article, the command \maketitle creates a title heading on the first page using the centered entries from the \title, \author, and, if present, \date and \thanks declarations. If the document class option titlepage has been given, the title appears on a separate page even for the article class.

An example of a title page in the standard LaTeX format is shown in Figure 3.2 on the previous page. Note that the current date appears automatically since the command \date is missing in the definition of the title page. This command may be used to put any desired text in place of the date.

For the unformatted title page produced with the titlepage environment, the commands \title and \author are left out and the entire title page is designed according to the author's specifications within the environment. To this end he or she may make use of all the structuring commands described in Chapter 4. In this case the printing of the title page is implemented at the end of the titlepage environment, so the command \maketitle is also left out.

Exercise 3.8: *Remove the declarations for changing the page format in Exercises 3.5–3.7. Add to your exercise text a title heading with the title 'Exercises', your name as author, and your address, together with a date entry in the form 'place, date'. To do this, write the following commands after* \begin{document}*:*

    \title{Exercises} \author{Your name\\Your address}
    \date{Your town, \today} \maketitle

Make sure that you have selected document class `article`. *After printing the document, change the document class command to*

> `\documentclass[titlepage]{article}`

to put the title information on to a title page instead of a title heading. Deactivate these commands by putting the comment character % at the beginning of each of the lines. In this way you avoid getting a title page in the following exercises but you can easily reactivate the commands simply by removing the % characters.

3.3.2   Abstract

The abstract is produced with the command

> `\begin{abstract}` *Text for the abstract* `\end{abstract}`

In document class `report`, the abstract appears on a separate page without a page number; in `article`, it comes after the title heading on the first page, unless the document class option `titlepage` has been selected, in which case it is also printed on a separate page. An abstract is not possible in document class `book`.

3.3.3   Sections

The following commands are available for producing automatic, sequential sectioning:

`\part`	`\chapter`	`\subsection`	`\paragraph`
	`\section`	`\subsubsection`	`\subparagraph`

With the exception of `\part`, these commands form a sectioning hierarchy. In document classes `book` and `report`, the highest sectioning level is `\chapter`. The chapters are divided into sections using the `\section` command, which are further subdivided by means of `\subsection`, and so on. In document class `article`, the hierarchy begins with `\section` since `\chapter` is not available.

The syntax of all these commands is

> `\sec_command[`*short title*`]{`*title*`}`   or
> `\sec_command*{`*title*`}`

In the first case, the section is given the next number in the sequence, which is then printed together with a heading using the text *title*. The text *short title* becomes the entry in the table of contents (Section 3.4) and the page head (provided that page style `headings` has been selected). If the optional *short title* is omitted, it is set equal to *title*; this is the normal situation unless *title* is too long to serve for the other entries.

In the second (*-form) case, no section number is printed and no entry in the table of contents is made (however, see Section 3.4.3).

The size of the title heading and the depth of the numbering depend on the position of the sectioning command within the hierarchy. For document class `article`, the `\section` command generates a single number (say 7), the `\subsection` command a double number with a period between the two parts (say 7.3), and so on.

In document classes `book` and `report`, the chapter headings are given a single number with the `\chapter` command, the `\section` command creates the double number, and so on. Furthermore, the command `\chapter` always starts a new page and prints **Chapter n** over the chapter title, where **n** is the current chapter number. At this point in the present book, we are in *Chapter 3, Section 3.3, Subsection 3.3.3*.

For each sectioning command there is an internal counter that is incremented by one every time that command is called, and reset to zero on every call to the next higher sectioning command. These counters are not altered by the *-forms, a fact that can lead to difficulties if standard and *-forms of the commands are mixed such that the *-forms are higher in the hierarchy than the standard forms. There are no problems, however, if the *-forms are always lower than the standard forms. The sequence

> `\section ... \subsection ... \subsubsection* ...`

numbers the headings for `\section` and `\subsection` while leaving the headings for `\subsubsection` without any numbering.

The sectioning command `\part` is a special case and does not affect the numbering of the other commands.

The automatic numbering of sections means that the numbers might not necessarily be known at the time of writing. The author may be writing them out of their final order, or might later introduce new sections or even remove some. If he or she wants to refer to a section number in the text, some mechanism other than typing the number explicitly will be needed. The LaTeX cross-reference system, described in detail in Section 8.3.1, accomplishes this task with the two basic commands

> `\label{name}`   `\ref{name}`

the first of which assigns a keyword *name* to the section number, while the second may be used as reference in the text for printing that number. The keyword *name* may be any combination of letters, numbers, or symbols. For example, in this book the command `\label{sec:xref}` has been typed in at the start of Section 8.3.1, so that this sentence contains the input text `at the start of Section \ref{sec:xref}`.

A second referencing command is `\pageref` for printing the page number where the corresponding `\label` is defined.

The referencing commands may be used in many other situations for labeling items that are numbered automatically, such as figures, tables, equations.

Every sectioning command is assigned a level number such that `\section` is always level 1, `\subsection` level 2, ... `\subparagraph` level 5. In document

⚠

class `article`, \part is level 0 while in `book` and `report` classes, \part is level −1 and \chapter becomes level 0. Section numbering is carried out down to the level given by the number `secnumdepth`. This limit is set to 2 for `book` and `report`, and to 3 for `article`. This means that for `book` and `report`, the section numbering extends only to the level of \subsection and for `article` to \subsubsection.

In order to extend (or reduce) the level of the section numbering, it is necessary to change the value of `secnumdepth`. This is done with the command

> \setcounter{secnumdepth}{*num*}

(The command \setcounter is explained in Section 7.1.) In `article`, *num* may take on values from 0 to 5, and in `book` and `report` from −1 to 5.

It is possible to change the initial value of a sectioning command within a document with the command

> \setcounter{*sec_name*}{*num*}

where *sec_name* is the name of the sectioning command without the preceding \. This procedure may be useful when individual sections are to be processed by LATEX as single files. For example,

> \setcounter{chapter}{2}

sets the \chapter counter to 2. The counter will be incremented on the next call to \chapter which then produces **Chapter 3**.

Occasionally it is necessary to change the font size or style in the section headline. This can be done by including the declarations described in Sections 4.1.2-4.1.6 in the section title text. For example,

> \section*{\Large\slshape Larger Slanted Font}

will print the section title 'Larger Slanted Font' in the size \Large and style \slshape. Such deviations from the standard fonts are *not* recommended since the changes affect only the title text itself, and not the preceding numbering.

3.3.4 Appendix

An appendix is introduced with the declaration

> \appendix

It has the effect of resetting the section counter for `article` and the chapter counter for `book` and `report` and changing the form of the numbering for these sectioning commands from numerals to capital letters A, B, ... Furthermore, the word 'Chapter' is replaced by 'Appendix' so that subsequent chapter headings are preceded by 'Appendix A', 'Appendix B', etc. The numbering of lower sectioning commands contains the letter in place of the chapter number, for example A.2.1.

3.3.5 Book structure

To simplify the structuring of a book, the commands

> $\boxed{2\varepsilon}$ \frontmatter
> *preface, table of contents*
> $\boxed{2\varepsilon}$ \mainmatter
> *main body of text*
> $\boxed{2\varepsilon}$ \backmatter
> *bibliography, index, colophon*

are provided in the book class. The \frontmatter command switches page numbering to Roman numerals and suppresses the numbering of chapters; \mainmatter resets the page numbering to 1 with Arabic numbers and reactivates the chapter numbering; this is once again turned off with \backmatter.

Exercise 3.9: *Insert at the beginning of your exercise text the command* \section{Title A} *and at some appropriate place near the middle* \section{Title B}. *Select some suitable text for Title A and Title B. Insert at appropriate places some* \subsection *commands with reasonable subtitles. Remove the commands included from Exercise 3.3:*

> \pagestyle{myheadings} \markright{Exercises}
> \pagenumbering{Roman}

and print the results.

Exercise 3.10: *Include the additional command* \chapter{Chapter title} *with your own appropriate Chapter title before your first* \section *command. Change the document class command to* \documentclass[twoside]{report} *and call the page style command* \pagestyle{headings} *in the preamble. Note the twofold effect of the sectioning commands in the headings and in the page headlines. Compare the results with the table in Section 3.2.1.*

Exercise 3.11: *Change the chapter command to*

> \chapter[Short form]{Chapter title}

by putting an abbreviated version of Chapter title for Short form. Now the page head contains the shortened title where the full chapter title previously appeared.

3.4   Table of contents

3.4.1 Automatic entries

LaTeX can prepare and print a table of contents automatically for the whole document. It will contain the section numbers and corresponding headings as given in the standard form of the sectioning commands, together with the page numbers on which they begin. The sectioning depth to which entries are made in the table of contents can be set in the preamble with the command

`\setcounter{tocdepth}{`*num*`}`

The value *num* has exactly the same meaning and effect as it does for the counter `secnumdepth` described above, by which the maximum level of automatic subsectioning is fixed. By default, the depth to which entries are included in the table of contents is the same as the standard level to which automatic sectioning is done: that is, to level `\subsection` for book and `report` and to level `\subsubsection` for `article`.

3.4.2   Printing the table of contents

The table of contents is generated and printed with the command

`\tableofcontents`

given at the location where the table of contents is to appear, which is normally after the title page and abstract.

This leads to a paradox, for the information in the table of contents is to be printed near the beginning of the document, information that cannot be known until the end. LATEX solves this problem as follows: the first time the document is processed, no table of contents can be included but instead LATEX opens a new file with the same name as the document file but with the extension `.toc`; the entries for the table of contents are written to this file during the rest of the processing.

The next time LATEX is run on this document, the `\tableofcontents` command causes the `.toc` file to be read and the table of contents is printed. As the processing continues, the `.toc` file is updated in case there have been major changes since the previous run. This means that the table of contents that is printed is always the one corresponding to the previous version of the document. For this reason, it may be necessary to run LATEX more than once on the final version.

3.4.3   Additional entries

The *-form sectioning commands are not entered automatically in the table of contents. To insert them, or any other additional entry, the commands

`\addcontentsline{toc}{`*sec_name*`}{`*entry text*`}`
`\addtocontents{toc}{`*entry_text*`}`

may be used.

With the first command, the entries will conform to the format of the table of contents, whereby `section` headings are indented more than those for `chapter`

but less than those for subsection. This is determined by the value of the
argument *sec_name*, which is the same as one of the sectioning commands
without the \ character (for example, section). The *entry_text* is inserted in the
table of contents along with the page number. This command is most useful to
enter unnumbered section headings into the table of contents. For example,

```
\section*{Author addresses}
\addcontentsline{toc}{section}{Author addresses}
```

The \addtocontents command puts any desired command or text into the
.toc file. This could be a formatting command, say \protect\newpage, which
takes effect when the table of contents is printed.

3.4.4   Other lists

!

In addition to the table of contents, lists of figures and tables can also be
generated and printed automatically by LaTeX. The commands to produce these
lists are

\listoffigures	reads and/or produces file .lof
\listoftables	reads and/or produces file .lot

The entries in these lists are made automatically by the \caption command
in the figure and table environments (see Section 6.4.4). Additional entries are
made with the same commands as for the table of contents, the general form of
which is

```
\addcontentsline{file}{format}{entry}
\addtocontents{file}{entry}
```

where *file* stands for one of the three types toc (*table of contents*), lof (*list of
figures*), or lot (*list of tables*). The argument *format* is one of the sectioning
commands for the table of contents, as described above, or figure for the list
of figures, or table for the list of tables. The argument *entry* stands for the text
that is to be inserted into the appropriate file.

Exercise 3.12: *In your exercise file, insert after the deactivated title page commands*

```
\pagenumbering{roman}
\tableofcontents \newpage
\pagenumbering{arabic}
```

Process your exercise file <u>twice</u> *with LaTeX and print out the second results.
Deactivate the above commands with % before doing the next run.*
*Note: if your section titles contain any commands, they might have to be preceded
by the* \protect *command to ensure that they are safely transferred to the
corresponding file.*

3.5   Fine-tuning text

3.5.1   Word and character spacing

The spacing between words and characters is normally set automatically by TEX, which not only makes use of the natural width of the characters but also takes into account alterations for certain character combinations. For example, an A followed by a V does not appear as AV but rather as AV; that is, they are moved together slightly for a more pleasing appearance. Interword spacing within one line is uniform, and is chosen so that the right and left ends match exactly with the side margins. This is called left and right *justification*. TEX also attempts to keep the word spacing for different lines as nearly the same as possible.

Words that end with a punctuation mark are given extra spacing, depending on the character: following a period '.' or exclamation mark '!', there is more space than after a comma ','. This corresponds to the rule in English typesetting that there should be extra spacing between sentences. In certain cases, the automatic procedures do not work properly, or it is desirable to override them, as described in the next sections.

Sentence termination and periods

TEX interprets a period following a lower case letter to be the end of a sentence where additional interword spacing is to be inserted. This leads to confusion with abbreviations such as *i. e.*, *Prof. Jones*, or *Phys. Rev.*, where the normal spacing is required. This can be achieved by using the characters ~ or \␣ instead of the normal blank. (The character ␣ is simply a symbol for the blank which is otherwise invisible.) Both these methods insert the normal interword spacing; in addition, ~ is a *protected space* that prevents the line from being broken at this point. The above examples should be typed in as i.~e., Prof.~Jones, and Phys.\ Rev., producing *i. e.*, *Prof. Jones*, and *Phys. Rev.* with the correct spacing and forcing the first two to be all on one line. In the third case, there is nothing wrong with putting *Phys.* and *Rev.* on different lines.

A period following an upper case letter is not interpreted as the end of a sentence, but as an abbreviation. If it really is the end of a sentence, then it is necessary to add \@ before the period in order to achieve the extra spacing. For example, this sentence ends with NASA. It is typed in as This sentence ends with NASA\@.

French spacing

The additional interword spacing between sentences can be switched off with the command \frenchspacing, which remains in effect until countermanded with \nonfrenchspacing. In this case, the command \@

is ignored and may be omitted. This paragraph has been printed with
\frenchspacing turned on, so that all word spacings within one line are
the same. It corresponds to the normal rule for non-English typesetting.

Character combinations "' and '"

A small spacing is produced with the command \,. This may be used, for
example, to separate the double quotes " and " from the corresponding
single quotes ' and ' when they appear together. For example, the text
''\,'Beginning' and 'End'\,'' produces "'Beginning' and 'End'".

Italic correction

In changing from a sloping typeface such as *italic* or *slanted* to a vertical
one, the last character may lean too far over into the following spacing,
making it appear too small. Extra spacing must be inserted here, depending on the letter (more for a *d* than for an *a*). TEX adds this extra spacing,
known as *italic correction*, when the command \/ is given. This should
always be included after the last letter of a sloping typeface before switching to a vertical one, for example {\slshape slanted\/} spacing for
slanted spacing. The command may be omitted when the last sloping
letter is followed by a period or comma.

The LATEX 2$_\varepsilon$ font commands \textsl and \textit (Section 4.1.4) automatically add italic corrections, which is another reason to prefer them
over the font declarations \slshape, \itshape. For example, instead of
{\slshape slanted\/}, one should type \textsl{slanted}. However,
it might sometimes be desirable to suppress the italic correction, something that may be accomplished with the command \nocorr at the end
of the sloping text.

Breaking ligatures

The same command \/ may be used to suppress ligatures, that is, letter
combinations that are printed as one character. The text shelf\/ful produces *shelfful* instead of the normal *shelfful*. As shown in Section 2.5.8,
TEX prints the combinations ff, fi, fl, ffi, and ffl as ligatures. By
typing f\/f\/i, the combination *ffi* is produced instead of the ligature
ffi.

The \/ command may also be used to suppress the reduced spacing
between certain letter combinations, such as AV or Te:

A\/V AV   T\/e Te   instead of   AV   Te

Inserting arbitrary spacing

Spacing of any desired size may be inserted into the text with the commands

> \hspace{*space*}
> \hspace*{*space*}

where *space* is the length specification for the amount of spacing, for example 1.5cm or 3em. (Recall that one *em* is the width of the letter M in the current typeface.)

This command puts blank space of width *space* at that point in the text where it appears. The standard form (without *) ignores the spacing if it occurs between two lines, just as normal blanks are removed at the beginning of lines. The *-form, on the other hand, inserts the spacing no matter where it occurs.

The length specification may be negative, in which case the command works as a backspace for overprinting characters with other ones.

A blank before or after the command will also be included:

`This is\hspace{1cm}1cm`	This is 1cm
`This is \hspace{1cm}1cm`	This is 1cm
`This is \hspace{1cm} 1cm`	This is 1cm

The command \hfill is an abbreviation for \hspace{\fill} (see Section 2.4.2). It inserts enough space at that point to force the text on either side to be pushed over to the left and right margins. With Left\hfill Right one produces

Left   Right

Multiple occurrences of \hfill within one line will each insert the same amount of spacing so that the line becomes left and right justified. For example, the text Left\hfill Center\hfill Right generates

Left   Center   Right

If \hfill comes at the beginning of a line, the spacing is suppressed in accordance with the behavior of the standard form for \hspace. If a rubber space is really to be added at the beginning or end of a line, \hspace*{\fill} must be used instead. However, LaTeX also offers a number of commands and environments to simplify most such applications (see Section 4.2.2).

A number of other fixed horizontal spacing commands are available:

> \quad   and   \qquad

The command \quad inserts a horizontal space equal to the current type size, that is, 10 pt for a 10 pt typeface, whereas \qquad inserts twice as much.

Inserting variable ...... and _____ sequences

Two commands that work exactly the same way as \hfill are

> \dotfill and \hrulefill

Instead of inserting empty space, these commands fill the gap with dots or a ruled line, as follows:

```
Start \dotfill\ Finish\\                                and
Left \hrulefill\ Center \hrulefill\ Right\\   produce
```

Start .... Finish
Left _____ Center _____ Right

Any combination of \hfill, \dotfill, and \hrulefill may be given on one line. If any of these commands appears more than once at one location, the corresponding filling will be printed that many more times than for a single occurrence.

```
Departure \dotfill\dotfill\dotfill\ 8:30 \hfill\hfill
Arrival \hrulefill\ 11:45\\
```

Departure .... 8:30   Arrival _____ 11:45

3.5.2   Line breaking

Breaking text into lines is done automatically in TEX and LATEX. However, there are times when a line break must be forced or encouraged, or when a line break is to be suppressed.

**The command **

A new line with or without additional line spacing can be achieved with the command \\. Its syntax is

> \\[*space*]
> *[*space*]

The optional argument *space* is a length that specifies how much additional line spacing is to be put between the lines. If it is necessary to start a new page, the additional line spacing is not included and the new page begins with the next line of text. The *-form prevents a new page from occurring between the two lines.

With *[10cm], the current line is ended and a vertical spacing of 10 cm is inserted before the next line, which is forced to be on the same page as the current line. If a page break is necessary, it will be made before the current line, which is then positioned at the top of the new page together with the 10 cm vertical spacing and the next text line.

The command \newline is identical to \\ without the option *space*. That is, a new line is started with no additional spacing and a page break is possible at that point.

Both commands may be given only within a paragraph, and not between them where they would be meaningless.

Further line-breaking commands

The command \linebreak is used to encourage or force a line break at a certain point in the text. Its form is

    \linebreak[*num*]

where *num* is an optional argument, a whole number between 0 and 4 that specifies how important a line break is. The command recommends a line break, and the higher the number the stronger the recommendation. A value of 0 allows a break where it otherwise would not occur (like in the middle of a word), whereas 4 compels a line break, as does \linebreak without *num*. The difference between this command and \\ or \newline is that the current line will be fully justified, that is, interword spacing will be added so that the text fills the line completely. With \\ and \newline, however, the line is filled with empty space after the last word and the interword spacing remains normal.

The opposite command

    \nolinebreak[*num*]

discourages a line break at the given position, with *num* specifying the degree of discouragement. Again, \nolinebreak without a *num* argument has the same effect as \nolinebreak[4], that is, a line break is absolutely impossible here.

Another way of forcing text to stay together on one line is with the command \mbox{*text*}. This is convenient for expressions such as 'Voyager-1' to stop a line break at the hyphen.

3.5.3 Paragraph spacing

The normal spacing between paragraphs is set by the length \parskip (Section 3.2.3), which may be altered from its original default value with the \setlength command:

    \setlength{\parskip}{*space*}

It is possible to add extra vertical spacing of amount *space* between particular paragraphs using the commands

    \vspace{*space*}
    \vspace*{*space*}

The *-form will add the extra space even when a new page occurs, or when the command appears at the top of a new page. The standard form ignores the extra vertical spacing in these situations.

If these commands are given within a paragraph, the extra space is inserted after the current line, which is right and left justified as usual.

The *space* parameter may even be negative, in order to move the following text higher up the page than where it would normally be printed.

The command \vfill is an abbreviation for \vspace{\fill} (see Section 2.4.2). This is the equivalent of \hfill for vertical spacing, inserting enough blank vertical space to make the top and bottom of the text match up exactly with the upper and lower margins. The comments on multiple occurrences of \hfill also apply to \vfill. If this command is given at the beginning of a page, it is ignored, just like the standard form of \vspace{\fill}. If a rubber space is to be put at the top of a page, the *-form \vspace*{\fill} must be used.

Further commands for increasing the spacing between paragraphs are

> \bigskip   \medskip   \smallskip

which add vertical spacing depending on the font size declared in the document class.

3.5.4 Paragraph indentation

The amount of paragraph indentation is set with the length \parindent (see Section 3.2.3), which can be changed with the \setlength command:

> \setlength{\parindent}{*space*}

The first line of each paragraph then has blank spacing of width *space* at the left side. To suppress indentation for one paragraph, or to force it where it would otherwise not occur, place

> \noindent   or   \indent

at the beginning of the paragraph to be affected.

|2ε| **Note:** the first paragraph of a section is not indented, not even with \indent. Include the package indentfirst (page 387) to alter this.

Instead of a blank line, the command \par may also be used to indicate the end of a paragraph.

3.5.5 Page breaking

Breaking text into pages occurs automatically in TeX and LaTeX, just as for line breaking. Here again, it may be necessary to interfere with the program's notion of where a break should take place.

Normal pages

The commands

>   \pagebreak[*num*]
>   \nopagebreak[*num*]

are the equivalents of \linebreak and \nolinebreak for page breaking. If \pagebreak appears between two paragraphs, a new page will be forced at that point. If it comes within a paragraph, the new page will be implemented after the current line is completed. This line will be right and left justified as usual.

The command \nopagebreak has the opposite effect: between paragraphs, it prevents a page break from occurring there, and within a paragraph, it stops a page break that might take place at the end of the current line.

Optional numbers between 0 and 4 express the degree of encouragement or discouragement for a page break. The analogy with the command \linebreak goes further: just as the line before the break is left and right justified with extra interword spacing, in the same way the page before the break is expanded with interline spacing to make it top and bottom justified.

The proper command to end a page in the middle, fill it with blank spacing, and go on to a new page is

>   \newpage

which is equivalent to \newline with regard to page breaking.

Exercise 3.13: Place one or two \newpage commands at appropriate spots in your exercise file. In the printout, the rest of the page will be empty. Now remove these commands and insert a \pagebreak command one or two lines after a natural page break. Afterwards, move the command a few lines before the normal break.

Pages with figures and tables

If the text contains tables, pictures, or reserved space for figures, these are inserted at the location of the corresponding command, provided that there is enough room for them on the current page. If there is not enough space, the text continues and the figure or table is stored to be put on a following page.

The command

>   \clearpage

ends the current page like \newpage and in addition outputs all the pending figures and tables on one or more extra pages. (See also Section 6.4.)

Two-column pages

If the document class option `twocolumn` has been chosen, or the command \twocolumn is in effect, then the two commands \pagebreak and \newpage end the current *column* and begin a new one, treating columns as pages. On the other hand, \clearpage and \cleardoublepage (see below) terminate the current page, inserting an empty right column if necessary.

Two-sided pages

An additional page-breaking command is available when the document option `twoside` has been selected:

> \cleardoublepage

which functions exactly the same as \clearpage (the current page is terminated, all pending figures and tables are output) but in addition, the next text will be put on to an *odd*-numbered page. If necessary, an empty page with an even number is printed to achieve this.

Controlling page breaks

LATEX 2_ε provides the possibility of increasing the height of the current page slightly with commands

> $\boxed{2_\varepsilon}$ \enlargethispage{*size*}
> $\boxed{2_\varepsilon}$ \enlargethispage*{*size*}

which add the length *size* to \textheight for this one page only. Sometimes the difference of a few points is all that is necessary to avoid a bad page break. The *-form of the command also shrinks any interline spacing as needed to maximize the amount of text on the page.

Further tips on page breaking

As mentioned above in Section 3.2.4, in document class book or option twoside the bottom edge of the page body is always the same on every page, but otherwise it may vary. This means that in these two cases extra interline spacing will be inserted at appropriate places, such as between paragraphs.

   This behavior is regulated by the internal declaration \flushbottom for book and twoside, and by \raggedbottom for all other situations. They are initially set by the internal class and option definitions, but may also be given by the user in the preamble or anywhere else in the text. They remain in effect until countermanded or until the end of the current environment.

It is not normally possible to have more text on a page than is allowed by the selected page length \texttheight (Section 3.2.4). However, if it is absolutely necessary to squeeze in one or two extra lines, vertical space must be cleared for them. This can be done by inserting shrinking rubber spacing with the \vspace command at an appropriate place, for example \vspace{0pt minus 2ex}. The best places are before and after *displays, enumerations, lists, tables, figures* (see Chapter 4).

!

A detailed description of the internal rules for page breaking cannot be given here, but it is sufficient to say that TeX issues so-called penalty points (with the command \penalty) at those places where a page break is to be discouraged, with *negative* points indicating an encouragement. Each line of a paragraph is given, say, 10 penalty points (\linepenalty=10), to make it more difficult to break the page within than between paragraphs.

The first and last lines of a paragraph are given 150 penalty points (that is, \clubpenalty=150 and \widowpenalty=150) to discourage (but not to forbid) page breaks that would leave a single line of a paragraph on one page.

As a last resort, one can change the penalty assignments to achieve a desired page break. For example, by setting \clubpenalty=450 it becomes three times more difficult to break the page after the first line of a paragraph compared to normal. Setting the penalty to 10 000 or more makes a page break *absolutely* impossible. Altering the penalty assignments, however, should only be undertaken when all else fails, since this destroys the balance with the rest of the document, and may conjure up even more monstrous page breaks.

3.6 Word division

When a line is to be right and left justified, it often turns out that the break cannot be made between whole words without either shoving the text too close together or inserting huge gaps between the words. It is then necessary to split a word. This fundamental task is performed by TeX, the underlying basis of LaTeX, by means of a word-dividing algorithm that works (almost) perfectly for English text, which is more than can be said for most authors. Nevertheless, even it makes mistakes at times which need to be corrected by human intervention.

If normal TeX/LaTeX is used for other languages, or if foreign words appear in English text, incorrect hyphenations are very likely to appear. (See Sections 3.6.5, 3.6.6, and D.1 for more discussions about LaTeX with other languages.) In these cases too, something must be done to override TeX's hyphenation rules, as described below.

3.6.1 Manual hyphenation

The simplest way to correct a wrongly divided word is to include a \- command at the right place within the word. The word *manuscript*, for example, will not be hyphenated at all, so if it causes problems with breaking a line, write it as man\-u\-script. This tells TEX to divide the word as necessary either as *man-uscript* or as *manu-script*, and to ignore its normal rules.

The \- command merely makes hyphenation possible at the indicated locations; it does not force it. If the author absolutely insists in dividing a word at a certain point, say between the *u* and *s* in *manuscript*, he or she can type manu-\linebreak script to achieve this. However, this brute force method is not recommended, because the line break will always occur here even if the text is later changed.

For English text, the spelling of a word remains the same when it is hyphenated, something that is not true in other languages. In traditional German, for example, if *ck* is split, it becomes *k-k*. TEX allows such behavior with the general hyphenation command

\discretionary{*before*}{*after*}{*without*}

where *before* and *after* are the letters (with hyphen) that come on either side of the break if division takes place, and *without* is the normal text with no hyphenation. Thus Boris Becker's name should be typed as

Boris Be\discretionary{k-}{k}{ck}er

something that one only wants to do in exceptional situations. Incidentally, the \- command is shorthand for \discretionary{-}{}{}.

Note: in the controversial reformed German spelling, *ck* is never split.

3.6.2   Hyphenation list

Words that are incorrectly hyphenated and that appear frequently within the document can be put into a *list of exceptions* in the preamble, to avoid laboriously inserting \- every time:

\hyphenation{*list*}

The *list* consists of a set of words, separated by blanks or new lines, with the allowed division points indicated by hyphens. For example,

\hyphenation{man-u-script com-pu-ter gym-na-sium
             coun-try-man re-sus-ci-tate ... }

The list may contain only words with the normal letters *a–z*, with no special characters or accents.

3.6.3   Suppressing hyphenation

Another means of avoiding bad word divisions is to turn hyphenation off, at least for a paragraph or two. Actually the command

\begin{sloppypar} *paragraph text* \end{sloppypar}

does not prevent word division, but does permit larger interword spacings without giving a warning message. This means that practically all lines are broken between words. It is also possible to put the command \sloppy in the preamble or in the current environment to reduce the number of word divisions in the whole document or within the environment scope. This is recommended when the line width is rather narrow.

When the command \sloppy is in effect, it is possible to undo it temporarily and to turn hyphenation back on with the command \fussy.

3.6.4   Line width and word division

!

It is appropriate to add some comments here about the effect of the line width on word division.

When TeX reaches the end of a paragraph, it attempts to organize the text into lines of equal width, taking into account the spacing requirements of Section 3.5.1, by breaking the text only between words; that is, without word division. If this fails, it will start looking for break points within the words themselves. Line breaking between words can occur much more easily if the line width is large, or if the average word length is small. Furthermore, for a given line width, the number of word divisions will be fewer if the font size is smaller.

With narrow line widths, right and left justification may not be possible even with word division, in which case TeX will need to insert more interword spacing than normally allowed. This can be accomplished with the sloppypar environment or \sloppy command described above. Justification will be achieved, but at the price of having the words pulled apart by an unacceptable amount. In both cases TeX prints a warning message, either Overfull \hbox (justification is impossible and a word will stick out beyond the right margin) or Underfull \hbox (word spacing too large). If neither situation is tolerable, the text will have to be adjusted, or suitable break points may be forced with the commands \linebreak and \hfill.

3.6.5   More about word division

!

The information that TeX needs to carry out word division is stored in special files containing lists of letter combinations used by the hyphenation algorithm. These lists are made up from statistical studies of the language under consideration. The files also contain lists of exceptions, those words that are incorrectly hyphenated by the algorithm. TeX first looks for a word in the exception list, splitting it according to the pattern it finds there; if it does not find it, it splits it using its algorithm. For each language, a different list of letter combinations and exceptions must be given. These lists are incorporated in the TeX/LaTeX formats when they are generated.

Originally Donald E. Knuth provided English hyphenation patterns in a file named hyphen.tex which is input automatically for Plain TeX. LaTeX 2.09 loaded a file lhyphen.tex which may be configured at each installation to input whatever hyphenation files are locally needed. With LaTeX 2_ε, the same result is achieved with the configuration file `hyphen.cfg` (Section D.5.2). Today, Knuth's hyphenation list is normally stored under the name `ushyphen.tex`, or something similar, to emphasize that it applies to American English.

One may examine the results of TeX's hyphenation algorithm on the screen with the command

> \showhyphens{*list of words*}

which prints out all the words in the *list of words* with the hyphenations that TeX chooses. For example, giving

> \showhyphens{interrelations penumbra summation}

will print to the screen

> in-ter-re-la-tions penum-bra sum-ma-tion

3.6.6   Word division with multilingual text

!

In TeX versions 2.99 or lower, only one set of letter combinations for hyphenation was allowed in one `plain.fmt` or `lplain.fmt` file. The LaTeX programs using these versions were applicable to a single language. It was not possible to mix languages in one document with multiple hyphenation rules.

From TeX version 3.0 on, multiple hyphenation lists may be included in the format, making it possible to switch hyphenation schemes within one document, using a new TeX command \language. This command may be used as part of language-specific adaptations to translate certain explicit English words in the output (such as 'Contents'), to simplify accents or punctuation, and to alter the definition of the date command \today. More information on multiple languages and \language is given in Section D.1.4.

Displayed Text

4

There are a variety of ways to display or emphasize the text: changing font style or font size, centering, indentation, marking the paragraphs, and so on. LaTeX supplies us with commands for the most common forms of display.

4.1 Changing font

In typography, a set of letters, numbers, and characters of a certain size and appearance is called a *font*. The standard font in LaTeX for the main body of text is an *upright, Roman* one of *medium* weight, in the size specified in the `\documentclass` statement at the start. The three possible basic sizes are 10, 11, and 12 pt, depending on the size options `10pt` (default), `11pt`, and `12pt`. (Recall, there are 72.27 points per inch or about 28.45 pt per cm.) The parenthesis characters ( ) extend the full height and depth of the font size.

The differences in the visual appearance of the three standard sizes are greater than would be expected from the ratios of the numbers:

This is an example of the 10 pt font. ( )

And this is the 11 pt font for comparison. ( )

And finally this is a sample of 12 pt font. ( )

4.1.1 Emphasis

In a typewritten manuscript the simplest way to emphasize text is with underlining. The typesetter will normally transform underlined text into *italics* for the printed version. Switching from standard to *emphasized* text is carried out in LaTeX with the declaration `\em` or with the command `\emph` $\boxed{2\varepsilon}$.

The `\em` declaration functions as the other font declarations described below: the change of font remains in effect until negated by another appropriate declaration (which can be \em itself), or until the end of the current *environment* (Section 2.2). An environment may also be created with a pair of curly braces {...}. This is the easiest way to *emphasize* short pieces of text, as for example:

```
This is the easiest way to {\em emphasize} short ...
```

A more logical method of *emphasizing* a word or two is provided in LATEX 2ε with the command \emph:

```
A more logical method of \emph{emphasizing} a word ...
```

Note carefully the difference between the *declaration* that remains in effect until the local environment is ended with the closing curly brace, and the command that operates on an argument enclosed in curly braces. Another more subtle difference is that the command \emph automatically inserts the *italic correction* (see Section 3.5.1), while with the declaration \em one must include it manually.

Both the declaration and the command switch to an emphasizing font. That means, if the current font is upright it switches to *italics, whereas if the text is already slanted, an* upright *font is selected.*

Nested emphasis is possible and is simple to understand:

```
The {\em first}, second, and {\em third font switch}
The {\em first, {\em second, and {\em third font switch}}}
```

both produce 'The *first,* second, and *third font switch*'.

4.1.2   Choice of font size

The following declarations are available in LATEX for changing the font size:

\tiny	smallest	\Large	larger
\scriptsize	very small	\LARGE	even larger
\footnotesize	smaller	\huge	still larger
\small	small	\Huge	largest
\normalsize	normal		
\large	large		

all of which are relative to the standard size selected in the document class option. In this book, the standard size is 10 pt, which is then the size selected with \normalsize.

The font size declarations behave as all other declarations: they make an immediate change that remains in effect until counteracted by another size declaration, or until the current environment comes to an end. If issued within curly braces {..}, the effect of the declaration extends only to the closing brace, as in a nameless environment:

```
normal {\large large \Large larger} normal again
```
normal large larger normal again

In LATEX 2.09, the font size declarations also reset all the other font attributes (Section 4.1.3) to their defaults, that is, Roman, upright, medium weight. By contrast, in LATEX 2$_\varepsilon$, these attributes remain unchanged. Compare:

2.09	`\sl slanted {\Large larger}`	*slanted* larger
2$_\varepsilon$	`\sl slanted {\Large larger}`	*slanted* *larger*

If `\documentstyle` is specified in place of `\documentclass`, these size declarations behave as they do in LATEX 2.09, for compatibility with older documents.

Changing the font size with one of the above commands also automatically changes the interline spacing. For every font size, there is a corresponding *natural* line spacing `\baselineskip`. This may be altered at any time. If the natural line spacing is 12 pt, the command `\setlength{\baselineskip}{15pt}` will increase it to 15 pt.

The value of `\baselineskip` that is effective at the end of the paragraph is used to make up the whole paragraph. This means that if there are several changes to `\baselineskip` within a paragraph, only the last value given will be taken into account.

With every change in font size, `\baselineskip` is reset to its natural value for that size. Any previous setting with `\setlength` will be nullified.

In order to create a change in the line spacing that is valid for all font sizes, one must make use of the factor `\baselinestretch`, which has a normal value of 1. The true interline spacing is really

```
\baselinestretch×\baselineskip
```

which maintains the same relative spacing for all font sizes. The user may change this spacing at any time with:

```
\renewcommand{\baselinestretch}{factor}
```

where *factor* is any decimal number. A value of 1.5 increases the interline spacing (baseline to baseline) by 50% over its natural size for all font sizes.

The new value of `\baselinestretch` does not take effect until the next change in font size. In order to implement a new value in the current font size, it is necessary to switch to another size and back again immediately. If the present font size is `\normalsize`, the sequence

```
\small\normalsize
```

will do the trick. Any size command may be used in place of `\small`.

Not all font styles are available in every size. If an impossible combination of font size and style is selected, LATEX makes a substitution and writes a warning to the transcript file.

4.1.3   Font attributes $\boxed{2_\varepsilon}$

The size of a font is only one of several *attributes* that may be used to describe it. With the New Font Selection Scheme (NFSS), which is an integral part of LaTeX 2_ε, it is possible to select fonts strictly by these attributes, as described in Section 8.5. However, for normal usage, there are some declarations and corresponding commands to simplify this procedure.

For the Computer Modern fonts provided with TeX and LaTeX, the following attributes and values exist:

Family: for the general overall style. Traditional typographical families have names like *Baskerville, Bodoni, Times Roman, Helvetica*, and so on. The standard LaTeX 2_ε installation provides three families with declarations

\rmfamily to switch (back) to a Roman font;
\ttfamily to switch to a typewriter font;
\sffamily to select a sans serif font.

Shape: for the form of the font. The shape declarations available with the standard installation are

\upshape to switch (back) to an upright font;
\itshape to select an *italic* shape;
\slshape to choose a font that is *slanted*;
\scshape to switch to CAPS AND SMALL CAPS.

Series: for the width and/or weight (boldness) of the font. The declarations possible are

\mdseries to switch (back) to medium weight;
\bfseries to select a **bold face** font.

These do not exhaust all the possible attribute settings, but they do cover the most standard ones, especially for the Computer Modern fonts. For other fonts, especially PostScript ones, additional attribute values exist. See Section 8.5 for more details.

These declarations are used just like any others, normally enclosed in a pair of curly braces {...}, such as {\scshape Romeo and Juliet} producing ROMEO AND JULIET. For longer sections of text, an environment is preferable:

\begin{*font_style*} ... *text in new font* ... \end{*font_style*}

This keeps better track of the beginning and end of the switch-over. For *font_style*, any of the above font commands may be used, leaving off the initial \.

Since changing any one attribute leaves the others as they were, all possible combinations may be obtained. (However, this does not mean that a font exists for each possible combination; if not, a substitution will be made.) If we select first a bold series with `\bfseries`, and then a slanted shape with `\slshape`, we obtain a bold, slanted font.

```
normal and {\bfseries bold and
    {\slshape slanted} and back} again.
```

produces: normal and **bold and *slanted* and back** again.

Finally, the declaration `\normalfont` resets all the attributes (except size) back to their defaults: Roman, upright, medium weight. It is often useful to issue this command just to be sure of the font in effect.

4.1.4 Font commands ⌷2ε⌷

For each of the font declarations listed above, there is a corresponding *font command* that sets its argument in a font with the specified attribute.

Family:	`\textrm{`*text*`}`	`\texttt{`*text*`}`	`\textsf{`*text*`}`
Shape:	`\textup{`*text*`}`	`\textit{`*text*`}`	`\textsl{`*text*`}`
	`\textsc{`*text*`}`		
Series:	`\textmd{`*text*`}`	`\textbf{`*text*`}`	
Default:	`\textnormal{`*text*`}`		
Emphasis:	`\emph{`*text*`}`		

Note that the `\emph` command is included here, corresponding to the declaration `\em`. The argument of `\textnormal` is set in the standard font selected with `\normalfont`.

The use of such commands to change the font for short pieces of text, or single words, is much more logical than placing a declaration inside an implied environment. The previous example now becomes

```
normal and \textbf{bold and \textsl{slanted}
    and back} again.
```

to make: normal and **bold and *slanted* and back** again.

As for the `\emph` command, these font commands automatically add any necessary *italic correction* (Section 3.5.1) when switching between upright and slanted/italic fonts. The correction may be suppressed by including `\nocorr` where it would otherwise appear, as

```
\textit{some italics\nocorr} without correction
{\slshape italics \nocorr\textup{without} correction}
```

Reminder: the font attribute declarations and commands are only available with LaTeX 2ε, not with version 2.09.

4.1.5 Old font declarations ⌷2.09⌷

For compatibility with LaTeX 2.09, the two-letter font declarations (originally part of TeX) have been retained.

\rm	Roman	\it	*Italic*	\sc	Small Caps
\bf	**Bold face**	\sl	*Slanted*	\sf	Sans Serif
\tt	Typewriter				

These have a different behavior from that of their newer counterparts: they rigidly select a particular font instead of altering only one attribute, albeit retaining the current size. It is as though a \normalfont declaration were issued with them.

> {\bf text} equals {\normalfont\bfseries text}

4.1.6 Additional fonts

!

It is likely that your computing center or your PC TeX package has even more fonts and sizes than those listed above. Consult the *Local Guide* or the package instructions. If so, they may be made available for use within a LaTeX document either by referring to them by name, or by their attributes, if they have been set up for NFSS.

To load a new font by name, the command

> \newfont{*fnt*}{*name* scaled *factor*}   or
> \newfont{*fnt*}{*name* at *size*}

is given, which assigns the font to the new command named *fnt*. In the first case, *factor* is a number 1000 times the scaling factor that is to be used to magnify or reduce the font from its basic or design size. In the second case, the font is scaled to be of the *size* specified. To install a slanted, sans serif font of size 20.74 pt, as \sss, we load cmssi17 at 20.74pt with

> \newfont{\sss}{cmssi17 at 20.74pt}

Now the declaration \sss switches to this font in the same way as \rm and \it change fonts, except that the baseline separation is not altered.

Alternatively, the new font declaration can be made by attributes with (see Section C.5.2)

> \DeclareFixedFont{\sss}{OT1}{cmss}{m}{sl}{20.74}

Indeed, if one wants to use the current encoding and \sffamily without knowing what they are, or without worrying so precisely what size must be stated, it is also possible to give

> \DeclareFixedFont{\sss}{\encodingdefault}{\sfdefault}
>   {m}{sl}{20}

(The defaults are explained in Section C.5.1.)

4.1.7 Character sets and symbols

!

The character sets that TeX and LaTeX print are provided with the program implementation. Depending on the version, there are between 400 and 800 font files for the different typefaces, sizes, and magnifications, each consisting of 128 or 256 single characters or symbols.

The individual character sets are each stored in their own files. The names of the 75 standard character files are listed in Appendix F, where many of them are also printed out.

Each symbol within a character set is addressed by means of a number between 0 and 127 (or 255). The command

> `\symbol{`*num*`}`

will produce that symbol with the internal identification number *num* in the current font. The symbol ¿ in the present font has the internal number 62 and can be printed with the command `\symbol{62}`. The identification number may also be given as an *octal* (prefix ') or *hexadecimal* (prefix ") number. Thus the symbol commands `\symbol{28}`, `\symbol{'34}`, and `\symbol{"1C}` are all identical, producing 'ø'.

The `\symbol` command may also be used to generate symbols for which no other command has been defined: for example, `{\ttfamily\symbol{'40}` `\symbol{'42} \symbol{'134}}` produces ␣ " \.

Section F.2.2 presents the assignments of the identification numbers with the characters for the different symbol families.

4.2   Centering and indenting

4.2.1   Centered text

The environment

> `\begin{center}` *line 1* `\\` *line 2* `\\` ... *line n* `\end{center}`

centers the sections of text that are separated by the `\\` command. (An optional additional line spacing may be inserted with `\\[`*len*`]`.) If the text is too long for one line, it is split over several lines using uniform word spacing, filling the whole line width as best it can, except for the last line. Word division does not occur.

Within an environment, the command `\centering` may be used to center the following text, again with `\\` as the line divider. The effect of the declaration lasts until the end of that environment.

A single line may be centered by typing its text as the argument of the TeX command `\centerline{`*text*`}`.

4.2.2   One-sided justification

The environments

> `\begin{flushleft}` *line 1* `\\` *line 2* `\\` ... *line 2* `\end{flushleft}`
> `\begin{flushright}` *line 1* `\\` *line 2* `\\` ... *line 2* `\end{flushright}`

produce text that is left (flushleft) or right (flushright) justified. If a section of text does not fit on to one line, it is spread over several with fixed word spacing, the same as for the center environment. Again, word division does not occur.

The same results may be produced within an environment with the declarations

\raggedright   replacing the flushleft   environment, and
\raggedleft   replacing the flushright environment.

4.2.3   Two-sided indentation

A section of text may be displayed by indenting it by an equal amount on both sides, with the environments

\begin{quote}   *text*   \end{quote}
\begin{quotation}   *text*   \end{quotation}

Additional vertical spacing is inserted above and below the displayed text to separate it visually from the normal text.

The text to be displayed may be of any length; it can be part of a sentence, a whole paragraph, or several paragraphs.

Paragraphs are separated as usual with an empty line, although no empty lines are needed at the beginning and end of the displayed text since additional vertical spacing is inserted here anyway.

The difference between the above two forms is thus:

In the quotation environment, paragraphs are marked by extra indentation of the first line, whereas in the quote environment, they are indicated with more vertical spacing between them.

The present text is produced within the quotation environment, while the sample above was done with the quote environment.

The quotation environment is only really meaningful when the regular text makes use of first-line indentation to show off new paragraphs.

4.2.4   Verse indentations

For indenting rhymes, poetry, verses, etc. on both sides, the environment

\begin{verse}   *poem*   \end{verse}

is more appropriate.

Stanzas are separated by blank lines
while the individual lines of the stanza are divided by \\.

If a line is too long for the reduced text width, it will be left
and right justified and continued on the next line, which is
indented even further.

The above indenting schemes may be nested inside one another. Within
a quote environment there may be another quote, quotation, or verse
environment. Each time, additional indentations are created on both sides
of the text and vertical spacing is added above and below; these quantities
however decrease as the depth of nesting increases. A maximum of six
such nestings is allowed.

Exercise 4.1: *Put some appropriate sections of text in your exercise file into the*
quote and quotation environments, that is, enclose these sections within

```
\begin{quote} . . . . . . . \end{quote}      or
\begin{quotation} . . . . . \end{quotation}
```

commands.

Exercise 4.2: *Make up a new file with the name poem.tex and type your favorite*
poem in the verse environment. Select 12pt as the standard font size and italic
as the typeface. Put the title of the poem before the verse environment in a
larger bold typeface, such as \Large\bfseries. Include the name of the poet
right justified.
Note: remember that you may include declarations to change the font style or
size within an environment and that these remain in effect only until the end of
that environment.

Exercise 4.3: *Make up another file with the name title.tex. Do you recall*
the titlepage environment for producing a free-form title page? If not, refer
to Section 3.3.1. Create a title page with this environment using font sizes and
styles of your choice, centering all the entries.
Note: within the titlepage environment you may of course make use of the
center environment, but it is also sufficient to give the \centering declaration
instead, since this will remain in effect only until the end of the titlepage
environment.
Choose the individual line spacings with the command \\[len] using an
appropriate value for the spacing len. Remember that vertical spacing before the
*first line of text must be entered with the *-form of the command \vspace*[len]*
(see Section 3.5.3).
Experiment with different font sizes and styles for the various parts of the
title page, such as title, author's name, address, until you are satisfied with the
results.
Compare your own title page with that of Exercise 3.8. If your creation appeals
to you more, include it in your standard exercise file by replacing the commands
\title, \author, \date, and \maketitle with the titlepage environment and
your own entries.

4.3   Lists

There are three environments available for producing formatted lists:

> \begin{itemize}   *list text* \end{itemize}
> \begin{enumerate}   *list text* \end{enumerate}
> \begin{description}   *list text* \end{description}

In each of these environments, the *list text* is indented from the left margin and a label, or marker, is included. What type of label is used depends on the selected list environment. The command to produce the label is \item.

4.3.1   Sample itemize

- The individual entries are indicated with a black dot, a so-called *bullet*, as the label.

- The text in the entries may be of any length. The label appears at the beginning of the first line of text.

- Successive entries are separated from one another by additional vertical spacing.

The above text was produced as follows:

```
\begin{itemize}
\item The individual entries are indicated with a black dot, a
      so-called \emph{bullet}, as the label.
\item The text in the entries may be of any length. The label
      appears at the beginning of the first line of text.
\item Successive entries are separated from one another by
      additional vertical spacing.
\end{itemize}
```

4.3.2   Sample enumerate

1. The labels consist of seqential numbers.

2. The numbering starts at 1 with every call to the enumerate environment.

The above example was generated with the following text:

```
\begin{enumerate}
\item The labels consist of sequential numbers.
\item The numbering starts at 1 with every call to the
      \texttt{enumerate} environment.
\end{enumerate}
```

4.3.3   Sample description

purpose This environment is appropriate when a number of words or expressions are to be defined.

example A keyword is used as the label and the entry contains a clarification or explanation.

other uses It may also be used as an author list in a bibliography.

The above sample was created using the following:

```
\begin{description}
\item[purpose] This environment is appropriate when a number of
    words or expressions are to be defined.
\item[example] A keyword is used as the label and the entry
    contains a clarification or explanation.
\item[other uses] It may also be used as an author list in a
    bibliography.
\end{description}
```

The \item[*option*] command contains an optional argument that appears in bold face as the label.

4.3.4   Nested lists

The above lists may be included within one another, either mixed or of one type, to a depth of four levels. The type of label used depends on the depth of the nesting. The indentation is always relative to the left margin of the enclosing list. A fourfold nesting of the itemize environment appears as follows:

- The label for the first level is a black dot, a *bullet*.
  - That of the second level is a long dash.
    * That of the third level is an asterisk.
      · And the label for the fourth level is a simple dot.
      · At the same time, the vertical spacing is decreased with increasing depth.
    * Back to the third level.
  - Back to the second level.
- And here we are at the first level of itemize once again.

Similarly for the enumerate environment, where the style of the numbering changes with the nesting level:

1. The numbering at the first level is with Arabic numerals followed by a period.

(a) At the second level, it is with lower case letters in parentheses.

    i. The third level is numbered with lower case Roman numerals with a period.

       A. At the fourth level, capital letters are used.

       B. The label style can be changed, as described in the next section.

    ii. Back to the third level.

(b) Back to the second level.

2. And the first level of enumerate again.

An example of a nested list with mixed types:

- The itemize label at the first level is a bullet.

    1. The numbering is with Arabic numerals since this is the first level of the enumerate environment.

       – This is the third level of the nesting, but the second itemize level.

       (a) And this is the fourth level of the overall nesting, but only the second of the enumerate environment.

       (b) Thus the numbering is with lower case letters in parentheses.

       – The label at this level is a long dash.

    2. Every list should contain at least two points.

- Blank lines ahead of an \item command have no effect.

The above mixed list was produced with the following text:

```
\begin{itemize}
  \item The {\tt itemize} label at the first level is a bullet.
  \begin{enumerate}
    \item The numbering is with Arabic numerals since this ...
    \begin{itemize}
      \item This is the third level of the nesting, but the ...
      \begin{enumerate}
        \item And this is the fourth level of the overall ...
        \item Thus the numbering is with lower case letters ...
      \end{enumerate}
      \item The label at this level is a long dash.
    \end{itemize}
    \item Every list should contain at least two points.
  \end{enumerate}

  \item Blank lines ahead of an ....
\end{itemize}
```

Exercise 4.4: *Produce a nested list using the* itemize *and* enumerate *environments as in the above example, but with a different sequence of these commands.*

Exercise 4.5: *Prepare a list of conference participants with their place of residence using the* description *environment, where the name of the participant appears as the argument in the* \item *command.*

Note: for all three types of lists, any text before the first \item *command will yield an error message on processing.*

4.3.5 Changing label style

The labels, or markers, used in the itemize and enumerate environments can be easily changed by means of the optional argument in the \item command. With \item[+] the label becomes +, and with \item[2.1:] it is 2.1:. The optional argument takes precedence over the standard label. For the enumerate environment, this means that the corresponding counter is *not* automatically incremented and the user must do the numbering manually.

The optional label appears right justified within the area reserved for the label. The width of this area is the amount of indentation at that level less the separation between label and text; this means that the left edge of the label area is flush with the left margin of the enclosing level.

It is also possible to change the standard labels for all or part of the document. The labels are generated with the internal commands

\labelitemi , \labelitemii , \labelitemiii , \labelitemiv
\labelenumi , \labelenumii , \labelenumiii , \labelenumiv

The endings i, ii, iii, and iv refer to the four possible levels.

These commands may be altered with \renewcommand. For example, to change the label of the third level of the itemize environment from * to +, give

    \renewcommand{\labelitemiii}{+}

Similarly the standard labels for the enumerate environment may be changed. However, here there is an additional complication that there is a counter for each enumerate level, named enumi, enumii, enumiii, and enumiv. As explained in Section 7.1.4, the value of a counter can be printed using one of the commands \arabic, \roman, \Roman, \alph, or \Alph, where the style of each command should be obvious from its name. That is, \Roman{xyz} prints the current value of the counter xyz in upper case Roman numerals, whereas \alph{xyz} prints it as a lower case letter (with *a* corresponding to 1 and *z* to 26).

These counters, together with the counter style commands, must be used in the redefinitions of the label commands. For example, to change the second-level label to Arabic numerals followed by '.)', it is necessary to give

```
\renewcommand{\labelenumii}{\arabic{enumii}.)}
```

which redefines \labelenumii to the value of counter enumii printed in Arabic, plus the characters '.)'. In this way, all the numbering levels may be changed. It is even possible to include more than one counter:

```
\renewcommand{\labelenumii}{\Alph{enumi}.\arabic{enumii}}
```

which will produce for every call to \item at level two the value of the counter enumi as a capital letter followed by the value of counter enumii as a number: that is, in the form A.1, A.2, ... , B.1, B.2, ... and so on.

If the new standard labels are to apply to the whole document, the redefining commands should be included in the preamble. Otherwise, they are valid only within the environment in which they appear.

Exercise 4.6: Change the standard labels for the itemize *environment into a long dash — (written ---) for the first level, to a medium dash – (--) for the second level, and to a hyphen - for the third level.*

Exercise 4.7: Change the standard labels for the enumerate *environment for the first level to (I), (II), ... , and for the second level to the Roman numerals of the first level followed by the number for the second level in the form I-1:, I-2:, ... , II-1:, II-2:, ...*

!

An alternative method of customizing the enumeration labels is with the enumerate package in the tools collection (Section D.3.3). Once this package has been loaded, the enumerate environment accepts an optional argument specifying the text of the label. The characters A a I i 1 represent the number in alphabetical, Roman, Arabic styles. If these characters appear elsewhere in the label text, they must be in {}. For example,

```
\begin{enumerate}[{Case} A]
 \item Witness tells the truth
 \item Witness is lying
\end{enumerate}
```
$\Rightarrow$

Case A Witness tells the truth

Case B Witness is lying

4.3.6 Bibliography

Academic publications normally include a list of references, or bibliography, containing the names of other works that are referred to within the text by means of a running number. Often the bibliography has not been finalized before the main text is started.

It would be a nuisance to have to go through the whole text and change all the reference numbering every time something was added to the bibliography. Therefore LaTeX is programmed not only to format the bibliography but also to keep track of its alterations and additions in order to modify the references in the text automatically.

The bibliography is produced with the environment

```
\begin{thebibliography}{sample_label}
   entries
\end{thebibliography}
```

The individual *entries* in the bibliography each begin with the command

```
\bibitem[label]{key} entry_text
```

Without the optional argument *label*, \bibitem produces a running number in square brackets as the label for the reference in the text. With *label*, one can give whatever indicator one wishes, such as an abbreviation of the author's name, or an arbitrary reference number. The mandatory argument *key* is a reference keyword that does not appear in the bibliography but is replaced in the text by the label. The keyword can be made up of any combination of letters, numbers, and symbols except commas.

The actual bibliography information is contained in *entry_text*, such as 'author, title, publisher, year, edition, page numbers', possibly in various typefaces. The text is indented after the first line by a width equal to that of the *sample_label*, so this should be as large as the longest label in the bibliography. For the standard application with running numbers, *sample_label* should be a dummy number with as many digits as the largest label (for example 99 if there are more than 10 but less than 100 entries).

The citation in the text itself is made with the command

```
\cite{key}
```

where *key* is the reference keyword that appears in the \bibitem command. For example, if the bibliography contains the entries

```
\begin{thebibliography}{99}
\bibitem{lamport} Leslie Lamport. \textsl{\LaTeX\ -- A Document
   Preparation System}. Addison--Wesley Co., Inc.,
   Reading, MA, 1985
   . . . . . . . .
\bibitem{knuth} Donald E. Knuth. \textsl{Computers and
   Typesetting Vol.\ A--E}. Addison--Wesley Co., Inc.,
   Reading, MA, 1984--1986
\bibitem[6a]{knuth:a} Vol A: \textsl{The \TeX\ book}, 1984
\bibitem[6b]{knuth:b} Vol B: \textsl{\TeX: The Program.}, 1986
   . . . . . . . .
\end{thebibliography}
```

then the text

```
For additional information about \LaTeX\ and \TeX\ see
\cite{lamport} and \cite{knuth,knuth:a}.
```

will produce: For additional information about LaTeX and TeX see [1] and [6, 6a].

Here `lamport`, `knuth`, and `knuth:a` have been chosen as keywords. The sample label is given as 99 since a two-digit number produces sufficient indentation for the standard form of `\bibitem`. The entry with the keyword `knuth` is the sixth in the list and thus it automatically receives the label [6]; in order for its sub-entries `knuth:a` ... `knuth:e` to be printed as [6a] ... [6e], it is necessary to set their optional *label* arguments to 6a ... 6e.

The results of the `thebibliography` environment are printed as the bibliography at the end of the document. For document classes `book` and `report`, the word **Bibliography** appears as a chapter title at the beginning, whereas for `article` the word **References** is written as a `section` heading. The above sample bibliography appears as:

[1] Leslie Lamport. *LATEX – A Document Preparation System.* Addison–Wesley Co., Inc., Reading, MA, 1985
. . . .

[6] Donald E. Knuth. *Computers and Typesetting Vol. A–E.* Addison–Wesley Co., Inc., Reading, MA, 1984–1986

[6a] Vol A: *The TEXbook*, 1984

[6b] Vol B: *TEX: The Program.*, 1986
. . . .

The actual bibliography style used in this book deviates from the LATEX standard: the citation in the text is made with the author's name and date of publication. No label at all appears in the bibliography listing. Section B.3.1 describes several extension packages that allow such a bibliography style to be achieved.

Exercise 4.8: *Produce a bibliography with the* `thebibliography` *environment, with a label consisting of the first three letters of the author's name followed by the last digits of the year of publication. If the same author has more than one work in a given year, add a running letter to distinguish them, for example* knu86c, knu86d. *With such labels, it is appropriate to make the reference keywords the same as the labels. The indentation depth should be 1.5 cm.*
Note: The indentation depth is determined by the argument sample_label *in the* `thebibliography` *environment. This is usually a dummy text with the right width. This width can be given precisely with* `\hspace{width}`.

Exercise 4.9: *Copy the* `thebibliography` *environment from the above exercise to the end of your standard exercise file (but ahead of the* `\end{document}` *command). Refer to the entries in the bibliography by inserting* `\cite` *commands into your text. Make sure that the keyword in the* `\cite` *command is written exactly as in the* `\bibitem` *command in the* `thebibliography` *environment.*

Usually the LATEX installation also contains the program BIBTEX, which can produce a bibliography automatically from the `\cite` (and `\nocite`) commands by referring to one or more databases of bibliography entries. This is described more fully in Section 8.3.3 and Appendix B. The latter also explains how bibliography databases may be made for application by many users.

4.4   Generalized lists

Lists such as those in the three environments `itemize`, `enumerate`, and `description` can be formed in a quite general way. The type of label and its width, the depth of indentation, spacings for paragraphs and labels, and so on, may be wholly or partially set by the user by means of the `list` environment:

$$\text{\textbackslash begin\{list\}\{\textit{stnd\_lbl}\}\{\textit{list\_decl}\} \textit{item\_list} \textbackslash end\{list\}}$$

Here *item_list* consists of the text for the listed entries, each of which begins with an `\item` command that generates the corresponding label.

The *stnd_lbl* contains the definition of the label to be produced by the `\item` command without an argument (see below).

The list parameters described in Section 4.4.2 are set by *list_decl* to whatever new values the user wishes.

4.4.1   Standard label

The first argument in the `list` environment defines the *stnd_lbl*, that is, the label that is produced by the `\item` command when it appears without an argument. In the case of an unchanging label, such as for the `itemize` environment, this is simply the desired symbol. If this is to be a mathematical symbol, it must be given as $symbol\_name$, enclosed in $ signs. For example, to select ⇒ as the label, *stnd_lbl* must be defined to be `$\Rightarrow$`.

However, the label is often required to contain a sequential numeration. For this purpose, a counter must be created with the `\newcounter{name}` command, where *name* is its designation. This command must appear before the first application of the counter in a `list` environment. Suppose a counter named `marker` has been defined for this use, then the argument *stnd_lbl* could be any of the commands for printing counters described in Section 4.3.5: for example, `\arabic{marker}` produces a running Arabic number.

Even more complex labels can be made up in this way. If the sequential labels are to be A-I, A-II, ... , *stnd_lbl* is set to `A--\Roman{marker}`.

Before a counter can function properly within the standard label, it must be associated with that list by including in the *list_decl* the command `\usecounter{counter}`, where *counter* is the name of the counter to be assigned (`marker` in the above example).

The standard label is actually generated by the command `\makelabel{label}`, which is called by the `\item` command. The user can redefine `\makelabel` with

!

the aid of the \renewcommand in the list declaration:

    \renewcommand{\makelabel}{*new_definition*}

If the standard label is defined in this manner, the corresponding entry in the list environment is left blank. This is because \makelabel is the more general command and overrides the other definition. Examples are presented in Section 7.5.9.

4.4.2   List style parameters

There are a number of style parameters used for formatting lists that are set by LaTeX to certain standard values. These values may be altered by the user in the *list_decl* for that particular list. The assignment is made in the usual way with the \setlength command. However, if the assignment is made outside the list environment, in most cases it will simply be ignored. This is because there are preset default values for each parameter at each level that can only be overridden by *list_decl*.

The style parameters are listed below and are also illustrated in Figure 4.1 on the opposite page, which is based on one taken from Lamport (1985, 1994).

\topsep

> is the vertical spacing in addition to \parskip that is inserted between the list and the enclosing text above and below. Its default value is set at each list level and cannot be globally redefined outside the *list_decl*.

\partopsep

> is the vertical spacing in addition to \topsep + \parskip that is inserted above and below the list when a blank line precedes the first or follows the last \item entry. It may be redefined globally, but only for the first and second levels.

\parsep

> is the vertical spacing between paragraphs of a single \item. Its default value is reset at each level, as for \topsep.

\itemsep

> is the vertical spacing in addition to \parsep that is inserted between two \item entries. As for \topsep and \parsep, its default value is reset at each level and cannot be globally changed.

\leftmargin

> is the distance from the left edge of the current environment to the left margin of the list text. There are default values for it at each level that may be globally redefined, as described in Section 4.4.6.

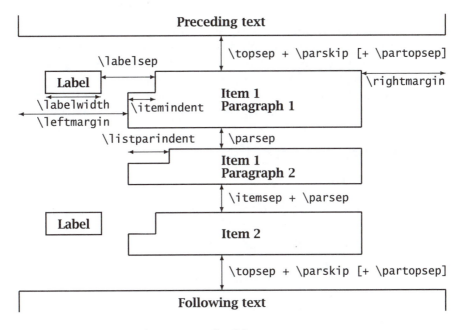

Figure 4.1: The list parameters

\rightmargin
> is the distance from the right edge of the current environment to the right margin of the list text. Its standard value is 0 pt, which can only be altered in *list_decl*.

\listparindent
> is the indentation depth of the first line of a paragraph within an \item with respect to the left margin of the list text. It is normally set to 0 pt so that no indentation occurs. This can only be changed in *list_decl*.

\labelwidth
> is the width of the box reserved for the label. The text of the label is printed right justified within this space. A new default value may be set globally which then applies to all list levels.

\labelsep
> is the spacing between the label box and the list text. A new value may be assigned globally, but it is only effective at the first level.

\itemindent
> is the distance by which the label and the first line of text in an

\item are indented to the right. It is normally set to 0 pt and so has no effect. This value can only be redefined in *list_decl*.

When changing the vertical spacings from their standard values, it is recommended that a rubber length (Section 2.4.2) be used.

The label created by the \item command normally appears right justified within a box of width \labelwidth. It is possible to make it left justified, as in the following list of parameters, by putting \hfill at the end of the definition of the standard label or in the \makelabel command.

4.4.3   Example of a user's list

List of Figures:

Figure 1:   *Page format with head, body, and foot, showing the meaning of the various elements involved.*

Figure 2:   *Format of a general list showing its elements.*

Figure 3:   *A demonstration of some of the possibilities for drawing pictures with L&TEX.*

This list was produced with the following input:

```
\newcounter{fig}
\begin{list}{\bfseries\upshape Figure \arabic{fig}:}
  {\usecounter{fig}
  \setlength{\labelwidth}{2cm}\setlength{\leftmargin}{2.6cm}
  \setlength{\labelsep}{0.5cm}\setlength{\rightmargin}{1cm}
  \setlength{\parsep}{0.5ex plus0.2ex minus0.1ex}
  \setlength{\itemsep}{0ex plus0.2ex} \slshape}
  \item Page format with head, body, and foot, showing the
        meaning of the various elements involved.
  \item Format of a general list showing its elements.
  \item A demonstration of some of the possibilities for
        drawing pictures with \LaTeX.
\end{list}
```

The command \newcounter{fig} sets up the counter fig. The standard label is defined to be the word **Figure** in upright, bold face, followed by the running Arabic number, terminated by :. This label is printed for each \item command.

The list declaration contains \usecounter{fig} as its first command, which makes the counter fig operational within the list. The width of the label box (\labelwidth) is set to 2.0 cm, the left margin of the list text (\leftmargin) to 2.6 cm, the distance between the label and the text (\labelsep) to 0.5 cm, and the right edge of the list (\rightmargin) is set to be 1 cm from that of the enclosing text.

The vertical spacing between paragraphs within an item (\parsep) is 0.5 ex but can be stretched an extra 0.2 ex or shrunk by 0.1 ex. The additional spacing between items (\itemsep) is 0 ex, stretchable to 0.2 ex.

Standard values are used for all the other list parameters. The last command in the list declaration is \slshape, which sets the list text in a *slanted* typeface.

Note: in LaTeX 2ε, if \upshape were not given in the label definition, the text of each \item would also be slanted, as ***Figure 1:***. In LaTeX 2.09, the declarations \bf and \sl must be used instead, and a bold, slanted font cannot be selected, even though it does exist.

4.4.4 Lists as new environments

If a particular type of list is employed several times within a document, it can become tiresome typing the same *stnd_lbl* and *list_decl* into the list environment every time. LaTeX offers the possibility of defining a given list as an environment under its own name. This is achieved by means of the \newenvironment command.

For example, the list in the above example can be stored so that it may be called at any time with the name figlist:

```
\newenvironment{figlist}{\begin{list}
    {\bfseries\upshape Figure \arabic{fig}:}
    {\usecounter{fig} ... {0ex plus0.2ex}\slshape}}
    {\end{list}}
```

It can then be called with

```
\begin{figlist} item_list \end{figlist}
```

so that it behaves as a predefined list environment.

Exercise 4.10: Define a new environment with the name sample *that produces a list in which every call to* \item *prints labels* Sample A, Sample B, *and so on. The labels are to be left justified within a box of width 20mm, and the distance between the label box and the item text is to be 2mm, with a total left margin of 22mm. The right edge of the text is to be moved in 5mm from that of the enclosing text. The extra vertical spacing between two items is to be* 1ex plus0.5ex minus0.4ex *in addition to the normal paragraph spacing. Secondary paragraphs within an item are to be indented by* 1em. *The normal paragraph separation should be* 0ex, *expandable to* 0.5ex.

> **!**

LaTeX itself makes frequent use of the list environment to define a number of other structures. For example, the quote environment is defined as

```
\newenvironment{quote}{\begin{list}{}
    {\setlength{\rightmargin}{\leftmargin}}
    \item[]}{\end{list}}
```

This environment is thus a list in which the value of \rightmargin is set to the current value of \leftmargin which has a default value of 2.5 em. The list itself consists of a single \item call with an *empty* label, a call that is automatically included in the definition of quote with the entry \item[].

In the same way, LaTeX defines the quotation and verse environments internally as special list environments. The left margins and the vertical spacings around the structures are left as the standard values for the list environment, and are therefore changed only when the standard values themselves are altered.

Finally, as an example of a possible user-defined special list we offer

\newenvironment{lquote}{\begin{list}{}{}\item[]}{\end{list}}

which creates an lquote environment that does nothing more than indent its enclosed text by the amount \leftmargin, with the right edge flush with that of the normal text, since \rightmargin has the standard value of 0 pt.

4.4.5   Trivial lists

!

LaTeX also contains a trivlist environment, with syntax

\begin{trivlist} *enclosed text* \end{trivlist}

in which the arguments *stnd_lbl* and *list_decl* are omitted. This is the same as a list environment for which the label is empty, \leftmargin, \labelwidth, and \itemindent are all assigned the value 0 pt, while \listparindent is set equal to \parindent and \parsep to \parskip.

LaTeX uses this environment to create further structures. For example, the call to the center environment generates internally the sequence

\begin{trivlist} \centering \item[] *enclosed text* \end{trivlist}

The environments flushleft and flushright are similarly defined.

4.4.6   Nested lists

Lists can be nested within one another with the list environments itemize, enumerate, and description, to a maximum depth of six. At each level, the new left margin is indented by the amount \leftmargin relative to that of the next higher one.

!

As mentioned earlier, it is only possible to change the standard values of a limited number of the list parameters with declarations in the preamble. One exception is the indentations of the left margins for the different nesting levels. These are set internally by the parameters \leftmargin$n$, where n stands for i, ii, iii, iv, v, or vi. These values can be changed by the user; for example, by declaring \setlength{\leftmarginiv}{12mm}, the left margin of the fourth-level list is shifted 12 mm from that of the third. These declarations must be made *outside* of the list environments and not in the *list_decl*.

At each level of list nesting, the internal macro \@list$n$ (n being i to vi) is called. This sets the value of \leftmargin equal to that of the corresponding \leftmargin$n$, unless \leftmargin is explicitly declared in the list environment. That is, there does not exist a single standard value for \leftmargin externally, but rather six different ones. The parameter \leftmargin has meaning only within a list environment.

4.5 Theorem-like declarations

In scientific literature one often has text structures like

Theorem 1 (Balzano–Weierstrass) *Every infinite set of bounded points possesses at least one maximum point.*

or

Axiom 4.1 *The natural numbers form a set S of distinct elements. For any two elements a, b, they are either identical, a = b, or different from one another, a ≠ b.*

Similar structures frequently appear with names such as *Definition, Corollary, Declaration, Lemma*, instead of *Theorem* or *Axiom*. What they have in common is that a keyword and a running number are printed in **bold face** and the corresponding text in *italic*.

Of course, these could be generated by the user by explicitly giving the type styles and appropriate number, but if a new structure of that type is later inserted in the middle of the text, the user would have the tedious job of renumbering all the following occurrences. With the command

\newtheorem{*struct_type*}{*struct_title*}[*in_counter*]

LATEX will keep track of the numbering automatically. Here *struct_type* is the user's arbitrary designation for the structure, while *struct_title* is the word that is printed in bold face followed by the running number (for example, **Theorem**). If the optional argument *in_counter* is missing, the numbering is carried out sequentially throughout the entire document. However, if the name of an existing counter, such as chapter, is given for *in_counter*, the numbering is reset every time that counter is augmented, and both are printed together, as in **Axiom 4.1** above.

The predefined structures are called with the command

\begin{*struct_type*}[*extra_title*]   *text*   \end{*struct_type*}

which also increments the necessary counter and generates the proper number. The above examples were produced with

```
\newtheorem{theorem}{Theorem} \newtheorem{axiom}{Axiom}[chapter]
. . . . . . . . . . . . . . .
\begin{theorem}[Balzano--Weierstrass] Every .... \end{theorem}
\begin{axiom} The natural numbers form .......... \end{axiom}
```

The optional *extra_title* also appears in bold face within parentheses ( ) following the running number.

Occasionally a structure is not numbered on its own but together with another structure. This can be included in the definition with another optional argument

`\newtheorem{`*struct_type*`}[`*num_like*`]{`*struct_name*`}`

where *num_like* is the name of an existing theorem structure that shares the same counter. Thus by defining `\newtheorem{subthrm}[theorem]` `{Sub-Theorem}`, the two structures `theorem` and `subthrm` will be numbered as a single series: **Theorem 1**, **Sub-Theorem 2**, **Sub-Theorem 3**, **Theorem 4**, and so on.

For more powerful theorem tools, see the $\mathcal{A}_{\mathcal{M}}\mathcal{S}$ `amsthm` package (Section E.4.1) and the `theorem` package in the tools collection (Section D.3.3).

4.6   Tabulator stops

4.6.1   Basics

On a typewriter it is possible to set *tabulator stops* at various positions within a line; then by pressing the tab key the print head or carriage jumps to the next tab location.

A similar possibility exists in LATEX with the `tabbing` environment:

`\begin{tabbing}`   *lines*   `\end{tabbing}`

One can think of the set tab stops as being numbered from left to right. At the beginning of the `tabbing` environment, no tabs are set, except for the left border, which is called the *zeroth* tab stop. The stops can be set at any spot within a line with the command `\=`, and a line is terminated by the command `\\`:

`Here is the \=first tab stop, followed by\= the second\\`

sets the first tab stop after the blank following the word *the*, and the second immediately after the word *by*.

After the tab stops have been set in this way, one can jump in the subsequent lines to each of the stops, starting from the left margin, with the command `\>`. A new line is started with the usual command `\\`.

Example:

Type	Quality	Color	Price
Paper	med.	white	low
Leather	good	brown	high
Card	bad	gray	med.

```
\begin{tabbing}
Type\qquad\= Quality\quad\=
Color\quad\= Price\\[0.8ex]
Paper   \> med. \> white \> low\\
Leather \> good \> brown \> high\\
Card    \> bad  \> gray  \> med.
\end{tabbing}
```

4.6.2 Sample line

It is often advantageous or even necessary to set the tab stops in a sample line that is not actually printed. It could contain, for example, the widest entries in the various columns that appear later, or the smallest intercolumn spacing between stops. The sample line may also contain \hspace commands to force the distance between stops to be a predetermined amount.

To suppress the printing of the sample line, it is terminated with the command \kill in place of \\.

> \hspace*{3cm}\=sample column \=\hspace{4cm}\= \kill

In addition to the left border, the above statement sets three tab stops:

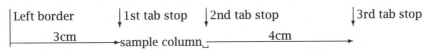

An \hspace command at the beginning of a sample line must be of the *-form, otherwise the inserted spacing will be deleted at the line margin.

4.6.3 Tab stops and the left margin

The left border of each line of the tabbing environment is at first identical with the left margin of the enclosing environment, and is designated the *zeroth* stop. By activating the 'tab key' \> at the start of a line, one sets the following text beginning at the first tab stop. However, the command \+ has the same effect, putting the left border permanently at the first stop, for all subsequent lines. With \+\+ at the beginning or end of a line, all the next lines will start two stops further along. There can be as many \+ commands in all as there are tab stops set on the line.

The command \- has the opposite effect: it shifts the left border for the following lines one stop to the left. It is not possible to set this border to be to the left of the *zeroth* stop.

The effect of the \+ commands may be overridden for a single line by putting \< at the start for each tab to be removed. This line then starts so many tabs to the left of the present border. With the next \\ command, the new line begins at the current left border determined by the total number of \+ and \- commands.

4.6.4 Further tabbing commands

Tab stops can be reset or added in every line. The command \= will add a stop if there have been sufficient \> commands to have jumped to the last stop, otherwise it will reset the next stop.

For example:

Old column 1	Old column 2		
Left column	Middle col	Extra col	
New col 1	New col 2	Old col 3	
Column 1	Column 2	Column 3	

```
Old column 1 \= Old column 2\\
Left column \> Middle col
   \= Extra col\\
New col 1 \= New col 2 \>
   Old col 3\\
Column 1\> Column 2 \> Column 3
\end{tabbing}
```

Occasionally it is desirable to be able to reset the tab stops and then to reuse the original ones later. The command \pushtabs accomplishes this by storing the current tabs and removing them from the active line. All the tab stops can then be set once again. The stored stops can be reactivated with the command \poptabs. The \pushtabs command may be given as many times as needed, but there must be the same number of \poptabs commands within any one tabbing environment.

It is possible to position text on a tab stop with *left_text* \' *right_text*, where *left_text* goes just before the current tab (or left border) with a bit of spacing, while *right_text* starts exactly at the stop. The amount of spacing between the *left_text* and the tab stop is determined by the tabbing parameter \tabbingsep. This may be changed by the user with the \setlength command as usual.

Text may be right justified up against the right border of a line with the command \' *text*. There must not be any more \> or \= commands in the remainder of the line.

The commands \=, \', and \' function as accent commands outside of the tabbing environment (Section 2.5.7). If these accents are actually needed within tabbing, they must be produced with \a=, \a', and \a' instead. For example, to produce ó, ò, or ō inside a tabbing environment, one must give \a'o, \a'o, or \a=o. The command \- also has another meaning outside of the tabbing environment (suggested word division) but since lines are not broken automatically within this environment, there is no need for an alternative form.

Here is an example illustrating all the tabbing commands:

Apples:	consumed by: people	
	horses	
	and sheep	
	reasonably juicy	
Grapefruits: a delicacy		
(see also: melons		
	pumpkins)	
Horses	feed on	apples

```
\begin{tabbing}
Grapefruits: \= \kill
Apples: \> consumed by: \= people\+\+\\
horses \\
and \' sheep\-\\
reasonably juicy\-\\
Grapefruits: \> a delicacy\\
\pushtabs
(see also: \= melons\\
   \> pumpkins)\\
\poptabs
Horses \> feed on \> apples
\end{tabbing}
```

4.6.5 Remarks on tabbing

TeX treats the `tabbing` environment like a normal paragraph, breaking a page if necessary between two lines within the environment. However, the commands `\newpage` and `\clearpage` are not allowed within it, and the command `\pagebreak` is simply ignored. If the user wishes to force a page break within the `tabbing` environment, there is a trick that he or she may employ: specify a very large interline spacing at the end of the line where the break should occur (for example, `\\[10cm]`). This forces the break and the spacing disappears at the start of the new page.

The line of text is effectively within a { } pair, so that any size or font declarations remain in force only for that one line. The text need not be put explicitly inside a pair of curly braces.

It is not possible to nest `tabbing` environments within one another.

Beware: the tab jump command `\>` always moves to the next logical tab stop. This could actually be a move backwards if the previous text is longer than the space available between the last two stops. This is in contrast to the way the tabulator works on a typewriter.

There is no automatic line breaking within the `tabbing` environment. Each line continues until terminated by a `\\` command. The text could extend beyond the right margin of the page. The user must take care that this does not happen.

The commands `\hfill`, `\hrulefill`, and `\dotfill` have no effect inside a `tabbing` environment, since no *stretching* takes place here.

Exercise 4.11: Generate the following table with the `tabbing` environment.

```
Project: Total Requirements = $900 000.00
         of which       1999 = $450 000.00
                        2000 = $350 000.00
                        2001 = $100 000.00

     1999 approved: $350 000.00  Deficiency: $100 000.00
     2000           $300 000.00              $150 000.00
     2001           $250 000.00  Surplus:    $150 000.00
     tentative      2000 = $100 000.00 for deficiency 1999
                    2001 = $  50 000.00              2000
                         + $100 000.00    excess for 1999 in 2000
Commitments         1999 = $100 000.00
                    2000 = $150 000.00          signed: H. André
```

Hint: the first line in the `tabbing` environment should read

> `Project: \=Total Requirements\= = \$900\,000.00 \+\\`

What is the effect of the `\+` command at the end of this line? How do you arrange, using these tab stops, for the years 1999, 2000, and 2001 in the second to fourth lines all to be positioned before the second tab stop? Which command should be at the end of the second line just before the terminator `\\`?

Lines 1–4 and 8–12 all use the same set of tab stops, even though there are additional stops set in the eighth line. With `\$1\=00\,000.00` *one can align the entry* `\$\>50\,000.00` *in the ninth line to match the decimal places of the lines above.*

Lines 5–7 have their own tab stops. Use the save and recall feature to store the preset tab stops and to bring them back. The left border of lines 5–7 correspond to the first stop of the first group. What command is at the end of the fourth line to ensure that the left border is reset to one stop earlier? How is the left border of the second-to-last line reset?

The last line contains 'signed: H. André' right justified. With what `tabbing` *command was this produced? Watch out for the accent é in this entry within the* `tabbing` *environment!*

4.7   Boxes

A *box* is a piece of text that TEX treats as a unit, like a single character. A *box* (along with the text within it) can be moved left, right, up, or down. Since the box is a unit, TEX cannot break it up again, even it was originally made up of smaller individual boxes. It is, however, possible to put those smaller boxes together as one pleases when constructing the overall box.

This is exactly what TEX does internally when it carries out the formatting: the individual characters are packed in *character* boxes, which are put together into *line* boxes horizontally with rubber lengths inserted between the words. The *line* boxes are stacked vertically into *paragraph* boxes, again with rubber lengths separating them. These then go into the *page body* box, which with the *head* and *foot* boxes constitutes the *page* box.

LATEX offers the user a choice of three *box types*: LR boxes, paragraph boxes, and rule boxes. The LR (left–right) box contains material that is ordered horizontally from *left* to *right*. A paragraph box will have its contents made into vertically stacked lines. A rule box is a rectangle filled solidly with black, usually for drawing horizontal and vertical lines.

4.7.1   LR boxes

To create boxes containing text in LR (left-to-right) mode the commands

> `\mbox{`*text*`}`   and   `\makebox[`*width*`][`*pos*`]{`*text*`}`
> `\fbox{`*text*`}`   and   `\framebox[`*width*`][`*pos*`]{`*text*`}`

are available. The two commands at the left produce an LR box with a width exactly equal to that of the *text* given between the braces { }. The `\fbox` command is the same as `\mbox` except that the $\boxed{text}$ is also framed.

With the two commands at the right, the width is predetermined by the optional length argument *width*. The other optional argument *pos* specifies how the text is positioned within the box. With no value given, the *text* is centered. Otherwise *pos* may be

l   to left justify the *text*,
r   to right justify it,
[2ε] s   to stretch it to fill up the full width.

Thus `\makebox[3.5cm]{centered text}` creates a box of width 3.5 cm in which the text is centered, as   centered text   , filled with white space, while with `\framebox[3.5cm][r]{right justified}` the text is right justified inside a framed box of width 3.5 cm: | right justified|. With LATEX 2ε, one may also give

> `\framebox[3.5cm][s]{stretched\dotfill text}`

to fill up the box, as |stretched . . . . text|, in which case some rubber length (Section 2.4.2) or other filler (page 48) must be added where the stretching is to occur.

If the *text* has a natural width that is larger than that specified in *width*, it will stick out of the box on the left, right, or both sides, depending on the choice of *pos*. For example,

> `\framebox[2mm]{centered}` produces   cen|t|ered

The above application may appear rather silly for `\framebox`, but it can indeed be very useful for `\makebox`. A width specification of 0 pt for `\makebox` can generate a centered, left, or right justified positioning of text in diagrams made with the `picture` environment (see Chapter 6 for examples). It may also be used to cause two pieces of text to overlap, as `\makebox[0pt][l]{/}S` prints a slash through an S, as $S\!\!\!/$.
Note: length specifications must always contain a dimensional unit, even when they are zero. Thus 0pt must be given for the width, not 0.

With LATEX 2ε, it is also possible to specify the *width* of an LR box relative to its natural dimensions (those produced by the simple `\mbox` command):

[2ε] `\width`   is the natural width of the box,
[2ε] `\height`   is the distance from baseline to top,
[2ε] `\depth`   is the distance from baseline to bottom,
[2ε] `\totalheight`   is `\height` plus `\depth`.

To make a framed box such that the width is six times the total height, containing centered text,

> `\framebox[6\totalheight]{Text}` | Text |

Note: these special length parameters only have meaning within the *width* specification of an LR box, or within the *height* specification of a paragraph box, as shown below. In any other context, they will produce an error message.

If a set piece of text is to appear in several places within the document, it can be stored by first giving the command

> `\newsavebox{\`*boxname*`}`

to create a box with the name `\`*boxname*. This name must conform to LATEX command name syntax (letters only) with an inital `\`. The name must not conflict with any existing LATEX command names. After such a box has been initiated, the commands

> `\sbox{\`*boxname*`}{`*text*`}` or
> `\savebox{\`*boxname*`}[`*width*`][`*pos*`]{`*text*`}`

will store the contents *text* for future use. The optional arguments *width* and *pos* have the same meanings as for `\makebox` and `\framebox`. Now with the command

> `\usebox{\`*boxname*`}`

the stored contents are inserted into the document text wherever desired, as a single unit.

The contents of an LR box may also be stored with the environment

> ②ε `\begin{lrbox}{`*boxname*`}`
>   *text*
> `\end{lrbox}`

This is equivalent to `\sbox{\`*boxname*`}{`*text*`}`. Its advantage is that it allows text within a user-defined environment (Section 7.4) to be stored for future use with `\usebox`.

4.7.2   Vertical shifting of LR boxes

The command

> `\raisebox{`*lift*`}[`*height*`][`*depth*`]{`*text*`}`

produces an `\mbox` with contents *text*, raised above the current baseline by an amount *lift*. The optional arguments tell LATEX to treat the box as though its extension above the baseline were *height* and that below were *depth*. Without these arguments, the box has its natural size determined by *text* and *lift*. Note that *lift*, *height*, and *depth* are lengths (Section 2.4.1). If *lift* is negative, the box is lowered below the baseline.

For example:

> `Baseline \raisebox{1ex}{high} and \raisebox{-1ex}{low}`
> `and back again`

produces: Baseline ^{high} and _{low} and back again.

The values for *height* and *depth* can be totally different from the actual ones of the *text*. Their effect is to determine how far away the previous and next lines of text should be from the current line, based on the heights and depths of all the boxes (characters are also boxes) in the line. By raising a box but specifying *height* to be the regular character size, the raised box will overprint the line above, and similarly for *depth* when a box is lowered.

4.7.3 Parboxes and minipages

Whole paragraphs can be put into separate *vertical* boxes (or *parboxes* in the LaTeX jargon) with the command

\parbox[*pos*]{*width*}{*text*}

or with the environment

\begin{minipage}[*pos*]{*width*}   *text* \end{minipage}

Both produce a vertical box of width *width*, in which the lines of text are stacked on top of each other as in normal paragraph mode.

The optional positioning argument *pos* can take on the values

b   to align the bottom edge of the box with the current baseline,
t   to align the top line of text with the current baseline.

Without any positioning argument, the parbox is centered vertically on the baseline of the external line of text.

The positioning argument is only meaningful when the \parbox command or the minipage environment occurs within a paragraph, for otherwise the current line and its baseline have no meaning. If the parbox is immediately preceded by a blank line, it begins a new paragraph. In this case, the vertical positioning of the parbox is made with reference to the following elements of the paragraph. These could be further parboxes. If the paragraph consists of only a single parbox or minipage, the positioning argument is meaningless and has no effect.

Examples:

```
\parbox{3.5cm}{\sloppy This is a 3.5 cm wide parbox. It is
    vertically centered on the}
\hfill CURRENT LINE \hfill
\parbox{5.5cm}{Narrow pages are hard to format. They usually
    produce many warning messages on the monitor. The command
    \texttt{\symbol{92}sloppy} can stop this.}
```

This is a 3.5 cm wide parbox. It is vertically centered on the   CURRENT LINE   Narrow pages are hard to format. They usually produce many warning messages on the monitor. The command \sloppy can stop this.

```
\begin{minipage}[b]{4.3cm}
The minipage environment creates a vertical box like the parbox
command. The bottom line of this minipage is aligned with the
\end{minipage}\hfill
\parbox{3.0cm}{middle of this narrow parbox, which in turn is
  aligned with}
\hfill
\begin{minipage}[t]{3.8cm}
the top line of the right hand minipage. It is recommended that
the user experiment with the positioning arguments to get used
to their effects.
\end{minipage}
```

The minipage environment creates a vertical box like the parbox command. The bottom line of this minipage is aligned with the   middle of this narrow parbox, which in turn is aligned with   the top line of the right hand minipage. It is recommended that the user experiment with the positioning arguments to get used to their effects.

In Section 4.7.7 we demonstrate how parboxes can be vertically stacked in any desired manner relative to one another.

The \parbox command produces a vertical box containing the *text* just like the minipage environment. However, the latter is more general. The *text* in a \parbox may not contain any of the centering, list, or other environments described in Sections 4.2–4.5. These may, on the other hand, appear within a minipage environment. That is, a minipage can include centered or indented text, as well as lists and tabbings.

4.7.4 Problems with vertical placement

⟦!⟧ Vertical positioning of minipages and parboxes can often lead to unexpected results, which can be explained by showing more graphically how a box is treated by LaTeX. Suppose we want to place two parboxes of different heights side by side, aligned on their first lines, and the two together set on the current line of text at the bottom. The 'obvious' way of doing this is

```
\begin{minipage}[b]{..}
    \parbox[t]{..}{..} \hfill  \parbox[t]{..}{..}
\end{minipage}
```

which does not work, for it produces instead the following results:

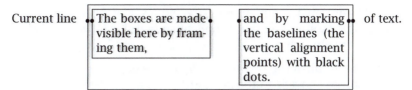

The reason for this is that each parbox or minipage is treated externally as a single character with its own height and depth above and below the baseline. As far as the outer minipage is concerned, it contains only two 'characters' on the same line, and that line is both the top and bottom one. Thus the bottom line of the outer minipage is indeed aligned with the line of text, but that bottom line is simultaneously the top line. The solution is to add a dummy second line to the outer box, as

```
\parbox[t]{..}{..} \hfill \parbox[t]{..}{..} \\ \mbox{}
```

The dummy line may not be entirely empty, hence the \mbox.

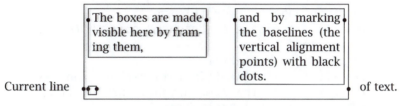

A similar problem occurs if two boxes are to be aligned with their bottom lines, and the pair aligned at the top with the current line of text. Here there are two possibilities to add a dummy *first* line.

```
\mbox{} \\                     aligns with the very top, or
\mbox{} \\[-\baselineskip]  aligns with the first text line.
```

An example of the first case is shown on page 186. Dummy lines are also needed for the solution of Exercise 4.12 on page 92.

4.7.5 Paragraph boxes of specific height $\boxed{2\varepsilon}$

In LaTeX 2ε, the syntax of the \parbox command and minipage environment has been extended to include two more optional arguments:

$\boxed{2\varepsilon}$ \parbox[*pos*][*height*][*inner_pos*]{*width*}{*text*}

$\boxed{2\varepsilon}$ \begin{minipage}[*pos*][*height*][*inner_pos*]{*width*}
    text
\end{minipage}

In both cases, *height* is a length specifying the height of the box; the parameters \height, \width, \depth, and \totalheight may be employed within the *height* argument in the same way as in the *width* argument of \makebox and \framebox (page 85).

The optional argument *inner_pos* states how the text is to be positioned *internally*, something that is only meaningful if *height* has been given. Its possible values are:

t to push the text to the top of the box,
b to shove it to the bottom,
c to center it vertically,
s to stretch it to fill up the whole box.

In the last case, rubber lengths (Section 2.4.2) should be present where the vertical stretching is to take place.

Note the difference between the external positioning argument *pos* and the internal one *inner_pos*: the former states how the box is to be aligned with the surrounding text, while the latter determines how the contents are placed within the box itself.

Example:

```
\begin{minipage}[t][2cm][t]{3cm}
  This is a minipage of height 2~cm with the text
  at the top.
\end{minipage}\hrulefill
\parbox[t][2cm][c]{3cm}{In this parbox, the text
  is centered on the same height.}\hrulefill
\begin{minipage}[t][2cm][b]{3cm}
  In this third paragraph box, the text is at the bottom.
\end{minipage}
```

This is a minipage————   ————
of height 2 cm
with the text at the   In this parbox, the
top.   text is centered on
                              the same height.   In this third para-
                                                        graph box, the text
                                                        is at the bottom.

The \hrulefill commands between the boxes show where the baselines are. All three boxes are the same size and differ only in their values of *inner_pos*.

4.7.6   Rule boxes

A rule box is a basically a filled-in black rectangle. The syntax for the general command is:

\rule[*lift*]{*width*}{*height*}

which produces a solid rectangle of width *width* and height *height*, raised above the baseline by an amount *lift*. Thus \rule{8mm}{3mm} generates ■. Without the optional argument *lift*, the rectangle is set on the baseline of the current line of text.

The parameters *lift*, *width*, and *height* are all lengths (Section 2.4.1). If *lift* has a negative value, the rectangle is set below the baseline.

It is also possible to have a rule box of zero width. This creates an invisible line with the given *height*. Such a construction is called a *strut* and is used to force a horizontal box to have a desired height or depth that is different from that of its contents. For this purpose, \vspace is inappropriate because it adds additional vertical space to that which is already there.

For example: \fbox{Text} produces ⬚Text⬚. In order to print ⬚Text⬚, one has to tell TeX that the box contents extend above and below the baseline by the desired amounts. This was done with \fbox{\rule[-2mm] {0cm}{6mm}Text}. What this says is that the text to be framed consists of 'an invisible bar beginning 2 mm below the baseline, 6 mm long, followed by the word *Text*'. The vertical bar indeed remains unseen, but it determines the upper and lower edges of the frame.

A rule box with zero height is also allowed, as an invisible horizontal line of given *width*; however, there seems to be no practical use for such a construction since horizontal displacements are readily achieved with the \hspace command.

4.7.7 Nested boxes

The box commands described above may be nested to any desired level. Including an LR box within a parbox or a minipage causes no obvious conceptual difficulties. The opposite, a parbox within an LR box, is also possible, and is easy to visualize if one keeps in mind that every box is a unit, treated by LaTeX as a single character of the corresponding size.

> A parbox inside an \fbox command has the effect that the entire parbox is framed. The present structure was made with
>
>   \fbox{\fbox{\parbox{10cm}{A parbox...}}}
>
> This is a parbox of width 10 cm inside a framebox inside a second framebox, which thus produces the double framing effect.

Enclosing a parbox inside a \raisebox allows vertical displacements of any desired amount. The two boxes here both have positioning [b], but the one at the right has been produced with:

                                                                a b c d e f g h i
    \raisebox{1cm}{\begin{minipage}[b]{2.5cm}   j k l m n o p q r
                a b c d e ... x y z\\   s t u v w x y z
                \underline{baseline}   <u>baseline</u>
            \end{minipage} }

which displaces it upwards by 1 cm.   <u>baseline</u>

A very useful structure is one in which `minipage` environments are positioned relative to one another inside an enclosing `minipage`. The positioning argument of the outside `minipage` can be used to align its contents as a unit with the neighboring text or boxes. An example of this is given in Exercise 4.12.

Finally, vertical boxes such as \parbox commands and `minipage` environments may be saved as the *text* in an \sbox or \savebox command, to be recalled later with \usebox, as described in Section 4.7.1.

4.7.8   Box style parameters

! There are two style parameters for the frame boxes \fbox and \framebox that may be reset by the user:

\fboxrule determines the thickness of the frame lines,

\fboxsep sets the amount of spacing between the frame and enclosed text.

New values are assigned to these length parameters in the usual LaTeX manner with the command \setlength: the line thickness for all the following \framebox and \fbox commands is set to 0.5 mm with \setlength{\fboxrule}{0.5mm}.

The scope of these settings also obeys the usual rule: if they are found in the preamble then they apply to the entire document; if they are within an environment then they are valid only until the end of that environment.

These parameters do *not* influence the \framebox command that is employed within the `picture` environment (Section 6.1.4) and which has different syntax and functionality from those of the normal \framebox command.

Exercise 4.12: How can the following nested structure be generated? (Note: font size is \footnotesize)

The first line of this 3.5 cm wide minipage or parbox is aligned with the first line of the neighboring minipage or parbox.

This 4.5 cm wide minipage or parbox is positioned so that its top line is at the same level as that of the box on the left, while its bottom line is even with that of the box on the right. The naïve notion that this arrangement may be achieved with the positioning arguments set to t, t, and b is incorrect. Why? What would this selection really produce?

The true solution involves the nesting of two of the three structures in an enclosing minipage, which is then separately aligned with the third one.

Note: there are two variants for the solution, depending on whether the left and middle structures are first enclosed in a minipage, or the middle and right ones. Try to work out both solutions. Incidentally, the third minipage is 3 cm wide.
Note: the problems of correctly aligning two side-by-side boxes as a pair on a line of text (Section 4.7.4) arises here once more. It will be necessary to add a dummy line to get the vertical alignment correct.

Exercise 4.13: Produce the framed structure shown below and store it with the command \sbox{\warning}{structure}. You will first have to create a box named \warning with the \newsavebox{\warning} command. Print this warning at various places in your exercise file by giving \usebox{\warning}.

> Vertical placement of minipages and parboxes can lead to surprising results which may be corrected by the use of dummy lines.

Note: the parbox width is 10 cm. There should be no difficulty producing the framed structure if one follows the previous example for the double framed box. Watch out when writing \sbox{\warning}{structure} that you have the correct number of closing braces at the end.

Next, change the values for the line thickness (\fboxrule) and frame spacing (\fboxsep) and print your results once more.

4.8 Tables

With the *box* elements and tabbing environment from the previous sections it would be possible to produce all sorts of framed and unframed tables. However, LATEX offers the user far more convenient ways to build such complicated structures.

4.8.1 Constructing tables

The environments tabular, tabular*, and array are the basic tools with which tables and matrices can be constructed. The syntax for these environments is

```
\begin{array}[pos]{cols}        rows  \end{array}
\begin{tabular}[pos]{cols}      rows  \end{tabular}
\begin{tabular*}{width}[pos]{cols}  rows  \end{tabular*}
```

The array environment can only be applied in *mathematical mode* (see Chapter 5). It is described here only because its syntax and the meaning of its arguments are exactly the same as those of the tabular environment. All three environments actually create a minipage. The meaning of the arguments is as follows:

pos Vertical positioning argument (see also the explanation of this argument for parboxes in Section 4.7.3). It can take on the values

   t   the top line of the table is aligned with the baseline of the current external line of text;

   b   the bottom line of the table is aligned with the external baseline;

       with no positioning argument given, the table is centered on the external baseline.

width This argument applies only to the `tabular*` environment and determines its overall width. In this case, the *cols* argument must contain the @-expression (see below) `@{\extracolsep{\fill}}` somewhere after the first entry. For the other two environments, the total width is fixed by the textual content.

cols The column formatting argument. There must be an entry for every column, as well as possible extra entries for the left and right borders of the table or for the intercolumn spacings.

The possible *column formatting symbols* are

l   the column contents are *left* justified;

r   the column contents are *right* justified;

c   the column contents are *centered*;

p{*wth*} the text in this column is set into lines of width *wth*, and the top line is aligned with the other columns. In fact, the text is set in a parbox with the command `\parbox[t]{`*wth*`}{`*column text*`}`;

*{*num*}{*cols*} the *column format* contained in *cols* is reproduced *num* times, so that `*{5}{|c}|` is the same as `|c|c|c|c|c|`.

The available *formatting symbols* for the left and right borders and for the intercolumn spacing are

| draws a vertical line;

|| draws two vertical lines next to each other;

@{*text*} this entry is referred to as an *@-expression*, and inserts *text* in every line of the table between the two columns where it appears.

    An @-expression removes the intercolumn spacing that is automatically put between each pair of columns. If white space is needed between the inserted text and the next column, this must be explicitly included with `\hspace{ }` within the *text* of the @-expression. If the intercolumn spacing between two particular columns is to be something other than the standard, this may be easily achieved by placing `@{\hspace{`*wth*`}}` between the appropriate columns in the formatting argument. This replaces the standard intercolumn spacing with the width *wth*.

    An `\extracolsep{`*wth*`}` within an @-expression will put extra spacing of amount *wth* between all the following columns, until countermanded by another `\extracolsep` command. In contrast to the standard spacing, this additional spacing is not removed by later @-expressions. In the `tabular*` environment, there must be a command `@{\extracolsep\fill}` somewhere in the column format so that all the subsequent intercolumn spacings can stretch out to fill the predefined table width.

If the left or right borders of the table do not consist of a vertical line, spacing is added there of an amount equal to half the normal intercolumn spacing. If this spacing is not wanted, it may be suppressed by including an empty @-expression @{} at the beginning or end of the column format.

rows contain the actual entries in the table, each horizontal row being terminated with \\. These rows consist of a sequence of column entries separated from each other by the & symbol. Thus each row in the table contains the same number of column entries as in the column definition *cols*. Some entries may be empty. The individual column entries are treated by LaTeX as though they were enclosed in braces { }, so that any changes in type style or size are restricted to that one column.

\hline This command may only appear before the first row or immediately after a row termination \\. It draws a horizontal line the full width of the table below the row that was just ended, or at the top of the table if it comes at the beginning.

Two \hline commands together draw two horizontal lines with a little space between them.

\cline{$n - m$} This command draws a horizontal line from the left side of column n to the right side of column m. Like \hline, it may only be given just after a row termination \\, and there may be more than one after another. The command \cline{1-3} \cline{5-7} draws two horizontal lines from column 1 to 3 and from column 5 to 7, below the row that was just ended. In each case, the full column widths are underlined.

\multicolumn{*num*}{*col*}{*text*} This command combines the following *num* columns into a single column with their total width including intercolumn spacings. The argument *col* contains exactly one of the positioning symbols l, r, or c, with possible @-expressions and vertical lines |. A value of 1 may be given for *num* when the positioning argument is to be changed for that column in one particular row.

In this context, a 'column' starts with a positioning symbol l, r, or c, and includes everything up to but excluding the next one. The first column also includes everything before the first positioning symbol. Thus |c@{}rl| contains three columns: the first is |c@{}, the second r, and the third l|.

The \multicolumn command may only come at the start of a row or right after a column separation symbol &.

\vline This command draws a vertical line with the height of the row at the location where it appears. In this way, vertical lines that do not extend the whole height of the table may be inserted within a column.

!   If a p-type column contains \raggedright or \centering, the \\ forces a new line *within the column entry* and not the end of the whole row. If this is the last column, use the \tabularnewline command instead, which always starts a new row. (Command added December 1, 1994.)

Since a table is a vertical box of the same sort as parbox and minipage, it may be positioned horizontally with other boxes or text (see examples in Section 4.7.3). In particular, the table must be enclosed within

\begin{center} *table* \end{center}

in order to center it on the page.

4.8.2   Table style parameters

!   There are a number of style parameters used in generating tables which LATEX sets to standard values. These may be altered by the user, either globally within the preamble or locally inside an environment. They should not be changed within the tabular environment itself.

\tabcolsep is half the width of the spacing that is inserted between columns in the tabular and tabular* environments;

\arraycolsep is the corresponding half intercolumn spacing for the array environment;

\arrayrulewidth is the thickness of the vertical and horizontal lines within a table;

\doublerulesep is the separation between the lines of a double rule.

Changes in these parameters can be made with the \setlength command as usual. For example, to make the line thickness to be 0.5 mm, give \setlength {\arrayrulewidth}{0.5mm}. Furthermore, the parameter

\arraystretch can be used to change the distance between the rows of a table. This is a multiplying factor, with a standard value of 1. A value of 1.5 means that the inter-row spacing is increased by 50%. A new value is set by redefining the parameter with the command

\renewcommand{\arraystretch}{*factor*}

4.8.3   Table examples

Creating tables is much easier in practice than it would seem from the above list of formatting possibilities. This is best illustrated with a few examples.

The simplest table consists of a row of columns in which the text entries are either centered or justified to one side. The column widths, the spacing between the columns, and thus the entire width of the table are automatically calculated.

Position	Club	Games	W	T	L	Goals	Points
1	Amesville Rockets	33	19	13	1	66:31	51:15
2	Borden Comets	33	18	9	6	65:37	45:21
3	Clarkson Chargers	33	17	7	9	70:44	41:25
4	Daysdon Bombers	33	14	10	9	66:50	38:28
5	Edgartown Devils	33	16	6	11	63:53	38:28
6	Freeburg Fighters	33	15	7	11	64:47	37:29
7	Gadsby Tigers	33	15	7	11	52:37	37:29
8	Harrisville Hotshots	33	12	11	10	62:58	35:31
9	Idleton Shovers	33	13	9	11	49:51	35:31
10	Jamestown Hornets	33	11	11	11	48:47	33:33
11	Kingston Cowboys	33	13	6	14	54:45	32:34
12	Lonsdale Stompers	33	12	8	13	50:57	32:34
13	Marsdon Heroes	33	9	13	11	50:42	31:35
14	Norburg Flames	33	10	8	15	50:68	28:38
15	Ollison Champions	33	8	9	16	42:49	25:41
16	Petersville Lancers	33	6	8	19	31:77	20:46
17	Quincy Giants	33	7	5	21	40:89	19:47
18	Ralston Regulars	33	3	11	19	37:74	17:49

The above table is made up of eight columns, the first of which is right justified, the second left justified, the third centered, the next three right justified again, and the last two centered. The column formatting argument in the `tabular` environment thus appears as

{rlcrrrcc}

The text to produce this table is

```
\begin{tabular}{rlcrrrcc}
Position & Club & Games & W & T & L & Goals & Points\\[0.5ex]
  1 & Amesville Rockets & 33 & 19 & 13 &  1 & 66:31 & 51:15 \\
  2 & Borden Comets     & 33 & 18 &  9 &  6 & 65:37 & 45:21 \\
... & .....             & .. & .. & .. & .. & ...   & ...   \\
 17 & Quincy Giants     & 33 &  7 &  5 & 21 & 40:89 & 19:47 \\
 18 & Ralston Regulars  & 33 &  3 & 11 & 19 & 37:74 & 17:49
\end{tabular}
```

In each row, the individual columns are separated from one another by the symbol & and the row itself is terminated with \\. The [0.5ex] at the end of the first row adds extra vertical spacing between the first two rows. The last row does not need the termination symbol since it is ended automatically by the \end{tabular} command.

The columns may be separated by vertical rules by including the symbol | in the column formatting argument. Changing the first line to

\begin{tabular}{r|l||c|rrr|c|c}

results in

Position	Club		Games	W	T	L	Goals	Points
1	Amesville Rockets		33	19	13	1	66:31	51:15
2	Borden Comets		33	18	9	6	65:37	45:21
⋮	⋮							⋮
17	Quincy Giants		33	7	5	21	40:89	19:47
18	Ralston Regulars		33	3	11	19	37:74	17:49

The same symbol | before the first or after the last column format
generates a vertical line on the outside edge of the table. Two symbols | |
produce a double vertical line. Horizontal lines over the whole width of
the table are created with the command \hline. They may only appear
right after a row termination \\ or at the very beginning of the table. Two
such commands \hline\hline draw a double horizontal line.

```
\begin{tabular}{|r|l||c|rrr|c|c|} \hline
Position & Club & Games & W & T & L & Goals & Points\\
 \hline\hline
 1  & Amesville Rockets & 33 & 19 & 13 &  1 & 66:31 & 51:15 \\
 \hline
 . . . . . . . . . . . . . . . . . . . . . . . . . . .
18  & Ralston Regulars  & 33 &  3 & 11 & 19 & 37:74 & 17:49 \\
 \hline
\end{tabular}
```

The table now appears as

Position	Club		Games	W	T	L	Goals	Points
1	Amesville Rockets		33	19	13	1	66:31	51:15
2	Borden Comets		33	18	9	6	65:37	45:21
⋮	⋮							⋮
17	Quincy Giants		33	7	5	21	40:89	19:47
18	Ralston Regulars		33	3	11	19	37:74	17:49

In this case, the row termination \\ must be given for the last row too
because of the presence of \hline at the end of the table.

In this example, all rows contain the same entry in the third column,
that is, 33. Such a common entry can be automatically inserted in the
column format as an @-expression of the form @{*text*}, which places *text*
between the neighboring columns. This could be accomplished for our
example by changing the column format to

{rl@{ 33 }rrrcc}   or   {|r|l||@{ 33 }|rrr|c|c|}

so that the text ' 33 ', blanks included, appears between the second and
third columns in every row. This produces the same table with slightly
different row entries: for example, the fourth row would now be given as

4   & Daysdon Bombers & 14 & 10 &   9 & 66:50 & 38:28 \\

The column format now consists of only seven column definitions, rlrrrcc. The previous third column c has been removed, and so each row contains one less column separation symbol &. The new third column, the number of games won, begins with the second & and is separated from the club name by the contents of @{ 33 }, which is entered automatically without any additional & symbol.

The last two columns give the relations between goals and points won and lost as a centered entry of the form *m:n*. The colons ':' are only coincidentally ordered exactly over one another since two-digit numbers appear in every case on both sides of the colon. If one entry had been 9:101, the colon would have been shifted slightly to the left as the entire entry was centered.

A vertical alignment of the ':' independent of the number of digits can also be achieved using an @-expression of the form r@{:}l in the column format. This means that a colon is placed in every row between a right and a left justified column. The column formatting argument in the example now becomes

{rl@{ 33 }rrrr@{:}lr@{:}l}   or
{|r|l||@{ 33 }|rrr|r@{:}l|r@{:}l|}

and the row entry is

4 & Daysdon Bombers & 14 & 10 &   9 & 66 & 50 & 38 & 28 \\

Each of the former c columns has been replaced by the two columns in r@{:}l. An @-expression inserts its text between the neighboring columns, removing the intercolumn spacing that would normally be there. Thus the r column is justified flush right with the ':' and the following l column flush left.

The same method can be employed when a column consists of numbers with decimal points and a varying number of digits.

The entries for the goal and point relationships are now made up of two columns positioned about the ':' symbol. This causes no problems for entering the number of goals won and lost or for the number of plus and minus points, since each entry has its own column. The column headings, however, are the words 'Goals' and 'Points', stretching over two columns each and without the colon. This is accomplished with the \multicolumn command, which merges selected columns in a particular row and redefines the column format. The first row of the unframed soccer table is then

Position& Club & W & T & L & \multicolumn{2}{c}{Goals}
                          & \multicolumn{2}{c}{Points}\\[0.5ex]

Here \multicolumn{2}{c}{Goals} means that the next two columns are to be combined into a centered column, containing the text 'Goals'. For the framed table, the new formatting argument in the \multicolumn commands must be {c|} since the vertical line symbol | was also removed when the old columns were combined. In deciding what belongs to a given column, use the rule that a column 'owns' everything up to but excluding the next r, l, or c.

The table of final results for our soccer league 1998/99 is to have the following title:

```
\begin{tabular}{|r|l||rrr|r@{:}l|r@{:}l||c|}\hline
  \multicolumn{10}{|c|}{\bfseries 1st Regional Soccer League ---
  Final Results 1998/99}\\ \hline
 &\itshape Club &\itshape W &\itshape T &\itshape L &
  \multicolumn{2}{c|}{\itshape Goals}
  & \multicolumn{2}{c||}{\itshape Points}
  & \itshape Remarks \\ \hline\hline
```

. . . .

	Club	W	T	L	Goals	Points	Remarks
							1st Regional Soccer League — Final Results 1998/99
1	Amesville Rockets	19	13	1	66:31	51:15	League Champs
2	Borden Comets	18	9	6	65:37	45:21	Trophy Winners
3	Clarkson Chargers	17	7	9	70:44	41:25	Candidates
4	Daysdon Bombers	14	10	9	66:50	38:28	for
5	Edgartown Devils	16	6	11	63:53	38:28	National
6	Freeburg Fighters	15	7	11	64:47	37:29	League
7	Gadsby Tigers	15	7	11	52:37	37:29	
8	Harrisville Hotshots	12	11	10	62:58	35:31	
9	Idleton Shovers	13	9	11	49:51	35:31	
10	Jamestown Hornets	11	11	11	48:47	33:33	
11	Kingston Cowboys	13	6	14	54:45	32:34	Medium Teams
12	Lonsdale Stompers	12	8	13	50:57	32:34	
13	Marsdon Heroes	9	13	11	50:42	31:35	
14	Norburg Flames	10	8	15	50:68	28:38	
15	Ollison Champions	8	9	16	42:49	25:41	
16	Petersville Lancers	6	8	19	31:77	20:46	Disbanding
17	Quincy Giants	7	5	21	40:89	19:47	Demoted
18	Ralston Regulars	3	11	19	37:74	17:49	

The horizontal lines for positions 3–5, 7–14, and 17 were made with the command \cline{1-9} while all the others used \hline:

```
11  & Kingston Cowboys  & 13 &  6 & 14 & 54&45 & 32&34 &
      Medium Teams \\ \cline{1-9}
```

The last two rows of the table deserve a comment. The remark 'Demoted' is vertically placed in the middle of the two rows. This is accomplished by typing

```
18  & Ralston Regulars  &  3 & 11 & 19 & 37&74 & 17&49
    & \raisebox{1.5ex}[0pt]{Demoted}\\ \hline
```

The \raisebox command lifts the text 'Demoted' by 1.5 ex. If the optional argument [Opt] had been left out, this lifting of the box would have increased the total height of the last row by 1.5 ex. This would have resulted in correspondingly more vertical spacing between the horizontal line of row 17 and the text of row 18. This additional spacing is suppressed by the optional argument *height* = [Opt]. (See Section 4.7.2 for a description of the \raisebox command.)

Occasionally one wants to increase the vertical spacing between horizontal lines and enclosed text. The soccer table would look better if the heading were thus:

1st Regional Soccer League — Final Results 1998/99						
Club	*W*	*T*	*L*	*Goals*	*Points*	*Remarks*

This is done by inserting an invisible vertical rule, a *strut* (Section 4.7.6), into the heading text:

```
\multicolumn{10}{|c|}{\rule[-3mm]{0mm}{8mm}\bfseries 1st
    Regional Soccer League ---  Final Results 1998/99}\\ \hline
```

The included rule has a width of 0 mm, which makes it invisible, extends 3 mm below the baseline, and is 8 mm high. It thus stretches 5 mm (8 − 3) above the baseline. It effectively pushes the horizontal lines away from the baseline in both directions. If a row consists of more than one column, it is sufficient to include a *strut* in only one of them since the size of the whole row is determined by the largest column.

Exercise 4.14: *Produce your own table for the final results of your favorite team sport in the same manner as for the soccer results above. Watch out that the colons ':' are properly aligned for the goals and points relationships.*

Exercise 4.15: *Generate the following timetable.*

		6.15–7.15 pm		7.20–8.20 pm		8.30–9.30 pm	
Day	Subj.	Teacher	Subj.	Teacher	Subj.	Teacher	
		Room		Room		Room	
Mon.	UNIX	Dr. Smith	Fortran	Ms. Clarke	Math.	Mr. Mills	
		Comp. Ctr		Hall A		Hall A	
Tues.	LATEX	Miss Baker	Fortran	Ms. Clarke	Math.	Mr. Mills	
		Conf. Room		Conf. Room		Hall A	
Wed.	UNIX	Dr. Smith	C	Dr. Jones	ComSci.	Dr. Jones	
		Comp. Ctr		Hall B		Hall B	
Fri.	LATEX	Miss Baker	C++	Ms. Clarke	canceled		
		Conf. Room		Conf. Room			

The entries 'Day' and 'Subj.' are raised in the same way as 'Demoted' was in the soccer table. To simplify its application, one can introduce a user-defined command with

$\newcommand{\rb}[1]{\raisebox{1.5ex}[0pt]{#1}}$ *(see Section 7.3.2)*

so that $\rb{entry}$ *behaves the same as* $\raisebox{1.5ex}[0pt]{entry}$*. This can be used, for example, as* $\rb{$ Mon.$}$ *or* $\rb{UNIX}$ *to elevate the entries by the necessary amount.*

In all the above examples, the entries in the individual columns are each a single line. Some tables contain certain columns with several lines of text that are somewhat separated from the rest of the row:

Model	Description	Price
FBD 200	**Desktop:** Intel Pentium 200 MMX, 32 MB RAM, 3.2 GB hard disk, 3.5" disk drive, 16 speed CD-ROM drive, 2 MB VGA, Windows 95, 15" monitor	883.70
FBD 233	**Desktop DeLuxe:** same as FBD 200 but with Pentium 233 MMX, 24 speed CD-ROM drive, 17" monitor	1376.40
FBT 266	**Mini Tower:** Intel Pentium II 266 MHz, 512 kB cache, 64 MB RAM, 4.2 GB hard disk, 3D graphics card, 32 speed CD-ROM drive, 3.5" disk drive, 19" monitor, Windows 98, Windows keyboard	2356.00

The above table is made up of three columns, the first left justified, the third right justified. The middle column contains several lines of text with a line width of 7.5 cm. This is generated with the column formatting symbol p{*width*}. The whole column formatting argument in this example is {lp{7.5cm}r}.

```
\begin{tabular}{lp{7.5cm}r}
  \bfseries Model & Description & \bfseries Price \\[1ex]
  FBD 200 &\small{\bfseries Desktop}: Intel Pentium
    200~MMX, 32~MB RAM, 3.2~GB hard disk, 3.5'' disk
    drive, 16 speed \mbox{CD-ROM} drive, 2~MB VGA,
    Windows~95, 15'' monitor & 883.70\\

       . . . . . . . . . . . . . . . . . . . . . . . . .
    Windows~98, Windows keyboard & 2356.00
\end{tabular}
```

The text for the middle column is simply typed in, being broken up into lines of width 8.0 cm automatically. The column is separated from the others with the & symbol in the usual way.

Warning: the line termination command \\ is ambiguous within a p column, for it can either start a whole new row, or if \raggedright or \centering have been given, it ends a line of text within that column entry. In this case, if this is the last column in the row, the only way to terminate the row is with \tabularnewline which always starts a new row.

Exercise 4.16: *Produce the following table.*

Course and Date	Brief Description	Prerequisites
Introduction to LSEDIT March 14 – 16	Logging on — explanation of the VMS file system — explanation and intensive application of the VMS editor LSEDIT — user modifications	none
Introduction to LaTeX March 21 – 25	Word processors and formatting programs — text and commands — environments — document and page styles — displayed text — math equations — simple user-defined structures	LSEDIT

The final example describes a blank form produced as a framed table. The difficulty here is to set the heights and widths of the empty boxes, since these are normally determined automatically by the text entries. The example shows how this may be accomplished with the help of *struts* and \hspace commands.

Budget Plan 1999–2001						
Project	Nr. ☐☐		Name ☐☐☐☐☐☐☐☐☐☐☐			
Year	1999		2000		2001	
	(Euros)	US $	(Euros)	US $	(Euros)	US $
Investment Costs						
Operating Costs						
Industrial Contracts						
Signature			Authorization			

```
\newsavebox{\k}\newsavebox{\kkk}
\sbox{\k}{\framebox[4mm]{\rule{0mm}{3mm}}}
\sbox{\kkk}{\usebox{\k}\usebox{\k}\usebox{\k}}
\begin{tabular} {|l|c|c|c|}\hline
  \multicolumn{4}{|c|}{\rule[-0.3cm]{0mm}{0.8cm}\bfseries
      Budget Plan 1999--2001}\\
  \hline\hline
  \rule[-0.4cm]{0mm}{1cm}Project
    & \multicolumn{3}{l|}{Nr. \usebox{\kkk}\hspace{0.5cm}
\vline\hspace{0.5cm}Name\usebox{\kkk}\usebox{\kkk}\usebox{\kkk}
  \usebox{\kk}}\\ \hline
\multicolumn{1}{|r|}{Year} & 1999 & 2000 & 2001 \\
\cline{2-4}
& (Euros) \vline\ US \$ & (Euros) \vline\ US \$
    & (Euros) \vline\ US \$ \\ \hline
Investment & \hspace{2.5cm}& \hspace{2.5cm}& \hspace{2.5cm} \\
```

```
Costs     & & & \\ \hline
Operating & & & \\
Costs     & & & \\ \hline
Industrial& & & \\
Contracts & & & \\ \hline
\multicolumn{4}{|l|}{\rule[-1.0cm]{0mm}{1.3cm}Signature
    \hspace{5cm}\vline~Authorization} \\ \hline
\end{tabular}
```

The first three lines are only indirectly related to the table construction. They arrange for three empty boxes [ ][ ][ ] to be drawn when the command \usebox{\kkk} is given (see Section 4.7.1).

Except for the command \hspace{2.5cm} to set the column widths of the last three columns and the command \vline to draw a vertical line within a column, this example contains nothing new that was not in the previous examples. It is only necessary to give a brief explanation of the last row in the table:

The command \multicolumn{4}{|l|} merges all four table columns into one, in which the text is set flush with the left margin. This text consists first of a strut \rule[-12mm]{0mm}{15mm} that says the height of the last row begins 12 mm below the baseline and is a total of 15 mm high. Then, beginning at the left margin, comes the word Signature, followed 5 cm later by \vline, a vertical line. The word Authorization is separated from the verical line by a blank space (~).

The above examples clearly illustrate how the column widths and row heights are automatically determined for tables. These sizes, however, may be influenced by *struts* and \hspace commands. In addition, the commands described in Section 4.8.2 permit the *intercolumn* and *inter-row* spacings as well as *line thickness* to be altered. For example,

```
\setlength{\tabcolsep}{5mm}
```

inserts 5 mm of spacing before and after every column; that is, it produces an intercolumn spacing of 10 mm. Section 4.8.2 gives more information about the use of these *table style* parameters.

4.8.4   Extension packages for tables

[!] As powerful as the tabular environment is, it does have many limitations. For this reason, there are a number of tools packages (Section D.3.3) that add additional features for constructing tables. To enable these packages, one must load them with \usepackage at the beginning of the document.

array extends the normal functionality of the tabular and array environments
by adding several column formatting arguments, and by allowing the user
to be able to define his or her own such arguments.

m{*wth*} produces a column of width *wth* which is aligned vertically in the
middle. (The standard p{*wth*} aligns the text with the top line.)

b{*wth*} is like p and m but aligns the text on the bottom line.

>{*decl*} inserts *decl* before the next column; thus >{\bfseries} sets the entire column in bold face without having to type \bfseries in each row.

<{*decl*} inserts *decl* after the last column; to have a centered column in math mode, give >{$}c<{$}.

!{*decl*} inserts *decl* between two columns without removing the normal intercolumn spacing, as for @{*decl*}.

With \newcolumntype{*type*}{*decl*} one can define new column specifiers for multiple applications. For example, to have C defined as a centered math column, give

> \newcolumntype{C}{>{$}c<{$}}

The height of all rows can be increased by setting \extrarowheight to some value with \setlength; this is useful to prevent horizontal lines from being too close to the text below them.

With \firsthline and \lasthline, horizontal lines can be issued before the first and after the last rows respectively without interfering with the vertical alignment of the table.

dcolumn loads the array package and defines a column specifier D to align a column of numbers on the decimal point. Its syntax is

> D{*in_char*}{*out_char*}{*number*}

where *in_char* is the input character for the decimal point (say .), *out_char* is the character that is output (say \cdot), and *number* is the maximum number of decimal places. If *number* is negative, the column is centered on the decimal point, otherwise it is right justified. Later versions allow *number* to specify the number of digits on both sides of the decimal point, for example as 3.2.

tabularx loads the array package and defines an environment tabularx which makes a table of a desired total width, like tabular*, but in which the columns expand, not the intercolumn spacings. The column specifier X is used to indicate the expandable columns, and is equivalent to p{*wth*} where *wth* is adjusted to the necessary size. For example,

> \begin{tabularx}{10cm}{c X 1 X} ... \end{tabularx}

produces a table of width 10 cm with columns 2 and 4 expandable.

delarray loads the array package and redefines the array environment so that it may be enclosed in braces that are automatically adjusted in size, as with the \left and \right commands of Section 5.4.1. The braces surround the column specifier. For example,

$$\begin{array}{[cc]}\ a \ \& \ b \ \backslash\backslash \ c \ \& \ d \ \end{array} \Rightarrow \begin{bmatrix} a & b \\ c & d \end{bmatrix}$$

longtable produces tables extending over several pages. It does not require the array package, but does recognize its extra features if loaded. The longtable environment takes the same column formatting argument as tabular and array, but has additional row entries at the start to determine:

- those rows that appear at the start of the table, terminated by \endfirsthead; this often includes the main \caption;

- those at the top of every continuation page, terminated by \endhead; these normally include an additional \caption and the column headers;

- those at the bottom of each page, terminated by \endfoot;

- and those rows at the end of the table, terminated by \endlastfoot.

An example of a long table is:

```
\begin{longtable}{|l|c|r|}
 \caption[Short title]{Demonstration of a long table}\\
 \hline
 Left & Center & Right \\
 \hline \endfirsthead
 \caption[]{\emph{continued}}\\
 \hline
 Left & Center & Right \\
 \hline \endhead
 \hline
 \multicolumn{3}{r}{\emph{continued on next page}}
 \endfoot
 \hline\endlastfoot
 Twenty-two & fifty & A hundred and eighty \\
 22 & 50 & 180 \\
 . . . . . .
\end{longtable}
```

The \caption command (Section 6.4.4) normally may only appear within table and figure environments but may also be used within longtable; however, when it is in a repeated row, it must have an empty optional argument [] to prevent multiple entries in the list of tables.

Up to four LATEX runs may be necessary to get the column widths right on all pages. Older versions had different ways of accomplishing this efficiently, so it is best to read the instructions in longtable.dtx for details.

4.8.5   Floating tables

The tabular environment produces a table at that place in the text where it appears, immediately after the previous text and followed by whatever comes afterwards. This causes no difficulty and is frequently just what one wants when the table fits on the page with the surrounding text. However, if the table is so long that it does not fit on the current page where it is defined, that page is ended and the next begins with the table,

Primary Energy Consumption

Energy Source	1975	1980	1986
Total Consumption			
(in million tons of BCU[a])	347.7	390.2	385.0
of which (percentages)			
petroleum	52.1	47.6	43.2
bituminous coal	19.1	19.8	20.0
brown coal	9.9	10.0	8.6
natural gas	14.2	16.5	15.1
nuclear energy	2.0	3.7	10.1
other[b]	2.7	2.3	3.0

[a]BCU = Bituminous Coal Unit (1 ton BCU corresponds to the heating equivalent of 1 ton of bituminous coal = 8140 kwh)
  [b]Wind, water, solar energy, etc.

Source: Energy Balance Study Group, Essen 1987.

followed by the subsequent text. This can lead to far too much blank space on the current page.

It would be better if the table were inserted into the immediate text position only if there is enough room for it on the present page, otherwise the text should just continue and the table be held in reserve until an appropriate place appears for it, say at the start of the next page. Since tables often are accompanied by headlines and/or captions, these should naturally be moved along with them.

LATEX does offer the possibility to *float* tables (as well as figures), including additional text, in just this manner. This is brought about with the environment

    \begin{table} *head_text*   *table*   *foot_text* \end{table}

where *table* stands for the entire table as defined in a `tabular` environment, *head_text* for whatever text appears above the table, and *foot_text* for that below. Widths, spacing, and positioning of the texts relative to the table are all matters for the user to arrange.

Independently of the enclosed text, everything that appears between \begin{table} and \end{table} is normally placed at the start of the current page. If a table already occupies the top of the page, an attempt is made to place it at the page bottom, if there is enough space for it. Otherwise, it will be placed on the next page, where further tables may be accumulated. The surrounding text is printed as though the table were not there. For further details about floats in general, including automatic sequential numbering, see Section 6.4.

The table at the top of this page was generated within the text at this location with the following (excluding the footnotes, which are described later in Section 4.10.4):

```
\begin{table} {\bfseries Primary Energy Consumption}\\[1ex]
  \begin{tabular*}{118mm}{@{}ll...rr@{}}
  . . . . . . . . . . . . . . . . . . . . .
\end{tabular*}\\[0.5ex]
\emph{Source:} Energy Balance Study Group, . . .
\end{table}
```

There are a number of formatting parameters that may be used in connection with the `table` environment, which are described together with those for figures in Section 6.4.

Exercise 4.17: *Complete the above text for the table on the previous page (without the footnotes). Pay attention to the following questions (check the explanations for the @-expressions in Section 4.8.1):*

1. *What is the effect of the @{} entries at the beginning and end of the formatting definition?*

2. *The* `tabular*` *environment generates a table with a given width, here 118 mm. What would be the effect of @{extracolsep{\fill}} at the beginning of the formatting definition?*

3. *Where in the formatting definition should @{extracolsep{\fill}} and the countermanding @{\hspace{1em}}@{\extracolsep{1em}} appear in order to format the table as it is printed here? How would the table appear if only @{\extracolsep{1em}} were given as countermand?*

4.9   Printing literal text

Occasionally it is necessary to print text exactly as it is typed, with all special characters, blanks, and line breaks appearing literally, unformatted, and in a typewriter font. Lines of computer code or samples of LATEX input text are examples of such literal text. This is accomplished with the environments

```
\begin{verbatim}    text  \end{verbatim}
\begin{verbatim*}   text  \end{verbatim*}
```

A new line is inserted before and after these environments.

With the *-form, blanks are printed with the symbol ␣ to make them visible.

As an example, on page 112 some input text is printed to demonstrate the use of footnotes in forbidden modes. This was done with

```
\begin{verbatim}
  \addtocounter{footnote}{-1}\footnotetext{Small insects}
  \stepcounter{footnote}\footnotetext{Large mammals}
\end{verbatim}
```

Literal text may also be printed within a line using the commands \verb and \verb*, as for example

```
\verb=\emph{words of text}=          \emph{words of text}
\verb*=\emph{words of text}=         \emph{words␣of␣text}
```

where the first character after \verb or \verb* (here =) is the delimiter, such that all text up to the next occurrence of that character is printed literally. This character may not appear in the literal text, obviously.

In contrast to the behavior in the verbatim environment, the literal text must be all on one line in the input text, otherwise an error message is printed 2ε. This is to indicate that you just might have forgotten to repeat the delimiting character.

Important: *neither the verbatim environment nor the \verb command may be used in an argument of any other command!*

Exercise 4.18: *Reproduce some input lines from this book as literal text.*

4.9.1 Extension packages for literal text

The standard package alltt (Section 8.8.3, page 241) provides an alltt environment that also prints its contents literally in a typewriter font, except that the characters \ { } retain their normal meaning. Thus LaTeX commands can be included within the literal text. For example,

```
\begin{alltt}
Underlining \underline{typewriter}
text is also possible.
Note that dollar ($) and
percent (%) signs are
treated \emph{literally}.
\end{alltt}
```

```
Underlining typewriter
text is also possible.
Note that dollar ($) and
percent (%) signs are
treated literally.
```

The standard package shortvrb (Section 8.8.3, page 242) offers a shorthand version of the \verb command. After declaring \MakeShortVerb{\|}, one can print short literal text with |text|. The command \DeleteShortVerb{\|} then restores the original meaning to |. Any arbitrary character may be temporarily turned into a literal switch this way.

One problem with the verbatim environment is that the entire literal text is input and stored before processing, something that can lead to memory overflows. The verbatim package in the tools collection (Section D.3.3, page 388) re-implements the environment to avoid this problem. The only drawback is that *there must not be any other text on the same line as the \end{verbatim}*.

The verbatim package offers two other extra features. It provides a comment environment that simply ignores its contents, as though each line started with a % sign. And it adds a command \verbatiminput{*filename*} to input the specified file as literal text. This is useful for listing actual computer programs rather than copying them into the LaTeX file.

4.10   Footnotes and marginal notes

4.10.1   Standard footnotes

Footnotes are generated with the command

> \footnote{*footnote_text*}

which comes immediately after the word requiring an explanation in a footnote. The text *footnote_text* appears as a footnote in a smaller typeface at the bottom of the page. The first line of the footnote is indented and is given the same footnote marker as that inserted in the main text. The first footnote on a page is separated from the rest of the page text by means of a short horizontal line.

The standard footnote marker is a small, raised number[1], which is sequentially numbered.

```
... raised number\footnote{The usual method of marking
footnotes in a typewritten ... same page.}, which is ...
```

The footnote numbering is incremented throughout the whole document for the `article` class, whereas it is reset to 1 for each new chapter in the `report` and `book` classes.

The \footnote command may only be given within the normal paragraph mode, and not within math or LR modes (Section 1.5.3). In practice, this means it may not appear within an LR box (Section 4.7.1) or a parbox (Section 4.7.3). However, it may be used within a `minipage` environment, in which case the footnote text is printed beneath the minipage and not at the bottom of the actual page.[2]

The \footnote command must immediately follow the word that is to receive the note, without any intervening blanks or spacing. A footnote at the end of a sentence can be given after the period, as in the last example above:

```
... of the actual page.\footnote{With nested ... wrong place.}
```

4.10.2   Non-standard footnotes

If the user wishes the footnote numbering to be reset to 1 for each \section command with the `article` class, this may be achieved with

> \setcounter{footnote}{0}

---

[1]The usual method of marking footnotes in a typewritten manuscript with *, **, etc., could also be done; however, since the page breaks are not known at the time of typing the text, there would be a problem of avoiding duplication of the symbols on the same page.

[2]With nested minipages, the footnote comes after the next \end{minipage} command, which could be at the wrong place.

just before or after a \section command.

The internal footnote counter has the name footnote. Each call to \footnote increments this counter by one and prints the new value in Arabic numbering as the footnote marker. A different style of marker can be implemented with the command

> \renewcommand{\thefootnote}{*number_style*{footnote}}

where *number_style* is one of the counter print commands described in Section 4.3.5: \arabic, \roman, \Roman, \alph, or \Alph. However, for the counter footnote, there is an additional counter print command available, \fnsymbol, which prints the counter values 1–9 as one of nine symbols:

> * † ‡ § ¶ ‖ ** †† ‡‡

It is up to the user to see that the footnote counter is reset to zero sometime before the tenth \footnote call.

An optional argument may be added to the \footnote command

> \footnote[*num*]{*footnote_text*}

where *num* is a positive integer that is used instead of the value of the footnote counter for the marker. In this case, the footnote counter is not incremented. For example**,

> \renewcommand{\thefootnote}{\fnsymbol{footnote}}
> For example\footnote[7]{The 7th symbol ... marker.},
> \renewcommand{\thefootnote}{\arabic{footnote}}

where the last line is necessary to restore the footnote marker style to its standard form. Otherwise, all future footnotes would be marked with symbols and not with numbers.

4.10.3 Footnotes in forbidden modes

A footnote marker can be inserted in the text with the command

> \footnotemark[*num*]

even where the \footnote command is normally not allowed, that is, in LR boxes, tables, and math mode. The marker is either the optional argument *num* or, if it is omitted, the incremented value of the footnote counter. The footnote itself is not generated. This must be done external to the forbidden mode with the command

> \footnotetext[*num*]{*footnote_text*}

**The 7th symbol appears as the footnote marker.

If the optional argument has been used for the footnote marker, the same *num* must be given as the option for the text command. Similarly, if no option was used for the marker, none may appear with the text. The footnote will be generated with the value of *num* or with that of the footnote counter.

This counter is incremented by a call to \footnotemark without an optional argument. The corresponding \footnotetext command, on the other hand, does not alter the counter.

If there are a number of \footnotemark commands without optional arguments appearing before the next \footnotetext command, it is necessary to adjust the counter with the command

> \addtocounter{footnote}{*dif*}

where *dif* is a negative number saying how many times the counter must be set back. Then before every \footnotetext command, the counter must be incremented by one. This can be done either with the command \addtocounter, with *dif*=1, or with the command

> \stepcounter{footnote}

which adds 1 to the given counter.

For example: | mosquitoes[3] and elephants[4] |

> For example: \fbox{mosquitoes\footnotemark\ and
>   elephants\footnotemark}

generates the footnote markers [3] and [4]. Now the counter has the value 4. In order for the first \footnotetext outside the framed box to operate with the correct counter value, it must first be decremented by one. The two footnote texts are made with

> \addtocounter{footnote}{-1}\footnotetext{Small insects}
> \stepcounter{footnote}\footnotetext{Large mammals}

immediately following the \fbox{} command. The footnote counter now has the same value as it did on leaving the \fbox.

4.10.4 Footnotes in minipages

As mentioned in Section 4.10.1, footnote commands are allowed inside the minipage environment. However, the footnote appears underneath the minipage, not below the main page.

---

[3] Small insects
[4] Large mammals

Footnote commands within a mini-page[a] have a different marker style. The footnote comes after the next \end{minipage} command.[b] Minipage footnotes have a counter separate from that of the main page, called mpfootnote, counting independently of footnote.

```
\begin{minipage}{6cm}
Footnote commands within
a minipage\footnote{The
marker is a raised
lower-case letter.} have
a different...
\end{minipage}
```

---

[a]The marker is a raised lower-case letter.
[b]Watch out for nested minipages.

!

　　Footnotes within a tabular environment can normally only be generated with the commands described above: \footnotemark within the table and \footnotetext outside the environment. However, if the tabular environment is inside a minipage, normal \footnote commands may also be used inside the table. The footnote appears below the table where the minipage comes to an end.

Exercise 4.19: *Produce a number of footnotes in your standard exercise file by inserting them where you think fit and by selecting some appropriate footnote text.*

Exercise 4.20: *Redefine the command \thefootnote so that the footnote markers become the symbols illustrated in Section 4.10.2. Add the redefinition to the preamble of your standard exercise file.*

Exercise 4.21: *Complete Exercise 4.17 so that the footnotes [a] and [b] appear as in the table on page 107.*

4.10.5　Marginal notes

Notes in the page margin are produced with the command

　　　\marginpar{*note_text*}

which puts the text *note_text* into the right-hand margin beginning at the level of the line where the command is given. The marginal note appearing here was generated with

This is a marginal note

```
... The marginal note \marginpar{This\\ is a\\ margin-\\al note}
    appearing here ...
```

　　The text is normally enclosed in a parbox of width 1.9 cm (0.75 in). Such a narrow box causes great difficulties with line breaking, which is why the lines are manually broken with the \\ command in the above example. Such a box is far more appropriate for marginal notes in the form of a single symbol, such as the arrow shown here.

⇐

　　Another common use for the marginal note is to draw attention to certain text passages by marking them with a vertical bar in the margin. This is often done to indicate changes in a text, for comparison with earlier versions after updated single sheets have been redistributed. The example marking this paragraph was made by including

```
\marginpar{\rule[-17.5mm]{1mm}{20mm}}
```

in the first line.

The width of the marginal note can be changed with a style parameter described in the next section. The user must watch out that the total page width does not become too big for the printer.

By default, marginal notes appear in the right-hand margin of the page, or in the outer margin when the `twoside` option has been selected. 'Outer' means the right margin for odd pages, and left for even ones. With the `twocolumn` option, they are placed in the outside margins: left for the left column and right for the right one.

⇒   This leads to a problem for marginal markings such as the arrow illustrated on the previous page. On this page, it must point in the opposite direction. In fact, its direction depends on which side of the page it is to appear, and that in turns depends on the page number or column. Since these are not known at the time of writing (and may even change with later revisions) it is necessary to have another solution. This is provided by the extended syntax of the `\marginpar` command

```
\marginpar[left_text]{right_text}
```

This form of the command contains two versions of the marginal text, *left_text* to go into the left margin, and *right_text* for the right margin, depending on which one is selected. Both the arrows on this and the previous page were generated with the same command

```
\marginpar[\hfill$\Longrightarrow$]{$\Longleftarrow$}
```

(The arrow commands are mathematical symbols that are fully explained in Section 5.3.5.)

!   Without the `\hfill` command in the above `\marginpar` example, the arrow in the left margin appears as it does at the side of this paragraph, too far over to the left. The reason for this is that the `\marginpar` command sets its contents flush with the left edge of the narrow margin box, made visible here with a frame. This left edge is aligned with the main text only when the note is put on the right ⇒   side; however, in the left margin, it is displaced from the main text. The `\hfill` command has the effect of setting the contents flush with the right edge of the margin box, which is then properly aligned with the main text.

A similar device was also applied in making the first marginal note in this section. The actual command that was given to produce it was

```
\marginpar[\flushright This\\ is a\\ margin-\\al note]
     {This\\ is a\\ margin-\\al note}
```

Here `\flushright` (Section 4.2.2) is equivalent to putting an `\hfill` on each line.

The standard positioning of the marginal notes can be switched with the command `\reversemarginpar`. Once this command has been given,   ⇐ marginal notes will appear in the left margin, or in the 'inner' margin for the `twoside` option. This command remains in effect until counter-manded by `\normalmarginpar`. For the `twocolumn` option, both these commands are inoperative.

!

A page break cannot occur within a marginal note. If the note is near the bottom of the page and there is not enough room for it, it extends below the last line of text. In this case, the text in the `\marginpar` command may include a `\vspace` command at the beginning to shove it upwards, or the note must be split across two pages with two separate `\marginpar` commands. Such manual corrections should only be undertaken after the whole document is finished, since subsequent alterations can upset the conditions under which they were made.

4.10.6 Style parameters for footnotes and marginal notes

!

There are two footnote style parameters that may be changed as needed, either in the preamble or locally within an environment.

`\footnotesep`
> The vertical spacing between two footnotes. This is a length that can be changed with the `\setlength` command.

`\footnoterule`
> The command that draws a horizontal line between the page text and the footnotes. It should not add any net vertical spacing. It may be changed, for example, by

> > `\renewcommand{\footnoterule}`
> > `{\rule{`*wth*`}{`*hght*`}\vspace{-`*hght*`}}`

> A value of 0 cm for the *hght* produces an invisible line of zero thickness.

The following style parameters may be changed to redefine how marginal notes appear:

`\marginparwidth`
> determines the width of the margin box;

`\marginparsep`
> sets the separation between the margin box and the edge of the main text;

`\marginparpush`
> is the smallest vertical distance between two marginal notes.

These parameters are all lengths and are assigned new values as usual with the `\setlength` command.

4.11 Comments within text

It is extremely useful to be able to include comments, explanations, re-minders, and so forth within the text, to tell the author or some other user later what the purpose of a certain construction was. These comments are not to be formatted with the rest of the text.

In TeX, the single character command % introduces a comment. When this character appears in the text, it and the rest of the line are ignored. If a comment is several lines long, each line must be prefixed with %.

The comment character % is also useful for deactivating commands temporarily. Putting a % character in front of such a command switches off the rest of the line. This is called 'commenting lines out'. Similarly, one may have alternative text versions in the source file, with some commented out, waiting until the author's mind is made up about which version will be used.

Finally, the % character has an important role to play in suppressing implied blanks at the end of a line. This is especially desirable in user definitions where unexpected blanks can creep in between otherwise invisible declarations with arguments. See Section 7.5.8.

Exercise 4.22: *Comment out the changes from Exercise 4.20 in your preamble. These commands may be reactivated later by removing the % character.*

!

As mentioned in Section 4.9.1, large sections of text may be turned into comments with the comment environment, provided by the verbatim package. Such a block comment is useful for explanations extending over several sentences or paragraphs, or to remove a large chunk of text temporarily. In both cases, it avoids the author's having to add % signs to every line manually.

5 Mathematical Formulas

Mathematics is the soul of TeX. It was because the setting of mathematical formulas is so complicated in normal printing, not to mention on a typewriter, that Donald Knuth invented his text formatting system. On the other hand, the soul of LaTeX is document design. Nevertheless, all the power of TeX's math setting is also available in LaTeX, offering an unbeatable combination.

In this chapter, we confine ourselves to the elements of mathematical typesetting available in standard LaTeX. The simplifications and additional elements provided by $\mathcal{A}_{\mathcal{M}}S$-LaTeX are reserved for Appendix E.

Mathematical formulas are produced by typing special descriptive text. This means that LaTeX must be informed that the *following text* is to be interpreted as a *mathematical formula*, and it must also be told when the *math text* has come to an end and normal text recommences. The processing of *math text* is carried out by switching to *math mode* (Section 1.5.3). Mathematical environments serve this purpose.

5.1 Mathematical environments

Mathematical formulas may occur within a line of text, as $(a + b)^2 = a^2 + 2ab + b^2$, or separated from the main text as

$$\int_0^\infty f(x)\,\mathrm{d}x \approx \sum_{i=1}^n w_i \mathrm{e}^{x_i} f(x_i)$$

These two types are distinguished by referring to them as *text* and *displayed formulas* respectively.

Text formulas, or equations, are generated with the environment

   \begin{math} *formula_text* \end{math}

Since text formulas are often very short, sometimes consisting of only a single character, a shorthand version is available as \(*formula_text* \).

If even that is too long, the very short form $formula\_text$ may also be employed. All three forms are effectively the same even though some internal differences may exist: for example, \(is fragile while $ is robust.

The contents of the formula, *formula_text*, consist of math constructs, which are described in the following sections.

Displayed formulas, or equations, are produced in the environments

```
\begin{displaymath}   formula_text  \end{displaymath}
\begin{equation}       formula_text  \end{equation}
```

The difference between these two is that the `equation` environment automatically adds a sequential equation number. The `displaymath` environment may be given with the shorthand forms \[... \] or $$ ... $$.

By default displayed formulas are centered horizontally with the equation number, if it is present, set flush with the right margin. By selecting the document class option `fleqn` (Section 3.1.1), the formulas are set left justified with an adjustable indentation. This option remains valid for the entire document whereas the amount of indentation may be changed at will with \setlength{\mathindent}{*indent*}, where *indent* is a length specification. Moreover, the document class option `leqno` sets the equation numbers flush with the left margin throughout the whole document.

Finally, multiline formulas can be created with the environments

```
\begin{eqnarray}    formula_text  \end{eqnarray}
\begin{eqnarray*}   formula_text  \end{eqnarray*}
```

where the standard form adds a sequential equation number for each line and the *-form is without equation numbers.

5.2   Main elements of math mode

5.2.1   Constants and variables

Numbers that appear within formulas are called *constants*, whereas *simple variables* are represented by single letters. The universal practice in mathematical typesetting is to put constants in Roman typeface and variables in *italics*. LaTeX adheres to this rule automatically in math mode. Blanks are totally ignored and are included in the input text simply to improve the appearance for the writer. Spacing between constants, variables, and operators like +, −, = are set automatically by LaTeX. For example $z=2a+3y$, $ z = 2 a + 3 y $ both produce $z = 2a + 3y$.

Mathematical symbols that are available on the keyboard are

$$+ \quad - \quad = \quad < \quad > \quad / \quad : \quad ! \quad ' \quad | \quad [\quad] \quad (\quad)$$

all of which may be used directly in formulas. The curly braces { } serve the purpose of logically combining parts of the formula and therefore cannot act as printable characters. To include braces in a formula, the same commands \{ and \} are used as in normal text.

$$M(s) < M(t) < |M| = m \qquad \text{\$M(s)<M(t)<|M| = m\$}$$
$$y'' = c\{f[y', y(x)] + g(x)\} \qquad \text{\$y'' = c\textbackslash\{f[y',y(x)] + g(x)\textbackslash\}\$}$$

Preparation: Create a new LATEX file with the name math.tex *containing at first only the commands* \documentclass{article}, \begin{document}, *and* \end {document}.

Exercise 5.1: *Produce the following text with your math exercise file:* 'The derivative of the indirect function $f[g(x)]$ is $\{f[g(x)]\}' = f'[g(x)]g'(x)$. For the second derivative of the product of $f(x)$ and $g(x)$ one has $[f(x)g(x)]'' = f''(x)g(x) + 2f'(x)g'(x) + f(x)g''(x)$.'
Note: higher derivatives are made with multiple ' *symbols:* \$y'''\$ *yields* y'''.

5.2.2 Exponents and indices

Mathematical formulas often contain exponents and indices, characters that are either raised or lowered relative to the main line of the formula, and printed in a smaller typeface. Although their mathematical meanings are different, superscripts and subscripts are typographically the same things as exponents and indices, respectively. It is even possible that exponents themselves have exponents or indices, and so on. These are produced by multiple applications of the raising and lowering operations.

LATEX and TEX make it possible to create any combination of exponents and indices with the correct type size in a simple manner: the character command ˆ sets the next character as an exponent (raised), while the character command _ sets it as an index (lowered).

$$x^2 \quad \text{x\textasciicircum2} \qquad a_n \quad \text{a\_n} \qquad x_i^n \quad \text{x\textasciicircum n\_i}$$

When exponents and indices occur together, their order is unimportant. The last example above could also have been given as x_iˆn.

If the exponent or index contains more than one character, the group of characters must be enclosed in braces { }:

$$x^{2n} \quad \text{x\textasciicircum\{2n\}} \qquad x_{2y} \quad \text{x\_\{2y\}} \qquad A_{i,j,k}^{-n!2} \quad \text{A\_\{i,j,k\}\textasciicircum\{-n!2\}}$$

Multiple raisings and lowerings are generated by applying ˆ and _ to the exponents and indices:

$$x^{y^2} \quad \text{x\textasciicircum\{y\textasciicircum2\}} \qquad x^{y_1} \quad \text{x\textasciicircum\{y\_1\}}$$
$$A_{j_{n,m}^{2n}}^{x_i^2} \qquad \text{A\textasciicircum\{x\_i\textasciicircum2\}\_\{j\textasciicircum\{2n\}\_\{n,m\}\}}$$

Note: The raising and lowering commands ˆ and _ are only permitted in math mode.

5.2.3   Fractions

Short fractions, especially within a text formula, are best represented using the slash character /, as in $(n+m)/2$ for $(n + m)/2$. For more complicated fractions, the command

\frac{*numerator*}{*denominator*}

is employed to write the *numerator* on top of the *denominator* with a horizontal fraction line of the right width between them.

$$\frac{1}{x+y}$$ \[\frac{1}{x+y} \]

$$\frac{a^2 - b^2}{a+b} = a - b$$ \[\frac{a^2 - b^2}{a+b} = a-b \]

Fractions may be nested to any depth within one another.

$$\frac{\frac{a}{x-y} + \frac{b}{x+y}}{1 + \frac{a-b}{a+b}}$$ \[\frac{\frac{a}{x-y} + \frac{b}{x+y}}
{1 + \frac{a-b}{a+b}}   \]

LATEX sets fractions within fractions in a smaller typeface. Section 5.5.2 describes how the automatic type sizes may be overridden if LATEX's choice is unsuitable.

5.2.4   Roots

Roots are printed with the command

\sqrt[*n*]{*arg*}

as in the example: $\sqrt[3]{8} = 2$ produces $\sqrt[3]{8} = 2$. If the optional argument n is omitted, the square root is generated: $\sqrt{a}$ yields $\sqrt{a}$.

The size and length of the root sign are automatically fitted to *arg*:
$\sqrt{x^2 + y^2 + 2xy} = x+y$ $\sqrt{x^2 + y^2 + 2xy} = x + y$, or

$$\sqrt[n]{\frac{x^n - y^n}{1 + u^{2n}}}$$ \[\sqrt[n]{\frac{x^n - y^n}{1 + u^{2n}}} \]

Roots may be nested inside one another to any depth:

$$\sqrt[3]{-q + \sqrt{q^2 + p^3}}$$ \[\sqrt[3]{-q + \sqrt{q^2 + p^3}} \]

5.2.5 Sums and integrals

Summation and integral signs are made with the two commands \sum and \int, which may appear in two different sizes depending on whether they occur in a text or displayed formula.

Sums and integrals very often possess upper and lower limits. These are printed with the exponent and index commands ^ and _. The positioning of the limits also depends on whether the formula is in text or displayed.

In a text formula \sum_{i=1}^n and \int_a^b produce $\sum_{i=1}^n$ and $\int_a^b$, whereas in a displayed formula they appear as at the left below:

$$\sum_{i=1}^n \quad \int_a^b$$

Some authors prefer the limits for the integral to be placed above and below the integral sign, the same as for summation. This is achieved with the command \limits immediately following the integral sign: \int\limits_{x=0}^{x=1}

$$\int_{x=0}^{x=1}$$

The rest of the formula text coming before and after the sum and integral signs is correctly aligned with them.

$$2 \sum_{i=1}^n a_i \int_a^b f_i(x) g_i(x) \, dx$$

\[2\sum_{i=1}^n a_i \int^b_a
f_i(x)g_i(x)\,\mathrm{d}x \]

!

Two points must be made regarding integrals such as $\int y \, dx$ and $\int f(z) \, dz$. First, there should be a little extra spacing between the differential operators dx and dz and the integrands preceding them. This is achieved with the small spacing command \,, mentioned in Section 3.5.1, not with a blank which is ignored in math mode. Second, the differential operator should be written upright, not italic, as explained in Section 5.4.10. This is accomplished by setting it within a \mathrm command (Section 5.4.2). Thus \int y\,\mathrm{d}x and \int f{z}\,\mathrm{d}z produce the desired results shown above, whereas \int y dx and \int f(z) dz yield $\int ydx$ and $\int f(z)dz$.

5.2.6 Continuation dots—ellipsis

Formulas occasionally contain a row of dots ... , meaning *and so on*. Simply typing a period three times in a row produces an undesirable result: ..., that is, the dots are too close together. Therefore LaTeX provides several commands

| \ldots | ... | *low dots* | \cdots | ⋯ | *center dots* |
| \vdots | ⋮ | *vertical dots* | \ddots | ⋱ | *diagonal dots* |

to space the dots correctly. The difference between the first two commands is best illustrated by the examples $a_0, a_1, \ldots, a_n$ and $a_0 + a_1 + \cdots + a_n$, which are produced with $a_0,a_1,\ldots,a_n$ for the first and $a_0 + a_1 + \cdots + a_n$ for the second.

The command \ldots is also available in normal text mode, whereas the other three are only allowed in math mode. In text mode, the command \dots may be used in place of \ldots with the same effect.

Exercise 5.2: Generate the following output:
The reduced cubic equation $y^3 + 3py + 2q = 0$ has one real and two complex solutions when $D = q^2 + p^3 > 0$. These are given by Cardan's formula as

$$y_1 = u + v, \quad y_2 = -\frac{u+v}{2} + \frac{i}{2}\sqrt{3}(u-v), \quad y_3 = -\frac{u+v}{2} - \frac{i}{2}\sqrt{3}(u-v)$$

where

$$u = \sqrt[3]{-q + \sqrt{q^2 + p^3}}, \qquad v = \sqrt[3]{-q - \sqrt{q^2 + p^3}}$$

Note: the spacings between the parts of the displayed equations are made with the spacing commands \quad and \qquad.

Exercise 5.3: Select the option fleqn *in the document class command and include the specification* \setlength{\mathindent}{2cm} *in the preamble. Redo the three* y *equations above, each as a separate displayed formula, using the* equation *environment instead of* displaymath *or* \[... \] *brackets.*

Exercise 5.4: Create the following text:
Each of the measurements $x_1 < x_2 < \cdots < x_r$ occurs $p_1, p_2, \ldots, p_r$ times. The mean value and standard deviation are then

$$x = \frac{1}{n}\sum_{i=1}^{r} p_i x_i, \qquad s = \sqrt{\frac{1}{n}\sum_{i=1}^{r} p_i(x_i - x)^2}$$

where $n = p_1 + p_2 + \cdots + p_r$.

Exercise 5.5: Although this equation looks very complicated, it should not present any great difficulties:

$$\int \frac{\sqrt{(ax+b)^3}}{x}\,dx = \frac{2\sqrt{(ax+b)^3}}{3} + 2b\sqrt{ax+b} + b^2\int \frac{dx}{x\sqrt{ax+b}}$$

The same applies to $\int_{-1}^{8}(dx/\sqrt[3]{x}) = \frac{3}{2}(8^{2/3} + 1^{2/3}) = 15/2$.

5.3   Mathematical symbols

There is a very wide range of symbols used in mathematical text, of which only a few are directly available from the keyboard. LaTeX provides almost every imaginable mathematical symbol that is commonly used. They are called with the symbol name prefixed with the command character \. The names themselves are derived from their mathematical meanings.

5.3.1 Greek letters

Lower case letters

α	\alpha	θ	\theta	o	o	τ	\tau
β	\beta	ϑ	\vartheta	π	\pi	υ	\upsilon
γ	\gamma	ι	\iota	ϖ	\varpi	ϕ	\phi
δ	\delta	κ	\kappa	ρ	\rho	φ	\varphi
ϵ	\epsilon	λ	\lambda	ϱ	\varrho	χ	\chi
ε	\varepsilon	μ	\mu	σ	\sigma	ψ	\psi
ζ	\zeta	ν	\nu	ς	\varsigma	ω	\omega
η	\eta	ξ	\xi				

Upper case letters

Γ	\Gamma	Λ	\Lambda	Σ	\Sigma	Ψ	\Psi
Δ	\Delta	Ξ	\Xi	Υ	\Upsilon	Ω	\Omega
Θ	\Theta	Π	\Pi	Φ	\Phi		

The Greek letters are made simply by putting the command character \ before the name of the letter. Upper case (capital) letters are distinguished by capitalizing the first letter of the name. Greek letters that do not appear in the above list are identical with some corresponding Latin letter. For example, upper case ρ is the same as Latin P and so needs no special symbol.

LATEX normally sets the upper case Greek letters in Roman (upright) type within a mathematical formula. If they need to be in italics, this can be brought about with the *math alphabet command* \mathnormal $\boxed{2\varepsilon}$: $\mathnormal{\Gamma\Pi\Phi}$ appears as $\Gamma\Pi\Phi$. (This replaces the math font *declaration* \mit $\boxed{2.09}$ which is used in LATEX 2.09 as $\mit \Gamma\Pi\Phi$.)

Greek letters may only be used in math mode. If they are needed in normal text, the command must be enclosed in $ ... $ signs.

5.3.2 Calligraphic letters

The following 26 *calligraphic* letters may also be used in math formulas:

$$\mathcal{A,B,C,D,E,F,G,H,I,J,K,L,M,N,O,P,Q,R,S,T,U,V,W,X,Y,Z}$$

These are called with the math alphabet command \mathcal $\boxed{2\varepsilon}$:

$\mathcal{A, B, C,...,Z}$

(replacing the 2.09 declaration \cal $\boxed{2.09}$).

5.3.3   Binary operators

Two mathematical quantities combined with one another to make a new quantity are said to be joined by a *binary operation*. The symbols that are available for use as binary operators are

±	\pm	∩	\cap	∘	\circ	○	\bigcirc
∓	\mp	∪	\cup	•	\bullet	□	\Box
×	\times	⊎	\uplus	◇	\diamond	◇	\Diamond
÷	\div	⊓	\sqcap	◁	\lhd	△	\bigtriangleup
·	\cdot	⊔	\sqcup	▷	\rhd	▽	\bigtriangledown
*	\ast	∨	\vee	⊴	\unlhd	◁	\triangleleft
⋆	\star	∧	\wedge	⊵	\unrhd	▷	\triangleright
†	\dagger	⊕	\oplus	⊘	\oslash	\	\setminus
‡	\ddagger	⊖	\ominus	⊙	\odot	≀	\wr
⊔⊔	\amalg	⊗	\otimes				

Note: the underlined symbol names in the above and following tables are only available in LaTeX 2ε if one of the packages latexsym (Section 8.8.3) or amsfonts (Section E.5.4) has been loaded.

5.3.4   Relations and their negations

When two mathematical quantities are compared, they are connected by a *relation*. The different types of relational symbols for the various comparisons are

≤	\le	\leq	≥	\ge	\geq	≠	\neq	~	\sim
≪	\ll		≫	\gg		≐	\doteq	≃	\simeq
⊂	\subset		⊃	\supset		≈	\approx	≍	\asymp
⊆	\subseteq		⊇	\supseteq		≅	\cong	⌣	\smile
⊏	\sqsubset		⊐	\sqsupset		≡	\equiv	⌢	\frown
⊑	\sqsubseteq		⊒	\sqsupseteq		∝	\propto	⋈	\bowtie
∈	\in		∋	\ni		≺	\prec	≻	\succ
⊢	\vdash		⊣	\dashv		⪯	\preceq	⪰	\succeq
⊨	\models		⊥	\perp		‖	\parallel |			\mid	

A number of the above symbols may be called by more than one name. For example, ≤ may be produced with either \le or \leq.

The opposite, or negated, meaning of the relation is indicated in mathematics with a slash / through the symbol: = and ≠ mean *equals* and *not equals*. For ≠ there is a special command \neq. However, in general one may put a slash through most of the above symbols by prefixing its name with \not. Thus \not\in yields ∉. The same is true for the keyboard characters: \not=, \not>, and \not< produce ≠, ≯, and ≮.

The following symbols may be negated in this manner. Note that the last two, \not\in and \notin, are not exactly the same: ∉ and ∉. The latter form is the preferred one.

≮	\not<	≯	\not>	≠	\not=
≰	\not\le	≱	\not\ge	≢	\not\equiv
⊀	\not\prec	⊁	\not\succ	≁	\not\sim
⋠	\not\preceq	⋡	\not\succeq	≄	\not\simeq
⊄	\not\subset	⊅	\not\supset	≉	\not\approx
⊈	\not\subseteq	⊉	\not\supseteq	≇	\not\cong
⋢	\not\sqsubseteq	⋣	\not\sqsupseteq	≭	\not\asymp
∉	\not\in	∉	\notin		

5.3.5 Arrows and pointers

Mathematical manuscripts often contain arrow symbols, also called *pointers*. The following multitude of arrow symbols are available:

←	\leftarrow \gets	⟵	\longleftarrow	↑	\uparrow
⇐	\Leftarrow	⟸	\Longleftarrow	⇑	\Uparrow
→	\rightarrow \to	⟶	\longrightarrow	↓	\downarrow
⇒	\Rightarrow	⟹	\Longrightarrow	⇓	\Downarrow
↔	\leftrightarrow	⟷	\longleftrightarrow	↕	\updownarrow
⇔	\Leftrightarrow	⟺	\Longleftrightarrow	⇕	\Updownarrow
↦	\mapsto	⟼	\longmapsto	↗	\nearrow
↩	\hookleftarrow	↪	\hookrightarrow	↘	\searrow
↼	\leftharpoonup	⇀	\rightharpoonup	↙	\swarrow
↽	\leftharpoondown	⇁	\rightharpoondown	↖	\nwarrow
⇌	\rightleftharpoons	⤳	\leadsto		

Here again the symbols → and ← may also be referred to under the names \to and \gets. Furthermore, the command \Longleftrightarrow may be substituted by \iff, although the latter (⟺) has a little more spacing on either side than the former (⟺).

5.3.6 Various other symbols

LaTeX contains by no means all possible symbols that may appear in mathematical texts. However, the following symbols are additional ones that LaTeX does make available. (Even more symbols are possible with the $\mathcal{AMS}$ symbol fonts and amssymb package, Section E.5.4.)

ℵ	\aleph	′	\prime	∀	\forall	□	\Box
ℏ	\hbar	∅	\emptyset	∃	\exists	◇	\Diamond
ι	\imath	∇	\nabla	¬	\neg	△	\triangle
ȷ	\jmath	√	\surd	♭	\flat	♣	\clubsuit
ℓ	\ell	∂	\partial	♮	\natural	♦	\diamondsuit
℘	\wp	⊤	\top	♯	\sharp	♥	\heartsuit
ℜ	\Re	⊥	\bot	‖	|	♠	\spadesuit
ℑ	\Im	⊢	\vdash	∠	\angle	⋈	\Join
℧	\mho	⊣	\dashv	\	\backslash	∞	\infty

5.3.7   Symbols with two sizes

The following symbols are printed in different sizes depending on whether they appear in text or displayed formulas:

Σ	$\sum$	\sum	$\cap$	$\bigcap$	\bigcap	$\odot$	$\bigodot$	\bigodot
$\int$	$\int$	\int	$\cup$	$\bigcup$	\bigcup	$\otimes$	$\bigotimes$	\bigotimes
$\oint$	$\oint$	\oint	$\sqcup$	$\bigsqcup$	\bigsqcup	$\oplus$	$\bigoplus$	\bigoplus
Π	$\prod$	\prod	$\vee$	$\bigvee$	\bigvee	$\uplus$	$\biguplus$	\biguplus
$\amalg$	$\coprod$	\coprod	$\wedge$	$\bigwedge$	\bigwedge			

The symbols \int and \sum have already been introduced in Section 5.2.5. There it was shown how these symbols may take on upper and lower limits; in the same way, all the above symbols may also be assigned upper and lower limits using the shifting commands ^ and _. The positioning of the limits varies for some symbols depending on whether they occur in text or displayed formulas. As indicated in Section 5.2.5, the command \limits forces the limits to be written above and below the symbol where they would otherwise be placed beside it. Similarly the complementary command \nolimits sets them beside the symbol when the standard positioning is above and below.

$$\oint_0^\infty \qquad \oint\limits_0^\infty$$ \[\oint^\infty_0 \oint\limits^\infty_0   \]

$$\prod_{\nu=0}^n \qquad \prod\nolimits_{\nu=0}^n$$ \[\prod^n_{\nu=0} \prod\nolimits^n_{\nu=0} \]

5.3.8   Function names

The universal standard for mathematical formulas is to set variable names in *italics* but the names of functions in Roman. If one were simply to write the function names *sin* or *log* in math mode, LATEX would interpret these as variables s i n and l o g and write them as sin and log. In order to tell LATEX that a function name is wanted, it is necessary to prefix the name with the command character \. The following functions are recognized by LATEX:

\arccos	\cosh	\det	\inf	\limsup	\Pr	\tan
\arcsin	\cot	\dim	\ker	\ln	\sec	\tanh
\arctan	\coth	\exp	\lg	\log	\sin	
\arg	\csc	\gcd	\lim	\max	\sinh	
\cos	\deg	\hom	\liminf	\min	\sup	

Some of these functions may also appear with limits attached to them. This is easily achieved by means of the $\lim_{x\to\infty}$ in text formulas and index command coming after the name $\lim_{x\to\infty}$ in displayed formulas of the function: `\lim_{x\to\infty}`.

The following function names may accept a limit with the lowering (index) command _:

> `\det  \gcd  \inf  \lim  \liminf  \limsup  \max  \min`
> `\Pr   \sup`

Finally, there are the function commands \bmod and \pmod{*arg*}, both of which produce the function *mod* in one of two forms:

`$ a \bmod b $`	$a \bmod b$	or as
`$ y \pmod{a+b} $`	$y \pmod{a+b}$.	

5.3.9 Mathematical accents

The following mathematical accents are available within math mode:

$\hat{a}$	`\hat{a}`	$\breve{a}$	`\breve{a}`	$\grave{a}$	`\grave{a}`	$\bar{a}$	`\bar{a}`
$\check{a}$	`\check{a}`	$\acute{a}$	`\acute{a}`	$\tilde{a}$	`\tilde{a}`	$\vec{a}$	`\vec{a}`
$\dot{a}$	`\dot{a}`	$\ddot{a}$	`\ddot{a}`	$\mathring{a}$	`\mathring{a}`		

The letters i and j should be printed without their dots when they are given an accent. To accomplish this, type the symbols \imath and \jmath instead of the letters, as in

> `$\vec{\imath} + \tilde{\jmath}$`: $\vec{\imath} + \tilde{\jmath}$

There are wider versions of \hat and \tilde available with the names \widehat and \widetilde. In this way, these accents may be placed over parts of a formula:

> $\widehat{1-x} = \widehat{-y}$ `$\widehat{1-x}=\widehat{-y}$`
> $\widetilde{xyz}$ `$\widetilde{xyz}$`

Exercise 5.6: *The union of two sets $\mathcal{A}$ and $\mathcal{B}$ is the set of all elements that are in at least one of the two sets, and is designated as $\mathcal{A} \cup \mathcal{B}$. This operation is commutative $\mathcal{A} \cup \mathcal{B} = \mathcal{B} \cup \mathcal{A}$ and associative $(\mathcal{A} \cup \mathcal{B}) \cup C = \mathcal{A} \cup (\mathcal{B} \cup C)$. If $\mathcal{A} \subseteq \mathcal{B}$, then $\mathcal{A} \cup \mathcal{B} = \mathcal{B}$. It then follows that $\mathcal{A} \cup \mathcal{A} = \mathcal{A}$, $\mathcal{A} \cup \{\emptyset\} = \mathcal{A}$ and $\mathcal{J} \cup \mathcal{A} = \mathcal{J}$.*

Exercise 5.7: *Applying l'Hôpital's rule, one has*

$$\lim_{x\to 0}\frac{\ln\sin\pi x}{\ln\sin x} = \lim_{x\to 0}\frac{\pi\frac{\cos\pi x}{\sin\pi x}}{\frac{\cos x}{\sin x}} = \lim_{x\to 0}\frac{\pi\tan x}{\tan\pi x} = \lim_{x\to 0}\frac{\pi/\cos^2 x}{\pi/\cos^2\pi x} = \lim_{x\to 0}\frac{\cos^2\pi x}{\cos^2 x} = 1$$

Exercise 5.8: The gamma function $\Gamma(x)$ is defined as

$$\Gamma(x) \equiv \lim_{n \to \infty} \prod_{v=0}^{n-1} \frac{n!\, n^{x-1}}{x+v} = \lim_{n \to \infty} \frac{n!\, n^{x-1}}{x(x+1)(x+2)\cdots(x+n-1)} \equiv \int_0^\infty e^{-t} t^{x-1}\, dt$$

The integral definition is valid only for $x > 0$ (2nd Euler integral).

Exercise 5.9: Remove the option `fleqn` from the document class command in Exercise 5.3 and redo the output.

Exercise 5.10:
$$\alpha \vec{x} = \vec{x}\alpha, \qquad \alpha\beta\vec{x} = \beta\alpha\vec{x}, \qquad (\alpha+\beta)\vec{x} = \alpha\vec{x} + \beta\vec{x}, \qquad \alpha(\vec{x}+\vec{y}) = \alpha\vec{x} + \alpha\vec{y}.$$
$$\vec{x}\vec{y} = \vec{y}\vec{x} \ \text{but} \ \vec{x}\times\vec{y} = -\vec{y}\times\vec{x}, \quad \vec{x}\vec{y} = 0 \ \text{for} \ \vec{x} \perp \vec{y}, \quad \vec{x}\times\vec{y} = 0, \ \text{for} \ \vec{x} \parallel \vec{y}.$$

Exercise 5.11: Reproduce Equations 5.1 and 5.2 from the next section.

5.4   Additional elements

The math elements described in the previous sections already permit the construction of very complex formulas, such as

$$\lim_{x \to 0} \frac{\sqrt{1+x}-1}{x} = \lim_{x \to 0} \frac{(\sqrt{1+x}-1)(\sqrt{1+x}+1)}{x(\sqrt{1+x}+1)} = \lim_{x \to 0} \frac{1}{\sqrt{1+x}+1} = \frac{1}{2} \quad (5.1)$$

$$\frac{\partial^2 U}{\partial x^2} + \frac{\partial^2 U}{\partial y^2} = 0 \ \Rightarrow \ U_M = \frac{1}{4\pi} \oint_\Sigma \frac{1}{r} \frac{\partial U}{\partial n}\, ds - \frac{1}{4\pi} \oint_\Sigma \frac{\partial \frac{1}{r}}{\partial n} U\, ds \quad (5.2)$$

$$I(z) = \sin(\frac{\pi}{2} z^2) \sum_{n=0}^\infty \frac{(-1)^n \pi^{2n}}{1 \cdot 3 \cdots (4n+1)} z^{4n+1} - \cos(\frac{\pi}{2} z^2) \sum_{n=0}^\infty \frac{(-1)^n \pi^{2n+1}}{1 \cdot 3 \cdots (4n+3)} z^{4n+3}$$
$$(5.3)$$

By reading the formulas from left to right there should be no difficulty in reconstructing the text that produced them. For example, the last equation is generated with

```
\begin{equation}
I(z) = \sin( \frac{\pi}{2} z^2 ) \sum_{n=0}^\infty
    \frac{ (-1)^n \pi^{2n} }{1 \cdot 3 \cdots (4n+1) } z^{4n+1}
    -\cos( \frac{\pi}{2} z^2 ) \sum_{n=0}^\infty
    \frac{ (-1)^n \pi^{2n+1} }{ 1 \cdot 3 \cdots (4n+3) } z^{4n+3}
\end{equation}
```

The above examples were made using the `equation` environment instead of the `displaymath` environment or its abbreviated form `\[ ... \]`, which has the effect of adding the equation numbers automatically. In the document classes `book` and `report`, equations are sequentially numbered within the chapter, the number being preceded by the chapter number and set within parentheses ( ), as illustrated above. For document class `article`, the equations are numbered sequentially throughout the entire document.

By default the equation number appears right justified and vertically centered with the equation. If there is not enough room for it on the same line, it is printed right justified below the equation. If the document class option `leqno` has been selected, the equation numbers are set left justified for the entire document.

The automatic numbering of equations means that the author may not know at the time of writing just what the equation number is. The LATEX cross-reference system described in Section 8.3.1 has already been explained for referring to section numbers (Section 3.3.3) and may also be used for equation numbers. By including a command `\label{`*name*`}` within the `equation` environment, one can print the unknown equation number in the text with the command `\ref{`*name*`}`, where *name* is a keyword consisting of any combination of letters, numbers, or symbols.

Examining Equation 5.3 more closely, one notices that the two parentheses pairs ( ) in cos() and sin() could stand to be somewhat larger. Furthermore, this equation just fills the line width, and if it were any longer, it would have to be broken at some appropriate spot and the parts positioned in a meaningful way relative to one another. None of the math elements described so far can accomplish these requirements.

Even something so simple as including some normal text within a formula has not yet been mentioned. The rest of this section addresses these problems.

Finally, there are times when one is not happy with the sizes that TEX has chosen, as for example in the last integral of Equation 5.2 where $\partial \dfrac{1}{r}$ would be more desirable than $\partial \frac{1}{r}$. This and other formatting aids, such as adjusting horizontal spacing between parts of formulas, are dealt with in Section 5.5.

5.4.1   Automatic sizing of bracket symbols

Mathematics often contains bracketing symbols, usually in pairs that enclose part of the formula. When printed, these bracket symbols should be the same size as the included partial formula. LATEX provides a pair of commands

> `\left`*lbrack*   *sub_form*   `\right`*rbrack*

to accomplish this. The command `\left` is placed immediately before the opening (left hand) bracket symbol *lbrack* while `\right` comes just before the closing (right hand) symbol *rbrack*.

$$\left[\int + \int \right]_{x=0}^{x=1}$$

```
\[ \left[ \int + \int \right]_{x=0}^{x=1} \]
```
The pair of brackets [ ] is adjusted to the size of the enclosed formula, as are the raised exponent and lowered index as well.

The commands \left and \right must appear as a pair. For every \left command there must be a corresponding \right command somewhere afterwards. The pairs may be nested. The first \left is paired with the last \right; the following \left with the second last \right, and so on. There must be the same number of \right as \left commands in a nesting.

The corresponding bracket symbols *lbrack* and *rbrack* may be perfectly arbitrary and do not need to be a logical pair.

$$\vec{x} + \vec{y} + \vec{z} = \left(\begin{matrix} a \\ b \end{matrix} \right[$$

This set of brackets is admittedly unusual but permissible.

```
\[ \vec{x} + \vec{y} + \vec{z} =
      \left( ... \right[ \]
```

Sometimes a formula contains only a single opening or closing bracket without a corresponding counterpart. However, the \left ... \right commands must still be given as a pair, but with a period '.' as an *invisible* bracket symbol.

$$y = \left\{ \begin{array}{r@{\quad:\quad}l} -1 & x < 0 \\ 0 & x = 0 \\ +1 & x > 0 \end{array} \right.$$

```
\[  y = \left\{ \begin{array}
        {r@{\quad:\quad}l}
      -1 & x<0 \\ 0 & x=0 \\ +1 & x>0
      \end{array} \right.     \]
```

The array environment in the above example is described in Section 4.8.1 and produces a table in math mode.

The \left ... \right commands may be applied to a total of 22 different symbols. These are

(	(	)	)	⌊	\lfloor	⌋	\rfloor
[	[	]	]	⌈	\lceil	⌉	\rceil
{	\{	}	\}	⟨	\langle	⟩	\rangle
|	|	‖	|	↑	\uparrow	⇑	\Uparrow
/	/	\	\backslash	↓	\downarrow	⇓	\Downarrow
				↕	\updownarrow	⇕	\Updownarrow

For example, \left| ... \right| produces two vertical bars adjusted in height to contain the enclosed formula text.

Exercise 5.12: *In Equation 5.3, generate* $\cos\left(\frac{\pi}{2}z^2\right)$ *and* $\sin\left(\frac{\pi}{2}z^2\right)$ *instead of* $\cos(\frac{\pi}{2}z^2)$ *and* $\sin(\frac{\pi}{2}z^2)$.

5.4.2 Ordinary text within a formula

It is often necessary to include some *normal* text within a formula, for example single words such as *and*, *or*, *if*, and so on. In this case one must switch to LR mode (Sections 1.5.3 and 4.7.1) while staying in math mode. This is carried out with the command \mbox{*normal text*} given inside the formula, together with horizontal spacing commands such as \quad or \hspace. For example:

$$X_n = X_k \qquad \text{if and only if} \qquad Y_n = Y_k \quad \text{and} \quad Z_n = Z_k$$

```
\[  X_n = X_k \qquad\mbox{if and only if}\qquad
    Y_n = Y_k \quad\mbox{and}\quad Z_n = Z_k          \]
```

In order to set a longer piece of text beside a displayed formula, as in some of the above examples, it is more appropriate to put both the formula and the text in their own parboxes or minipages, placed side by side with the proper vertical positioning.

On the other hand, if letters from text fonts are required as mathematical symbols, they should be entered with the LATEX 2_ε *math alphabet commands*:

$$\mathrm{\boxed{2\varepsilon}} \qquad \mathtt{\boxed{2\varepsilon}} \qquad \mathbf{\boxed{2\varepsilon}}$$
$$\mathsf{\boxed{2\varepsilon}} \qquad \mathit{\boxed{2\varepsilon}} \qquad \mathcal{\boxed{2\varepsilon}}$$

We have already met \mathcal in Section 5.3.2 on calligraphic letters. All these commands function the same way: they set their argument in the corresponding font.

$$\mathbf{B}^0(x) \qquad T_j^i \qquad \text{\$\mathbf{B}^0(x)\$ \quad \$\mathsf{T}^i\_j\$}$$

The command \mathnormal in Section 5.3.1 also belongs to this group. The difference between it and \mathit is that it sets its argument in the regular *math* italic font, while the latter uses the normal *text* italic. The letters are the same, but the spacing is different.

```
$\mathnormal{differ} \ne \mathit{differ}$
```
$differ \ne differ$

All the math alphabet commands set their text in math mode, which means that spaces are ignored as usual. This is not the case for text placed in an \mbox.

5.4.3 Matrices and arrays

$$\begin{matrix} a_{11} & a_{12} & \cdots & a_{1n} \\ \vdots & \vdots & \ddots & \vdots \\ a_{n1} & a_{n2} & \cdots & a_{nn} \end{matrix}$$

Structures like the one at the left are the basis for matrices, determinants, system of equations, and so on. They will all be referred to here as *arrays*.

Arrays are produced by means of the `array` environment, whose syntax and construction are described in Section 4.8.1 on tables. The `array` environment generates a table in math mode, that is, the column entries are interpreted as formula text. For example:

$$
\begin{array}{*{3}{c@{\;:+\;:}}c@{\;=\;}c}
a_{11}x_1 + a_{12}x_2 + \cdots + a_{1n}x_n = b_1 \\
a_{22}x_1 + a_{22}x_2 + \cdots + a_{2n}x_n = b_2 \\
\cdots\cdots\cdots\cdots\cdots\cdots\cdots \\
a_{n1}x_1 + a_{n2}x_2 + \cdots + a_{nn}x_n = b_n
\end{array}
$$

```
\[  \begin{array}{*{3}{c@{\:+\:}}c@{\;=\;}c}
      a_{11}x_1 & a_{12}x_2 & \cdots & a_{1n}x_n & b_1 \\
      a_{22}x_1 & a_{22}x_2 & \cdots & a_{2n}x_n & b_2 \\
        \multicolumn{5}{c}{\dotfill}                    \\
      a_{n1}x_1 & a_{n2}x_2 & \cdots & a_{nn}x_n & b_n
   \end{array}      \]
```

> **!**

As a reminder of the table construction elements (Section 4.8.1): `@{t}` inserts the contents of t between the adjacent columns. In the above example, this is `\:+\:` and `\;+\;`. The commands `\:` and `\;` have not yet been introduced but they produce small horizontal spacing in math mode (Section 5.5.1). `*{3}{c@{\:+\:}}` is an abbreviation for three repetitions of the column definition `c@{\:+\:}`. `c` defines the column to be one of centered text. `\multicolumn{5}{c}` says that the next five columns are to be merged and replaced by one with centered text. `\dotfill` fills the column with dots. The above system of equations could be produced somewhat more simply with

```
\begin{array}{c@{\:+\:}c@{\:+\cdots+\;}c@{\;=\;}c}
```

It is possible to nest `array` environments:

$$
\left(\left| \begin{array}{cc} x_{11} & x_{12} \\ x_{21} & x_{22} \end{array} \right| \atop \begin{matrix} y \\ z \end{matrix} \right)
$$

```
\[  \left( \begin{array}{c}
        \left| \begin{array}{cc}
        x_{11} & x_{12} \\ x_{21} & x_{22}
          \end{array} \right| \\
        y \\ z \end{array} \right)        \]
```

The outermost array consists of one column with centered text (c). The first entry in this column is also an array, with two centered columns. This array is surrounded left and right by vertical lines with adjusted sizes.

The `array` environment is structurally the same as a vertical box. This means that it is treated as a single character within the surrounding environment, so that it may be coupled with other symbols and construction elements.

$$
\sum_{p_1<p_2<\cdots<p_{n-k}}^{(1,2,\dots,n)} \Delta \quad \begin{matrix} p_1 p_2 \cdots p_{n-k} \\ p_1 p_2 \cdots p_{n-k} \end{matrix} \quad \sum_{q_1<q_2<\cdots q_k} \left| \begin{array}{cccc} a_{q_1 q_1} & a_{q_1 q_2} & \cdots & a_{q_1 q_k} \\ a_{q_2 q_1} & a_{q_2 q_2} & \cdots & a_{q_2 q_k} \\ \cdots\cdots\cdots\cdots\cdots \\ a_{q_k q_1} & a_{q_k q_2} & \cdots & a_{q_k q_k} \end{array} \right|
$$

```
\[ \sum_{p_1<p_2<\cdots<p_{n-k}}^{(1,2,\ldots,n)}
   \Delta_{\begin{array}{l}
           p_1p_2\cdots p_{n-k} \\ p_1p_2\cdots p_{n-k}
       \end{array}}
   \sum_{q_1<q_2<\cdots q_k} \left| \begin{array}{llcl}
               a_{q_1q_1} & a_{q_1q_2} & \cdots & a_{q_1q_k} \\
               a_{q_2q_1} & a_{q_2q_2} & \cdots & a_{q_2q_k} \\
               \multicolumn{4}{c}\dotfill\\
               a_{q_kq_1} & a_{q_kq_2} & \cdots & a_{q_kq_k}
           \end{array} \right|       \]
```

In this example, an `array` environment is used as an index on the Δ. However, the indices appear too large with respect to the rest of the formula. Section 5.4.6 presents a better solution for array indices.

As for all table environments, an optional vertical positioning parameter `b` or `t` may be included with the `array` environment. The syntax and results are described in Sections 4.7.3 and 4.8.1. This argument is included only if the array is to be positioned vertically relative to its top or bottom line rather than its center.

$$x - \begin{array}{c} a_1 \\ \vdots \\ a_n \end{array} - u - v \quad \begin{array}{cc} 10 \\ 12 \\ u+v & -120 \end{array}$$

```
\[ x - \begin{array}{c}
        a_1 \\ \vdots \\ a_n \end{array}
     - \begin{array}[t]{cl}
        u - v & 10\\
        u + v & \begin{array}[b]{r}
               12\\-120   \end{array}
     \end{array}                      \]
```

We suggest that the reader try to deduce how the various arrays are structured with the help of the generating text on the right.

Exercise 5.13: The solution for the system of equations

$$F(x,y) = 0 \quad \text{and} \quad \begin{vmatrix} F''_{xx} & F''_{xy} & F'_x \\ F''_{yx} & F''_{yy} & F'_y \\ F'_x & F'_y & 0 \end{vmatrix} = 0$$

yields the coordinates for the possible inflection points of $F(x,y) = 0$.
Note: the above displayed formula consists of two sub-formulas, between which the word 'and' plus extra spacing of amount \quad are inserted. Instead of enclosing the `array` *environment within size-adjusted vertical lines with* \left| ...
\right|*, one may use a formatting argument* {| ... |} *(Section 4.8.1) to produce the vertical lines. Such a structure is called a* determinant *in mathematics.*

Exercise 5.14: The shortest distance between two straight lines represented by the equations

$$\frac{x - x_1}{l_1} = \frac{y - y_1}{m_1} = \frac{z - z_1}{n_1} \quad \text{and} \quad \frac{x - x_2}{l_2} = \frac{y - y_2}{m_2} = \frac{z - z_2}{n_2}$$

is given by the expression

$$\frac{\pm \begin{vmatrix} x_1 - x_2 & y_1 - y_2 & z_1 - z_2 \\ l_1 & m_1 & n_1 \\ l_2 & m_2 & n_2 \end{vmatrix}}{\sqrt{\begin{vmatrix} l_1 & m_1 \\ l_2 & m_2 \end{vmatrix}^2 + \begin{vmatrix} m_1 & n_1 \\ m_2 & n_2 \end{vmatrix}^2 + \begin{vmatrix} n_1 & l_1 \\ n_2 & l_2 \end{vmatrix}^2}}$$

If the numerator is zero, the two lines meet somewhere.
Note: we do not recommend using {|cc|} in the formatting argument of the three determinants in the denominator under the root sign. Here the \left| ... \right| pair should be applied. Try out both possibilities for yourself and compare the results.

Exercise 5.15: *Laurent expansion:* using $c_n = \frac{1}{2\pi i} \oint (\zeta - a)^{-n-1} f(\zeta)\, d\zeta$, for every function $f(z)$ the following representation is valid ($n = 0, \pm1, \pm2, \ldots$)

$$f(x) = \sum_{n=-\infty}^{+\infty} c_n (z-a)^n = \begin{cases} c_0 + c_1(z-a) + c_2(z-a)^2 + \cdots + c_n(z-a)^n + \cdots \\ \qquad + c_{-1}(z-a)^{-1} + c_{-2}(z-a)^{-2} + \cdots \\ \qquad\qquad + c_{-n}(z-a)^{-n} + \cdots \end{cases}$$

Tip: the right-hand side of the equation can be created with an array environment consisting of only one column. What is its formatting argument?

5.4.4 Lines above and below formulas

The commands

\overline{*sub_form*}   and   \underline{*sub_form*}

can be used to draw lines over or under a formula or sub-formula. They may be nested to any level:

$$\overline{\overline{a}^2 + xy + \overline{\overline{z}}}$$

```
\[ \overline{\overline{a}^2 + \underline{xy}
         + \overline{\overline{z}}} \]
```

The command \underline may also be employed in normal text mode to underline text, whereas \overline is allowed only in math mode.
Exactly analogous to these are the two commands

\overbrace{*sub_form*}   and   \underbrace{*sub_form*}

which put horizontal curly braces above or below the sub-formula.

$$\overbrace{a + \underbrace{b + c} + d}$$

```
\overbrace{a + \underbrace{b+c} + d}
```

In displayed formulas, these commands may have exponents or indices attached to them. The (raised) exponent is set above the *overbrace* while the (lowered) index is placed below the *underbrace*.

$$
\underbrace{a + \overbrace{b + \cdots + y}^{123} + z}_{\alpha\beta\gamma}
$$

```
\[  \underbrace{a + \overbrace{b + \cdots +
     y}^{123} + z}_{\alpha\beta\gamma}   \]
```

Exercise 5.16: The total number of permutations of n elements taken m at a time (symbol P_n^m) is

$$
P_n^m = \prod_{i=0}^{m-1} (n - i) = \underbrace{n(n - 1)(n - 2)\ldots(n - m + 1)}_{\text{total of } m \text{ factors}} = \frac{n!}{(n - m)!}
$$

5.4.5 Stacked symbols

The command

$\quad$ \stackrel{*upper_sym*}{*lower_sym*}

places the symbol *upper_sym* centered on top of *lower_sym*, whereby the symbol on top is set in a smaller typeface.

$\vec{x} \stackrel{\mathrm{def}}{=} (x_1, \ldots x_n)$ `$ \vec{x} \stackrel{\mathrm{def}}{=} (x_1,...$`

$A \stackrel{\alpha'}{\longrightarrow} B \stackrel{\beta'}{\longleftarrow} C$ `$ A \stackrel{\alpha'}{\longrightarrow} B ... $`

By making use of math font size commands (Section 5.5.2) it is possible to construct new symbols with this command. For example, some authors prefer to have the \le symbol appear as $\overset{<}{=}$ instead of $\leq$. This is achieved by combining $<$ and $=$ with $\stackrel{\textstyle<}{=}$. If the command \textstyle were omitted, the symbol would be printed as $\overset{<}{=}$.

5.4.6 Additional TEX commands for math

The TEX math commands \atop and \choose are useful additions to the set of commands and may be applied within any LATEX document. (In fact, all TEX math commands except \eqalign, \eqalignno, and \leqaligno may be used in a LATEX manuscript.) Their syntax is

$\quad$ {*top* \atop *bottom*}
$\quad$ {*top* \choose *bottom*}

Both commands produce a structure that looks like a fraction without the dividing line. With the \choose command, this structure is also enclosed within round brackets (in mathematics this is called a *binomial coefficient*).

$$
\binom{n + 1}{k} = \binom{n}{k} + \binom{n}{k - 1}
$$

```
\[ {n+1 \choose k} =
   {n \choose k} + {n \choose k-1} \]
```

$$\prod_{j\geq0}\left(\sum_{k\geq0} a_{jk}z^k\right) = \sum_{n\geq0} z^n \left(\sum_{\substack{k_0,k_1,\ldots\geq0 \\ k_0+k_1+\cdots=0}} a_{0k_0}a_{1k_1}\cdots\right)$$

```
\[ \prod_{j\ge0}\left( \sum_{k\ge0} a_{jk}z^k \right) =
\sum_{n\ge0} z^n \left(\sum_{k_0,k_1\ldots\ge0 \atop
    k_0+k_1+\cdots=0} a_{0k_0} a_{1k_1}\ldots \right)    \]
```

Similar structures can be generated with the LATEX environments

```
\begin{array}{c}   upper_line \\   lower_line \end{array}        (atop)
\left(\begin{array}{c} upper \\   lower \end{array}\right) (choose)
```

The difference between these array structures and those of the TEX commands is that the former are always printed in the size and style of normal text formulas, whereas the latter will have varying sizes depending on where they appear within the formula.

For comparison

$$\Delta_{\substack{p_1p_2\cdots p_{n-k} \\ p_1p_2\cdots p_{n-k}}}$$ The index array is produced using \atop

$$\Delta \quad \substack{p_1p_2\cdots p_{n-k} \\ p_1p_2\cdots p_{n-k}}$$ The index array is produced using \array

The above TEX commands may also be employed to produce small matrices within text formulas, such as $\left(\begin{smallmatrix}1&0\\0&1\end{smallmatrix}\right)$ or $\left(\begin{smallmatrix}a&b&c\\l&m&n\end{smallmatrix}\right)$. Here the first matrix was typed in with

```
${1\,0\choose0\,1}$
```

and the second with

```
$\left({a\atop l}{b\atop m}{c\atop n}\right)$
```

The syntax of these Plain TEX commands is radically different from that normally used by LATEX 2_ε. See page 197 for a way to correct this.

!

5.4.7 Multiline equations

A multiline equation is one that is developed over several lines in which the relation symbols (for example, = or ≤) in each line are all vertically aligned with each other. For this purpose, the environments

```
\begin{eqnarray}   line_1\\   ... \\ line_n end{eqnarray}
\begin{eqnarray*} line_1\\   ... \\ line_n end{eqnarray*}
```

are used to set several lines of formulas or equations in displayed math mode. The individual lines of the equation or formula are separated from one another by \\. Each entry line has the form

$$\textit{left\_formula} \quad \& \quad \textit{mid\_formula} \quad \& \quad \textit{right\_formula} \setminus\setminus$$

When printed, all the *left_formula*s appear right justified in a left column, the *right_formula*s left justified in a right column, and the *mid_formula*s centered in between. The column separation character & designates the various parts of the formula. Normally the *mid_formula* is a single math character, the relation operator mentioned above. The individual lines thus have the same behavior as they would in a \begin{array}{rcl} ... \end{array} environment.

The difference between the array and eqnarray environments is that in the latter the lines are set as displayed formulas. This means that for those symbols listed in Section 5.3.7 the larger form will be selected, and that the numerator and denominator of fractions will be in normal size. On the other hand, for the array environment the column entries will be set as text formulas, the smaller form of these symbols will be chosen and the parts of the fraction will appear in a smaller type size.

The standard form of the eqnarray environment adds an automatic sequential equation number, which is missing in the *-form. To suppress the equation number for a single line in the standard form, add the command \nonumber just before the line termination \\.

The equation numbers may be referred to in the text with the command \ref{*name*} once the keyword *name* has been assigned to the equation number with the \label{*name*} command somewhere in the line of the equation. See Section 8.3.1 for more details.

Examples:

$$
\begin{array}{rcl}
(x+y)(x-y) & = & x^2 - xy + xy - y^2 \\
 & = & x^2 - y^2 \\
(x+y)^2 & = & x^2 + 2xy + y^2
\end{array}
$$

(5.4)

(5.5)

```
\begin{eqnarray}
    (x+y)(x-y) & = & x^2-xy+xy-y^2 \nonumber\\
               & = & x^2 - y^2 \\
    (x+y)^2    & = & x^2 + 2xy + y^2
\end{eqnarray}
```

$$
\begin{array}{rcl}
x_n u_1 + \cdots + x_{n+t-1} u_t & = & x_n u_1 + (a x_n + c) u_2 + \cdots \\
 & & + \left(a^{t-1} x_n + c(a^{t-2} + \cdots + 1) \right) u_t \\
 & = & (u_1 + a u_2 + \cdots + a^{t-1} u_t) x_n + h(u_1, \ldots, u_t)
\end{array}
$$

```
\begin{eqnarray*}
x_nu_1 + \cdots + x_{n+t-1}u_t & = & x_nu_1 + (ax_n + c)u_2 +
                                                        \cdots\\
   &     & + \left(a^{t-1}x_n + c(a^{t-2} + \cdots+1)\right)u_t\\
   & = & (u_1 + au_2 + \cdots + a^{t-1}u_t)x_n + h(u_1,\ldots,u_t)
\end{eqnarray*}
```

The second example requires some explanation. In the second line there is a \left(... \right) pair for the automatic sizing of the (). Such a pair may only appear within a single line; that is, it may *not* be broken by a line terminator \\! If automatic bracket sizing is to occur in a multiline equation, it may only be found within a single line.

If bracket pairs must appear on different lines, one can try using the construction \left(... \right. \\ \left. ... \right). In the first line, the \left(is paired with the *invisible* bracket \right., while the second begins with the invisible \left. which is paired with the closing \right). However, this will only work satisfactorily if both parts of the equation have roughly the same height so that the two automatic sizings yield much the same results. Section 5.5.3 describes how to select bracket sizes manually in the event that the automatic method fails.

The + sign at the beginning of the second line also requires a remark. The signs + and − have two meanings in mathematics: between two math quantities they act as a coupling (*binary operator*), but coming before a math symbol they serve as a sign designation (positive or negative). LATEX stresses this difference by inserting different spacing in the two cases (for example, compare $+b$ with $a + b$).

$$y \;=\; a+b+c+d$$
$$+e+f+g$$
$$+h+i+j$$

If a long formula is broken into several lines and one of them begins with + or −, LATEX regards it as a sign designation and moves it closer to the next character.

The solution is to introduce an *invisible* character of *zero* width at the beginning of such a line. This may be the empty structure {}. Compare the effects of && +e+f+g and &&{}+h+i+j in the above equation.

Since the extra spacing is always inserted between + and an opening parenthesis (, it was not necessary to give the empty structure {} at the beginning of the second line of the eqnarray* example above.

Occasionally it is better to break long multiline equations as follows:

$$w + x + y + z =$$
$$a + b + c + d + e + f +$$
$$g + h + i + j + k + l$$

that is, the second and subsequent lines are not aligned with the equals sign but are left justified with a certain indentation from the beginning of the first line.

```
\begin{eqnarray*}
\lefteqn{w+x+y+z = } \\
 & & a+b+c+d+e+f+  \\
 & & g+h+i+j+k+l
\end{eqnarray*}
```

The command \lefteqn{w+x+y+z =} \\ in the first line has the effect that the contents of the argument are indeed printed out, but that LATEX considers them to have *zero* width. The left-hand column then contains only intercolumn spacing, which produces the indentation for the rest of the lines.

The indentation depth may be altered by inserting \hspace{*depth*} between the \lefteqn{ ... } and the line termination \\. A positive value for *depth* increases the indentation, a negative value decreases it.

Exercise 5.17: *The following equations are to be broken as shown:*

$$\arcsin x \quad = \quad -\arcsin(-x) = \frac{\pi}{2} - \arccos x = \left[\arccos \sqrt{1 - x^2}\right]$$

$$= \quad \arctan \frac{x}{\sqrt{1 - x^2}} = \left[\operatorname{arccot} \frac{\sqrt{1 - x^2}}{x}\right] \tag{5.6}$$

$$f(x + h, y + k) = f(x, y) + \left\{\frac{\partial f(x, y)}{\partial x} h + \frac{\partial f(x, y)}{\partial y} k\right\}$$

$$+ \frac{1}{2} \left\{\frac{\partial^2 f(x, y)}{\partial x^2} h^2 + 2 \frac{\partial^2 f(x, y)}{\partial x \partial y} kh + \frac{\partial^2 f(x, y)}{\partial y^2} k^2\right\} \tag{5.7}$$

$$+ \frac{1}{6}\{\cdots\} + \cdots + \frac{1}{n!}\{\cdots\} + R_n$$

Note on possible error messages:

Long formulas with many sets of logical brackets, especially deeply nested ones, will almost always contain errors at first. The cause is often brackets that have been incorrectly ordered or overlooked.

If LaTeX produces error messages during formula processing, which the beginner is not able to interpret correctly (error messages are described in detail in Chapter 9), he or she should check the bracket pairing very carefully for the formula text. Some text editors can do the search for the matching bracket, which greatly simplifies the task.

If the error is not found in this way, one can try pressing the return key at the error message in order to proceed with the processing. The resulting printout can indicate where things may have gone wrong.

Exercise 5.18: *The* eqnarray *environment inserts additional spacing where the column separation character* & *appears. This is undesirable when an equation is to be broken aligned on* + *or* − *within a long summation. For example:*
The inverse function of the polynomial expansion $y = f(x) = ax + bx^2 + cx^3 + dx^4 + ex^5 + fx^6 + \cdots$ $(a \neq 0)$ begins with the elements

$$x = \varphi(y) = \frac{1}{a} y \quad - \quad \frac{b}{a^3} y^2 + \frac{1}{a^5}(2b^2 - ac)y^3$$

$$+ \quad \frac{1}{a^7}(5abc - z^2 d - fb^3)y^4$$

$$+ \quad \frac{1}{a^9}(6a^2 bd + 3a^2 c^2 + 14b^4 - a^3 e - 21ab^2 c)y^5$$

$$+ \quad \frac{1}{a^{11}}(7a^3 be + 7a^3 cd + 84ab^3 c - a^4 f -$$

$$28a^2 b^2 d - 28a^2 bc^2 - 43b^5)y^6 + \cdots$$

Select a value for the declaration \arraycolsep (Section 4.8.2) such that the distance between the + and − and the break points are as near as possible to those in the rest of the formula.

5.4.8   Framed or side-by-side formulas

!

Displayed formulas or equations may be put into vertical boxes of appropriate width, that is, in a \parbox command or minipage environment. Within the vertical box, the formulas are horizontally centered or left justified with indentation \mathindent according to the selected document class option.

Vertical boxes may be positioned relative to one another just like single characters (Sections 4.7.3 and 4.7.7). In this way the user may place displayed formulas or equations side by side.

$$\begin{aligned}\alpha &= f(z) \quad (5.8)\\ \beta &= f(z^2) \quad (5.9)\\ \gamma &= f(z^3) \ (5.10)\end{aligned}$$

$$\begin{aligned} x &= \alpha^2 - \beta^2\\ y &= 2\alpha\beta \end{aligned}$$

The left-hand set of equations is set in a \parbox of width 4 cm, the right-hand set in one of width 2.5 cm, while this text is inside a minipage of width 4.5 cm.

```
\parbox{4cm}{\begin{eqnarray} \alpha &=& f(z)...\end{eqnarray}}
\hfill  \parbox{2.5cm}{\begin{eqnarray*}
  x &=& \alpha^2 - \beta^2\\ y &=& 2\alpha\beta \end{eqnarray*}}
\hfill  \begin{minipage}{4.5cm} The left-hand ... \end{minipage}
```

!

Vertical boxes can also be useful when equation numbers are placed in an unconventional manner. The eqnarray environment generates an equation number for every line, which may be suppressed with \nonumber. To add a vertically centered equation number to a set of equations, for example,

$$\begin{aligned} P(x) &= a_0 + a_1 x + a_2 x^2 + \cdots + a_n x^n\\ P(-x) &= a_0 - a_1 x + a_2 x^2 - \cdots + (-1)^n a_n x^n \end{aligned} \qquad (5.11)$$

the following text may be given:

```
\parbox{10cm}{\begin{eqnarray*} ... \end{eqnarray*}} \hfill
\parbox{1cm}{\begin{eqnarray}\end{eqnarray}}
```

The actual set of equations is produced here in the eqnarray* environment, within a vertical box of width 10 cm, followed by an empty eqnarray environment in a box of width 1 cm that generates the equation number. Both boxes are vertically aligned along their center lines.

!

Emphasizing formulas by framing requires no new construction elements. It is sufficient to put them into an \fbox (Section 4.7.7). Text formulas $\boxed{a+b}$ are simply framed with \fbox{$a+b$}. For displayed formulas, \displaystyle (Section 5.5.2) must be issued, else they are set as text formulas.

$$\boxed{\int_0^\infty f(x)\, \mathrm{d}x \approx \sum_{i=1}^{n} w_i e^{x_i} f(x_i)}$$

is produced with

```
\[\fbox{$\displaystyle \int^\infty_0 f(x)\,\mathrm{d}x ..$}\]
```

An alternative method to frame displayed equations is with the $\mathcal{A}_{\mathcal{M}}\mathcal{S}$-LaTeX \boxed command, page 419.

5.4.9 Chemical formulas and bold face in math formulas

In mathematics it is sometimes necessary to set individual characters or parts of the formula in bold face. This can be achieved simply with the math alphabet command \mathbf that we met in Section 5.4.2:

$\mathbf{S^{-1}TS = dg(\omega_1,\ldots,\omega_n) = \Lambda}$

produces $\mathbf{S^{-1}TS = dg(\omega_1,\dots,\omega_n) = \Lambda}$.

In this example, the entire formula has been set as the argument of \mathbf so that everything should be set in bold face. In fact, only numbers, lower and upper case Latin letters, and upper case Greek letters are set in **bold Roman** with \mathbf. Lower case Greek letters and other math symbols appear in the normal *math* font.

If only part of the formula is to be set in bold face, that part must be given as the argument of the \mathbf command.

$\mathbf{2\sqrt{x}/y} = z$ $2\sqrt{\mathbf{x}}/y = z$

The math font style command \boldmath will set all characters in bold face, with the following exceptions:

- raised and lowered symbols (exponents and indices)
- the characters $+$ $:$ $;$ $!$ $?$ $($ $)$ $[$ $]$
- symbols that exist in two sizes (Section 5.3.7)

The \boldmath declaration may not appear in math mode. It must be called before switching to math mode or within a parbox or minipage. The countercommand \unboldmath resets the math fonts back to the normal ones.

$$\oint\limits_C V\,\mathrm{d}\tau = \oint\limits_\Sigma \nabla\times V\,d\sigma$$

```
\boldmath \[ \oint\limits_C V
\,\mathrm{d}\tau =\oint\limits_\Sigma
\nabla\times V\,d\sigma \] \unboldmath
```

If \boldmath has been turned on outside the math mode, it may be temporarily turned off inside with \mbox{\unboldmath$ ... $}.

\boldmath\(P = \mbox{\unboldmath$m$}b\)\unboldmath

yields: $\boldsymbol{P} = m\boldsymbol{b}$. Similarly, \boldmath can be temporarily turned on within math mode with the structure \mbox{\boldmath$ ... $}: $W_r = \int M\,d\varphi = r^2 m\omega^2/2$

\(W_r = \int\mbox{\boldmath$M\,\mathrm{d}\varphi$} =..\)

An alternative method of printing single symbols in bold face is provided by the bm package described on page 386.

Chemical formulas are normally set in Roman type, not in italics as for mathematical formulas. This may be brought about by setting the formula as the argument of the font command \mathrm:

$\mathrm{Fe_2^{2+}Cr_20_4}$ $\mathrm{Fe_2^{2+}Cr_2O_4}$

5.4.10 International typesetting standards

The International Standards Organization (ISO) has established the recognized conventions for typesetting mathematics, the essential elements of which are presented in an article by Beccari (1997), along with a description of how they may be realized with LaTeX. Some of these rules have already been mentioned and demonstrated in the examples. Here are the major points.

1. Simple variables are represented by italic letters, as $a\ b\ c\ x\ y\ z$.

2. Vectors are written in bold face italic, as $\boldsymbol{B}\ \boldsymbol{v}\ \boldsymbol{\omega}$.

3. Tensors of 2nd order and matrices may appear in a sans serif font, as $\mathsf{M}\ \mathsf{D}\ \mathsf{I}$.

4. The special numbers e, i, π, as well as the differential operator d, are to be *written in an upright font* to emphasize that they are not variables.

5. A measurement consisting of a number plus a dimension is an indivisible unit, with a smaller than normal space between them, as 5.3 km and 62 kg. The dimension is in an upright font.

Point 1 is fulfilled automatically by LaTeX. Point 5 is easily achieved by inserting the small space `\,` command between the number and dimension, as `5.3\,km` and `62\,kg`. Using the protected space `~` instead is very common practice among LaTeX users, which ensures that the two parts are not split, but with regular, not small spacing: 5.3 km and 62 kg.

Point 2 is not satisfied with the `\vec` command, which produces $\vec{B}$. Nor does `\mathbf` help, for this yields **B**, in an upright font. The best solution is to use the `\boldsymbol` command from the $\mathcal{A}_{\mathcal{M}}\mathcal{S}$ package amsbsy (Section E.3.1) or the `\bm` command from the tools package bm (Section D.3.3). Otherwise one must resort to defining

```
\renewcommand{\vec}[1]{\mbox{\boldmath$#1$}}
```

for a revised `\vec` command.

Similarly point 3 can be met with the math alphabet command `\mathsf`. However, Beccari does point out that even tensor variables should probably be italic. This is more difficult, but it can be accomplished. See example 5 on page 200.

Point 4 is the one that is most often violated, especially for the differential d. We have demonstrated it in the examples in this book and have shown how it may be achieved with the `\mathrm` math alphabet command. However, to simplify the application, it is recommended to create some user-defined commands, such as

```
\newcommand{\me}{\mathrm{e}}       for math e
\newcommand{\mi}{\mathrm{i}}       for math i
\newcommand{\dif}{\mathrm{d}}      for differential
```

An upright π is not so easy since this is not provided in the usual math fonts.

With these new commands, the equation on page 117 is more conveniently set with

```
\[ \int^{\infty}_0 f(x)\,\dif x \approx
          \sum^n_{i=1}w_i \me^{x_i} f(x_i) \]
```

5.5   Fine-tuning mathematics

5.5.1   Horizontal spacing

Even though TeX has a thorough knowledge of the rules of mathematical typesetting, it cannot hope to understand the mathematical meaning. For example, $y\,dx$ normally means the joining of the variable y with the differential operator dx, and this joining is designated by a small space between the two. However, TeX removes the blank in the entry y dx and prints ydx, the product of three variables y, d, and x. At this point, LaTeX needs some fine-tuning assistance.

Small amounts of horizontal spacing can be added in math mode with the commands

\,	small space	= 3/18 of a quad
\:	medium space	= 4/18 of a quad
\;	large space	= 5/18 of a quad
\!	negative space	= −3/18 of a quad

In the following examples, the third column contains the results without the additional horizontal spacing command.

`$\sqrt{2}\,x$`	$\sqrt{2}\,x$	$\sqrt{2}x$
`$\sqrt{\,\log x}$`	$\sqrt{\,\log x}$	$\sqrt{\log x}$
`$0\left(1/\sqrt{n}\,\right)$`	$O\left(1/\sqrt{n}\,\right)$	$O\left(1/\sqrt{n}\right)$
`$[\,0,1)$`	$[\,0,1)$	$[0,1)$
`$\log n\,(\log\log n)^2$`	$\log n\,(\log\log n)^2$	$\log n(\log\log n)^2$
`$x^2\!/2$`	$x^2\!/2$	$x^2/2$
`$n/\!\log n$`	$n/\!\log n$	$n/\log n$
`$\Gamma_{\!2}+\Delta^{\!2}$`	$\Gamma_{\!2}+\Delta^{\!2}$	$\Gamma_2+\Delta^2$
`$R_i{}^j{}_{\!kl}$`	$R_i{}^j{}_{\!kl}$	$R_i{}^j{}_{kl}$
`$\int_0^x\!\int_0^y`	$\int_0^x\!\int_0^y dF(u,v)$	$\int_0^x\int_0^y dF(u,v)$
`\mathrm{d}F(u,v)$`		
`\[ \int\!\!\!\int_D`	$\iint_D dx\,dy$	$\int\int_D dxdy$
`\mathrm{d}x\,\mathrm{d}y \]`		

Note: in the third last example `R_i{}^j` is so constructed that an invisible character of zero width comes after the index of R_i, and it is this dummy character that receives the following exponent. The result is $R_i{}^j$, instead of R_i^j which is produced by `R_i^j`.

There are no hard and fast rules for applying the math spacing commands. Some candidates are the differential operator, small root signs in text formulas followed by a variable, the dividing sign /, and multiple integral signs. The above examples illustrate many suitable situations.

5.5.2   Selecting font size in formulas

It is possible to alter the font sizes that TeX selects for the various parts of the formula. First we must explain what sizes are available in math mode and what TeX's selection rules are.

In math mode there are four font sizes that may be chosen, their actual sizes being relative to the basic font size of the document class:

\displaystyle	D	Normal size for displayed formulas
\textstyle	T	Normal size for text formulas
\scriptstyle	S	Normal size for first sub-, superscript
\scriptscriptstyle	SS	Normal size for later sub-, superscripts

From now on we shall make use of the symbolic abbreviations D, T, S, and SS. When math mode is switched on, the active font size becomes D for displayed and T for text formulas. Their only difference lies in those symbols that appear in two sizes, plus the corresponding style of those superscripts and subscripts (Section 5.3.7). The larger symbols belong to D, the smaller to T.

Starting from these base sizes, various math elements will be set in other sizes. Once another size has been chosen for an element, it remains the active size within that element.

The table below shows the selection rules:

Active size	Fractions upper	lower	Super-, subscripts
D	T	T	S
T	S	S	S
S	SS	SS	SS
SS	SS	SS	SS

If the active size is D, size T is selected for both parts of the fraction. That is, for both numerator and denominator T becomes the active size. If they in turn contain further fractions, these will be set in S. If the active size is D or T, superscripts and subscripts (exponents and indices) appear in S; within them S is the active size and any fractions or shiftings inside them will be set in SS.

The TeX elements { \atop } and { \choose } are treated as fractions.

The active font size inside an array environment is T.

The smallest available math font size is SS. Once it has been reached, no further reduction is possible, so that all subsequent superscripts and subscripts appear in SS as well.

From the table one easily sees that

\[a_0 + \frac{1}{a_1 + \frac{1}{a_2
    + \frac{1}{a_3 + \frac{1}{a_4}}}} \]

$$a_0 + \cfrac{1}{a_1 + \cfrac{1}{a_2 + \cfrac{1}{a_3 + \frac{1}{a_4}}}}$$

must appear as shown at the right.

Such a math structure, called a *continued fraction*, is normally printed as

$$a_0 + \cfrac{1}{a_1 + \cfrac{1}{a_2 + \cfrac{1}{a_3 + \cfrac{1}{a_4}}}}$$

```
\[ a_0 + \frac{1}{\displaystyle a_1
   + \frac{1}{\displaystyle a_2
   + \frac{1}{\displaystyle a_3
   + \frac{1}{a_4}}}}          \]
```

By explicitly giving the font size within each element, one makes that size active rather than relying on the internally selected size. In this example, D is chosen for each denominator. Thus the next fraction acts as though it were outermost. The `\displaystyle` command may be omitted in the last fraction. (Can you see why?)

The selection table allows one to predict precisely which math font size will be applied at any part of the formula so that an explicit specification may be made if necessary. The effects of such a choice may also be calculated for the following math elements.

In the examples below, the right-hand column shows how the formulas would appear if the math font size is not explicitly specified.

$$\frac{\dfrac{a}{x-y} + \dfrac{b}{x+y}}{1 + \dfrac{a-b}{a+b}}$$

```
\[ \frac{\displaystyle\frac{a}{x-y}
          +\frac{b}{x+y}}
   {\displaystyle 1+\frac{a-b}{a+b}} \]
```

$$\frac{\frac{a}{x-y} + \frac{b}{x+y}}{1 + \frac{a-b}{a+b}}$$

$$e^{-\frac{x_i-x_j}{n^i+n^j}}$$

```
\[    e^{\textstyle -
   \frac{x_i-x_j}{n^i+n^j}},   \]
```

$$e^{-\frac{x_i-x_j}{n^i+n^j}}$$

$$\left(\begin{pmatrix} ab \\ cd \end{pmatrix}\quad \dfrac{e+f}{g-h} \atop 0 \quad \left| \dfrac{ij}{kl} \right| \right)$$

```
\[ \left(\begin{array}{cc}
   \displaystyle{ab\choose cd} &
   \displaystyle\frac{e+f}{g-h}\\
   0 & \displaystyle \left|
        {ij\atop kl} \right|
   \end{array}\right)           \]
```

$$\left(\begin{pmatrix} ab \\ cd \end{pmatrix}\quad \frac{e+f}{g-h} \atop 0 \quad \left| \frac{ij}{kl} \right| \right)$$

If an explicit size specification is to be given frequently within a document—say within every entry in an `array` environment—then considerable writing effort may be avoided by adding to the preamble

```
\newcommand{\D}{\displaystyle}\newcommand{\T}{\textstyle}...
```

In this way the size command may be given simply by typing `\D` or `\T`, etc.

The math font size selection rules given above are in fact a simplification. For those readers who wish to know the exact details, we now present the complete description.

For each of the four math font sizes D, T, S, SS there is a modified version D', T', S', SS'. The difference is that with D, T, S, SS the superscripts (exponents) are somewhat higher than they are with D', T', S', SS'. Compare the positions of the exponent 2 above and below the line: $\dfrac{x^2}{x^2}$. Otherwise the primed and unprimed font sizes are identical. The true selection rules are given in the table at the right.

Active size	Fractions upper	lower	Super- scripts	Sub- scripts
D	T	T'	S	S'
D'	T'	T'	S'	S'
T	S	S'	S	S'
T'	S'	S'	S'	S'
S, SS	SS	SS'	SS	SS'
S', SS'	SS'	SS'	SS'	SS'

When the font size is explicitly stated within a numerator or superscript, the unprimed version is selected; and in the denominators and subscripts, the primed fonts.

5.5.3   Manual sizing of bracket symbols

The \left ... \right commands preceding one of the 22 bracket symbols listed in Section 5.4.1 adjust the symbol size automatically according to the height of the enclosed formula text. However, it is possible to select a size explicitly by placing one of the TeX commands \big, \Big, \bigg, or \Bigg before the bracket symbol.

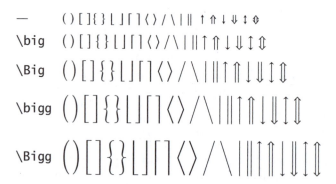

In contrast to the \left ... \right pairs, the explicit bracket size commands do not have to be contained within one line of a multiline formula. The opening and closing brackets may appear on different lines. This applies also to the commands described in the next paragraph.

!

Another two sets of bracket size commands exist with the names \bigl ... \Biggl and \bigr ... \Biggr. These cause the following bracket symbol to be treated internally as an opening or closing bracket. The practical difference between these and the normal commands is insignificant for LaTeX.

Further bracket size commands are \bigm ... \Biggm. With them, the following bracket symbol acts like a relation operator in that larger horizontal spacing is inserted between it and the neighboring parts of the formula.

$$[(a+b)\,|\,(c+d)]$$

```
\[ \big[ (a+b) \big| (c+d) \big] \]
\[ \bigl[ (a+b) \bigl| (c+d) \bigr] \]
\[ \bigm[ (a+b) \bigm| (c+d) \bigm] \]
```

$$[(a+b)\,|\,(c+d)]$$

$$[\,(a+b)\,|\,(c+d)\,]$$

5.5.4 Math style parameters

!

The mathematical style parameters listed below are set to standard values by LaTeX. The user may alter them at any time with the \setlength command in the usual manner.

\arraycolsep
    Half the width of the intercolumn spacing for the array environment (see also Section 4.8.2).

\jot   Extra vertical spacing that appears between the rows of multiline equations in the eqnarray and eqnarray* environments.

\mathindent
    The amount of indentation for math formulas when the document class option fleqn has been selected.

\abovedisplayskip
    The vertical spacing above displayed formulas when the left side of the formula is closer to the left margin than the end of the preceding line of text. Such a formula is designated *long*.

\belowdisplayskip
    The vertical spacing inserted below a *long* displayed formula.

\abovedisplayshortskip
    Vertical spacing added above a *short* displayed formula. This is one in which the left edge is to the right of the end of the preceding line of text.

\belowdisplayshortskip
    Vertical spacing inserted after a *short* displayed formula.

\topsep
    The above four spacings are not used with the document class option fleqn where instead \topsep is inserted above and below displayed formulas (see Section 4.4.2).

All the above parameters, except \jot, should be rubber lengths (Section 2.4.2).

5.5.5 Some further advice

Sometimes authors desire horizontal and vertical alignments of their formulas that just cannot be achieved using the means described so far. They should consider placing their formulas inside horizontal or vertical boxes which may then be positioned wherever they please.

Similarly the `array` environment together with explicit size declarations and the table construction and style elements from Sections 4.8.1 and 4.8.2 should be able to accomplish just about any horizontal and vertical alignment.

Exercise 5.19: *Generate this continued fraction. Note: in contrast to the example on page 144, the 1 here in the numerator appears left justified. Hint: do you remember the command* \hfill?

$$a_0 + \cfrac{1}{a_1 + \cfrac{1}{a_2 + \cfrac{1}{a_3 + \cfrac{1}{a_4}}}}$$

Exercise 5.20: *Produce the following set of equations with the* `array` *environment.*

$$\sin 2\alpha = 2 \sin \alpha \cos \alpha, \qquad \cos 2\alpha = \cos^2 \alpha - \sin^2 \alpha$$
$$\sin 3\alpha = 3 \sin \alpha - 4 \sin^3 \alpha \qquad \cos 3\alpha = 3 \cos^3 \alpha - 3 \cos \alpha$$
$$\sin 4\alpha = 8 \cos^3 \alpha \sin \alpha - 4 \cos \alpha \sin \alpha \quad \cos 4\alpha = 8 \cos^4 \alpha - 8 \cos^2 \alpha + 1$$

Hint: recall that @{...} *expressions in the formatting field of the* `array` *environment can be used to insert horizontal spacing and/or mathematical text between the columns (see the first example in Section 5.4.3).*

Exercise 5.21: *Create the following output with the* `array` *environment.*

Equations for the tangential plane and surface normal		
Equation for the surface	Tangential plane	Surface normal
$F(x,y,z) = 0$	$\dfrac{\partial F}{\partial x}(X-x) + \dfrac{\partial F}{\partial y}(Y-y)$ $+ \dfrac{\partial F}{\partial z}(Z-z) = 0$	$\dfrac{X-x}{\dfrac{\partial F}{\partial x}} = \dfrac{Y-y}{\dfrac{\partial F}{\partial y}} = \dfrac{Z-z}{\dfrac{\partial F}{\partial z}}$
$z = f(x,y)$	$Z - z = p(X-x) + q(Y-y)$	$\dfrac{X-x}{p} = \dfrac{Y-y}{q} = \dfrac{Z-z}{-1}$
$x = x(u,v)$ $y = y(u,v)$ $z = z(u,v)$	$\begin{vmatrix} X-x & Y-y & Z-z \\ \dfrac{\partial x}{\partial u} & \dfrac{\partial y}{\partial u} & \dfrac{\partial z}{\partial u} \\ \dfrac{\partial x}{\partial v} & \dfrac{\partial y}{\partial v} & \dfrac{\partial z}{\partial v} \end{vmatrix} = 0$	$\dfrac{X-x}{\begin{vmatrix} \frac{\partial y}{\partial u} & \frac{\partial z}{\partial u} \\ \frac{\partial y}{\partial v} & \frac{\partial z}{\partial v} \end{vmatrix}} = \dfrac{Y-y}{\begin{vmatrix} \frac{\partial z}{\partial u} & \frac{\partial x}{\partial u} \\ \frac{\partial z}{\partial v} & \frac{\partial x}{\partial v} \end{vmatrix}}$ $= \dfrac{Z-z}{\begin{vmatrix} \frac{\partial x}{\partial u} & \frac{\partial y}{\partial u} \\ \frac{\partial x}{\partial v} & \frac{\partial y}{\partial v} \end{vmatrix}}$
$r = r(u,v)$	$(R - r)(r_1 \times r_2) = 0$ or $\quad (R - r)N = 0$	$\cdot \quad R = r + \lambda(r_1 \times r_2)$ or $\quad R = r + \lambda N$
In this Table x, y, z and r are the coordinates and the radius vector of a fixed point M on the curve; X, Y, Z and R are the coordinates and radius vector of a point on the tangential plane or surface normal with reference to M; furthermore $p = \frac{\partial z}{\partial x}$, $q = \frac{\partial z}{\partial y}$ and $r_1 = \partial r / \partial u$, $r_2 = \partial r / \partial v$.		

Note: If you succeed in reproducing this mathematical table you should have no more problems with positioning formulas and their parts!
    Hints for the solution:

1. *Define abbreviations such as* \D *for* \displaystyle *and* \bm *for* \boldmath *and possibly even* \ba *and* \ea *for* \begin{array} *and* \end{array}.

2. *Build the table in stages. Start with the head and only proceed further when it looks reasonable. Watch out that normal text within the* array *environment must be included inside an* \mbox.

3. *Continue with the first mathematical row. Here the first entry in the second column is also an* array *environment that is aligned with the rest of the row by means of the* [t] *positioning argument. Do not forget to activate the size* \D *at the necessary places in the inner structures. Insert a possible strut (Section 4.7.6) to generate the right distance from the head. The distance to the next row can be adjusted with a length specification added to the row terminating command* \\[..].

4. *After this row has been completed successfully, the next should pose no difficulties.*

5. *In the third mathematical row, both the first column and the left side of the second column consist of* array *environments. The third column contains three fractions whose denominators may be produced either with* array *or with the* TEX {...\atop...} *command; the fractions themselves are placed in two rows with another* array *environment.*

6. *The second and third columns of the last mathematical row again contain* array *environments. Some of the sub-formulas in this row appear in* \boldmath. *Recall that this command may only be invoked within text mode, that is, within an* \mbox.

7. *In the last row, all three columns of the outer* array *environment are merged together and the text is set within a parbox of appropriate width:*

    \multicolumn{3}{|c|}{\parbox{..}{......}}

5.6 Beyond standard LATEX

The typesetting elements presented in this chapter should be able to handle most of the everyday problems of mathematical composition. For more complicated situations, the user might consider taking advantage of the math extensions offered by the American Mathematical Society, $\mathcal{AMS}$. This set of packages, first developed for Plain TEX users, is available for LATEX as well. Known as $\mathcal{AMS}$-LATEX, it includes many extra commands, further environments to supplement eqnarray, and many additional symbols. An overview of this system is presented in Appendix E.

    In any event, typesetting mathematics is a very complicated matter. It was for this reason that Donald Knuth selected the $ sign as the switch to math mode: in the old days of movable lead type, setting mathematical formulas was an expensive business because of the extra workload it entailed. Even today, electronic typesetting is far more involved for math than for simple text.

6 Illustrations

Diagrams, photographs, screen shots, plots of experimental data, all these are non-text material that are standard parts of today's documents, whether of a scientific nature or not. Since LaTeX deals only with the textual aspect of these documents, it must provide some means of including such graphical material that has been prepared by other programs.

Standard LaTeX does actually contain the means to make somewhat primitive drawings on its own. The word 'primitive' should not be considered derogatory, for simple building blocks are the basic units for constructing very complicated, sophisticated structures. Additional packages do exist to exploit the LaTeX picture elements very extensively; however, we shall not deal with them in this book since there are more elegant means today for placing graphical material inside a LaTeX document.

The LaTeX Graphics Companion (Goossens *et al.*, 1997) provides an excellent presentation of most of these methods. They include packages that work directly inside LaTeX, and which are thus independent of the output driver, as well as those that are driver-specific, usually based on PostScript. In this book, we describe only the LaTeX native picture environment (Section 6.1) and the importation of graphic files produced by external software (Section 6.2). We then show how colored text can be included (Section 6.3). Finally we discuss how figures (and tables) can be made to float to the most appropriate location (Section 6.4).

6.1 Drawing pictures with LaTeX

6.1.1 Picture coordinates

The picture building blocks can only be put in place once a *coordinate system* has been established for that picture. This consists of a *reference point* or *origin*, and two mutually perpendicular *coordinate axes*, as well as a *length unit* for the coordinates. The origin is the lower left corner of the picture and the axes are its lower and left edges. These edges are referred to as the *x-axis* (lower) and *y-axis* (left).

Once the unit of length (UL) has been specified, every point within the picture area can be uniquely referred to with two decimal numbers: the first is the number of length units along the x-axis, the second the number along the y-axis.

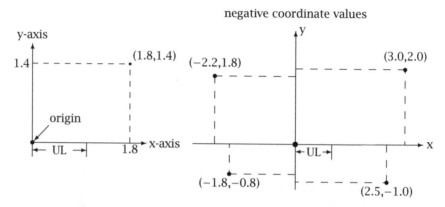

The coordinate numbers are generally positive, meaning that the point lies to the right and above the reference point. Since this point is the lower left corner of the picture, all other points should be more to the right and higher than it. However, negative values are also possible. A negative x value (a negative number for the first member of a coordinate pair) defines a point *left* of the origin, while a negative y value (the second member of the pair is negative) specifies a point below the reference point.

The unit length is selected with the command

```
\setlength{\unitlength}{length}
```

In the left-hand example above, the unit length was set to a value of 1.5 cm with the command `\setlength{\unitlength}{1.5cm}`. The point (1.8,1.4) then lies 1.8 times the unit length (= 2.7 cm) to the right, and 1.4 times (= 2.1 cm) above the origin. In the right-hand example, the unit length is set to 1 cm.

The unit of length is normally set to a convenient size such as 1 cm, 1 mm, or 1 in and the picture is built up accordingly. Once it has been completed, it is possible to rescale the whole thing simply by changing the value of the length unit. A picture that was originally designed with `\unitlength` set to 1 cm can be enlarged by a factor 1.2 by redefining its length unit to be 1.2 cm.

6.1.2 The picture environment

Pictures are constructed within the `picture` environment which is started with

```
\begin{picture}(x_dimen,y_dimen)
  picture_commands
\end{picture}
```

where (*x_dimen*, *y_dimen*) is a pair of numbers that specifies the size (dimensions) of the picture in the *x-direction* (horizontal) and *y-direction* (vertical). This pair of numbers is enclosed in *round* parentheses! The unit of length is that previously selected by \unitlength.

```
\setlength{\unitlength}{1.5cm}
\begin{picture}(4,5) ... ... ... \end{picture}
```

produces a picture that is 4 length units wide and 5 units high. Since the length unit has been set to 1.5 cm, the actual size is 6 cm wide and 7.5 cm high.

Picture commands are those commands described below that are used to produce and position the individual picture elements. These are the only commands that are allowed within the picture environment, other than font style and size declarations (Section 4.1) and the line thickness commands \thicklines and \thinlines. These last determine which of the two available line thicknesses will become current for drawing lines: one may switch back and forth as one pleases. Initially *thin* lines are active.

The value of the parameter \unitlength must not be altered within the picture environment, for it must remain the same for the entire picture. It may, however, be changed between pictures.

If the \unitlength specification together with the picture environment are enclosed within another environment, such as \begin{center} ... \end{center}, then that value of \unitlength is valid only until the end of the environment. A picture environment without a preceding \unitlength command uses the standard value of 1 pt.

6.1.3   The positioning commands

Picture elements are generated and positioned by means of the two commands \put and \multiput, which have the syntaxes:

```
\put(x_coord,y_coord){pic_elem}
\multiput(x_coord,y_coord)(x_incr,y_incr){num}{pic_elem}
```

The *pic_elem* is one of the picture element commands described in the next section. The arguments (*x_coord*, *y_coord*) are the *placement coordinates*, designating the location of the picture element within the picture coordinate system, in units of \unitlength. If this is 1 cm, then (2.5,3.6) means that the element is to be positioned 2.5 cm to the right and 3.6 cm above the lower left corner of the picture.

The \multiput command generates the same picture element *num* times, moving it (*x_incr*, *y_incr*) each time. Thus the element is drawn at

(*x_coord*, *y_coord*), (*x_coord* + *x_incr*, *y_coord* + *y_incr*),
(*x_coord* + 2*x_incr*, *y_coord* + 2*y_incr*), ... up to
(*x_coord* + [*num* − 1]*x_incr*, *y_coord* + [*num* − 1]*y_incr*)

The coordinate pair (*x_coord*, *y_coord*) is incremented by (*x_incr*, *y_incr*) for each successive placement. The values of the incrementing pair may be positive or negative.

Thus \multiput(2.5,3.6)(0.5,-0.6){5}{*pic_elem*} produces the *pic_elem* a total of five times, first at the location (2.5,3.6) and then at (3.0,3.0), (3.5,2.4), (4.0,1.8), and finally (4.5,1.2).

Note that the numbers for the *coordinate* and *increment* pairs are given in round parentheses (,) and that the two numbers within each pair are separated by a comma. The *num* and *pic_elem* entries, on the other hand, are enclosed in curly brackets { } as usual.

Warning: since the comma separates the two numbers in a coordinate pair, it may *not* be used in place of a decimal point. *For coordinate entries, decimal numbers must be written with a period, not a comma.*

6.1.4   Picture element commands

Text within pictures

The simplest picture element of all is a piece of text, positioned at the desired location within the picture. This is accomplished by putting text in place of *pic_elem* in the \put or \multiput command.

An arrow

(1.8,1.2)

The arrow points to the location (1.8,1.2). The command \put(1.8,1.2){An arrow} inserts the text 'An arrow' so that its lower left corner is at the specified position.

The text as picture element may also be packed into a \parbox or a minipage environment, and the reference point for the coordinate entry in the \put command depends on the positioning arguments of that box:

\parbox[b]{..}{..}	\parbox{32mm}{...}	\parbox[t]{..}{..}
Reference point is the lower left corner of the last line in the parbox.	For a standard parbox, the reference point is the vertical center of the left edge.	Reference point is the lower left corner of the top line in the parbox.

Exercise 6.1: Produce a picture 100 mm wide and 50 mm high with \unitlength equal to 1 mm. Place the given texts at the following locations: (0,0) 'The First Picture', (90,47) 'upper left', (70,40) 'somewhere upper right', and put a parbox of width 60 mm at (25,25) containing 'A separate exercise file with the name picture.tex *should be created for the exercises in this Chapter.'*

Exercise 6.2: *Repeat the picture processing with a value of 1.5 mm for* \unitlength, *and with positioning arguments* t *and* b *for the parbox.*

Picture boxes—rectangles

The box commands \framebox, \makebox, and \savebox (Section 4.7.1) are available in the picture environment but with an extended syntax. In addition, there is another box command \dashbox:

> \makebox(*x_dimen*,*y_dimen*)[*pos*]{*text*}
> \framebox(*x_dimen*,*y_dimen*)[*pos*]{*text*}
> \dashbox{*dash_len*}(*x_dimen*,*y_dimen*)[*pos*]{*text*}

The *dimensional* pair (*x_dimen*, *y_dimen*) defines the width and height of the rectangular box in units of \unitlength. The positioning argument *pos* determines how the *text* is located within the box. It may take on values:

[t] *top*   The input text appears—centered horizontally—*below* the upper edge of the box.

[b] *bottom*   The input text appears—centered horizontally—*above* the lower edge of the box.

[l] *left*   The input text *begins*—centered vertically—at the *left* edge of the box.

[r] *right*   The input text *ends*—centered vertically—at the *right* edge of the box.

2ε [s] *stretch*   The input text is stretched horizontally to fill up the box, and centered vertically.

Without the optional argument *pos*, the input text is centered vertically and horizontally within the box.

These positional values may be combined two at a time:

[tl] *top left*   The text appears at the *upper left*.

[tr] *top right*   The text appears at the *upper right*.

[bl] *bottom left*   The text appears at the *lower left*.

[br] *bottom right*   The text appears at the *lower right*.

The order of the values is unimportant: tl has the same effect as lt.

These box commands are to be used as *pic_elem* in the placement commands \put and \multiput. The box is so placed that its lower left corner is at the position given by the coordinate pair in the placement command.

\put(1.5,1.2){\framebox(2.5,1.2){center}}

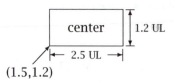

The arrow indicates the point (1.5,1.2) which is the position of the lower left corner of the rectangle with width 2.5 units and height 1.2 units. The text 'center' is centered both horizontally and vertically. UL = 0.8 cm.

The effect of the text positioning argument is made clear with the following examples (UL = 1 cm):

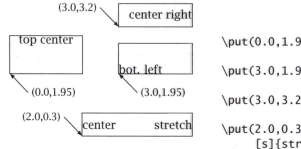

\put(0.0,1.95){\framebox(2,1.0)
                [t]{top center}}
\put(3.0,1.95){\framebox(2,0.8)
                [lb]{bot. left}}
\put(3.0,3.2){\framebox(2,0.6)
                [r]{center right}}
\put(2.0,0.3){\framebox(2,0.6)
                [s]{stretch\hfill center}}

The picture element \makebox is exactly the same as the \framebox command but without the rectangular frame. It is most often employed with the dimensional pair (0,0) in order to place text at a desired location. (See Section 4.7.1 for the effect of *zero* width boxes on the enclosed text.)

\put(3,1.6){\makebox(0,0){center center}}
\put(2,0.5){\makebox(0,0)[tr]{top right}}
\put(4,1.0){\makebox(0,0)[b]{bot center}}
\put(2,2.8){\makebox(0,0)[l]{flush left}}

(2.0,2.8) →flush left
(3.0,1.6)
center center
(2.0,0.5)   bot center
top right   (4.0,1.0)

The combination [lb] positions the text in exactly the same way as simply typing the text in as *pic_elem* without a box, as shown on page 154.

The picture element \dashbox also produces a framed box, but with a *dashed* line around it. The argument *dash_len* specifies the *dash length*.

\put(1.0,0.75){\dashbox{0.2}(4,1){dashed frame}}

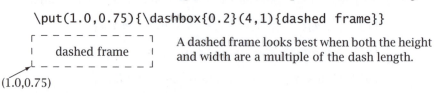

A dashed frame looks best when both the height and width are a multiple of the dash length.

(1.0,0.75)

Even in the above picture box commands, the entered *text* may be put into a vertical box (\parbox or minipage). Since vertical boxes themselves possess an optional positioning argument b or t, which must not conflict with that of the picture box, the following rule must be observed:

If a picture box command contains the positioning argument b or t, then the same value must be applied to the enclosed vertical box. If the picture box command has no positioning argument or only r or l, then the vertical box must be used in the standard (no argument) form.

The positioning argument of a picture box has the same effect on the enclosed vertical box as it does on a line of text, in that the whole box is treated as a single unit.

Exercise 6.3: Reproduce the organization table below with the boxes and included text but without the horizontal and vertical lines and arrows. These will be part of the next exercise.
Hint: first draw the boxes on a piece of squared paper so that the edges of the boxes lie on the printed rules. Select the unit of length to be the rule spacing of the paper. Take as the origin the lower left corner of an imaginary frame surrounding the entire diagram.
Note: you will soon be able to generate your own squared paper with whatever rule spacing you wish!

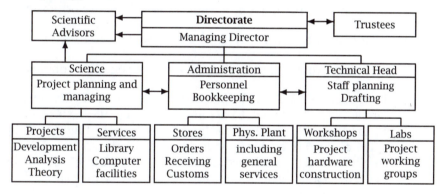

Straight lines

In the `picture` environment, LaTeX can draw straight lines of any length, horizontally and vertically as well as at a limited number of angles. The syntax for this picture element reads

$$\text{\textbackslash line}(\Delta x, \Delta y)\{length\}$$

For horizontal and vertical lines, *length* specifies how long the line is to be in length units. For lines at an angle, it has a somewhat more complicated meaning, as is explained below. The line begins at that spot given by the placement coordinates in the `\put` or `\multiput` command.

```
\thicklines
\put(0,0){\line(1,0){6}}
\put(0,0){\line(0,1){1}}
\put(6,0){\line(0,1){0.5}}
```

(0,0)   (7.5,0)

The angle at which the line is drawn is given by the *slope pair* $(\Delta x, \Delta y)$. The slope pair $(1,0)$, in which $\Delta x = 1$ and $\Delta y = 0$, produces a *horizontal* line, while the pair $(0,1)$ leads to a *vertical* line. This is illustrated in the above example.

In general $(\Delta x, \Delta y)$ has the following meaning:

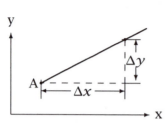

Beginning at a point A on the line and moving a distance Δx in the x direction (horizontally), Δy is the distance one must move in the y direction (vertically) in order to rejoin the line.

By specifying a slope pair $(\Delta x, \Delta y)$, a line is drawn at just that angle to fulfill the above conditions.

As mentioned already, the number of different slopes available is limited. This is because Δx and Δy may only take on values according to certain rules:

1. The number must be a whole integer (negative or positive).

2. Only the values 0, 1, ... , 6 are allowed.

3. The two numbers in the pair may not contain a common divisor.

Pairs such as $(3.5,1.2)$ (rule 1) and $(7,0)$ (rule 2) are thus forbidden. Similarly $(2,2)$ and $(3,6)$ are invalid pairs by rule 3, since both numbers in the first pair are divisible by 2, and those in the second by 3. The same angles are achieved with the pairs $(1,1)$ and $(1,2)$ respectively. In all there are 25 allowed slope pairs, including $(1,0)$ for horizontal and $(0,1)$ for vertical lines, as one can verify by writing down all the possibilities.

In addition, the numbers in the slope pair may be positive or negative, for example, $(0,-1)$, $(-2,-5)$. A negative Δx in the above diagram means motion to the left, and a negative Δy is for motion downwards. Thus \put(2,3){\line(0,-1){2.5}} draws a line starting at point (2,3) and going 2.5 length units vertically downwards.

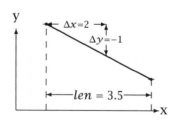

For lines at an angle, the argument *length* determines the projected distance along the x-axis. This is made clearer with the help of the diagram on the left.

\put(1.0,2.75){\line(2,-1){3.5}}

If one draws dashed lines straight down from the two end points, then that part of the x-axis between them is the projection of the line on to the x-axis.

Sloping lines must have a minimum length of about 10 pt or 3.5 mm, otherwise a warning is issued and no line is drawn.

Arrows

The *arrow* picture element is made with the command

$$\text{\\vector}(\Delta x, \Delta y)\{length\}$$

which functions exactly the same as the \line command as far as the arguments and their limitations are concerned. The command draws a line from the placement location given in the \put or \multiput command and places an arrow head at the end.

Just as for lines, arrows too must have a length of at least 10 pt or 3.5 mm. Rules 1–3 apply to Δx and Δy as well, with the further restriction that the allowed numbers are limited to 0, 1, ... , 4. This makes a total of 13 possible angles for arrows, not counting changes in sign.

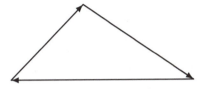

```
\begin{picture}(5,2)\thicklines
  \put(5,0){\vector(-1,0){5}}
  \put(0,0){\vector(1,1){2}}
  \put(2,2){\vector(3,-2){3}}
\end{picture}
```

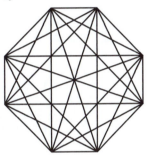

Exercise 6.4: Complete the diagram in Exercise 6.3 by including the missing horizontal and vertical lines and arrows.

Exercise 6.5: Generate the figure at the right. The corner points are (0,5), (0,10), (5,15), (10,15), (15,10), (15,5), (10,0), and (5,0) and the length unit is 0.1 inch.

Circles

The *circle* picture element is produced with the commands

```
\circle{diameter}
\circle*{diameter}
```

With the *-form of the command, a solid filled-in circle is printed rather than just an outline as for the standard form. Only certain sizes are available so LATEX selects the one closest to the specified *diameter* entry. If the size is too small, a warning is output to the monitor and no circle is printed.

```
\begin{picture}(3,1.6)
  \put(1,1){\circle*{0.2}}
  \put(1,1){\circle{1.2}}
  \put(1,1){\vector(0,1){0.6}}
  \put(2.5,1){\circle*{0.5}}
\end{picture}
```

The placement location in the corresponding \put command refers to the center of the circle.

Ovals and rounded corners

The term *oval* is used here to mean a rectangle whose corners have been replaced by quarter circles; the largest possible radius is chosen for the circles such that the sides join together smoothly. The command to produce them is

\oval(*x_dimen*,*y_dimen*)[*part*]

The placement location in the corresponding \put command refers to the center of the oval.

\put(3.0,0.75){\oval(4.0,1.5)}

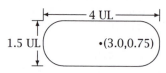

Here we have set the values *x_dimen*=4.0 UL and *y_dimen*=1.5 UL, whereby the unit length UL has been selected to be 0.8 cm. The center of the oval is at the placement coordinates in the \put command, at (3.0,0.75).

The optional argument *part* takes on values t, b, l, or r, for making half ovals.

\put(1.75,4.2){\oval(3.5,1.2)[b]}
\put(1.75,2.6){\oval(3.5,1.2)[t]}
\put(1.75,1.0){\oval(3.5,1.2)[l]}
\put(3.25,1.2){\oval(3.5,1.2)[r]}

The width and height specifications for half ovals are always those of the *entire* figure even though only part of it is being drawn. Similarly, the placement coordinates in the corresponding \put command refer to the center of the complete oval. (Unit length here is UL = 1 cm.)

The argument *part* may also be one of the combinations tl, tr, bl, or br to generate a quarter oval. The order of the two letters is unimportant, so that lt, rt, lb, and rb are equally valid.

\put(2.0,2.5){\oval(3.0,1.0)[tl]}
\put(2.5,2.5){\oval(3.0,1.0)[tr]}
\put(1.0,1.5){\oval(1.0,2.0)[bl]}
\put(3.5,1.5){\oval(1.0,2.0)[br]}

Once again the size specifications refer to the *entire* oval and not just to the part that is drawn, and the placement coordinates in the \put command refer to the center of the complete oval.

Quarter and half circles may also be drawn as partial ovals with equal width and height, but only up to a certain size. The following examples demonstrate that sections of circles are possible up to a size of about 1.5 cm.

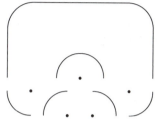

```
\put(2.0,1.0){\oval(4.0,4.0)[t]}
\put(2.0,1.0){\oval(1.5,1.5)[t]}

\put(0.75,0.75){\oval(1.5,1.5)[bl]}
\put(1.75,0.0){\oval(1.5,1.5)[tl]}
\put(2.25,0.0){\oval(1.5,1.5)[tr]}
\put(3.25,0.75){\oval(1.5,1.5)[br]}
```

Sections of ovals may be combined with other picture elements. The placement coordinates in the \put command, however, require some serious consideration to get them positioned properly.

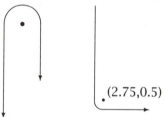

```
\put(0.5,2.5){\oval(1.0,1.0)[t]}
\put(0.0,2.5){\vector(0,-1){2.5}}
\put(1.0,2.5){\vector(0,-1){1.5}}
\put(0.5,2.5){\circle*{0.1}}

\put(2.5,0.5){\line(0,1){2.5}}
\put(2.75,0.5){\oval(0.5,0.5)[bl]}
\put(2.75,0.25){\vector(1,0){1.25}}
```

In all the above examples, the centers of the ovals have been marked with a black dot to indicate where they are. They are not normally a part of the \oval picture element.

UL=1 mm

Exercise 6.6: Although the object pictured here is not that attractive, it does make an excellent exercise for the picture *environment. Hint: sizing and positioning can be worked out best by overlaying the illustration with transparent graph paper.*

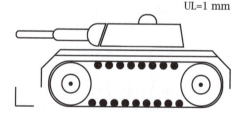

Vertically stacked text

It is sometimes necessary to write text vertically in a diagram, as in the example here at the right. This is carried out with the command

$\quad$ \shortstack[*pos*]{*col*}

The positioning argument can take on values l, r, or c. The standard is c for centered. The command is similar to a tabular environment with only one column. The text is entered as *col*, each row being separated from the next by \\.

y
a
x
i
s

The \shortstack command is most frequently implemented for placing short lines of text inside a framed box, or for stacking single letters vertically. The individual rows are separated from each other with the smallest possible vertical spacing. This means that rows with letters sticking up or down (like *h* and *y*) will have larger apparent gaps between them than rows without such letters. Adding \strut to each line, as in the third example, equalizes the vertical spacing.

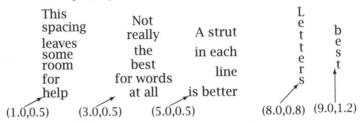

The placement coordinates of the corresponding \put command refer to the lower left corner of an imaginary box that contains the vertically stacked text. The first of the above texts is left justified, the second centered, and the third right justified. The last two on the right are centered. They were entered with

```
\put(1.0,0.5){\shortstack[l]{This\\spacing\\leaves\\some\\...}}
\put(3.0,0.5){\shortstack{Not\\really\\the\\best\\ ...}}
\put(5.0,0.5){\shortstack[r]{A strut\strut\\in each\strut\\...}}
\put(8.0,0.8){\shortstack{L\\e\\t\\t\\e\\r\\s}}
\put(9.0,1.2){\shortstack{b\\e\\s\\t}}
```

The \shortstack command may also be used outside of the `picture` environment within normal text. One possible application is for marginal notes, as in Section 4.10.5.

Framed text

The \framebox command generates a frame of a predetermined size in which text may be inserted at various positions (page 155). In text mode, there is the command \fbox for drawing a frame around text that fits it exactly (Section 4.7.1). This command is also available in the `picture` environment.

The amount of spacing between the box frame and the enclosed text is given by the parameter \fboxsep. The placement of an \fbox by means of the \put command occurs in an unexpected manner, as shown below:

```
\begin{picture}(5,2)
\setlength{\fboxsep}{0.25cm}
\put(0,0){\framebox(5,2){}}
\put(1,1){\fbox{fitted frame}}
\end{picture}
```

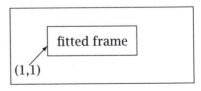

The additional frame spacing is often unwanted in a diagram, especially if the frame surrounds a picture object rather than text. In this case, the command

> \frame{*pic_elem*}

is used instead. The placement coordinate of the \put command then refers to the lower left corner as usual.

```
\put(0.0,0.5){\frame{TEXT}}
\put(1.5,0.0){\frame{\shortstack{W\\O\\R\\D}}}
```

The contents of the \frame command can be any of the previous picture elements, and need not be merely text. However, in many cases the output comes out wrong.

```
\put(0,0){\frame{\vector(1,1){1.0}}}
\put(2,0){\frame{\circle{1.0}}}
```

The first example produces the correct result, while the second has failed. In such cases, one can try putting the picture object inside a \makebox of suitable size and positioning as argument for the \frame command. However, it would then make more sense to use the \framebox command itself in place of \frame{\makebox...}.

Curved lines

Curved lines may be drawn in the picture environment with the commands

2.09 \bezier{*num*}(x_1,y_1)(x_2,y_2)(x_3,y_3)

2ε \qbezier[*num*](x_1,y_1)(x_2,y_2)(x_3,y_3)

which draw a quadratic *Bézier* curve from point (x_1, y_1) to (x_3, y_3) with (x_2, y_2) as the extra Bézier point. The curve is actually drawn as $num + 1$ dots. The only difference between \bezier and \qbezier is that, for the latter, the number of points *num* is an optional argument; if it is omitted, its value will be calculated to produce a solid-looking line. The \bezier command has been retained only for compatibility with older LATEX 2.09 documents.

The meaning of the extra point can be illustrated with the example at the right. The input is

```
\begin{picture}(40,20)
\qbezier(0,0)(20,20)(40,10)
\end{picture}
```

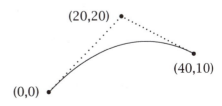

The curve is drawn from (0,0) to (40,10) such that the tangents at the endpoints (the dotted lines) intersect at the extra Bézier point (20,20). Another way of stating this is that, as one moves from the first to the third point, one begins by heading directly towards the second point, and on arrival at the destination, one is moving directly away from that second point again. The dotted lines in the above example, which are not drawn by the \bezier functions, illustrate this.

A dotted curve may be drawn by specifying the number of points *num*; some experimentation may be required to get just the right effect.

Note: the \bezier command was not an integral part of LaTeX 2.09, but was contained in an extra file bezier.sty which had to be read in by including bezier in the list of options to \documentstyle. In LaTeX 2_ε, an empty file of the same name is provided for compatibility.

Line thickness

There is a choice of two line thicknesses for the picture elements \circle, \oval, and \vector, as well as for sloping lines. These may be selected with

    \thicklines   or   \thinlines

Each of these declarations remains in effect until countermanded by a call to the other one, or until the end of the environment in which it was invoked. Initially, \thinlines is in effect.

The thickness of the horizontal and vertical lines may be set with the declaration

    \linethickness{*thickness*}

to any desired size. The argument *thickness* is a positive length specification. With \linethickness{1.5mm}, all subsequent horizontal and vertical lines will have a thickness of 1.5 mm.

Since frame boxes are constructed out of horizontal and vertical lines, the command \linethickness also affects the \framebox and \dashbox commands.

Saving parts of pictures

It is possible to store a combination of picture elements as a sub-picture under a certain name, and to recall the whole set as often as one wants without having to reissue the individual commands every time.

First each sub-picture must have a name reserved for it with

    \newsavebox{*sub_pic_name*}

which creates a box with the name *sub_pic_name* for storing the picture. Afterwards, the sub-picture is saved with

\savebox{*sub_pic_name*}(*x_dimen*,*y_dimen*)[*pos*]{*sub_pic*}

where the arguments (*x_dimen*,*y_dimen*) and *pos* have the same meaning as for \makebox on page 155.

If the picture commands *sub_pic* are simply a piece of text, this command is exactly the same as the \makebox command except that the text is not printed but stored under the name *sub_pic_name*. The *sub_pic* may be set down anywhere within the main picture as a separate picture element.

\usebox{*sub_pic_name*}

```
\newsavebox{\sub}
\savebox{\sub}(2,1)[br]{\small Sub-Pic}
....
\put(0.7,0.0){\frame{\usebox{\sub}}}
\put(3.0,1.0){\frame{\usebox{\sub}}}
```

This example may not seem very practical, since the \savebox and \usebox commands could have been replaced by a \framebox together with \multiput to achieve the same result with even less effort. However, the main advantage of these two commands is not for multiple setting of text, but rather for more complex *sub_pic* compositions.

It should be pointed out that a \savebox command may be given outside of the picture environment, and even within the preamble. Such a sub-picture is then available in all picture environments throughout the document. However, if a \savebox is defined within an environment, it keeps its contents only until that environment comes to an end.

The picture elements inside a \savebox will be sized according to the value of \unitlength in effect at the time that the box is constructed. It will not be rescaled by a later change in \unitlength.

6.1.5   Shifting a picture environment

The generalized syntax of the picture environment contains a further coordinate pair as an optional argument

\begin{picture}(*x_dimen*,*y_dimen*)(*x_offset*,*y_offset*)
                    picture_commands \end{picture}

In this form, (*x_offset*,*y_offset*) specifies the coordinates of the lower left corner. This means that for all \put commands in the environment, the amounts *x_offset* and *y_offset* are effectively subtracted from the placement coordinates, so that the entire picture is shifted by *x_offset* to the *left* and by *y_offset downwards*.

6.1.6   Making graph paper $\boxed{2_\varepsilon}$

The package graphpap adds a new command for drawing gridded paper:

$\boxed{2_\varepsilon}$ \graphpaper[*num*](*x*,*y*)(*lx*,*ly*)

which places a grid with its lower left corner at (x, y), which is *lx* units wide and *ly* units high. Grid lines are drawn for every *num* units, with every fifth one thicker and labeled. If *num* is not specified, it is assumed to be 10. All arguments must be integers, not decimal fractions.

For example, \graphpaper(50,50)(200,100) produces

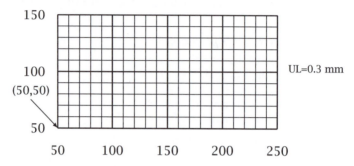

Exercise 6.7: *Produce a sheet of graph paper 10 cm by 15 cm with 2 mm separation between the lines. This sheet may be used to determine the positioning of picture elements when planning a new diagram.*

6.2   The graphics packages $\boxed{2_\varepsilon}$

Rather than trying to draw complicated figures with the limited LATEX picture environment, one may take advantage of other, more powerful software packages to produce a graphics file. This might be a representation of experimental data, a photograph, or some other graphical material that cannot be rendered by simple line drawings. In such cases, one wants to be able to *import* the external graphics file into the LATEX document.

Even under LATEX 2.09 there were methods for including PostScript graphics, and to take advantage of other PostScript features, such as scaling, rotation, and color. However, the necessary packages were designed for specific DVI drivers, using different syntaxes at the user level. The resulting LATEX document files threatened to become tied to one particular driver.

With LATEX 2_ε, a standardized set of commands is available for all drivers by means of the packages graphics and color. The specialized coding to translate the user commands into instructions for the specific drivers is placed in certain .def files which are loaded by options to these packages. Thus one only needs to change the option to switch to another driver; the main text remains unchanged.

Strictly speaking, these packages, referred to as the *graphics collection*, are extensions to the basic LATEX installation. We describe them here, however, rather than in Appendix D because of their growing importance in producing professional-looking documents and camera-ready copy. Provided by Sebastian Rahtz and David Carlisle and many other contributors, they should be part of every LaTeX installation. They are located on the CTAN servers in the directory `graphics` (diagram on page 401). One can obtain a manual for their use by processing the accompanying file `grfguide.tex`. Even more details on these packages and on advanced applications of PostScript to LaTeX are to be found in the book *The LaTeX Graphics Companion* (Goossens *et al.*, 1997).

The commands defined by these packages are the building blocks for other packages that either emulate the older driver-specific ones or provide a more comfortable syntax for these functions. As long as these other packages are based on `graphics` and `color`, they should be equally compatible with all the supported drivers.

These features are not limited to PostScript drivers. As long as a driver can support the inclusion and manipulation (scaling, rotation) of graphics, and/or the use of color, a `.def` file can be written to enable it to make use of the standardized graphics and color commands.

Driver names that may be used as options are

dvipdf	dviwindo	pctexhp⬟□⇑	tcidvi
dvips	emtex⬟□⇑	pctexps	textures
dvipsone	oztex	pctexwin⬟□⇑	truetex□⇑
dviwin⬟□⇑	pctex32	pdftex	xdvi

(Limited functionality: ⬟no color support; □no scaling; ⇑no rotation.)

New drivers are being added at each update, thus one should consult the current documentation for the latest status.

A driver option *must* be specified when loading the `graphics` and `color` packages, for example, as

    `\usepackage[dvips]{graphics,color}`

A local default driver may be established by means of the configurations files described in Section 6.2.4.

6.2.1   Importing external graphics

We wish to have a graphics file produced by some other program included in the document, possibly scaled to a desired size or rotated by 90°. One wants to do by computer what used to be done with scissors and glue.

There are two packages available for importing and manipulating external graphics files: the more basic `graphics` and the more extended `graphicx`. They both have identical functionality, differing only in syntax.

In addition to the driver names, there are some other options that may be invoked when the packages `graphics` or `graphicx` are loaded:

draft   does not import but places a framed box where the graphic would appear, with the file name printed inside; this speeds up the processing considerably when one is only working on the text;

final   counteracts draft; needed when the draft option has been issued globally in \documentclass;

hidescale
          leaves blank space where scaled text should be;

hiderotate
          leaves blank space where rotated text should be; this and the option hidescale are useful if the previewer cannot handle scaling or rotation;

hiresbb
          look for bounding box values in %%HiResBoundingBox instead of the normal %%BoundingBox line.

Importing with the graphics package

The basic importing command with the graphics package is

$$\includegraphics[llx,lly][urx,ury]{file\_name}$$

where *llx, lly* are the coordinates of the lower left corner, and *urx, ury* those of the upper right corner of the *bounding box* containing the part of the picture that is to be included. In other words, they say where the scissors are to be applied. Units may be specified (like [3cm, 2in]) but if they are omitted, big points (bp, 72 per inch, 28.3464 ... per cm) are assumed. If only one optional argument is given, it is the upper right corner, and the lower left is assumed to be [0,0].

If no bounding box coordinates are given, the driver will obtain them some other way, depending on the type of graphics file. For example, for the very common *encapsulated PostScript* files with extension .eps, the bounding box information is extracted from the graphics file itself. The figure at the right is stored in such a file and is included simply with the command

$$\includegraphics{clock.eps}$$

With \includegraphics*, the figure is clipped, so that any drawing outside the specified bounding box is suppressed. This is useful if only part of a figure is to be reproduced. It is also vital with some perverse figures that paint the whole page white!

Scaling

The graphics file in the above commands is transferred to the document in its original size. In order to rescale it, there are two commands available:

> \scalebox{*h_scale*}[*v_scale*]{*text*}

which applies horizontal and vertical scale factors to the contents *text*; if *v_scale* is omitted, it is the same as *h_scale*;

> \resizebox{*h_length*}{*v_length*}{*text*}

adjusts the figure to fit into the specified horizontal and vertical sizes; if either length is given as !, the one scale factor is used for both dimensions. A *-form allows *v_length* to refer to the height + depth of the box, rather than just to the height. In both cases, the contents *text* may be an \includegraphics command, but it may also be any arbitrary text.

Reflection

The contents of a box may be reflected horizontally with

> \reflectbox{*text*}

Rotation

Rotation of a box about the left-hand end of its baseline is done with

> \rotatebox{*angle*}{*text*}

where *angle* is in degrees, and the rotation is counterclockwise.

To illustrate this, we have scaled the previous clock figure to a height of 2 cm and then rotated it by 30°, using the commands

\rotatebox{30}{\resizebox{!}{2cm}{%
    \includegraphics{clock.eps}}}

For demonstration purposes, we have added framed boxes around the figure before and after the rotation. It is the inner (tilted) box that is 2 cm high, while the overall outer box is somewhat higher and broader due to its inclined contents. Without these frames the figure seems to have extraneous space above and to the left, something that can be very puzzling even to experienced users. (The frames would not normally appear with the above commands.)

Exercise 6.8: Copy the lines at the right to a file named demo.eps and then include it in a LaTeX document, with some normal text above and below it. The result should appear as:

```
%!PS-Adobe-3.0 EPSF-3.0
%%BoundingBox: 169 158 233 242
220 200 moveto
200 200 20 0 360 arc
170 170 moveto
230 220 lineto
170 210 lineto
225 160 lineto
205 240 lineto
170 170 lineto
stroke
showpage
```

Importing, scaling, rotating with the graphicx package

If one selects the graphicx rather than the graphics package, a different interface is available for both importing and rotation, one making use of keys and values:

\includegraphics[*key=value*, ...]{*file_name*}

The keys are of two types: those that take a numerical value, and those that are flags with the values true or false. Simply giving the name of a flag without a value is equivalent to setting it to true. Possible keys and their values are:

scale= *number*; enters the number by which the figure size should be magnified over its natural size;

width= *length*; specifies the width to which the figure should be scaled; if height not given, it is scaled with the same factor as the width;

height= *length*; specifies the height to which the figure should be scaled; if width is not given, it is scaled with the same factor as the height;

totalheight= *length*; like height but specifies the height plus depth of the figure; should always be used in place of height if the figure has been rotated;

keepaspectratio (=true/false); if both height and width are specified, this flag ensures that the original height/width ratio remains unchanged; the figure will not exceed either of the given dimensions;

angle= *number*; the angle by which the figure is to be rotated counterclockwise, in degrees; any height or width specifications coming before this key are also rotated, so that the height becomes the width, while the width becomes either the height (positive angle) or depth (negative angle);

origin= *loc*; determines the point about which the rotation occurs; default is bl for bottom left corner; also possible are c for center, t for top, r for right, and B for baseline; any sensible combination, such as tr, is allowed;

draft (=true/false); like the draft package option but applied to the one graphics file; the figure is not imported, but rather a framed box of the correct size is printed containing the name of the file;

clip (=true/false); suppresses the printing of any graphic outside the bounding box;

bb= *llx lly urx ury*; enters the coordinates of the bounding box manually, if they are missing or incorrect in the graphics file, or to be deliberately altered; the specifications are four lengths separated by blanks; units may be given, but if omitted, big points (bp) are assumed;

viewport= *llx lly urx ury*; specifies the bounding box but relative to the lower left corner of the existing one; useful for correcting the bounding box, or (with clip) to select only a portion of the whole figure;

trim= *dllx dlly durx dury*; reduces the existing bounding box by the amounts specified;

hiresbb (=true/false); like the hiresbb package option but applied to the one graphics file; reads bounding box information from the %%HiResBoundingBox line in the graphics file.

The keys are all optional; they are included as needed. Their order is not important other than that angle can change previous height and width meanings. The sets of key/values are separated from each other by commas.

With the key/value syntax, the tilted, scaled graphic on page 169 is produced with

```
\includegraphics[height=2cm,angle=30]{clock.eps}
```

For compatibility with the graphics package, there also exists an \includegraphics* version that clips the imported figure; this is equivalent to including the key clip.

With the graphicx package, the \rotatebox command is similarly redefined to accept the optional key origin.

Exercise 6.9: Include the graphics file demo.eps from Exercise 6.8 scaled by a factor of 2 and rotated by 45° using the \includegraphics command with the graphicx package. Experiment with various keys, in particular with height and width together.

6.2.2   Additional graphics packages

The epsfig package

Sebastian Rahtz has provided a package epsfig that not only updates
the earlier (2.09) version, but also re-implements Rokicki's epsf package
by means of the graphics commands. This is helpful for users who are
accustomed to those syntaxes. For epsf, this is

> \epsfysize=*y_size*   or   \epsfxsize=*x_size*
> \epsf[*llx lly urx ury*]{*file_name*}

The epsfig package also defines an importing command that makes
use of the regular keys and values to enter its parameters:

> \epsfig{file=*file_name*,*key=value*, ... }

For compatibility with some older versions, there is a \psfig command
which is synonymous with the above.

The epsfig package is included in the bundle of graphics packages
and drivers.

The lscape package

Another extra package in the graphics bundle is lscape, by David
Carlisle. This defines a landscape environment that prints its contents
rotated 90° on a page for itself. Head and footlines remain as normal.
This is intended primarily for inserting figures that are in landscape mode,
that is, wider than they are high.

The rotating package

The rotating package by Sebastian Rahtz and Leonor Barroca tries to
make the interface for rotation somewhat simpler. It defines

> \begin{sideways} *text* \end{sideways}
> \begin{turn}{*angle*} *text* \end{turn}
> \begin{rotate}{*angle*} *text* \end{rotate}
> \turnbox{*angle*}{*text*}

where sideways rotates *text* by 90°, turn by an arbitrary angle. The
environment rotate and command turnbox are equivalent: they rotate
but in a box of zero size, so that the contents overlap the surroundings.

This package is not part of the graphics bundle and must be obtained
separately.

6.2.3   Troubleshooting

How importation works

!

In order to understand better what can go wrong when a graphics file is imported, and to know what to do about it, it is important to realize how the interplay between LaTeX and the driver program functions.

LaTeX has no idea what is in the graphics file; for it, the figure is simply a box of a given height, width, and depth, as indeed are all the characters that LaTeX processes. The information on the graphic's natural size is somehow obtained, by reading %%BoundingBox line in the graphics file itself, or in some other file, or through the optional entries in the \includegraphics command or equivalent. After scaling and rotating, LaTeX knows the final size that it must reserve in the output text for the figure.

What is then written to the .dvi file is the name of the graphics file and information on how it should be transformed. Just how this information is coded depends on the graphics driver selected. When the printer driver program processes the .dvi file, it interprets these special instructions, reads in the specified graphics file, performs the transformations, and places the result where LaTeX has said it should go. The end result is that the area inside the designated bounding box coincides with the box that LaTeX has reserved for it. If the bounding box information is incorrect, the figure is obviously going to be misplaced.

The most common type of graphics file to be imported is *encapsulated PostScript*. Files adhering to this standard are intended to be included within other PostScript files, and so must not contain certain PostScript commands that reset the whole graphics page. Most importantly, it must contain a comment line (ignored by PostScript itself) of the form:

%%BoundingBox: *llx lly urx ury*

giving the coordinates of the lower left and upper right corners of the bounding box. The units are never specified, always being big points.

Having pointed out how the importation takes place, we can now discuss what can go wrong along this chain of processes.

Problems with importation

The most common problems encountered when importing encapsulated PostScript files are listed here.

No bounding box. If the bounding box information is totally missing from the graphics file, LaTeX issues the error message

```
! LaTeX Error: Cannot determine size of graphic
                       in ... (no BoundingBox).
```

The solution is to determine the bounding box coordinates somehow (see next point) and to include them, either in the \includegraphics command, or by editing the graphics file itself. However, if there really is no bounding box information in the file, it is unlikely to conform to the encapsulated standard and will cause other problems.

The placement is incorrect. Both LaTeX and the driver process without error messages, but the figure is either displaced from the expected position, or is far too small.

Most likely the bounding box information is incorrect. Many applications that produce PostScript files are too lazy to calculate the true bounding box, or they think they are generating a whole page with a figure somewhere in the middle. In either case, the bounding box corresponds to the full page even though the printed figure occupies only a portion of it.

Find the true bounding box by one of the following methods:

1. Print the figure, mark the lower left and upper right corners of the box containing the figure, and measure their distances from the left and bottom edges. Enter these distances in the `\includegraphics` command, or edit the PostScript file. In the latter case, convert to big points.

   Difficulties with this are that some encapsulated PostScript files cannot be printed on their own, and that the left and bottom edges of the paper need not be the exact lines from which the printer really measures.

2. Include the figure in a short LaTeX file such as

   ```
   \setlength{\fboxsep}{-\fboxrule}
   \fbox{\includegraphics{test.eps}}
   ```

   and print the output. The apparent bounding box appears as a framed box. Measure the true bounding box relative to the left and bottom edges of the apparent one, and enter the values in the `\includegraphics` command with the `viewport=` key. If necessary, scale the figure down to fit on the page, but then remember to increase the measurements by the same scale factor.

3. Use the GhostView program to fix up the bounding box, either automatically or manually. This is the most convenient method if you have this utility for viewing and manipulating PostScript files.

Immovable graphic. It does not shift, nor scale nor rotate, no matter what is specified. In this case, it violates the encapsulated PostScript rules and contains some global plotting commands. Graphics produced by word processing programs are notorious for this. Often the offending command is `setpagedevice`.

There is little that can be done to correct this, other than trying to regenerate the graphics file with an option for encapsulated PostScript.

Judicious editing can remove the troublesome lines, but this could result in the file becoming totally unreadable.

6.2.4 Configuring graphics importation

Although the graphics syntax has been standardized, and most of the driver-specific coding hidden in the .def files, there are still a number of items that must be set up for any particular installation and operating system. These are most conveniently placed in the local configuration file `graphics.cfg` which is read in if it is present.

Default driver

The choice of driver option must always be specified, but a local default can be established by placing the line

> \ExecuteOptions{*driver*}

into the configuration file, where *driver* is one of the allowed driver options. This may be overridden by any explicit driver option specified in the document file itself.

The rest of the configuring commands in this section can be issued either in the `graphics.cfg` file, or in the document.

Search path for graphics files

One can specify the directories where LATEX is to look for graphics files with

> \graphicspath{*dir_list*}

where *dir_list* is a list of directory names, each enclosed in brackets { }, with no other separator. The syntax of the local operating system must be used. Without this command, LATEX searches for graphics files in the same directories as for all other TEX files. Example,

\graphicspath{{figs/}{eps/}}	*for Unix, DOS, Windows*
\graphicspath{{:figs:}{:eps:}}	*for Macintosh*
\graphicspath{{[.figs]}{[.eps]}}	*for VMS*

Default extensions

A list of default extensions for the graphics files can be defined with

> \DeclareGraphicsExtensions{*ext_list*}

This means that only the root name of the file must be given and LATEX will attempt to find it by attaching all the possible extensions. For PostScript drivers, the *ext_list* is usually set to .eps, .ps. At our installation, we also include the non-standard extension .psc. Note that the above command does not add to the list of extensions but rewrites it anew; if you wish to add to the list, you must include all the allowed extensions in the one declaration.

Graphics types

Defining the extensions is only part of the task: one must also associate each extension with a graphics type so that LaTeX knows how to process it. PostScript recognizes only one type, eps, encapsulated PostScript, but there do exist other types such as bmp and pcx for other drivers. For the non-standard .psc extension above, we must also give

```
\DeclareGraphicsRule{.psc}{eps}{}{}
```

to inform LaTeX that this extension belongs to type eps. The other two (empty) arguments specify that the bounding box information is to be read from the file itself, and that no other program needs to be applied to the file.

Compressing graphics files

Since PostScript files are often extremely large, it makes sense to try to compress them with either the zip or gzip programs. In such a case, the .eps file is replaced by a file with extension .zip, or .eps.gz, or .eps-gz. Two problems now arise: first, LaTeX cannot read such files to obtain the bounding box information, and secondly, the driver needs to unpack such a file to include it in the final output. This can be accomplished with, for example,

```
\DeclareGraphicsRule{.eps.gz}{eps}{.eps.bb}{`gunzip -c #1}
```

which establishes the graphics type as eps, with the bounding box information in the file of the same name and extension .eps.bb, and that the operating system command gunzip -c must be applied to the file (represented as #1). The single quote ` is required to indicate a system command. The %%BoundingBox line of the original file must be copied and stored in the .eps.bb file.

Such decompression rules are system dependent and thus need to be configured for the local installation. For example, under the VMS operating system, the gzip program produces files with extension .eps-gz and decompression is performed with gzip -d rather than with gunzip. The corresponding rule becomes

```
\DeclareGraphicsRule{.eps-gz}{eps}{.bb}{`gzip -d -c #1}
```

6.3   Adding color $\boxed{2_\varepsilon}$

The color package accepts the same driver options listed on page 167 for the graphics package. In addition, it also recognizes the options:

monochrome

> to convert all color commands to black and white, for previewers that cannot handle color;

dvipsnames

> makes the named color model of dvips (Section 6.3.1) available to other drivers;

nodvipsnames
>   disables the named model for dvips, to save memory;

usenames
>   loads all the named colors as defined ones; again, see Section 6.3.1 for details.

!

A local configuration file color.cfg can be set up in the same way as for the graphics package. The default driver option is specified in exactly the same way as in Section 6.2.4.

Colors are specified either by a defined name, or by the form

> [*model*]{*specs*}

where *model* is one of rgb (red, green, blue), cmyk (cyan, magenta, yellow, black), gray, or named. The *specs* is a list of numbers from 0 to 1 giving the strengths of the components in the model. Thus [rgb]{1,0,0} defines red, [cmyk]{0,0,1,0} yellow. The gray model takes only one number. The named model accesses colors by internal names that were originally built into the dvips driver, but which may now be used by some other drivers too. This model is described in Section 6.3.1.

A color can be defined with

> \definecolor{*name*}{*model*}{*specs*}

and then the *name* may be used in all the following color commands. Certain colors are automatically predefined for all drivers: red, green, blue, yellow, cyan, magenta, black, white.

In the following color commands, *col_spec* is either the name of a defined color, like {blue}, or [*model*]{*spec*}, like [rgb]{0,1,0}.

\pagecolor *col_spec* sets the background color for the current and following pages;

\color *col_spec* is a declaration to switch to setting text in the given color;

\textcolor *col_spec*{*text*} sets the text of its argument in the given color;

\colorbox *col_spec*{*text*} sets its argument in a box with the given color as background;

\fcolorbox *col_spec1* *col_spec2*{*text*} like \colorbox, with a frame of *col_spec1* around a box of background color *col_spec2*; the two specifications must either both be defined ones, or both use the same model, which is given only once; for example, \fcolorbox{red}{green}{Text} sets 'Text' in the current text color on a green background with a red frame;

\normalcolor switches to the color that was active at the end of the preamble. Thus placing a \color command in the preamble can change the standard color for the whole document. This is the equivalent to \normalfont for font selection.

Normally one would try to define all the colors needed as names for the *col_spec* entries. This simplifies changing the color definition everywhere should fine-tuning be required after the initial printed results are seen. The same color definition can produce quite different effects on different printers. Even the display on the monitor is no reliable guide as to how the output will appear on paper.

6.3.1   The named color model

One very useful color model is called named and is based on the 68 predefined internal colors of the dvips PostScript driver. Sample names are BurntOrange or DarkOrchid. This model can be activated for other drivers with the option dvipsnames, in which case one can define colors as, for example

```
\definecolor{titlecol}{named}{DarkOrchid}
```

The color titlecol can then be used as *col_spec* in the various color commands.

The named colors can be defined with their own names if one invokes the option usenames, which effectively declares

```
\definecolor{BurntOrange}{named}{BurntOrange}
```

and so on, for all 68 colors.

It is possible to generate a palette of the named colors by processing the following short LaTeX file and sending the output to the desired printer.

```
\documentclass[12pt,a4paper]{article}
\usepackage[dvips]{color}
\usepackage{multicol}
\pagestyle{empty}
\setlength{\oddsidemargin}{0pt}
\setlength{\textwidth}{16cm}
\setlength{\textheight}{22cm}
\setlength{\parindent}{0pt}
\setlength{\parskip}{0pt}

\begin{document}
\renewcommand*{\DefineNamedColor}[4]{%
    \textcolor[named]{#2}{\rule{7mm}{7mm}}\quad
    \texttt{#2}\strut\\}
```

```
\begin{center}\Large Named colors in \texttt{dvipsnam.def}
\end{center}
\begin{multicols}{3}
\input{dvipsnam.def}
\end{multicols}
\end{document}
```

Remember, each printer can reproduce the colors differently, so it is important to test this table with every color printer that might be used.

Exercise 6.10: Copy the above lines to a file named palette.tex and process it. View the output on a color monitor or send it to a color printer.

6.4   Floating tables and figures

Whether a figure is produced in the picture environment or imported with \includegraphics, it is inserted in the text where the drawing or importing commands are issued, coming between the previous and following texts. This can present the same difficulties as for tables, described in Section 4.8.5: if the figure is so high that it no longer fits on the current page, that page is prematurely terminated and the figure is placed at the top of the next page, with too much empty space left on the original page.

In Section 4.8.5 we mentioned how this problem is solved for tables. Here we now give a complete description of the *float* procedures that apply to both figures and tables.

6.4.1   Float placement

LaTeX does make it possible to *float* figures and tables, together with their headlines and captions, to an appropriate location without interrupting the text. This is invoked with the environments

```
\begin{figure}[where]   figure  \end{figure}
\begin{figure*}[where]  figure  \end{figure*}
\begin{table}[where]    table   \end{table}
\begin{table*}[where]   table   \end{table*}
```

The *-forms apply only to the two-column page format and insert the figure or table across both columns instead of the normal single column. They function exactly the same as the standard forms when the page format is single column.

In the above syntax, *figure* and *table* are the texts for the contents of the float, either a picture or tabular environment, or an \includegraphics command, together with a possible \caption command, as described later in Section 6.4.4.

The argument *where* specifies the allowed locations for the figure or table. There are several possibilities so that *where* consists of from zero to four letters with the following meanings:

h *here*: the float may appear at that point in the text where the environment is typed in; this is not permitted for the *-form;

t *top*: the float may appear at the top of the current page, provided there is enough room for both it and the previous text; if this is not the case, it is added at the top of the next page; the subsequent text continues on the current page until the next normal page break; (for two-column format, read *column* in place of *page*);

b *bottom*: the float may be placed at the bottom of the page; the subsequent text continues until the room left on the current page is just enough for the float; if there is already insufficient room, the float will be put at the bottom of the next page; this is not permitted for the *-form;

p *page*: the float may be put on a special page (or column) containing only figures and/or tables;

$\boxed{2_\varepsilon}$! used together with any combination of the other letters, suspends all the spacing and number restrictions described in Section 6.4.3.

The argument values may be combined to allow several possibilities. If none is given, LaTeX assumes the standard combination tbp.

The placement arguments permit certain possibilities for locating the float, but the actual insertion takes place *at the earliest possible point* in accordance with the following rules:

• no float appears on a page prior to that where it is defined;

• figures and tables are output in the order in which they were defined in the text, so that no float appears before a previously defined float of the same type; figures and tables may, however, be mixed in their output sequence; in two-column format, the double column *-floats may also appear out of sequence;

• floats will be located only at one of the allowed positions given by the placement argument *where*; without an argument, the standard combination tbp is used;

• unless the ! is included in *where*, the positioning obeys the limitations of the style parameters described in Section 6.4.3;

- for the combination ht, the argument h takes priority; the float is inserted at the point of its definition even if there is enough room for it at the top of the page.

When any of the commands \clearpage, \cleardoublepage, or \end{document} is given, all floats that have not yet been output will be printed on a separate page or column regardless of their placement arguments.

Sometimes LaTeX gets stuck on a float which holds up the entire queue until the end of the document. One way to clear the queue is to issue \clearpage right after the troublesome float. However, that would insert a new page at that point, something that may not be desired. The package afterpage in the tools collection (Section D.3.3) provides the command \afterpage which executes its argument at the end of the current page. Thus \afterpage{\clearpage} solves this problem.

6.4.2 Postponing floats

Occasionally one wants to prevent floats from appearing on a certain page, for example at the top of a title page. (LaTeX automatically corrects that case.) However, there are other situations where a float should be suppressed temporarily. One might want it at the top of a page, but not before the start of the section that refers to it. The command

$\boxed{2\varepsilon}$ \suppressfloats[*loc*]

sees to it that *for the current page only* no further floats of the specified placement *loc* should appear. If the optional *loc* is omitted, all floats are suppressed; otherwise *loc* may be either t or b, but not both.

Note that \suppressfloats does not suppress all floats for the current page, but only *further* ones that come between the issuing of this command and the end of the page. Thus it is possible for floats from a previous section still to appear on the page.

Alternatively, the package flafter (Section 8.8.3) may be loaded to ensure that *all* floats appear only after their position in the text.

The command \suppressfloats and the location parameter !, which are both new to LaTeX 2_ε, are attempts to give the author more control over the sometimes capricious actions of float placement.

6.4.3 Style parameters for floats

There are a number of style parameters that influence the placement of floats, which may be altered by the user as desired.

topnumber
        The maximum number of floats that may appear at the top of a page.

bottomnumber
> The maximum number of floats that may appear at the bottom of a page.

totalnumber
> The maximum number of floats that may appear on any page regardless of position.

dbltopnumber
> The same as topnumber but for floats that extend over both columns in two-column page format.

The above parameters are all *counters* and may be reset to new values with the command \setcounter{*ctr*}{*num*}, where *ctr* is the name of the counter and *num* the new value that it is to take on.

\topfraction
> A decimal number that specifies what fraction of the page may be used for floats at the top.

\bottomfraction
> A decimal number that specifies what fraction of the page may be used for floats at the bottom.

\textfraction
> The fraction of a page that must be filled with text. This is a minimum, so that the fraction available for floats, whether top or bottom, can never be more than 1−\textfraction.

\floatpagefraction
> The smallest fraction of a float page that is to be filled with floats before a new page is called.

\dbltopfraction
> The same as \topfraction but for double column floats in two-column page format.

\dblfloatpagefraction
> The same as \floatpagefraction but for double column floats in two-column page format.

These style parameters are altered with \renewcommand{*cmd*}{*frac*} where *cmd* stands for the parameter name and *frac* for the new decimal value, which in every case must be less than 1.

\floatsep
> The vertical spacing between floats appearing either at the top or at the bottom of a page.

\textfloatsep
> The vertical spacing between floats and text, for both top and bottom floats.

\intextsep
> The vertical spacing above and below a float that appears in the middle of a text page with the h placement argument.

\dblfloatsep
> The same as \floatsep but for double column floats in two-column page format.

`\dbltextfloatsep`
> The same as `\textfloatsep` but for double column floats in two-column page format.

This group of style declarations are rubber lengths that may be changed with the `\setlength` command (Section 2.4.2).

[2ε] `\topfigrule`
> A command that is executed after a float at the top of a page. It may be used to add a rule to separate the float from the main text. Whatever it adds must have zero height.

[2ε] `\botfigrule`
> Similar to `\topfigrule`, but is executed before a float that appears at the bottom of a page.

[2ε] `\dblfigrule`
> Similar to `\topfigrule`, but for double column floats.

These three commands normally do nothing, but they may be redefined if necessary. For example, to add a rule of thickness 0.4 pt below a top float,

```
\renewcommand{\topfigrule}{\vspace*{-3pt}
    \rule{\columnwidth}{0.4pt}\vspace{2.6pt} }
```

Because of the negative argument in `\vspace*`, the total vertical spacing is zero, as required.

All the one-column style parameters also function within the two-column page format, but they apply only to floats that fill up one column.

If the style parameters are set to new values within the preamble, they apply from the first page onwards. However, if they are changed within a document, they do not take effect until the next page.

6.4.4 Float captions

A figure caption or table title is produced with the command

> `\caption[`*short_title*`]{`*caption_text*`}`

inside the `figure` or `table` environment. The *caption_text* is the text that is printed with the float and may be fairly long. The *short_title* is optional and is the text that appears in the list of figures or tables (Section 3.4.4). If it is missing, it is set equal to *caption_text*. The *short_title* should be given if the *caption_text* is longer than about 300 characters, or if it is more than one line long.

In the `table` environment, the `\caption` command generates a title of the form 'Table *n*: *caption_text*', and in the `figure` environment 'Figure *n*: *caption_text*', where *n* is a sequential number that is automatically incremented. In document class `article`, the figures and tables are numbered from 1 through to the end of the document. For the `report` and `book` classes, they are numbered within each chapter in the form *c.n*, where *c*

Table 6.1: Computer Center Budget for 1999

Nr.	Item	51505	52201	53998	Total
1.1	Maintenance	130 000		15 000	145 000
1.2	Network costs	5 000		23 000	28 000
1.3	Repairs	25 000	6 000		31 000
1.4	Expendables		68 000		68 000
1.	Total	160 000	74 000	38 000	272 000

is the current chapter number and n is the sequential number reset to 1 at the start of each chapter. Figures and tables are numbered independently of one another.

The \caption command may be omitted if numbering is unwanted, since any text included in the *float* environment will accompany the contents. The advantages of \caption over simple text are the automatic numbering and entries in the lists of figures and tables. However, a manual entry in these lists may be made as described in Section 3.4.4 using

\addcontentsline   and   \addtocontents

A *title*, or *headline*, is produced with \caption when the command comes at the beginning of the material in the *float* environment: the number and text are printed above the table or figure. A *caption* is added below the object if the command comes after all the other float commands. In other words, the \caption is just another item within the float and whether its text appears at the top (as *title*) or below (as *caption*) depends on how the user places it.

The *caption_text* will be centered if it is shorter than one line, otherwise it is set as a normal paragraph. The total width may be adjusted to that of the table or figure by placing the command inside a parbox or minipage. For example,

\parbox{*width*}{\caption{*caption_text*}}

The following examples contain further examples of how text, table, and picture commands may be combined in floats.

6.4.5   Float examples

The first two tables are produced with the following texts:

```
\begin{table} \caption{Computer Center Budget for 1999}
  \begin{tabular}{|l|l||r|r|r|r|} ... ...  \end{tabular}
\end{table}
```

(This text was typed in before the last paragraph of the previous section. The second table is entered here in the current text.)

Table 6.2: Estimates for 1999. *A continuation of the previous budget is no longer practical since, with the installation of the new computing system in 1998, the operating conditions have been completely overhauled.*

Nr.	Item	51505	52201	53998	Total
1.1	Maintenance	240 000			240 000
1.2	Line costs	12 000	8 000	36 000	56 000
1.3	Training			50 000	50 000
1.4	Expansion	80 000	3 000		83 000
1.5	Expendables		42 000		42 000
1.	Total	332 000	53 000	86 000	471 000

```
\begin{table}
  \caption{\textbf{Estimates for 1999} {\em A continuation...}}
    \begin{tabular}{|l|l||r|r|r|r|} ... ... ... \end{tabular}
\end{table}
```

Because the placement argument is missing from the `table` environments, the standard values `tbp` are used. The first was placed at the top on the facing page because it was typed in early enough (during the last section) that there was still room for it there. Then the command `\suppressfloats` (Section 6.4.2) was issued to prevent any more floats appearing on the same page. The second table is therefore forced to float to the top of this page.

Narrow figures or tables may be set beside each other, as shown in the following example (which appears at the bottom of this page).

```
\begin{figure}[b]
\unitlength1cm
\begin{minipage}[t]{5.0cm}
\begin{picture}(5.0,2.5) ... ... \end{picture}\par
\caption{Left}
\end{minipage}
\hfill
```

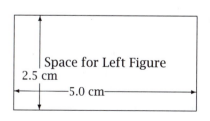

Figure 6.1: Left

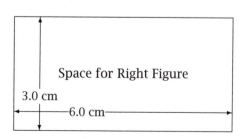

Figure 6.2: Right

```
\begin{minipage}[t]{6.0cm}
\begin{picture}(6.0,3.0) ... ... \end{picture}\par
\caption{Right}
\end{minipage}
\end{figure}
```

The two figures along with their captions are each set in a minipage environment of widths 5 and 6 cm. The minipages are separated from one another by an \hfill space. The positioning argument t has the effect that the minipages are aligned along their first lines (Section 4.7.3). The entire structure within the figure environment floats as a single entity.

The question might now arise as to why, since the two figures have unequal heights and are supposedly aligned vertically along their top lines, their bottom edges are at the same level. The explanation is that a picture environment establishes an LR box (Section 4.7.1) to contain all the picture commands, and that is viewed by LaTeX as a single line of output text with the baseline at the bottom edge of the picture. In both minipages, the picture environment is the first entry and is therefore the first logical line of text. It is these baselines that are taken for the vertical alignment of the minipages.

If the two pictures were to be aligned along their top edges, it would be necessary to include a dummy first line (Section 4.7.4) in each minipage before the picture environments. This could be something like \mbox{}, for example.

The application of box commands within a float permits completely free positioning. If the caption text is to appear, say, beside the table or figure, instead of above or below it, the objects may be put into minipages or parboxes with suitable alignment arguments. Here is an example:

```
\begin{table}[b]
  \centerline{\bfseries Results and Seat Distribution of the...}
  \mbox{\small
  \begin{minipage}[b]{7.7cm}
    \begin{minipage}[t]{4.4cm}
```

Results and Seat Distribution of the 1998 General Election

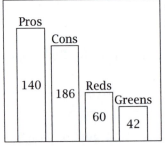

Seat Distribution
in the Assembly of
1998–2002

Results:
1998 General
Election

Party	Votes	%
Pros	15 031 287	37.7
Cons	12 637 418	31.7
Reds	6 499 210	16.3
Greens	4 486 514	11.3
Others	1 201 078	3.0

```
        \mbox{}\\ \unitlength0.75cm
        \begin{picture}(5.75,5.0) ... ... ... \end{picture}
     \end{minipage} \hfill
     \parbox[t]{3.2cm}{\makebox[0cm]{}\\{\bfseries Seat ...} ...}
  \end{minipage}
  \hspace{-3cm}
  \begin{minipage}[b]{7cm}
     \parbox[b]{2.5cm}{{\bfseries Results:}\\ General Election..}
     \hfill
     \begin{tabular}[b]{|l||r|r|} ... ... \end{tabular}
  \end{minipage} }
\end{table}
```

! Here the vertical boxes are nested inside one another. The left side, consisting of graphics and a title at the upper right, is contained in a minipage of width 7.7 cm, in which the picture is in another minipage of width 4.4 cm while the text is in a parbox of width 3.2 cm. Both are aligned with the top lines. A dummy line containing \mbox{} (Section 4.7.4) is added before the first picture to provide a top line with which the second parbox can be aligned.

The right-hand side consists of a minipage of width 7 cm, shifted to the left by 3 cm, aligned with the left-hand minipage along the bottom lines. It contains a parbox of width 2.5 cm for the text as well as a table, which is automatically a vertical box. Both of these are aligned with the bottom lines.

6.4.6   References to figures and tables in text

The automatic numbering of tables and figures has the consequence that the author does not know the numbers at the time of writing. Since he or she would like to be able to refer to these objects by number, as 'see Figure 3' or 'Table 5 illustrates', another means of referencing must be found. It is not sufficient to keep track of the number of \caption calls that have been made, since the document may not be written in the order in which it finally appears, and new figures or tables may be inserted during revision, or some removed.

These problems are solved with the LaTeX cross-reference system, described in more detail in Section 8.3.1. The basic commands are

\label{*name*}   \ref{*name*}

which assign a keyword *name* to the figure or table number that may be used as reference in the text. The keyword *name* may be any combination of letters, numbers or symbols. The assignment is made with the \label command given anywhere within the caption text of the \caption command; in the main text, the command \ref inserts the number that is associated with its keyword.

This is best shown with an example. The budget table on page 184 was actually written as

```
\caption{\label{budget95} Computer Center ... }
```

so that `Table \ref{budget95}` produces 'Table 6.1' when used in the main text.

There is a second referencing command `\pageref` to generate the page number where the referenced object is to be found. For example, a few lines up, the input text `on page \pageref{budget95}` was used to create the text '... on page 184'.

7

User Customization

LaTeX allows the user to define his or her own commands and environments. However, since these make extensive use of the LaTeX counters and lengths, we will first present a more detailed discussion of these objects and how they may be manipulated.

7.1 Counters

7.1.1 LaTeX counters

LaTeX manages a number of counters by giving them initial values at the start and by changing these values when certain commands are called. Most of these counters have the same name as the commands that alter them:

part	chapter	paragraph	figure	enumi
	section	subparagraph	table	enumii
	subsection	page	footnote	enumiii
	subsubsection	equation	mpfootnote	enumiv

The meanings of most of these counters are obvious from their names and need no further explanation. The counters enumi ... enumiv refer to the four levels of the **enumerate** environment (Sections 4.3.4 and 4.3.5), while the counter mpfootnote controls the footnote numbering within the minipage environment (Section 4.10.4).

In addition to these counters, there may be further counters created by the \newtheorem command, having the same name as the *struct_type* argument (Section 4.5). For the two examples of the \newtheorem command on page 79, this book also contains the counters theorem and axiom.

The value of a counter is an integer number, usually non-negative. A command may output several numbers at once: the current \subsection

189

command outputs 7.1.1, which addresses three counters in all. For example, the \subsection command increments the value of the subsection counter by one, and prints the values of the chapter, section, and subsection counters, separated by periods. At the same time, this command sets the subsubsection counter to zero.

7.1.2   User-defined counters

The user may create new counters with the command

$$\newcounter\{counter\_name\}\ [in\_counter]$$

where *counter_name* is the name of the newly established counter. This may be any combination of letters that is not already the name of an existing counter. Thus none of the names of the LaTeX counters listed above may be used as *counter_name* nor any name of a previously defined user counter. The optional argument *in_counter* is the name of another counter that already exists (LaTeX or user defined) and has the effect that the newly defined counter is reset to zero whenever *in_counter* is incremented by one of the commands \stepcounter or \refstepcounter (see below).

When a counter has been created with \newcounter, its initial value is zero.

The \newcounter command may not appear in any file that is read in with the \include command (Section 8.1.2). It is therefore best to put all \newcounter commands into the preamble.

7.1.3   Changing counter values

Every counter, whether LaTeX or user defined, can have its value altered with the following commands:

\setcounter{*counter*}{*num*}
> This command is self-explanatory: the specified *counter* is assigned the integer value *num*.

\addtocounter{*counter*}{*num*}
> With this command, the value of the *counter* is increased by the integer *num*, which may be positive or negative.

\stepcounter{*counter*}
> The value of the *counter* is increased by one, and at the same time all its sub-counters (that is, those that have this counter as their *in_counter*) are reset to zero (see above).

\refstepcounter{*counter*}
> This command has the same effect as \stepcounter but also makes *counter* the current counter for the cross-referencing command \label (see Section 8.3.1).

This last command may be applied, for example, within the figure or table environments when the \caption command is missing and yet there is to be a reference in the text to the figure or table number using the \ref command. Then \refstepcounter{figure} or \refstepcounter {table} is given within that float environment to bring the corresponding counter to the right value and to allow that value to be assigned a keyword with the \label command (Section 8.3.1).

The value of a counter may be treated as a number with the command

> \value{*counter*}

which does not change the value at all. This command is mostly used in connection with \setcounter or \addtocounter. For example, if the user-defined counter mypage has been created, it may be set to the same value as the page counter page by giving \setcounter{mypage}{\value {page}}.

Normally the \protect command, which prevents fragile commands from being broken up during transport, may also be included in front of robust commands without any harm being done. The \value is the exception to this rule. It may never be preceded by the \protect command.

7.1.4 Printing counter values

The numerical value in a counter can be printed with the commands

\arabic{*counter*}	as an Arabic number,
\Roman{*counter*}	as a capital Roman numeral,
\roman{*counter*}	as a lower case Roman numeral,
\alph{*counter*}	as a lower case letter,
\Alph{*counter*}	as a capital letter,
\fnsymbol{*counter*}	as a footnote symbol.

For the commands \alph and \Alph, the numbers 1 … 26 correspond to the letters *a* … *z* and A … Z. It is up to the user to ensure that the counter value lies within this range. For \fnsymbol, the numbers 1 … 9 are output as the symbols * † ‡ § ¶ ‖ ** †† ‡‡. Here again, the user must take care that the counter does not reach a value of 10 or more.

For each counter, a command of the form

> \the*counter*

is also available, consisting of \the immediately followed by the name of that counter, such as \thepage. This type of command is initially identical to \arabic{*counter*}, but may redefined to be composed of several counter commands. In the document classes book and report, for example, the command \thesection is defined in terms of both the chapter and section numbers: `\arabic{chapter}.\arabic{section}`. Here is the result of printing `\thesection` at this point: 7.1.

The automatic printing of counter values such as the page, equation, or sectioning numbers is accomplished by means of calls to the appropriate `\the`*counter* commands. If a different format for some automatic numbering is desired, say alphabetical equation numbers, the definition of the corresponding `\the`*counter* command can be altered using the methods described in Section 7.3.

Exercise 7.1: *Take your standard* `exercise.tex` *file and print out the final values of the LATEX counters with* `\arabic{counter}` *commands at the end. Change some the values with the* `\setcounter` *and* `\addtocounter` *commands and print the values out once more.*

7.2   Lengths

It has been constantly pointed out in all the descriptions of the length parameters such as `\parskip` or `\textwidth` that new values may be assigned with the `\setlength` command. Some of these parameters expect rubber length values that may stretch or shrink. These are mainly parameters that produce vertical spacing. The types of length units, both fixed and rubber, are described in detail in Section 2.4. This will not be repeated here, but rather additional commands for assigning and handling lengths are discussed in this section.

The standard LATEX method for assigning a value to a length parameter is with the command

> `\setlength{\`*length_cmd*`}{`*length_spec*`}`

where *length_spec* may be a length specification (with units) or another length parameter. In the latter case, `\`*length_cmd* takes on the current value of that other parameter. Thus with `\setlength{\rightmargin}{\leftmargin}` the right-hand margin in a `list` environment is set to the same value as that of the left-hand margin.

Lengths may be increased with

> `\addtolength{\`*length_cmd*`}{`*length_spec*`}`

which adds *length_spec* to the value of the length parameter `\`*length_cmd*. A negative value for *length_spec* decreases `\`*length_cmd* by that amount. Once again, another length parameter may be used for *length_spec*, with or without a preceding minus sign, and its value will be added or subtracted. A decimal number just before a length parameter multiplies its value by that quantity: `0.5\textwidth` means half the width of the text column and `2\parskip` twice the inter-paragraph spacing.

With the command

$$\setminus\texttt{settowidth}\{\setminus length\_cmd\}\{text\}$$

the length parameter *length_cmd* is set equal to the natural length of a piece of text set in LR mode (normal left-to-right).

Similarly the commands

$\boxed{2\varepsilon}$ \\settoheight{*length_cmd*}{*text*}

$\boxed{2\varepsilon}$ \\settodepth{*length_cmd*}{*text*}

set the *length_cmd* equal to the height and depth of the *text* above and below the baseline, respectively.

Finally, the command

$$\setminus\texttt{stretch}\{decimal\_num\}$$

yields a rubber length that is *decimal_num* times as stretchable as \\fill (Section 2.4.2).

A user-defined length parameter is created with

$$\setminus\texttt{newlength}\{\setminus new\_len\_cmd\}$$

which establishes *new_len_cmd* as a length with a value of 0pt. All the above commands may be used to manipulate its value further.

The command

$$\setminus\texttt{addvspace}\{length\_spec\}$$

inserts extra vertical spacing of the amount *length_spec* at that point where it appears. If more than one such command is given, the total inserted spacing will be that of the largest argument and not the sum of them all. This command may only be given *between* paragraphs. Its application for user-defined commands and environments lies in the generation of structures that should behave like paragraphs.

7.3   User-defined commands

New commands may be defined or redefined under LaTeX with the commands

$$\setminus\texttt{newcommand}\{\setminus com\_name\}[narg][opt]\{def\}$$
$$\setminus\texttt{renewcommand}\{\setminus com\_name\}[narg][opt]\{def\}$$

The first version is used to define a command *com_name* that does not yet exist. Its name may be any combination of letters that do not form the name of another command. The second version redefines an already existing command *com_name*. In both cases, an error message is printed if the incorrect variant is called. The first optional argument *narg* is a number between 1 and 9 specifying how many arguments the new or altered command is to have. A second optional argument *opt* gives the default value for an optional argument that the new command may take. The actual definition of the command is contained in the text *def*. (In LaTeX 2.09 the [*opt*] optional argument was not available.)

7.3.1   Commands without arguments

We will first illustrate the use of the \newcommand without the optional argument [*narg*]. This form is applied when a fixed combination of LaTeX or user commands is to be repeated frequently as a command with its own name. For example, the structure $x_1, \ldots, x_n$, called an x-vector, often occurs in mathematical formulas and is formed in math mode with x_1,\ldots,x_n. Typing

> \newcommand{\xvec}{x_1,\ldots,x_n}

creates a new command named \xvec that may be called and used just like any other command. When called, it inserts the sequence of text and commands, in this case x_1,\ldots,x_n, into the current text exactly as if one had typed it oneself. In fact, this is precisely what happens: when \xvec is called, it is its definition that goes into the LaTeX processing.

Since the new command \xvec contains a math command (the subscript command _), it may only be called within math mode. Thus $\xvec$ is required to produce $x_1, \ldots, x_n$ in text mode. It might therefore seem a good idea to include the switching to math mode in the definition itself, as

> \newcommand{\xvec}{$x_1,\ldots,x_n$}

Now \xvec yields $x_1, \ldots, x_n$. However, this command may only be applied in text and never in math mode. There is a trick to enable the command to be called in both modes: define it as

> 2ε \newcommand{\xvec}{\ensuremath{x_1,\ldots,x_n}}

Now both \xvec and $\xvec$ are allowed, both with the same result.

(In LaTeX 2.09, the command \ensuremath does not exist. Instead, one must use \mbox{$...$}. In text mode, the \mbox is ignored, but in math mode, it temporarily switches to text mode, from which the $ signs activate math mode. The two methods do not produce identical results, and \ensuremath is to be preferred in LaTeX 2ε.)

The above text mode example was actually written as \xvec{}, since TeX treats it as a command without arguments, terminating its name with the first non-letter that it finds. If this character is a blank, it only ends the command name and does not insert interword spacing (Section 2.1). Thus \xvec and ... produces '$x_1, \ldots, x_n$and ... ' without any spacing between. This problem is solved by adding a space command \␣ or the empty structure {} after the command name, in this case as \xvec\␣ or \xvec{}.

It would also have been possible to include the blank in the definition of \xvec, as {\ensuremath{x_1,\ldots,x_n} }. Now the blank following the command name is still removed, but the command itself inserts one to make up for it. However, this is not recommended practice since this 'programmed' blank will always be present, even when some punctuation or other symbol directly follows the command. (A better solution is provided by the `xspace` package in the tools collection, on page 388.)

The different versions of the above example were all illustrated with the command `\newcommand`, although in fact this command may only be used once to initiate a user-defined command that does not already exist. Once `\xvec` has been created, a revised definition can only be given with the command `\renewcommand`. This was indeed done with the second and subsequent definitions of `\xvec`.

Following this example, the user may employ `\newcommand` (or alternatively `\renewcommand`) to combine any set of commands and text to form a command under a new name that may then be called whenever necessary. In this way, a considerable amount of typing can be avoided and the possibility of error reduced, especially for complex mathematical structures.

If it is not known whether or not a command with the chosen name already exists, one may use instead

$\boxed{2_\varepsilon}$ `\providecommand{\`*com_name*`}[`*narg*`][`*opt*`]{`*def*`}`

which has the same syntax as `\newcommand` and `\renewcommand`. The difference is that if the command already exists, the new definition will be ignored. The opposite effect (overwriting the current definition of a command without knowing if it already exists) can be achieved by first calling `\providecommand` to ensure the command exists, and then issuing the true definition with `\renewcommand`. However, this should be done only with great care!

Exercise 7.2: *Define the commands* `\iint`*,* `\iiint`*, and* `\idotsint`*, to make the multiple integrals shown at the right as displayed formulas, or as the following text formulas:* $\iint, \iiint, \int \cdots \int$

$$\iint \quad \iiint \quad \int \cdots \int$$

Exercise 7.3: *Change the* `\thechapter`*,* `\thesection`*, and* `\thesubsection` *commands so that for the document classes* book *and* report *the chapter numbering is done with capital letters, such as B, the section numbering with capital Roman numerals after the chapter letter, in the form B-III, and the subsection numbering with lower case Roman numerals following a comma: B-III,v.*
Hint: the original versions of these commands in book *and* report *are defined as*

```
\newcommand{\thechapter}{\arabic{chapter}}
\newcommand{\thesection}{\thechapter.\arabic{section}}
\newcommand{\thesubsection}{\thesection.\arabic{subsection}}
```

Now apply `\renewcommand` *to make the required changes.*

7.3.2   Commands with arguments

In addition to the structure $x_1, \ldots, x_n$, there are equivalent vectors $y_1, \ldots, y_n$ and $z_1, \ldots, z_n$ in mathematics. It would be possible to define commands \yvec and \zvec following the pattern for \xvec. However, it is also possible to define a generalized vector command and to specify the variable part as an argument. In the present example, the variable part is the letter x, y, or z. A command with *one* variable part is created with the optional argument [1]. For example,

```
\newcommand{\avec}[1]{\ensuremath{#1_1,\ldots,#1_n}}
```

defines the general vector command \avec{*arg*}. Calling \avec{x} yields $x_1, \ldots, x_n$ while invoking \avec{y} prints out $y_1, \ldots, y_n$. The character #1 in the command definition is a dummy argument representing the text of *arg* that replaces all occurrences of #1 when the command is called. By imagining an x or a y at each location of the #1 in the definition, the desired structures in each case can be recognized.

The digit 1 in the dummy argument #1 seems at first to be rather pointless. In fact, for a command with only one argument it really has no meaning. However, its function becomes more obvious for commands with *multiple* arguments. Let us say, for example, that we want a command to generate structures such as $u_1, \ldots, u_m$ as well as $v_1, \ldots, v_n$. This requires two arguments, one to specify the letter u, v, etc., and a second to determine the last subscript n, m, and so on. Such a command is created with

```
\newcommand{\anvec}[2]{\ensuremath{#1_1,\ldots,#1_#2}}
```

and invoked as \anvec{u}{n} for $u_1, \ldots, u_n$ and as \anvec{v}{m} for $u_1, \ldots, u_m$. The optional argument [2] for \newcommand says that the command being defined contains two arguments; in the definition part, #1 is replaced by the first argument and #2 by the second. By imagining u or v in place of #1 and n or m where #2 stands, one can see how the command \anvec{*arg1*}{*arg2*} operates.

This pattern may be carried on for even more arguments. With

```
\newcommand{\subvec}[3]{\ensuremath{#1_#2,\ldots,#1_#3}}
```

a command \subvec is defined with three arguments. It should be clear from the definition that calling \subvec{a}{i}{j} produces $a_i, \ldots, a_j$.

A command argument that consists of only a single character need not be put into curly brackets { } but may be given directly. If it is the first argument, it must be separated from the command name with a blank, as usual. Thus the sequence \subvec aik is the same as \subvec{a}{i}{k}, and \subvec x1n produces the same structure $x_1, \ldots, x_n$ as our first user-defined example \xvec.

Arguments must be enclosed in curly brackets { } when they contain more than one character, since the brackets indicate that the contents are to be treated as a unit. Thus \subvec{A}{ij}{lk} prints out $A_ij,\ldots,A_lk$. The three replacement arguments are A for #1, ij for #2, and lk for #3.

Why does \subvec{A}{ij}{lk} produce $A_ij,\ldots,A_lk$ and not the expected result $A_{ij},\ldots,A_{lk}$? The answer is that, although the arguments within curly brackets are set as units into the definition text, the brackets themselves are not. The command text after replacement is \ensuremath{A_ij,\ldots,A_lk} so that only the first characters following the subscript symbols _ are actually lowered. In order that both letters be lowered, they must be seen as a unit within the command text, that is as $A_{ij},\ldots,A_{lk}$. This may be achieved with the \subvec command by including an extra set of brackets in the arguments: \subvec{A}{{ij}}{{lk}}. A better solution, however, is to put the brackets in the definition to begin with:

\renewcommand{\subvec}[3]{\ensuremath{#1_{#2},\ldots,#1_{#3}}}

which will always produce the desired result with only a single set of brackets per argument: \subvec{A}{ij}{lk} prints $A_{ij},\ldots,A_{lk}$.

7.3.3 Commands with an optional argument $\boxed{2\varepsilon}$

$\boxed{!}$

As we have seen, many LaTeX commands may take *optional* arguments, including the command \newcommand itself. In LaTeX 2ε, it is also possible to define user commands with one optional argument. The advantage of this is that, although an argument is provided, in most applications it will usually take some standard value that need not be given explicitly.

As an example, the user-defined vector command \subvec in the last section has three arguments, for the letter and for the first and last subscripts. However, it may be that the letter is normally x, so it makes sense to include it as an optional argument, which is to be specified only for a different letter. This is accomplished with

\renewcommand{\subvec}[3][x]
    {\ensuremath{#1_{#2},\ldots,#1_{#3}}}

The difference between this and the previous definition is the addition of [x] after the [3] argument. This states that the *first* of the three arguments is to be optional, and its standard value is x. Now \subvec{i}{j} prints $x_i,\ldots,x_j$ while \subvec[a]{1}{n} produces $a_1,\ldots,a_n$.

There may only be one optional argument in the user-defined command, and it will always be the first one, the #1 in the definition.

7.3.4 Additional examples of user-defined commands

In the above explanation of user-defined commands, a very simple case of a vector structure was taken as an example. We would now like to demonstrate some more complex situations, in which counters, lengths, and even some special TeX commands are applied.

Example 1: In Section 5.4.6, the TeX commands \atop and \choose were presented as useful mathematical commands even for LaTeX applications.

Unfortunately, the syntax of these commands deviates considerably from that of the similar LaTeX command \frac. However,

```
\newcommand{\latop}[2]{{#1\atop#2}}          and
\newcommand{\lchoose}[2]{{#1\choose#2}}
```

define two commands \latop and \lchoose that yield the same results with a syntax like that of LaTeX: \latop{*upper*}{*lower*}.

Example 2: The command \defbox{*sample_text*} is to set a box width equal to the length of the text *sample_text*. A subsequent call to the command \textbox{*text*} centers *text* within a frame with the same width as *sample_text*.

```
\newlength{\wdth}
\newcommand{\defbox}[1]{\settowidth{\wdth}{#1}}
\newcommand{\textbox}[1]{\framebox[\wdth]{#1}}
```

First, a new length parameter \wdth is created, then \defbox is defined so that \wdth is set equal to the length of its argument (Section 7.2), and finally \textbox makes a framed box of that same width containing its argument, centered. (Do not name the length parameter \width, for this already exists, Section 4.7.5.)

```
as wide as this text\\
\defbox{as wide as this text}\textbox{}\\
\textbox{text}\\
\textbox{longer text}
```

as wide as this text

text
longer text

Example 3: A footnote command \myftnote is to be created that behaves as the normal command \footnote{*text*} in putting *text* into a footnote, but instead of using numbers as the marker, it should take the symbols * † ‡ § ¶ ‖ ** †† ‡‡ one after the other, starting again with the symbol * on each new page. First a new counter must be established that will be reset to zero every time the page counter is incremented. This is done with (see Section 7.1.2)

```
\newcounter{myfn}[page]
```

making a user-defined counter myfn that is set to zero every time the page counter is incremented.* Next the command

```
\renewcommand{\thefootnote}{\fnsymbol{footnote}}
```

redefines the footnote marker to be that symbol in the sequence given by the counter footnote (Sections 4.10.2 and 7.1.4). Now the actual new footnote command can be constructed with

---

*Actually the page counter is not incremented exactly at the end of the page; LaTeX reads in the whole paragraph before it decides if and where a page break should occur. This can cause problems with resetting the myfn counter near the top of a page.

```
\newcommand{\myftnote}[1]{\setcounter{footnote}{\value{myfn}}%
   \footnote{#1}\stepcounter{myfn}}
```

yielding the desired results. The user-defined command \myftnote possesses one argument, which is passed to the LaTeX \footnote command after the LaTeX counter footnote has been set equal to the value of the user counter myfn. Once the command \footnote has been executed, the counter myfn is then incremented by one with \stepcounter{myfn}. This counter, however, is reset to zero whenever the page counter is incremented, that is, whenever a new page begins.

The footnote on the previous page was generated with the command \myftnote as described. It is now used here* and again here†, demonstrating how the symbols have been reset on a new page.

Example 4: A command \alpheqn is to be set up so that once it has been called, the subsequent equations will all have the same number but be followed by letters *a*, *b*, ... , separated by a hyphen '-' from the number. The command \reseteqn restores the numbering scheme to its original style. Thus a sequence of equation numbers could be 4, 5, 6-a, 6-b, 7.

```
\newcounter{saveeqn}
\newcommand{\alpheqn}{\setcounter{saveeqn}{\value{equation}}%
   \stepcounter{saveeqn}\setcounter{equation}{0}%
   \renewcommand{\theequation}
      {\mbox{\arabic{saveeqn}-\alph{equation}}}}
\newcommand{\reseteqn}{\setcounter{equation}{\value{saveeqn}}%
   \renewcommand{\theequation}{\arabic{equation}}}
```

The example should be easy to comprehend with the help of the commands and counters in Section 7.1. The current value of counter equation is saved in the counter saveeqn and then incremented, while equation itself is set to zero. The form of the equation marker, \theequation, is redefined using these two counters. The equation numbering routines will operate on equation as usual, leaving saveeqn unchanged. The resetting command \reseteqn puts the value of saveeqn back into equation and restores the definition of \theequation.

This example is only appropriate for the document class article. For report and book, the definition of \theequation is

```
\arabic{chapter}.\arabic{equation}
```

The necessary modifications are left as an exercise for the reader.

The \mbox command in the first \renewcommand{\theequation} defining the combined equation number is necessary because the result will be printed in *math mode*, where the hyphen '-' is interpreted as a binary operator (minus sign) with extra spacing between it and its two 'operands', \arabic{saveeqn}

---

*another footnote
†and yet another footnote

and \alph{equation}. Thus $6 - a$ would be output instead of 6-a. The \mbox command causes a temporary switch out of math into LR mode.

Example 5: In Section 5.4.10, the international standards for mathematical typesetting are outlined, in which tensor and matrix variables should be set in a sans serif typeface, preferably italic or slanted. The math alphabet command \mathsf accomplishes this only for an upright font. The solution is to create a new math alphabet \mathsfsl for use in a \tensor command.

```
\DeclareMathAlphabet{\mathsfsl}{OT1}{cmss}{m}{sl}
\newcommand{\tensor}[1]{\mathsfsl{#1}}
```

The NFSS system of font attributes is presented in Section 8.5.1, and the \DeclareMathAlphabet command is explained in Section C.5.3.

The Computer Modern sans serif fonts do not possess an italic form which is why the slanted shape is taken instead. If one were to select italic with {i} in place of {sl}, LATEX would automatically substitute the slanted version.

In Section 5.4.10, several new commands are suggested to simplify typing math according to the ISO standard. These are:

```
\newcommand{\me}{\mathrm{e}}
\newcommand{\mi}{\mathrm{i}}
\newcommand{\dif}{\mathrm{d}}
\renewcommand{\vec}[1]{\mbox{\boldmath$#1$}}
%\renewcommand{\vec}[1]{\boldsymbol{#1}}    (amsbsy package)
%\renewcommand{\vec}[1]{\bm{#1}}            (bm package)
```

These permit the *constants* e, i, and differential operator d to be printed upright, and not italic. Three versions of the replacement \vec command are offered, depending on which extension packages are available.

If these commands are used frequently in many different documents, we recommend saving them in a separate file, say isomath.tex, to be input at the start of those documents, with \input{isomath}.

Exercise 7.4: Define LATEX commands \Lbrack and \Lbrace in the same manner as in Example 1 for \latop and \lchoose corresponding to the TEX commands \brack and \brace. These TEX commands behave as \choose from Section 5.4.6 except that they enclose their contents in square brackets [\Lbrack] or curly braces {\Lbrace}. (Note that the names \lbrack and \lbrace are already defined and so should not be used for these commands.)

Exercise 7.5: Generalize Example 4 with a command \vareqn{num}{type} to make the subsequent equation numbers have the value num followed by a running number in square brackets printed as \alph ... \Roman, as given by the argument type. For example, 33[A], 33[B], would result from calling \vareqn{33}{\Alph}.

Exercise 7.6: *Generalize the integral commands in Exercise 7.2 to include an argument to represent the area of integration set centered below the entire symbol. Thus* \iint{(D)}, \iiint{V}, *and* \idotsint{G} *should produce:*

$$\iint\limits_{(D)} \quad \iiint\limits_{V} \quad \int\cdots\int\limits_{G}$$

Hint: the second command can be made simply with a subscript on the middle integral (but see \limits *in Section 5.2.5). For the other two, negative horizontal shifting of the subscripted symbol is needed using* \hspace{-..}.

Exercise 7.7: *The* \tensor *command defined in Example 5 above sets tensor variables in a slanted, sans serif font. If matrices are to be set instead in an upright, sans serif font, define a command* \mtrx *to produce this.*

Note: do not name this command \matrix *because that already exists to produce matrices, not their variable names.*

7.3.5 Conditional text

Practiced TEX users will be familiar with the conditional commands that are available, both for TEX and LATEX. However, their usage is not always straightforward and often requires extensive knowledge of TEX's deeper principles. Leslie Lamport has provided a package named ifthen, extended by David Carlisle for LATEX 2_ε, which not only simplifies their application, but also gives them a LATEX syntax.

The package is loaded as usual with the command

$\boxed{2_\varepsilon}$ \usepackage{ifthen}

in the preamble. It then makes available the two commands \ifthenelse and \whiledo, which have the following syntaxes:

\ifthenelse{*test*}{*then_text*}{*else_text*}
\whiledo{*test*}{*do_text*}

In both cases, *test* is a logical statement (explained below); for the first command, *then_text* or *else_text* is inserted into the text depending on whether *test* is ⟨*true*⟩ or ⟨*false*⟩. For the second command, the *do_text* is inserted (executed) as long as *test* evaluates to ⟨*true*⟩. (The *do_text* must alter the inputs to *test* or it will never stop!) The texts may also contain commands, or even define or redefine commands.

There are four types of basic logical statements which may be combined to form more complicated ones.

Testing numbers

To compare two numbers or commands that evaluate to numbers, simply put one of the relational operators <, =, or > between them, which stand for *less than*, *equals*, and *greater than*, respectively. The value of a counter may be tested by putting its name as the argument of the \value command. Examples:

```
\newcommand{\three}{3}
\ifthenelse {\three = 3} {O.K.} {What?}
\ifthenelse {\value{page} < 100 }
    {Page xx} {Page xxx}
```

The first case prints 'O.K.' since \three is equal to 3; in the second case *Page xx* is printed if the current page number is less than 100, else *Page xxx*. (The blanks above are added for clarity, since spaces between arguments are always ignored.)

Whether a number is even or odd can be tested with \isodd $\boxed{2\varepsilon}$

```
\ifthenelse {\isodd{\value{page}}
    {odd} {even}
```

Testing text

To test if two commands evaluate to the same piece of text, or if a command is defined as a certain *string* of text, use

```
\equal{string1}{string2}
```

where *string1* and *string2* are texts or commands that reduce to text. For example, with

```
\ifthenelse {\equal{\name}{Fred}} {Fredrick} {??}
```

the text Fredrick is inserted if \name has been defined to be Fred (with \newcommand), otherwise two question marks are printed.

Testing lengths

Another logical statement compares two lengths.

$\boxed{2\varepsilon}$ \lengthtest{*relation*}

where *relation* consists of two lengths or length commands separated by a relational operator <, =, or >. For example,

```
\newlength{\horiz} \newlength{\vert}
\newlength{\min}
. . . . . . .
\ifthenelse {\lengthtest{\horiz > \vert}}
    {\setlength{\min}{\vert}} {\setlength{\min}{\horiz}}
```

sets \min to be the smaller of \horiz and \vert.

Testing switches

A *boolean switch* is a parameter that is either ⟨*true*⟩ or ⟨*false*⟩, also called a *flag*. Three commands exist to handle them:

2ε	\newboolean{*string*}	creates a new switch
2ε	\setboolean{*string*}{*value*}	assigns a value true or false
2ε	\boolean{*string*}	tests its value

The last of these is used as *test* in \ifthenelse and \whiledo.

There are a number of internal LaTeX switches that may also be tested (but never reset!). The most useful of these are @twoside and @twocolumn for checking if two-side or two-column modes are active. Since they contain the character @, they may only be used inside a class or package file (Appendix C).

Combining logical statements

Any of the above logical statements may be combined to form a more complex statement by means of the logical operators

> \and   \or   \not   \(\)

which should be straightforward to anyone familiar with boolean logic. For example, to set \textwidth to 10 cm if two-column mode is active or if \paperwidth is greater than 15 cm and the page counter is below 100,

```
\ifthenelse {\lengthtest{\textwidth > 10cm} \or
     \( \lengthtest{\paperwidth > 15cm} \and
         \value{page} < 100 \) }
     {\setlength{\textwidth}{10cm}} {}
```

will accomplish this.

Such conditionals are fairly complicated, and they are most appropriate for defining new commands that may have alternative actions, or for including in package and class files (Appendix C).

A relatively simple example: suppose one is uncertain whether British or American spelling is wanted by the publisher. One can write the document with both included in the text, with a switch in the preamble to make the final selection.

```
\newboolean{US}
\setboolean{US}{true} %For American spelling
%\setboolean{US}{false} %For British spelling
\newcommand{\USUK}[2]{\ifthenelse{\boolean{US}}{#1}{#2}}
```

Now \USUK is a command that prints its first or second argument according to the setting of the flag US. Thus in the text one may write

```
... the \USUK{color}{colour} of music ...
```

which will yield American spelling if `\setboolean{US}{true}` has been specified, otherwise the British version. In fact, different sections of the work could have different spelling simply by changing the value of the switch at the appropriate point. (This book was written in this way, which was a good thing, because the editor did change his mind about spelling style after the manuscript was finished.)

As an example of `\whiledo`: one wishes to write some text n times, where both the text and n are to be variable.

```
\newcounter{mycount}
\newcommand{\replicate}[2]{\setcounter{mycount}{#1}
    \whiledo{\value{mycount}>0}{#2\addtocounter{mycount}{-1}}}
```

Now `\replicate{30}{?}` will print 30 question marks.

7.4   User-defined environments

Environments may be created or changed with the commands

> `\newenvironment{`*env_name*`}[`*narg*`][`*opt*`]{`*beg_def*`}{`*end_def*`}`
> `\renewenvironment{`*env_name*`}[`*narg*`][`*opt*`]{`*beg_def*`}{`*end_def*`}`

where the arguments have the following meanings:

env_name: the name of the environment; for `\newenvironment`, it may not be the same as any existing environment or command name, whether LATEX or user-defined. For `\renewenvironment`, on the other hand, there must already be an environment bearing this name. Any changes to LATEX environments should only be undertaken if the user knows what he or she is doing.

narg: a number between 1 and 9 that states how many arguments the environment is to have; if the optional argument *narg* is omitted, the environment is to have no arguments.

[2ε] *opt:* the default text for the first argument (#1) if it is to be optional; this behaves the same as for `\(re)newcommand` (page 193).

beg_def: the *initial* text to be inserted when `\begin{`*env_name*`}` is called; if this text contains entries of the form #*n*, with $n = 1, \ldots, narg$, then when the environment is started with the call

> `\begin{`*env_name*`}{`*arg_1*`}...{`*arg_n*`}...`

each occurrence of #*n* within *beg_def* is replaced by the text of the argument *arg_n*.

end_def: the *final* text that is inserted when \end{*env_name*} is called; here the dummy arguments #*n* are not allowed since they are only to appear in the *beg_def* text.

7.4.1   Environments without arguments

Just as for user-defined commands, environments without the optional argument *narg* will be illustrated first. A user-defined environment named sitquote is created with

```
\newenvironment{sitquote}{\begin{quote}\small
    \itshape}{\end{quote}}
```

which sets the text appearing between \begin{sitquote} text \end {sitquote} in the typeface \small\itshape, and indented on both sides from the main margins, as demonstrated here.

In this case, *beg_def* consists of the command sequence \begin{quote} \small\itshape while *end_def* is simply \end{quote}. Now the call

```
\begin{sitquote} text \end{sitquote}    is the same as
\begin{quote}\small\itshape text \end{quote}
```

which produces the desired result.

    This example does not appear to be very practical since the same effect can be achieved with less typing by adding \small\itshape at the beginning of the quote environment. However, if such a structure were to occur very often within a document, this new environment would be useful in saving some typing and reducing the chance of writing the command sequence \small\itshape incorrectly.

    Let us expand the previous example somewhat as follows:

```
\newcounter{com}
\newenvironment{comment}
{\noindent\slshape Comment:\begin{quote}\small\itshape}
{\stepcounter{com}\hfill(\arabic{com})\end{quote}}
```

where now *beg_def* contains the text and commands

```
\noindent\slshape Comment:\begin{quote}\small\itshape
```

and *end_def* is

```
\stepcounter{com}\hfill(\arabic{com})\end{quote}
```

where com is a user counter created by the \newcounter command. Now since the command \begin{comment} inserts the text *beg_def* at the start of the environment and \end{comment} the text *end_def* at the finish, it should be clear that

```
\begin{comment} This is a comment.
   Comments should ...
   ... in round parentheses.
\end{comment}
```

will generate the following type of structure:

Comment:

> *This is a comment. Comments should be preceded by the word Com-ment: the text being in a small, italic typeface, indented on both sides from the main margins. Each comment receives a running comment number at the lower right in round parentheses.*   (1)

The reader should examine the sequence of commands with the re-placement text of this example to see precisely what the effect of this environment is. Two weaknesses should become apparent: what would happen if \begin{comment} were called in the middle of a line of text without a blank line before, and what would take place if the last line of the comment text were so long that there was no more room for the running comment number on the same line?

The following revision removes these two problems:

```
\renewenvironment{comment}
{\begin{sloppypar}\noindent\slshape Comment:
   \begin{quote}\small\itshape}
{\stepcounter{com}\hspace*{\fill}(\arabic{com})\end{quote}
   \end{sloppypar}}
```

With the \begin{sloppypar} command, the call to the environment always starts a new paragraph in which no *overfull* line can occur upon line breaking. If the comment number does not fit on the last line of text, a new line begins with the number right justified because of the command \hspace*{\fill}. Again the reader should examine carefully just what is inserted into the processing between the \begin{comment} and the \end{comment}.

7.4.2   Environments with arguments

Passing arguments over to an environment is carried out exactly as for commands. As an example, the comment environment will be modified so that the name of the person making the comment is added after the word *Comment:*, and this name will be an argument when the environment is invoked.

```
\renewenvironment{comment}[1]
{\begin{sloppypar}\noindent\slshape Comment: #1
      \begin{quote}\small\itshape}
{\stepcounter{com}\hspace*{\fill}(\arabic{com})%
      \end{quote}\end{sloppypar}}
```

The text

```
\begin{comment}{Helmut Kopka} This is a modified ...
   ... environment argument \end{comment}
```

now produces

Comment: Helmut Kopka

> *This is a modified comment. Comments should be preceded by the word Comment:, followed by the name of the commenter, with the text of the comment being in a small, italic typeface, indented on both sides from the main margins. Each comment receives a running comment number at the lower right in round parentheses. The name of the commenter is transferred as an environment argument.* (2)

This example will now be modified once again by interchanging the comment number and the name of the commenter. Placing the running number after the word *Comment:* is no problem, for it is simply necessary to insert those commands from {*end_def*} at the location of the #1 dummy argument. However, putting the symbol #1 where the comment number used to be will produce an error message during the LATEX processing since this violates the syntax of the \newenvironment command: '*No dummy arguments shall appear within the {end_def}.*' If the dummy argument comes after \begin{quote}, the name will be printed at the wrong place, at the beginning of the comment text.

There is a trick to solve this problem:

```
\newsavebox{\comname}
\renewenvironment{comment}[1]
{\begin{sloppypar}\noindent\stepcounter{com}\slshape
   Comment \arabic{com}\sbox{\comname}{#1}
   \begin{quote}\small\itshape}
{\hspace*{\fill}\usebox{\comname}\end{quote}\end{sloppypar}}
```

The commands \newsavebox, \sbox, and \usebox are all described in Section 4.7.1. Here \comname is the name of a box that has been created with \newsavebox, in which the first argument, the commenter's name, is stored. With this new definition, a comment now appears as follows:

Comment 3

> *In this form, every comment is assigned a sequential number after the word Comment. The comment text appears as before, while the name of the commenter is entered as the environment argument and is placed at the right of the last line.* Helmut Kopka

The implementation of more than one argument for environments is the same as for commands and needs no further explanation.

7.4.3   Environments with an optional argument $\boxed{2_\varepsilon}$

Similarly, environments may be defined with an optional argument, just
as commands. To take our last example once more, if we feel that most
comments will be made by Helmut Kopka, we could alter the first line of
the definition to

```
\renewenvironment{comment}[1][Helmut Kopka]{..}{..}
```

Now it is only necessary to specify the name of the commenter if it is
someone else. With

```
\begin{comment}[Patrick W. Daly]More than ...
... appropriate.\end{comment}
```

we obtain

Comment 4

> *More than one person may want to make a comment, but perhaps
> one person makes more than others do. An optional argument for the
> name is then appropriate.*   Patrick W. Daly

Exercise 7.8: *Extend the definition of the comment environment so that a page
break cannot occur either between the heading 'Comment n' and the comment
text or between the comment text and the commenter's name.*

Exercise 7.9: *Create a new environment making use of the* minipage *environment, to be named* varpage *and possessing one argument that is a sample text
to determine the width of the minipage. The call*

```
\begin{varbox}{'As wide as this sample text'}
. . . . . . . . . . . . . . . . . . . .\end{varbox}
```

*should pack the enclosed text into a minipage that has a width equal to that of
the text 'As wide as this sample text'.*
Hint: a user-defined length parameter must first be established, say \varwidth.
See Section 7.2 for details on assigning lengths to text widths.

Exercise 7.10: *Generate an environment named* varlist *with two arguments
that behaves as a generalization of the sample list in Section 4.4.3. The first
argument is to be the item word that is printed on each call to* \item; *the second
is the numbering style of the item numeration. For example, with the call*

```
\begin{varlist}{Sample}{\Alph} . . . . . \end{varlist}
```

every \item *command within the environment should produce the sequence
'Sample A', 'Sample B', ... The indentation should be 1 cm larger than the width
of the item word, which itself should be left justified within the label box.*
Hint: once again a user-defined length, say \itemwidth, *is necessary for this
solution. After the length of the item word has been stored with* \settowidth,
the indentation may be set with

$$\setlength{\leftmargin}{\itemwidth}$$

to be equal to the width of the item word, and then with

$$\addtolength{\leftmargin}{1cm}$$

to be 1 cm larger. Length assignments for \labelwidth *and* \labelsep *may be similarly set to appropriate values.*
All further details may be obtained from Section 4.4.

7.5 Some comments on user-defined structures

This section contains some general remarks on the creation and use of user-defined LaTeX structures. They are not strict rules but merely reflect ideas that we have obtained through our own experiences. Every user will develop his or her own methods of application according to personal requirements.

7.5.1 Saving user-defined structures

The creation of user-defined structures can save the user a great deal of work. Frequently used commands and/or text should be stored individually with the \newsavebox, \newcommand or \newenvironment definitions written to a separate file. This file may then be read in with the \input command and made available in any other document file.

With time, a very large number of such stored structures may accumulate. To put them all into a *single* file will slow down the processing and may make it difficult to keep track of the contents and command names. It is therefore recommended that the user structures be sorted according to areas of application and stored in a number of *separate* files.

7.5.2 Abbreviating structures

A simple application of user-defined commands is to shorten the names of LaTeX structures. For example, with

```
\newcommand{\be}{\begin{enumerate}}
\newcommand{\ee}{\end{enumerate}}
```

it is sufficient to write \be to start the LaTeX environment enumerate, and \ee to end it.

Such abbreviations can avoid a considerable amount of writing, but they are not so suitable for a large collection of stored commands since it is easy to forget the meaning behind the command name. The author of LaTeX, Leslie Lamport, has deliberately selected names that clearly indicate the function of the command or environment. Such unambiguous, unabridged names are easier to remember than unclear, arbitrary abbreviations. Nevertheless, some abbreviations may be practical within a single document, provided they do not become too numerous. It is a question of the individual user's style and his or her typing capabilities.

7.5.3 Identical command and counter names

In the previous examples a number of counters were introduced for application to specific commands or environments, such as `myfn` for the command `\myfootnote` on page 199, or `com` for the environment `comment` on page 205. In these cases, the counter and the corresponding command or environment were given different names. This was not necessary: *counters may have a name that is identical to that of a command or environment.* LaTeX knows from the context whether the name refers to a counter or to a command/environment.

The different names were chosen in the examples in order to avoid confusion for the beginner. In fact, it is reasonable to give a counter the same name as the command or environment with which it is coupled, a practice that LaTeX itself constantly employs (Section 7.1.1). In the above-mentioned examples, the counters could just as easily have been named `myfootnote` and `comment` to emphasize their interdependence with the command and environment of the same names.

7.5.4 Scope of user definitions

User structures that are defined within the preamble are valid for the entire document. Command and environment definitions that are made inside an environment remain in effect only until its end. Even their names are unknown to LaTeX outside the environment in which they were defined. Thus if they are to be defined once again in another environment, it is `\newcommand` or `\newenvironment` that must be used and not the `\renew` versions.

For command and environment names that have been globally defined, that is, in the preamble, all further new definitions of these names must be made with the `\renew` version. However, these subsequent definitions will apply only locally within the environment in which they are declared; outside, the previous global definitions will be valid once more.

The same applies to structure definitions within nested environments. A definition in the outer environment is effective within all inner ones, but a new definition must be made at the deeper levels with `\renewcommand` or `\renewenvironment`. On leaving the inner environment, the new definition will no longer be operative, but rather the old one will be re-established.

Warning: structures that have been created with `\newsavebox` or with

\newcounter are globally defined. If they are given within an environment, their definitions remain effective even outside that environment. Similarly \setcounter functions globally, although \savebox does not.

7.5.5 Order of definitions

User-defined structures may be nested within one another. If one user definition contains another user-defined structure, it is often the case that the inner one has already been defined. However, this is not necessary. *User definitions may contain other user structures that are defined afterwards.* What is important is that the other structure is defined before the first command is invoked.

For example, a normal sequence of definitions would be

> \newcommand{\A}{*defa*}
> \newcommand{\B}{*defb*}
> \newcommand{\C}{\A \B}

where \C is defined in terms of \A and \B. However, it is also permissible to write

> \newcommand{\C}{\A \B}
> *normal text, but without calling* \C
> \newcommand{\B}{*defb*}
> \newcommand{\A}{*defa*}
> *further text with any number of calls to* \A, \B *and* \C

7.5.6 Transferring arguments

The dummy arguments in a command or environment definition may be used as arguments for other commands within that definition. For example, if the commands \A and \B each take one argument, the command definition

> \newcommand{\C}[3]{\A{#1}#2\B{#3}}

is a permissible one. Here the first and third arguments of \C are transferred to the commands \A and \B, and only the second argument enters the definition directly. All combinations of direct and transferred arguments are allowed. Here is a more concrete example:

> \newcommand{\sumvec}[4]{\anvec{#3}{#4} = #1_1+#2_1,\ldots
> ,#1_#4+#2_#4}

The call $\sumvec xyzn$ produces $z_1,\ldots,z_n = x_1 + y_1,\ldots,x_n + y_n$, where \anvec is the command defined in Section 7.3.2.

7.5.7   Nested definitions

!

User definitions may be nested inside one another. A structure such as

```
\newcommand{\outer}{{\newcommand{\inner}...}}
```

is permissible. The command that is defined with the name {\inner} is valid
and known only within the command {\outer}, according to the remarks on the
scope of definitions on page 210. Although the TeX macros make copious use of
nested command definitions to limit the lifetime of temporary commands, it is
not recommended to nest LaTeX definitions excessively since it is too easy to lose
track of the bracketing. A forgotten bracket pair will produce an error message
on the *second* call to the outer command, since the inner definition still exists as
a leftover from the first call. Nevertheless, here is an example:

```
\newcommand{\twentylove}
  {{\newcommand{\fivelove}
    {{{\newcommand{\onelove}
       {I love \LaTeX!}%
       \onelove\ \onelove\ \onelove\ \onelove\ \onelove}}}
     \fivelove\\ \fivelove\\ \fivelove\\ \fivelove}}
```

The entry My opinion of \LaTeX:\\ \twentylove now produces:

My opinion of LaTeX:

I love LaTeX! I love LaTeX! I love LaTeX! I love LaTeX! I love LaTeX!
I love LaTeX! I love LaTeX! I love LaTeX! I love LaTeX! I love LaTeX!
I love LaTeX! I love LaTeX! I love LaTeX! I love LaTeX! I love LaTeX!
I love LaTeX! I love LaTeX! I love LaTeX! I love LaTeX! I love LaTeX!

Indenting the lines of the definition, as shown above, can help to keep track of
the nesting levels; each line of the same level starts in the same column, ignoring
braces.

!

If both the *inner* and *outer* definitions are to be provided with arguments,
the symbols for the inner and outer dummy arguments must be distinguished.
The symbols for the *inner* definition are ##1 ... ##9, while those for the *outer*
one are the normal #1 ... #9. For example:

```
\newcommand{\thing}[1]{{\newcommand{\color}[2]{The ##1 is ##2.}
  \color{#1}{red} \color{#1}{green} \color{#1}{blue}}}
```

The entry The colors of the objects are\\
  \thing{dress}\\ \thing{book}\\ \thing{car}   produces
The colors of the objects are

The dress is red. The dress is green. The dress is blue.
The book is red. The book is green. The book is blue.
The car is red. The car is green. The car is blue.

The separate definition and calling as in Section 7.5.5 is easier to follow and
would be:

```
\newcommand{\thing[1]}{\color{#1}{red} \color{#1}{green}
                \color{#1}{blue}
\newcommand{\color}[2]{The #1 is #2.}
```

7.5.8 Unwanted spacing

Occasionally user-defined structures generate spacing where none was expected, or more spacing than was desired. This is almost always due to blanks or new lines in the definition, included only to improve the legibility of the input text but interpreted as spacing when the structure is invoked.

For example, if the % character had been left out of the first line of the \myftnote definition on page 199, a line feed would have been added to the command text at that point, and converted into a blank. This blank would be inserted between the previous word which should receive the footnote marker and the call to the \footnote command that actually generates the marker, with the result that it would be displaced from that word (for example, wrong * instead of right*).

! At this point we should like to remind the reader that many LaTeX commands are *invisible*, in that they do not produce any text at that point where they are called. If such an invisible command is given separated by blanks from the surrounding text, it is possible that two blanks will appear.

> For example \rule{0pt}{0pt} produces

'For example produces' the interword spacing twice. Invisible commands without arguments do not present this problem since the trailing blank acts solely as a command name terminator and disappears. Furthermore, the following LaTeX commands and environments always remove the subsequent blanks even when arguments are present.

\pagebreak	\linebreak	\label	\glossary	\vspace	figure
\nopagebreak	\nolinebreak	\index	\marginpar		table

7.5.9 Two final examples

! The normal description environment (Section 4.3.3) sets the label in bold face, and indents all lines following the first one by a fixed amount. For describing command names in computer code, it might be more desirable to print the labels in typewriter font (\texttt command or \ttfamily declaration) and to have the subsequent lines indented by the amount of the longest label. Let us define a new environment ttscript for this purpose that is invoked thus:

> \begin{ttscript}{*sample_text*}   *list_text*   \end{ttscript}

The text of the argument *sample_text* determines the left indentation as its width with the font command \texttt. The *list_text* consists of \item[*label_text*] commands followed by the corresponding explanatory text, every line of which is indented from the left margin by the width of *sample_text*. The *label_text* appears printed with \texttt left justified within the label box. The ttscript environment is defined as

```
\newenvironment{ttscript}[1]{%
  \begin{list}{}{%
```

```
    \settowidth{\labelwidth}{\texttt{#1}}
    \setlength{\leftmargin}{\labelwidth}
    \addtolength{\leftmargin}{\labelsep}
    \setlength{\parsep}{0.5ex plus0.2ex minus0.2ex}
    \setlength{\itemsep}{0.3ex}
    \renewcommand{\makelabel}[1]{\texttt{##1\hfill}}}}
  {\end{list}}
```

which may be easily adapted to other font styles for the label text, renamed, and stored as a user-defined structure for general applications.

The first seven lines of the above definition should not present any problems in understanding. The environment definition contains one argument that is transferred to

```
    \settowidth{\labelwidth}{\texttt{#1}}
```

in order to set the width of the label. The left margin is then set with

```
    \setlength{\leftmargin}{\labelwidth}
```

to be the same as the width of the label \labelwidth, and finally

```
    \addtolength{\leftmargin}{\labelsep}
```

increments the left margin by the amount of the label separation \labelsep.

The sixth and seventh lines set \parsep and \itemsep to appropriate values that may be changed as needed by the user. Other list parameters may also be included if desired.

The last line redefining the \makelabel command needs some further explanation. This command was briefly described in Section 4.4.1. It is valid only within the list environment and is used by each call to \item to form the actual label. Its normal argument is the *optional* argument to the \item command. With the above redefinition, that label is now set with the font command \texttt, followed by \hfill to left justify the whole within the label box. Because of the nesting of the redefinition within the definition of the ttscript environment, it is necessary for the dummy argument to be given as ##1 instead of #1, as explained in the previous section.

As an example, by giving the text

```
    \begin{ttscript}{description}
    \item[list] environments are useful for . . .
    \item[enumerate] environments number . . .
    \item[itemize] environments mark the . . .
    \item[description] environments label . . .
    \end{ttscript}
```

we can generate the following description of the LaTeX listing environments:

list   environments are useful for generating lists in which the various items are set off from one another for greater clarity;

enumerate   environments number their items consecutively, starting at 1;

itemize   environments mark the various items with a distinguishing symbol, often a bullet •;

`description` environments label the items with some text in bold face type, for greater stress.

Exercise 7.11: Generalize the `ttscript` environment into a universal `varscript` environment with two arguments such that the second one determines the font style of the item text.

User-defined commands and environments may contain up to nine arguments. The more arguments a structure possesses, the more flexible it is. On the other hand, the calls to the commands become more complex and awkward, since not only the number but also the sequence of arguments must be strictly maintained.

The sample user-defined environment `figlist` on page 77 in Section 4.4.4 contains no arguments. When it is called, every `\item` command produces **Figure 1:**, **Figure 2:**, etc. The text is indented 2.6 cm from the left and 2 cm from the right margin of the surrounding text. Additional list parameters like `\labelsep`, `\parsep`, and `\itemsep` are set to fixed values in the list declaration. However, with the definition

```
\newcounter{itemnum} \newlength{\addnum}
\newenvironment{genlist}[8]
  {\begin{list}{\textbf{#1 \arabic{itemnum}:}}
    {\usecounter{itemnum}
    \settowidth{\labelwidth}{\textbf{#1}}
    \settowidth{\addnum}{\textbf{\ \arabic{itemnum}: }}
    \addtolength{\labelwidth}{\addnum}
    \setlength{\labelsep}{#2}
    \setlength{\leftmargin}{\labelwidth}
    \addtolength{\leftmargin}{\labelsep}
    \setlength{\rightmargin}{#3}
    \setlength{\listparindent}{#4}
    \setlength{\parsep}{#5}
    \setlength{\itemsep}{#6}
    \setlength{\topsep}{#7}#8}}
  {\end{list}}
```

a generalized list is created in which the first argument determines the uniform item name printed with every `\item` command together with the sequential number. The left indentation is set to the width of the item name plus the size of `\labelsep`, given as the second argument. The next five arguments determine the lengths of various list parameters. The last argument specifies the font declaration for the text within the environment. The syntax for this new environment is

```
\begin{genlist}{itemname}{labelsep}{rightmargin}{listparindent}
  {parsep}{itemsep}{topsep}{fontstyle}
\item text
  ... ...
\item text
\end{genlist}
```

and the call

```
\begin{genlist}{Sample}{4mm}{1cm}{0pt}{1ex plus0.5ex}
    {0pt}{0pt}{\slshape}
  \item no value
  \item The enclosed coupon is worth a value of \$2.00 towards
        the purchase of your next order.
\end{genlist}
```

produces

Sample 1:   *no value*

Sample 2:   *The enclosed coupon is worth a value of $2.00 towards the*
               purchase of your next order.

In this example, the length entries must be given with their units. It is also possible to include the units directly in the definition so that only pure numbers need to be given in the call. Similarly the rubber length in the vertical spacing could be generated by an algorithm within the definition:

```
\setlength{\parsep}{#5ex plus0.3#5ex minus0.5#5ex}
```

which expects a pure number as the fifth argument. If this entry were 2, `\parsep` would be set equal to `2ex plus0.6ex minus1ex`. Once again, there is a trade-off between simplified entries and the complexity of remembering what they mean, especially if the various length arguments all have different dimensions and algorithms.

Advanced Features

8

This chapter describes those features of LATEX that justify the name 'document preparation system', whereas previous chapters have concentrated more on text processing. The term *advanced* is perhaps misleading, since these features are essential for producing large, complex documents in an efficient manner. The subjects included here are the splitting of a document into several files, selective processing of part of a document, cross-references to sections, figures, and equations, automated production of bibliographies, indices, and glossaries, handling different sets of fonts, and preparing presentation material.

8.1 Processing parts of a document

As often pointed out, a LATEX document consists of a preamble and the actual text part. Short documents, such as those that a beginner might have, are written to a single file by means of a text editor, and might be corrected after the first trial printing. As the user gathers experience and confidence, the LATEX documents will rapidly expand in length until one is faced with the task of producing an entire book of more than a hundred pages.

Such long documents could theoretically be kept in one file, although that would make the whole operation increasingly clumsy. The file editor functions less efficiently on longer files and the LATEX processing takes correspondingly more time. A better idea is to split the work into several files which LATEX then merges during the processing.

8.1.1 The \input command

The contents of another file may be read into a LATEX document with the command

  \input{*filename*}

where the name of the other file is *filename*.tex. It is only necessary to specify the root name of the other file, leaving off the extension .tex. During the LaTeX processing, the text contained in this second file is read in at that location in the first file where the command is given.

The result of the \input command is the same as if the contents of the file *filename*.tex had been typed into the document file at that position. The command may be given anywhere in the document, either in the preamble or within the text part.

Since the \input command may be given in the preamble, it is possible to put the whole preamble text itself into a separate file. The actual LaTeX processing file could even be reduced to simply \begin{document} ... \end{document} with a number of \input commands. A preamble file makes sense if one has a series of documents all of the same type requiring a common preamble. This also simplifies any later change to the specifications that must be made in *all* the documents. Different preamble files may be prepared for various types of processing and may then be selected with \input{*proc_type*}.

A file that is read in by means of the \input command may also contain further \input commands. The nesting depth is limited only by the capacities of the computer.

In order to obtain a listing of all extra files read in, give the command

$\boxed{2\varepsilon}$ \listfiles

in the preamble. The list appears both on the computer monitor and in the transcript file, at the end of the processing run. The version numbers and any other loading information are also printed. This provides a check as to which files have been input. See Section C.2.9 for more details.

Exercise 8.1: Put the preamble of your standard exercise file exercise.tex *into a separate file* preamble.tex. *Split the text part into three files* exer1.tex, exer2.tex, *and* exer3.tex. *What should the main file now contain to ensure that LaTeX processes the whole exercise text?*

8.1.2   The \include command

Splitting the document into several files may be practical for writing and editing, but when the files are merged with the \input commands, it is still the entire document that is processed. Even if only one file contains a small correction, all files will be read in and processed once again. It would therefore be desirable to be able to reprocess only the one corrected file.

One rough-and-ready method is to write a temporary main file containing only the preamble (which may be read in) and an \input command to read in that specific file. The disadvantage is that all automatic numbering of page numbers, sections, figures, equations, etc. will start from 1, since all the information from the previous files will be missing. Furthermore, all cross-references from other files will be absent.

A much better method is to employ the LaTeX command

> \include{*filename*}

which is only allowed within the text part of the document, together with

> \includeonly{*file_list*}

in the preamble, containing a list *file_list* of those files that are to be read in. The file names are separated by commas and the extension .tex is left off.

If *filename* is in the *file_list*, or if \includeonly is missing from the preamble, the command \include{*filename*} is identical to

> \clearpage \input{*filename*} \clearpage

However, if *filename* is not contained within the *file_list*, \include is equivalent to \clearpage and the file contents are not read in.

The command \include is less general than the \input command, since it always begins a new page. The document must therefore be split into files at those points where a new page occurs, such as between chapters. Another limitation is that \include commands may not be nested: they may only appear in the main processing file. However, an \input command may be given within a file that is \included.

The great advantage of the \include command is that the additional information about page, section, and equation numbers will be supplied by the \includeonly command so that the selective processing takes place with the correct values of these counters. Cross-reference information from the other files is also available so that the \ref and \pageref commands (Section 8.3.1) yield the correct results. All these values will have been determined during a previous processing of the entire document.

If the changes in the file that is being selectively processed lead to an increase or reduction in the number of pages, the following files will also have to be reprocessed to correct their page numbers. The same is true if sections are added or removed, or if the number of equations, footnotes, figures, etc. is altered.

For example, suppose *file_3* ends on page 17, but after the selective processing it now extends to page 22. The following *file_4* still begins on page 18, and all further files also have their original starting page numbers. If *file_4* is now selectively processed, it will receive the correct number for its first page, 23, based on the stored information in the revised *file_3*. So far so good. However, if instead *file_6* were to be selectively processed right after *file_3*, it would receive its starting page number from *file_5*, which has not yet been corrected and would be in error by 5. The same applies to all other structure counters. Their correct values can only be guaranteed when the files have been reprocessed in their proper order.

In spite of these restrictions, the \include command is extremely useful for large documents, saving considerable computation time. Longer documents are normally written and edited in many stages. The \include command permits one to reprocess selective alterations in a short time, even if the numbering systems temporarily go awry. This can be repaired later on with a complete reprocessing of the entire document, by deactivating the \includeonly command in the preamble.

A file that is read in with \include may not contain any \newcounter declarations. This is not much of a restriction, since they should normally be given in the preamble.

Each chapter of this book was written to a separate file with names gtl1.tex, gtl2.tex, ... The processing file itself contains the text

```
\documentclass{book}
 . . . . . . . . . . . . . . .
 \includeonly{...}
\begin{document}
 \frontmatter \include{toc}
 \mainmatter
 \include{gtl1} . . . \include{gtl7} . . .
 \backmatter \printindex
\end{document}
```

where the file toc.tex contains simply

```
\setcounter{page}{7}
\tableofcontents \listoftables \listoffigures
```

By making the appropriate entry in the \includeonly command, it is possible to process the various chapters selectively: for example, by giving \includeonly{toc,gtl3} one processes the table of contents and Chapter 3.

8.1.3 Monitor input and output

There are times when it is desirable that LATEX write a message to the computer monitor during processing. This can be achieved with the command

> \typeout{*message*}

where *message* stands for the text that is to appear on the screen. This text is written when the LATEX processing reaches this command. At the same time, the message is written to the .log file (Section 8.9).

If *message* contains a user-defined command, it will be interpreted and its translation appears on the screen. The same thing applies to

LATEX commands. This could have dire consequences if the commands, either user or LATEX, are not really printable. To print the command name literally, add the command \protect in front of it.

The command

> \typein[\com_name]{message}

also writes the *message* text to the monitor, but then it waits for the user to enter a line of text from the keyboard, terminated by a carriage return. If the optional argument \com_name is missing, the line of text is inserted directly into the processing. In this way, one could, for example, reuse the same text for a letter for several addressees, entering the name each time from the keyboard. Suppose the text contains

> Dear \typein{Name:}\\ ...

then what appears on the screen is

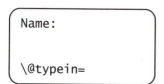

At this point, one enters the name of the recipient. If on successive processings 'George', 'Fred', and 'Mary' are entered, the result will be a set of identical letters differing only in their salutations as 'Dear George', 'Dear Fred', and 'Dear Mary'.

If the \typein command contains the optional argument \com_name, this is treated the same as

> \typeout{message} \newcommand{\com_name}{entered definition}

In this case the definition is stored under the command name \com_name interactively and may be invoked and executed in the rest of the document like any other LATEX command.

With some experience in LATEX methodology, it soon becomes obvious that *interactive* processing with the \typein command can be very practical. For example, if the preamble contains

> \typein[\files]{Which files?}
> \includeonly{\files}

the following appears on the monitor screen:

```
Which files?


\files=
```

and LATEX waits for the user to type in the names of one or more files (separated by commas) to be processed. This avoids having to modify the main processing file with an editor every time.

This book was produced with precisely this addition to the main file shown on the previous page.

A similar procedure may be employed when a form letter is to be sent to various recipients. One can enter the name and address and even the salutation interactively. Complete forms may be processed by LATEX in this way, with the entries being made from the keyboard. All the order forms in the authors' institute are produced in this manner, for example.

Warning: the \typein command may not be used as the argument of another LATEX command! It may, however, be given in environments such as minipage.

Exercise 8.2: *Change the main file from Exercise 8.1 so that the files* exer1.tex, exer2.tex, *and* exer3.tex *may be read in with the* \include *command. Arrange that you can determine interactively which of the files is to be processed.*

Exercise 8.3: *Generate output of the form*

Certificate

Olympic Spring Games
Walterville 1992

Finger Wrestling

Gold	A. T. Glitter	AUR	7999.9	Points
Silver	S. Lining	ARG	7777.7	Points
Bronze	H. D. Tarnish	CUP	7250.0	Points

so that the following enquiries appear on the screen one after the other

Message	Command=	Input
Sport:	\@typein =	Finger Wrestling
Unit:	\unit =	Points
Gold:	\@typein =	A. T. Glitter
Country:	\@typein =	AUR
Value:	\@typein =	7999.9
Silver:	\@typein =	S. Lining
. . .	. . . =	. . .

and the appropriate entries are made interactively. The third column contains the answers necessary to produce the above sample output. Repeat the program with different entries. Let your imagination run wild.

8.2   Including TEX commands in LATEX

The rapid increase in TEX applications is primarily due to the availability of LATEX, something that many users are unaware of. They consider LATEX to be a separate executable program, an alternative to TEX. In reality, LATEX is merely an interface between the user and the TEX processing program that simplifies the TEX operation enormously. It *translates* the input logical structures into functional TEX commands and executes them through internal calls to TEX.

Thus it is possible to give pure TeX commands within LaTeX. This applies to all the TeX primitive commands as a matter of course. In addition to the roughly 300 primitive TeX commands, there are another 600 or so macros defined in the Plain TeX format. The formal difference between running LaTeX and Plain TeX is that the format file `latex.fmt` (or `lplain.fmt` for LaTeX 2.09) is loaded instead of the original TeX format file `plain.fmt`. However, since the LaTeX formats contain most, but not all, of the macro definitions in Plain TeX, the majority of the TeX macros may also be invoked within LaTeX.

Those TeX macros that are not permissible in LaTeX are listed in Section G.3. These are the only TeX commands that may not be used or that function differently in LaTeX.

Notwithstanding the possibility of employing (Plain) TeX commands in a LaTeX document, it is recommended to stay with the high-level commands wherever possible. This is the best guarantee for portability and durability. However, since many refined features (or tricks) can only be realized at the TeX level, it would be foolish to ban such practices altogether. See Section C.1.4 for a further discussion on coding guidelines, and Section C.2.12 for some useful TeX commands.

8.3 In-text references

In a longer text, one often wants to refer to other chapters, sections, tables, and figures, or to pages where some other description has been given. One also needs an index of certain keywords that occur throughout the document. In the days before electronic text processing, such cross-references and indices meant an enormous amount of extra work for the author or secretary. Today the computer can alleviate much of this burden.

In the old days, referring to earlier parts of the text by page number was tedious but feasible. Referencing future parts that had not yet been written had to be limited to section numbers since the page numbers were not yet known, or space had to be left for the page number to be entered at a later date.

The production of a book is usually a progressive and constantly shifting process. The manuscript may not necessarily be written in the order in which it is to appear, and even once the initial draft version is finished, there will be major changes due to new considerations by the author or to knowledgeable advice from reviewers. Revisions, deletions, and insertions of entire sections or even chapters are commonplace, not to mention possible reordering of parts of the text.

LaTeX relegates all the problems associated with such major alterations to the past. No matter what changes the author makes, the information necessary for the cross-references and keyword index is stored for use at any point in the text.

8.3.1 Cross-references

As already pointed out in several places, the command

\label{*marker*}

is used to set a marker at that point in the text where it appears, which may be referred to at other places. The handle identifying the marker is the text *marker*, which may be any combination of letters, numbers, and characters (except those special symbols representing single character commands: \ # $ % & ˜ ˆ _ { }).

The page number on which the \label command was issued can be printed with

\pageref{*marker*}

at any point earlier or later in the document.

If the \label command is given after a sectioning command, or within an equation, eqnarray, or enumerate environment, or inside the argument of a \caption within the figure or table environments, the command

\ref{*marker*}

prints that number of the section, equation, figure, table, or enumeration that was current when *marker* was defined, and in the correct format. For enumerate, the same number is printed as is generated by the \item command where the \label appeared. Reference may also be made to theorem-like structures created by a \newtheorem command if the \label is given within the text of that theorem command. For example, a \label with marker text bo-wei was issued in the text of the Bolzano-Weierstrass theorem on page 79:

```
\begin{theorem}[Balzano--Weierstrass]
                \label{bo-wei}...\end{theorem}
```

so that the input text

```
Theorem~\ref{bo-wei} on page~\pageref{bo-wei}
```

generates the output 'Theorem 1 on page 79'. Similarly, the input text

```
for Table~\ref{budget95} on page~\pageref{budget95},
see also Section~\ref{sec:figref}
```

produces 'for Table 6.1 on page 184, see also Section 6.4.6', since that section contains the marker \label{sec:figref}.

The marker command \label may be given within the argument of a sectioning command, but may also be issued anywhere within the text of that section. The number printed by the corresponding \ref will be that of the innermost section where the \label appeared. To avoid possible confusion, it is recommended to place the \label immediately after the sectioning command to which it is to refer.

A list of all label markers, their translations, and page numbers may be produced by processing the file lablst.tex, provided with the LaTeX installation.

!

It is useful to know how the cross-referencing information is managed. What the \label command actually does is to write the marker text together with the current value of the appropriate counter(s) and the current page number to an auxiliary file with the root name of the document file plus the extension .aux. The \ref and \pageref commands obtain their information from this .aux file, which is read in at the beginning by the \begin{document} command. The same situation as for the table of contents (Section 3.4.2) also applies here: on the first run, the .aux file does not exist so that no cross-reference information can be output; instead the information is gathered and written to a new .aux file at the end of the run. A warning message is printed at the end if the auxiliary information has changed and a new run is necessary.

8.3.2 Extension packages for cross-referencing

!

There are two packages in the tools collection (Section D.3.3) that offer additional cross-referencing features. The xr package allows one to refer to \labels defined in another LaTeX file. This is useful for related documents that make cross-references to each other. With \externaldocument{*filename*}, all the markers in the specified file become available in the current document, to be printed with \ref{*marker*} as usual. Difficulties arise if there are duplicate marker names in any of the external or current files. To circumvent this problem, an optional argument can be given, for example as \externaldocument[x-]{*filename*}, in which case all the markers in the external file will be prefixed with x-. Thus the external marker intro is referred to with \ref{x-intro}.

The other package, varioref, defines a command \vref{*marker*} which is basically equivalent to '\ref{*marker*} on page \pageref{*marker*}.' However, if the *marker* is on the current, previous, or next page, it prints appropriate text. Thus Fig. \vref{f1} can produce 'Fig. 5', 'Fig. 5 on the preceding page', 'Fig. 5 on the next page', 'Fig. 5 on the facing page', or 'Fig. 5 on page 24'. The command \vpageref prints only the page part of the text, or 'on this page' if the *marker* is on the current page. It may also take two optional arguments: the texts for the current page and non-current pages. For example, the \vpageref[above example][example]{f1} prints 'the above example' or 'the example on the next page'. All texts can be customized and many language options exist to translate the texts. More details can be found by processing the file varioref.dtx.

8.3.3   References to the bibliography

The creation of a bibliography and references (citations) to it have already been described in Section 4.3.6, but are repeated here to complete the discussion on cross-referencing. The bibliography is produced as follows:

```
\begin{thebibliography}{sample_label}
  \bibitem[label_1]{key_1} entry_text_1
  \bibitem[label_2]{key_2} entry_text_2
      . . . . . . . . . . . . . . . . .
\end{thebibliography}
```

The meaning of the arguments is explained in Section 4.3.6 and will not be repeated here, except for the reference key. This marker *key* plays the same role as *marker* for the \label command. Again, any combination of letters, numbers, and characters is allowed, except commas in this case. Reference is made in the text to the bibliography entry with

```
\cite[extra]{key}
```

which sets the corresponding \bibitem label in square brackets. The input text

```
For additional information about \LaTeX\ and \TeX\
see~\cite{lamport} and \cite{knuth,knuth:a}.
```

together with the sample bibliography in Section 4.3.6 produces as output:

> For additional information about LATEX and TEX see [1] and [6, 6a].

The reference markers lamport, knuth, and knuth:a are translated into the corresponding citation labels from the bibliography.

If the optional argument *extra* is included in the \cite command, this text is added after the label(s) but still inside the square brackets.

```
The creation of a bibliographic database is described in
\cite[Appendix B]{lamport}, while the program \BibTeX\
itself is explained in \cite[pages 74, 75]{lamport}.
```

The creation of a bibliographic database is described in [1, Appendix B], while the program BIBTEX itself is explained in [1, pages 74, 75].

The thebibliography environment allows the author to set up a bibliography with \bibitem commands for the individual entries. There is a separate program BIBTEX that will generate the bibliography automatically by searching one or more bibliographic databases for the *key* markers given in the \cite commands. The keys must of course agree with those in the databases.

Entries to which no reference is made in the text may be included in the bibliography. This is achieved by putting

\nocite{*key_i, key_j, ...* }

anywhere in the text, where *key_i, key_j, ...* are the keys for the extra entries. The bibliography itself is now generated with the command

\bibliography{*database_a, database_b, ...* }

where *database_a, database_b, ...* are the root names of the files containing the bibliographic databases to be searched. The extension for these files is .bib.

Appendix B describes how bibliographic databases may be set up and how the program BibTeX may be used properly. Basically, BibTeX must be called as a separate program with the root name of the document file after the first run on LaTeX. Afterwards, LaTeX must be run again *at least twice*, the first time to set up the bibliography and to define the relationships between keys and labels, the second time to insert the label text in place of the keys in the \cite commands. BibTeX need only be run again later if new references are added or old ones removed by means of their keys.

8.3.4 Keyword index

LaTeX itself does not produce a keyword index automatically as it does a table of contents, but it can prepare input data for a program that does generate such an index, in a form that LaTeX can use on a later run.

A keyword index is formatted in the environment

\begin{theindex} *index_entries* \end{theindex}

which switches to a two-column page format with a running head *INDEX*. The first page of the index carries the heading **Index** in the same size as a chapter heading for the book and report document classes, and as a section heading for the article class. (More precisely, the actual word printed is contained in the command \indexname which may be redefined for other languages.) The individual entries are made with commands

\item \subitem \subsubitem and \indexspace

followed by the keywords and their page numbers. For example,

```
commands, 18                          \item commands, 18
    arguments, 19, 101                \subitem arguments, 19, 101
        multiple, 103, 104            \subsubitem multiple, 103, 104
        replacement symbol, 20        \subsubitem replacement symbol, 20
    as environments, 42               \subitem as environments, 42
    used as arguments for sec-        \subitem used as arguments for
        tioning commands, 41, 42          sectioning commands, 41, 42
                                      \indexspace
displayed text, 21-32                 \item displayed text, 21--32
```

If the text entry is too long for one line, it is broken and continued on the next line, indented deeper than all other lines, as in the above example 'used as argument for sectioning commands, 41, 42'. The command \indexspace leaves a blank line in the index.

The theindex environment only sets up a suitable format for the index. The entries themselves, as well as their page numbers, are generated by the MakeIndex program described in Section 8.4. This program requires input information from the LATEX file in the form of unsorted keywords and page numbers. The author enters this information in the text file with the command

\index{*index_entry*}

where *index_entry* is the text to be entered into the index. It may contain any combination of letters, numbers, and symbols *including even command characters and blanks.* This means that even commands may be included in the *index_entry*, such as \index{\LaTeX\ logo}. Even the one command that may otherwise never be used as an argument, \verb, may be included. However, if *index_entry* does contain a command, \index may not be used as an argument of another command.

Normally all the \index commands are ignored by LATEX and do absolutely nothing. They are activated only when the preamble contains the command

\makeindex

in which case, a file with the document's root name plus the extension .idx is opened. Now the \index commands write *index_entry* and the current page number to this file in the form

\indexentry{*index_entry*}{*pagenumber*}

It is this .idx file that is the input to the MakeIndex program. There are special symbols that may be used in the *index_entry* to indicate main and subitems and other refinements; these are explained in the next section.

In the LATEX installation, there is the file idx.tex which can improve the readability of the .idx files. By processing the idx.tex file (that is, calling latex idx), the user is prompted for the name of the .idx file to list:

```
********************************
* Enter idx file's first Name. *
********************************

\filename=
```

After the root name of the .idx file has been entered from the keyboard, a two-column page format is output containing the *index_entry* texts listed under page headings of the form **Page n**. No further information is obtained through such a formatted listing, but it is far easier to work with than a straightforward output of the .idx file alone.

Even if the \index commands have no effect without the \makeindex command in the preamble, it is a good idea to include them in the text from the initial development of the document. The \makeindex command can be included later when the final version of the document is ready. At this point, the MakeIndex program can be run to produce the theindex environment from the data in the .idx file just as BIBTEX produces thebibliography environments. This tool is described in the next section.

> ！

The LaTeX installation also contains the package showidx, which prints the index entries from the \index commands as marginal notes, beginning at the top of the page on which they occur. This is useful for going through a preliminary version of the document to check that all index entries are really on the proper page or whether additional entries have to be made. The package is loaded with the \usepackage command.

If showidx is used, it is a good idea to increase the width of the marginal note box with the declaration \marginparwidth (Section 4.10.6) in the preamble. Unfortunately, the present book format is not suitable for such a demonstration.

8.3.5 Glossary

A 'glossary' is a special index of terms and phrases alphabetically ordered together with their explanations. To help set up a glossary, LaTeX offers the commands

> \makeglossary   in the preamble and
> \glossary{*glossary_entry*}   in the text part

which function just like the commands for making up a keyword index. The entries are written to a file with extension .glo after the command \makeglossary has been given in the preamble. The form of these file entries from each \glossary command is

> \glossaryentry{*glossary_entry*}{*pagenumber*}

The information the .glo file can be used to establish a glossary. However, there is no equivalent to the theindex environment for a glossary, but a recommended structure is the description environment (Section 4.3.3) or a special list environment (Section 4.4).

8.4   MakeIndex—a keyword processor

The tiresome drudgery of making up the theindex environment for a keyword index can be dispensed with if the program MakeIndex is available. This program was written by Pehong Chen with the support of Leslie Lamport. We will give only an abbreviated description of its use here. More details are in the documentation accompanying the program package.

The program MakeIndex processes the .idx file, producing as output another file with the root name of the document and the extension .ind, containing the complete theindex environment. The program is called by

> makeindex *root_name*.idx or simply makeindex *root_name*

A subsequent LATEX processing outputs the index at the location of the \printindex command which, together with the \see command, is defined in the package file makeidx.sty. Thus the production of an index in this way requires the package file makeidx to be loaded with the \usepackage command.

The MakeIndex program expects the \index entries to be of one of the three forms:

> \index{*main_entry*}
> \index{*main_entry*!*sub_entry*}
> \index{*main_entry*!*sub_entry*!*sub_sub_entry*}

The individual main and sub-entries may contain any combination of characters with the exceptions of !, @, and |. The exclamation point is the divider between the various entry fields. If the \index command contains only a main entry, this will become the text for an \item command. The main entries will be in alphabetical order.

If the \index command contains a main and a sub-entry, the text *sub_entry* will be assigned to a \subitem command underneath the corresponding *main_entry*. The texts for the \subitem will also be in alphabetical order. Similarly the text of a *sub_sub_entry* will appear following a \subsubitem command, alphabetically ordered, below the appropriate *sub_entry* text.

The main and sub-entries may also contain special characters and even LATEX commands that are to be neglected during alphabetization. This is indicated by an entry of the form *lex_entry*@*print_entry*, in which *lex_entry* is used for the alphabetical ordering while *print_entry* is the actual text to be output. For example, in the keyword index in this book, which was generated with MakeIndex, all the command names appear as input text and are ordered without the preceding \ character. These entries were made in the text as \index{put@\verb=\put=}.

An entry may be terminated with the character sequences |(or |) to designate the beginning and end of a range of page numbers. For example,

> \index{picture!commands|(}   on page 152
> \index{picture!commands|)}   on page 164

produces the sub-entry 'commands' to the main entry 'picture' with page numbers given as 152–164.

Instead of having the page number printed after the entry, a reference may be made to another index entry. For example, with

> \index{space|see{blank}}

the entry 'space, *see* blank' is made in the index. (More precisely, it is the text stored in the command \seename that is printed in place of *see*; this may be changed for other languages.)

The three characters !, @, and | therefore have special functions for the MakeIndex program. In order to print these characters literally as text without their functions, the *quote* character " must precede them. For example, "! represents a literal exclamation point and not the entry divider.

The quote character itself is thus a fourth special symbol and must be entered literally as "". However, there is a special rule in MakeIndex syntax that says a quote character preceded by a backslash will be interpreted as part of a command: thus \" may be used to put a German umlaut in an entry (as in \index{Knappen, J\"org}). This special rule can lead to additional problems at times: in this book the index entry \! had to be typed as "\"!.

It is possible to specify varying fonts for the page number. For example, in the index of this book, page numbers are set in bold face to indicate the page where a command is defined or first explained. This is achieved with an entry of the form

> \index{blank}   on page 11, and
> \index{blank|textbf}   on page 20

which in the second case puts the page number for this entry as the argument of a command \textbf. The line in the theindex environment becomes

> \item blank, 11, \textbf{20}

Note: the vertical bar in the \index entry is not a typing error, but must replace the backslash as the LATEX command symbol under these circumstances.

The alphabetical ordering of MakeIndex normally follows the standard ASCII code, first *symbols*, then *digits*, and finally *letters*, where upper case comes before lower case. Blanks are included as symbols. There are a number of options that allow these rules to be changed. How the options are invoked when the program is called depends on the computer type, but we assume here a hyphen preceding the option letter, as in

> makeindex -g -l *root_name*

The most important options are:

-l   Letter ordering: blanks are ignored when sorting;

-c   Compress blanks: multiple and leading blanks are removed as they are in normal LATEX;

-g   German ordering: following the German rules in which *symbols* precede *letters* (*lower* before *upper* case) which precede *numbers*; the sequences "a, "o, "u, and "s (the codes for *ä, ö, ü,* and *ß* in the usual LATEX adaptions for German) are treated as though they were ae, oe, ue, and ss, which is standard German practice;

-s   <u>S</u>tyle specification: allows the name of an index formatting file to be included to redefine the functioning of MakeIndex.

The -s option reads in an index style file containing commands to define both the input and output of the MakeIndex program. For example, it is possible to change the special symbols !, @, |, and " so that different characters execute their functions and they themselves revert to being pure text.

The style-defining file consists of a list of pairs in the form *keyword attribute*. The attribute is either one character in single quotes (for example, 'z'), or a character string in double quotes (for example, "a string"). The most important keywords, together with their default definitions, are:

quote   '"' defines the quote symbol;

level   '!' defines the entry separation symbol;

actual '@' defines the lexical switch symbol;

encap   '|' defines the dummy command symbol for page formatting.

There are many more keywords for defining complicated output structures. These are described in the documentation that should be included with the MakeIndex program package.

The file makeindex.tex contains a short manual by Leslie Lamport (but without mentioning style definitions).

8.5   The New Font Selection Scheme (NFSS) $\boxed{2_\varepsilon}$

When TeX and LaTeX were originally invented, the fonts available for them were very limited in number. For this reason, LaTeX 2.09 had an inflexible system of defining the fonts that were to be used, since it was not obvious that one might want to change them. The association between the high-level font commands such as \large and \bf and the external font name that is ultimately selected was rigidly fixed in the file lfonts.tex which went into making up the lplain format.

Today there are many additional fonts available, some of which should be used alongside the standard CM fonts, and others that should replace them altogether. For example, the Cyrillic fonts of Section E.5.2 should be added parallel to the Latin fonts, but to make them operate automatically under the LaTeX size commands was a complex procedure. (We know: we have done it!) Similarly, installing PostScript fonts involved calling intricate interface macros, which admittedly were provided with the PostScript driver package, but which represented an enormous effort on the part of some LaTeX programmer.

Another problem with LATEX 2.09 was the behavior of the font style and size commands (Sections 4.1.5 and 4.1.2). Each of the font declarations \rm, \bf, \sc, \sl, \it, \sf, and \tt activates a particular font, depending on the currently selected size. Each of these declarations is mutually exclusive. Thus the combination \bf\it is exactly the same as \it. There is no way to select a bold italic font in this scheme. Furthermore, every size declaration from \tiny to \Huge automatically switches to \rm, so that \bf\large does *not* produce a bold large font as one might expect. Finally, the use of declarations rather than commands with arguments violates the basic philosophy of LATEX. That is, to emphasize a *single* word, it makes more sense to type \emph{single} than {\em single\/}. (Experienced LATEX users may deny this, but that is only because they have deeply rooted habits.)

In 1989 Frank Mittelbach and Rainer Schöpf proposed a New Font Selection Scheme (NFSS) for LATEX and a preliminary test package was ready in early 1990. A second release (NFSS2) was published mid-1993 with many substantial changes. With the official release of LATEX 2_ε in June 1994, NFSS has become firmly entrenched in the new standard. The new font declarations and commands are described in Sections 4.1.3 and 4.1.4. Here we explain the usage in more detail; in Section C.5 we describe how new sets of fonts may be installed.

8.5.1 Font attributes under NFSS

According to the NFSS scheme, every character set can be classified by five attributes called *encoding*, *family*, *series*, *shape*, and *size*, which may be selected with the commands

\fontencoding{*encode*} \fontfamily{*fam*} \fontseries{*wt_wth*}
\fontshape{*form*}   and   \fontsize{*sz*}{*line_sp*}

The *encode* attribute was new to the second release of NFSS. It specifies the layout of the characters within the font. Possible values for it are listed in Table 8.1 on the next page. It is unlikely that one would want to change encoding within a document, except to activate Cyrillic fonts perhaps. This feature is more for LATEX programmers to install new fonts on the system.

The argument *fam* in \fontfamily denotes a basic set of font properties. For the Computer Modern fonts, listed in Section F.2.2, all the serif fonts belong to the family cmr. The family cmss includes all the sans serif fonts while cmtt contains the typewriter fonts. A number of special decorative fonts are the only members of their families. Table 8.3 on page 235 lists the CM fonts according to the family and other attributes.

Note: The font attribute *family* has no relationship to the primitive TEX concept of the same name. A TEX family consists of three fonts of

Table 8.1: The NFSS encoding schemes

Encoding	Description	Sample Font	Page
OT1	Original text fonts from Knuth	cmr10	440
OT2	Univ. Washington Cyrillic fonts	wncyr10	445
T1	The Cork (DC/EC) fonts	ecrm1000	449
OML	TEX math letter fonts	cmmi10	443
OMS	TEX math symbol fonts	cmsy10	443
OMX	TEX math extended fonts	cmex10	444
U	Unknown coding	—	—

different sizes for use in math formulas as normal characters, and first- and second-level indices.

The argument *wt_wth* in \fontseries designates the *weight* (=*bold-ness*) and *width* of the characters. These are specified by 1 to 4 letters as shown in Table 8.2.

Table 8.2: The NFSS *series* attributes

Weight class		Width class		
Ultralight	ul	Ultracondensed	50%	uc
Extralight	el	Extracondensed	62.5%	ec
Light	l	Condensed	75%	c
Semilight	sl	Semicondensed	87.5%	sc
Medium (normal)	m	Medium	100%	m
Semibold	sb	Semiexpanded	112.5%	sx
Bold	b	Expanded	125%	x
Extrabold	eb	Extraexpanded	150%	ex
Ultrabold	ub	Ultraexpanded	200%	ux

The argument for \fontseries{*wt_wth*} consists of the letter or letters for the weight, followed by those for the width. Thus ebsc indicates weight *extrabold* and width *semicondensed* while bx means weight *bold* and width *expanded*. The letter m is omitted when combined with any non-normal weight or width; if both are to be normal, it is sufficient to give m alone.

In \fontshape, the argument *form* is one of the letter combinations n, it, sl, or sc for selecting normal (or upright), italic, slanted, or small caps.

The \fontsize attribute command takes two arguments, the first *sz* being the point size of the font (without the dimension pt explicitly given) and the second *line_sp* being the vertical spacing from one baseline to the next. The second argument becomes the new value of \baselineskip (Section 3.2.3). For example, \fontsize{12}{15} selects a font size of 12 pt with interline spacing of 15 pt. (The second argument may be given a dimension, such as 15pt, but pt is assumed if no dimension is stated.)

Once all five attributes have been set, the font itself is selected with the command \selectfont. The new feature here is that the various attributes are independent of one another. Changing one of them does not alter the others. For example, if the selection

\fontfamily{cmr} \fontseries{bx} \fontshape{n} \fontsize{12}{15}

has been made for an upright, bold, expanded, Roman font of size 12 pt and interline spacing 15 pt, then when \fontfamily{cmss} is later selected for a sans serif font, the attributes weight and width bx, form n, and size 12(15pt) remain in effect when the next \selectfont is issued.

Alternatively, all attributes but the size may be specified and the font activated immediately with the command

\usefont{*code*}{*family*}{*series*}{*shape*}

Table 8.3: Attributes of the Computer Modern fonts

Series	Shape(s)	Examples of external names
Computer Modern Roman — (\fontfamily{cmr})		
m	n, it, sl, sc, u	cmr10, cmti10, cmsl10, cmcsc10, cmu10
bx	n, it, sl	cmbx10, cmbxti10, cmbxsl10
b	n	cmb10
Computer Modern Sans Serif — (\fontfamily{cmss})		
m	n, sl	cmss10, cmssi10
bx	n	cmssbx10
sbc	n	cmssdc10
Computer Modern Typewriter — (\fontfamily{cmtt})		
m	n, it, sl, sc	cmtt10, cmitt10, cmsltt10, cmtcsc10

The above table (by F. Mittelbach and R. Schöpf) lists the classification of the Computer Modern character sets according to the attributes \fontfamily, \fontseries, and \fontshape. That there are so many attribute combinations without a corresponding CM font may appear to be a weakness in the NFSS system, but it must be recalled that it is designed for the future. It may also be employed with the PostScript fonts, which are becoming ever more popular, to exploit their complete variability.

Formally it is possible to set any combination of attributes; however, there may not exist any font matching all the attributes selected. If that is the case, then when \selectfont is called, LaTeX issues a warning stating which font has been activated in its place. The font size attribute of the \fontsize command may normally take on values of 5, 6, 7, 8, 9, 10, 10.95, 12, 14.4, 17.28, 20.74, and 24.88, but other values may also be added. The second argument, the interline spacing, may take on any value since it is not something intrinsic to the font itself.

With the \begin{document} command, LaTeX sets the five attributes to certain preset default values. These are normally standard encoding OT1, family cmr, medium series m, normal shape n, and the base size selected. The user may change these initial values within the preamble, or they might be set differently by a special option, say when a PostScript font has been selected.

8.5.2   Simplified font selection

The attribute commands \fontencoding, \fontfamily, \fontseries, \fontshape, and \fontsize, together with the command \selectfont, are the basic tools in the New Font Selection Scheme. The user need not employ these commands directly, but rather may make use of the higher-level declarations presented in Sections 4.1.2 and 4.1.3. In fact, a font declaration like \itshape is defined as \fontshape{it}\selectfont.

The high-level commands to select font sizes are:

\tiny	(5pt)	\normalsize	(10pt)	\LARGE	(17.28pt)
\scriptsize	(7pt)	\large	(12pt)	\huge	(20.74pt)
\footnotesize	(8pt)	\Large	(14.4pt)	\Huge	(24.88pt)
\small	(9pt)				

The sizes listed for the commands are those when 10pt (the default) has been selected as the basic size option in the \documentclass command; for 11pt and 12pt, they will all scale accordingly.

The family declarations and their standard family attribute values are:

\rmfamily (cmr)   \sffamily (cmss)   \ttfamily (cmtt)

which are the three Computer Modern families, Roman, Sans Serif, and Typewriter (Section F.2.2).

The series declarations and their initial attribute values are:

\mdseries (m)   \bfseries (bx)

meaning that only a medium and a bold extended series are provided as standard.

Finally, the shape declarations and their attribute values are:

```
\upshape   (n)        \itshape   (it)
\slshape   (sl)        \scshape   (sc)
```

to select upright, *slanted*, *italic*, and SMALL CAPS.

Note that there are no high-level declarations for the encoding attributes. This is because there is normally no need to change encoding within a document. An exception might be to use Cyrillic fonts (coding OT2), in which case one could define

```
\newcommand{\cyr}{\fontencoding{OT2}\selectfont}
\newcommand{\lat}{\fontencoding{OT1}\selectfont}
```

to be able to switch back and forth more conveniently.

The family, shape, and series attributes may be reset to their standard values at any time with the \normalfont command, which also activates that font in the current size.

For each of the above font attribute declarations there is also a corresponding font command (Section 4.1.4) that sets its argument in that font. Thus \textit{text} is almost the same as {\itshape text}, the only difference being that the command also contains the *italic correction* automatically. The complete list of such commands is:

Family:	\textrm	\textsf	\texttt	
Series:	\textmd	\textbf		
Shape:	\textup	\textit	\textsl	\textsc
Other:	\emph	\textnormal		

The \emph command is described in Section 4.1.1; \textnormal sets its argument in \normalfont.

8.6   Typing special symbols directly $\boxed{2_\varepsilon}$

$\boxed{!}$ The commands for producing the special characters and accented letters listed in Sections 2.5.6 and 2.5.7 may be suitable for typing isolated 'foreign' words, but become quite tedious for inputting large amounts of text making regular use of such characters. Most computer systems provide non-English keyboards with appropriate fonts for typing these national variants directly. Unfortunately, the coding of such extra symbols is by no means standard, depending very much on the computer system.

For example, the text Gauß meets Ampère entered with an MS-DOS editor (code page 437 or 850) appears in a Windows application (code page 1252) as Gauá meets AmpŠre and on a Macintosh as Gau· meets Ampäre. Since LATEX is intended to run on all systems, it simply ignores all such extra character codes on the grounds that they are not properly defined.

The inputenc package solves this problem. It not only informs LATEX which input coding scheme is being used, it also tells it what to do with the extra characters. One invokes it with

Table 8.4: Input coding schemes for `inputenc` package

ascii	7-bit ASCII encoding, characters 32–127 only
latin1	ISO Latin-1 encoding (Western Europe)
latin2	ISO Latin-2 encoding (Eastern Europe)
latin3	ISO Latin-3 encoding (Catalan, Esperanto, Galacian, Maltese)
latin5	ISO Latin-5 encoding (Turkish)
decmulti	DEC Multinational encoding
cp850	IBM 850 code page (Western Europe)
cp852	IBM 852 code page (Eastern Europe)
cp437	IBM 437 code page (North America)
cp437de	Variant on 437, with German ß replacing Greek β in position 225
cp865	IBM 865 code page (Scandinavia)
applemac	Macintosh encoding
next	NeXt encoding
ansinew	Windows ANSI encoding
cp1252	Windows 1252 code page (same as `ansinew`)
cp1250	Windows 1250 code page, for Central and Eastern Europe

> `\usepackage[`*code*`]{inputenc}`

where *code* has one of the values listed in Table 8.4 (more are added with each LATEX update).

The files *code*`.def` define the extended characters with the commands

> `\DeclareInputText{`*pos*`}{`*text*`}`   or
> `\DeclareInputMath{`*pos*`}{`*math*`}`

which assign *text* or *math* coding to the character *pos*. For example, `latin1.def` contains

> `\DeclareInputText{198}{\AE}`

to translate the input character code 198 to the LATEX command `\AE`.

Documents making use of this package are fully portable to other computer systems. A document file produced with a DOS editor may look very strange to a human user reading it on a Macintosh, but when the Macintosh LATEX processes it, the proper DOS interpretations will be applied so that the end result is what the author intended.

Some TEX installations automatically contain such conversions for their own local input coding scheme; files written under such a system without the `inputenc` package will be processed correctly on their own system, but will generate garbage on another one. Thus this package should always be added when special symbols are directly typed in, especially if there is any possibility that the file may be processed on another computer system.

8.7   Alternatives for special symbols $\boxed{2\varepsilon}$

$\boxed{!}$ A number of symbols in the text fonts cannot be addressed explicitly, but only as ligatures, for example ¿ as ?'. Other symbols that one might want in text

Table 8.5: Alternative commands for special symbols in text mode

Command	Symbol	Replaces	Release
	Ligatures		
\textemdash	—	---	*1994/12/01*
\textendash	–	--	*"*
\textexclamdown	¡	!'	*"*
\textquestiondown	¿	?'	*"*
\textquotedblleft	"	' '	*"*
\textquotedblright	"	' '	*"*
\textquoteleft	'	'	*"*
\textquoteright	'	'	*"*
	Math symbols		
\textbullet	•	$\bullet$	*1994/12/01*
\textperiodcentered	·	$\cdot$	*"*
\textbackslash	\	$\setminus$	*1995/12/01*
\textbar			$|$	*"*
\textless	<	$<$	*"*
\textgreater	>	$>$	*"*
	Miscellaneous		
\textvisiblespace	␣	\verb*+ +	*1994/12/01*
\textasciicircum	^	\verb+^+	*1995/12/01*
\textasciitilde	~	\verb+~+	*"*
^{123}	123	^{123}	*1995/06/01*
	Special symbols		
\textcompwordmark		(ligature break)	*1994/12/01*
\textregistered	®		*1995/12/01*
\texttrademark	™		*"*
\textcircled{x}	Ⓧ		*1994/12/01*

are only available in math mode. LaTeX 2_ε provides some \text.. commands, listed in Table 8.5, to print these and other characters directly. (Note that these commands were made available in different releases of LaTeX 2_ε.)

8.8   Additional standard files

The basic LaTeX format contains most of the basic features needed for producing regular documents. However, there are many additional features that are indispensable for certain applications, and may be included by means of the \usepackage command (Section 3.1.2). Hundreds of such packages have been written by various users and made available over computer networks (Section D.6); others have been developed by members of the LaTeX3 Project Team and are offered parallel to the standard installation; finally, a number are part of that installation itself.

8.8.1   Special documents

The standard installation includes a number of special 'documents', files with the .tex extension. These are:

[2.09] small.tex a short sample document in LaTeX 2.09;

[2.09] sample.tex a longer sample document in LaTeX 2.09;

[2ε] small2e.tex a short sample document in LaTeX 2_ε;

[2ε] sample2e.tex a longer sample document in LaTeX 2_ε;

lablst.tex (Section 8.3.1) for printing out all the cross-reference labels, with translations and page numbers; it asks interactively the name of the document that is to be so listed;

idx.tex (Section 8.3.4) lists all the \index entries for a document, ordered in two columns, by page; it asks interactively the name of the document that is to be so listed;

testpage.tex to test the positioning of the text on a page; this is more a test of the printer driver, to make sure that the margins are correct and that the scaling comes out properly; in LaTeX 2.09, this file is for American letter paper only, while for LaTeX 2_ε, one is asked interactively for the desired paper size;

[2ε] nfssfont.tex is a document to print out a font table, sample text, and various other tests; it asks for the name of the font interactively; the command \help prints out a list of commands that can be given;

[2ε] docstrip.tex (Section D.3.1) this is more a program than a document; it is the basic tool for unpacking operating files from their documented source files; it not only removes comments, but also includes alternative coding depending on various options.

8.8.2   Extra classes [2ε]

In addition to the standard classes that have already been discussed (book, report, article, letter) the standard installation also contains

proc for producing camera-ready copy of conference proceedings; it is a variant of the two-column article class; it contains one extra command: \copyrightspace, which leaves blank space at the bottom of the first column for typing in a copyright notice;

minimal a bare-bones class for testing and debugging;

slides (Section 8.10) the former SₗₗTₑX of LATₑX 2.09, for producing presentation material;

ltxdoc (Section D.3.2) to produce class and package documentation.

8.8.3   Standard packages

Those packages that are part of the standard installation of LATₑX should be considered universal, in that one can expect them always to be available. The other packages, even those of the LATₑX3 Team, must be specially obtained.

The standard packages, some of which are described elsewhere, are:

alltt (Section 4.9.1) enables the alltt environment which is much like verbatim except that \ { } behave as normal; this means that commands included in the text are executed and not printed literally;

_{2.09} bezier (Section 6.1.4) provides the possibility to draw quadratic curves in the picture environment; this feature is now internal to LATₑX 2$_\varepsilon$, but a dummy package file is available for compatibility with older documents;

_{2ε} doc (Section D.3.2) provides many special features for documenting LATₑX packages;

_{2ε} exscale allows the mathematical symbols that appear in different sizes to scale with the font size option given in the \documentclass command; normally the size of these symbols is fixed, independent of the basic size of the text;

_{2ε} flafter (Section 6.4.2, page 181) ensures that floating objects (figures and tables) do not appear before their position in the text;

_{2ε} fontenc (Section C.5.6, page 359) declares font encodings by loading their .def files; the names of the encodings are listed as options;

_{2ε} graphpap (Section 6.1.6) provides the command \graphpaper which produces coordinate grids in the picture environment;

ifthen (Section 7.3.5) allows text or command definitions depending on the state of various conditionals; one may test the state of a boolean switch, the value of a counter or length, or the text definition of a command;

_{2ε} inputenc (Section 8.6) allows a LATₑX document file to be prepared with a system-dependent input encoding scheme for special characters;

$\boxed{2_\varepsilon}$ latexsym loads the 11 special LATEX symbols in the lasy fonts (Section F.2.2), which are no longer internal to LATEX 2_ε; this file is automatically read in in compatibility mode;

makeidx (Section 8.4) provides the commands that may be used with the MakeIndex program to generate a keyword index;

showidx (Section 8.3.4) makes the \index entries visible by printing them as marginal notes;

$\boxed{2_\varepsilon}$ shortvrb provides \MakeShortVerb and \DeleteShortVerb, to designate (and remove) a certain character as the equivalent of \verb for printing literal text. The character, preceded by a backslash, is given as an argument to these commands. After \MakeShortVerb{\|} has been issued, |\cmd{x}| is equivalent to \verb|\cmd{x}|, producing \cmd{x};

$\boxed{2_\varepsilon}$ syntonly provides the command \syntaxonly which, when given in the preamble, suppresses all output to the .dvi file, but continues to issue errors and warnings; the job runs about four times faster than normal; this is used to check the processing when output is not immediately needed;

$\boxed{2_\varepsilon}$ t1enc makes the T1 font encoding (Section F.4.3) the standard, for use with DC or EC fonts;

$\boxed{2_\varepsilon}$ textcomp makes the special symbols in the text companion fonts available in text mode (Section F.4.4); available as of December 1, 1997;

$\boxed{2_\varepsilon}$ tracefnt is a diagnostic tool for checking NFSS font selections; the options given with \usepackage control its action:

errorshow suppresses all warnings and information messages to the monitor, which are still sent to the transcript file; only error messages are printed to the monitor;

warningshow prints warnings and error messages on the monitor; this behavior is much the same as when the package is not used at all;

infoshow (default option) sends font selection information to the monitor, which is normally only written to the transcript file;

debugshow writes considerably more information to the transcript file about every change of font; can produce a very large transcript;

pausing turns all warnings into error messages so that processing temporarily stops, awaiting user response;

loading shows the loading of external fonts.

8.8.4 Accessory software

On the network servers where LaTeX can be found (Section D.6), there are a number of parallel directories offering even more accessory packages than those in the standard installation. Although not part of the basic LaTeX kernel, they are considered essential to any complete installation. Some of them are described later in Appendix D on extensions to the standard.

amslatex is a set of packages supported by the American Mathematical Society for advanced mathematical typesetting, Appendix E.

babel provides support for multilingual LaTeX, described in Section D.1.3.

graphics a bundle of packages for including, rotating, manipulating external figures and diagrams, color, described in Section 6.2.

cyrillic contains support for various Cyrillic font encodings as well as for a large number of input encoding schemes.

psnfss provides packages for installing certain PostScript Type 1 fonts with NFSS, as illustrated in Section D.2.1.

tools is a set of useful packages written by members of the LaTeX3 Project Team, outlined in Section D.3.3.

Most of the above packages come with full documentation, making use of the documented source code technique described in Section D.3.2. The source file is processed under LaTeX to obtain the manual and/or detailed explanation of the coding, while an installation file is run to extract various files (.sty, .fd, and so on) from the source file.

8.8.5 Contributed packages

The LaTeX3 Project Team members are not the only ones who have written class and package files for LaTeX. On the network servers (Section D.6) there are hundreds of private packages that have been made public. A number of the most useful are described in *The LaTeX Companion* (Goossens *et al.*, 1994).

The ability to write one's own classes and packages (explained in more detail in Appendix C) is one of the great strengths of LaTeX. Rather than expecting the originators to anticipate *everything* that might be required of the text processor, the real users can establish their specialist needs, find the solutions, and make them readily available for others. The basic program is not overloaded with applications that are used extensively only by a limited number of people; furthermore, it is those users with the practical experience who produce the new extensions.

8.9   The various LATEX files

!

A number of different files are used during the LATEX processing: some are read in while others are created to store information for the next run. They all consist of two parts:

> *root_name*.*extension*

For every LATEX document, there is a *main* file, whose root name is used when the program LATEX is called. It normally possesses the extension .tex, as do any other files that it might read in with \input or \include commands. (If they have a different extension, it must be given explicitly.)

The other files that are created during a LATEX processing run normally have the same root name as the main file but a variety of different extensions, depending on their functions. If the main file contains \include commands, there will be additional files created having the same root names as the included .tex files but with the extension .aux.

Some of these files are created with every LATEX run, whereas others appear only when certain LATEX commands have been issued, such as \tableofcontents or \makeindex. The .aux files as well as those produced by special LATEX commands may be suppressed by including the command

> \nofiles

in the preamble. This command is useful if a document is being constantly corrected and reprocessed, so that the information in the special files is not yet finalized and may therefore be dispensed with. Files created at the TEX level of processing will always be generated.

Other extensions describe files that contain formatting information, or additional instructions (coding), or databases. The root part of these file names is not associated with any document file. An example of this type is the article.cls file that defines the article class.

The rest of this section contains a list of the LATEX file extensions with a short description of the role they play in the LATEX processing.

.aux   This is the auxiliary file written by LATEX, containing information for cross-references as well as some commands necessary for the table of contents and other lists. There will be one .aux file created (or re-created) for the main file, in addition to one for every file read in by an \include command.

When a document is processed by LATEX for the first time, there is no .aux file and the cross-referencing commands cannot function. The message

> No file *main_name*.aux.

is written to the monitor in this case. The .aux files are read in by the command \begin{document}. As they are read in, their names appear on the monitor screen in the form

> (*main_name*.aux) (*sub1_name*.aux) ...

New .aux files are written for every LATEX processing run. The new .aux file for the main file begins with the \begin{document} command, whereas those for the \included files are initiated by the corresponding \include command, and are closed when the .tex file is fully read in. The main .aux file is closed by the \end{document} command.

The auxiliary .aux files may be suppressed by issuing \nofiles in the preamble.

.bbl This file is not created by LATEX, but by the program BIBTEX. It has the same root name as the main file. BIBTEX actually only reads the .aux file for its information. The .bbl file is read into the next LATEX processing run by the command \bibliography and produces the list of literature references.

.bib Bibliographic databases have the extension .bib. BIBTEX reads them to extract the information it needs to generate the .bbl file. The root name describes the database and not (necessarily) any text file. The database name is included in the text by means of the \bibliography command.

.blg This is the transcript file from a BIBTEX run. The root name is the same as that of the input file.

.bst This is a *bibliography style file* that is used as input to BIBTEX to determine the format of the bibliography. The name of the .bst file is included in the text by means of the \bibliographystyle command.

2ε .cfg Some classes permit local *configurations* for paper size or other requirements by putting specifications into a .cfg file of the same root name as the class. The ltxdoc class (Section D.3.2) is one of these.

2ε .clo This is a *class option* file in LATEX 2_ε, containing the coding for certain options that might apply to more than one class. There is no special command to input them.

2ε .cls This is a *class* file in LATEX 2_ε, defining the overall format of the document. It is read in by the \documentclass command. A .cls file may also be input from another one with the \LoadClass command.

2ε .def Additional definition files of various types carry the .def extension. Examples are t1enc.def (to define the T1 encoding) and latex209.def (for compatibility mode between LATEX 2_ε and 2.09).

2ε .dtx These are documented source files for the LATEX 2_ε installation. If a .dtx file is processed directly, the documentation is printed. This is an explanatory manual and/or a detailed description of the coding. The class, package, .ltx, .fd or other file(s) are extracted by processing the .dtx file with DocStrip (Section D.3.1), usually by running LATEX or TEX on a supplied .ins file.

.dvi This is the TEX output file, containing the processed text in a format that is independent of the output printer, hence a 'device-independent' file. It too is generated by every TEX process run and cannot be suppressed by \nofiles. The message

    Output written on *main_name*.dvi(*n* pages, *m* bytes).

is written to the monitor when this file is closed.

When certain errors occur, the .dvi file is not written, in which case the monitor message is

```
No pages of output
```

The .dvi file must be further processed by a *printer driver* program that converts it into the special commands for the desired output device (printer).

[2ε] .fd   Files containing the NFSS commands associating external font names with font attributes (Section C.5.6) are *font description files*. The root names consist of the encoding and family designations, such as ot1cmr.fd. When a font of a given family is requested for the first time, the corresponding .fd file is automatically read in, if it exists.

[2ε] .fdd   Documented sources for .fd files have the extension .fdd, and are nothing more than a special type of .dtx file. They may be processed for the documentation, or have the .fd files extracted by DocStrip.

.fmt   This extension belongs to a TeX *format* file that has been created with the special program initex (Section 1.2.1). Format files contain all the basic instructions in a compact coding for quick loading. LaTeX is nothing less than TeX run with the format latex.fmt [2ε] or lplain.fmt [2.09]. (Plain TeX uses the format file plain.fmt.)

.glo   This file is only generated when the command \makeglossary exists in the preamble. It has the same root name as the main file and contains only \glossaryentry commands that have been produced by \glossary commands in the input text. This file will be suppressed in spite of the presence of \makeglossary if the command \nofiles is issued in the preamble.

.gls   This file is the glossary equivalent of the .ind file. It too is generated by the MakeIndex program, but with the .glo file as input. The input and output extensions must be explicitly given. There are no standard LaTeX commands to read in the .gls file. The doc package uses it to record a change history.

.idx   This file is only generated when the command \makeindex exists in the preamble. It has the same root name as the main file and contains only \indexentry commands that have been produced by \index commands in the input text. This file will be suppressed in spite of the presence of \makeindex if the command \nofiles is issued in the preamble.

.ilg   This is the transcript file from a MakeIndex run. The root name is the same as that of the input file.

.ind   This file is generated by the MakeIndex program, with the .idx file as input. It is read in by the \printindex command which is defined by the makeidx package. It contains a theindex environment, built up from the entries in the input file.

[2ε] .ins   To facilitate the DocStrip run on a .dtx file, the necessary commands (and options) are put into an *installation* file. The extraction is then accomplished simply by processing (with TeX or LaTeX) the appropriate .ins file.

.ist This is an *index style file*, containing style settings for MakeIndex. The root name reflects the style and has no relationship to any text file. For the doc package, the files gind.ist and gglo.ist are provided for specialized index formatting.

.lis This is an alternative to .log on some systems.

.lof This file contains the information for the list of figures. It behaves exactly like the .toc file except that it is opened by the command \listoffigures instead of by \tableofcontents.

.log This file contains the protocol of the TeX processing run, that is, all the messages that were sent to the monitor during the run as well as additional information that can be of use to TeX experts when errors occur. This file is always generated, even when \nofiles has been issued. The message

    Transcript written on *main_name*.log.

is written to the monitor when this file is closed.

.lot This file contains the information for the list of tables. It behaves exactly like the .toc file except that it is opened by the command \listoftables instead of by \tableofcontents.

.ltx As an alternative to .tex to indicate that this is a LaTeX and not a Plain TeX file, this good idea has never become truly popular, probably because one must always give .ltx explicitly when processing.

For LaTeX 2ε, certain files that are inputs to the initex program bear the extension .ltx. They form the basis for generating the format latex.fmt. The main such file is named latex.ltx.

.mf METAFONT source files (Sections F.1 and F.3) contain drawing instructions for the characters in a font. The METAFONT program converts these to bitmap .pk files in a given size and resolution for DVI driver programs.

.pk Packed pixel, or bitmap, files (Section F.1) contain images of the characters in a font. These are read in by the DVI drivers to print the characters either on paper or on the computer monitor.

.sty This is a *package* file in LaTeX 2ε, to be loaded with \usepackage. A package contains additional LaTeX commands to define new features or alter existing ones. It should not contain any printable text. It may be input by another package with the command \RequirePackage.

In LaTeX 2.09, the .sty files were *main style* files (equivalent to classes) as well as *style option* files (packages). The two have quite different properties although they both have the same extension. They were both read in by the \documentstyle command.

.tex The file containing the user's input text should have the extension .tex. For every LaTeX document there is at least one .tex file. If there is only one .tex file, it is also the *main* file that determines the root name of all the others. If there are \input or \include commands within the input text, there are other files belonging to the document with different root names but all with the extension .tex. The main file is the one whose name appears when the program LATEX is called. It contains the highest level \input commands and is the only file that may have \include commands.

.tfm TEX font metric files (Section F.1) contain all the font information that TEX needs to process a document: height, width, depth of each letter, ligature information, italic correction, and so on. The only thing that is missing is what the character looks like. This is provided by .pk files.

.toc This file contains the information for the table of contents. The command \tableofcontents reads in the .toc file, if it exists, and then outputs the table of contents. At the same time, a new .toc file is opened where all the contents information for the current processing run is written. This .toc file is closed by the \end{document} command, so that if the LATEX run is interrupted before completion, the .toc file is lost.

A .toc file is only generated if the document contains a \tableofcontents command and if the \nofiles command has not been issued. It has the same root name as the main file.

.vf Virtual font files (Section F.1) are read in in place of .pk files for fonts that do not really exist independently. For each character, there are instructions for the driver program, usually to take character *nn* from real font *xx*.

8.10 Presentation material (SLITEX)

LATEX 2.09 provided a parallel version of itself called SLITEX for producing presentation material such as colored slides or viewgraphs. This special variant makes use of a different set of fonts that is more appropriate for projection than the regular book style of letters. However, this required another TEX *format* (Section 1.2.1) since the older LATEX version programs its whole set of fonts explicitly into the format file. The name *SLITEX* was used to distinguish it from regular LATEX. Running SLITEX rather than LATEX meant invoking the TEX program with the splain instead of the lplain format.

Under LATEX 2$_\varepsilon$, with the New Font Selection Scheme, there is no longer any problem in replacing all the fonts within the one LATEX format. Since SLITEX has ceased to be something similar to, but separate from, LATEX, there is no further need for a distinguishing name. Instead, we refer to the class slides for generating presentation material. This class may be selected with regular LATEX 2$_\varepsilon$ just like any other class, whereas with LATEX 2.09, the slides style could only be chosen with SLITEX, and in fact was the only style that it would accept.

We will use the word *SLITEX* only to refer to the LATEX 2.09 version, which has some important differences from the LATEX 2$_\varepsilon$ slides class.

8.10.1 The slides class

By *presentation material*, we mean slides or viewgraphs that are to be projected before an audience. One could of course make up the text for such sheets with the normal LaTeX commands, but there would be problems getting the font size right, arranging overlays, making sure that the text does not change pages unexpectedly, and the question of whether book-style fonts are suitable for projection. The LaTeX class slides attempts to solve these.

A different set of fonts is used by this class, ones in which the lower case letters are relatively larger than in the normal fonts, and with a much bigger base size. This has the effect of limiting the amount of text that fits on one page/slide, but this is in fact very good practice for presentation material. The text really should be restricted to keywords and abbreviated sentences. A full page of normal text projected on to a screen will not be read by the audience.

Slides should also make use of color. On this point, SʟɪTeX is rather out of date. When LaTeX was first created in the mid-1980s, there were no reasonably priced color printers available. Instead, a system of *color layers* was used, whereby the text for each color was printed separately (in black and white) to be later copied in color to transparencies that would then be superimposed. Today there are better methods to produce colored slides directly. Although color support is not part of the LaTeX 2_ε kernel, there is a color package that does provide color commands for certain drivers (Section 6.3). This package may be used everywhere in LaTeX 2_ε, not just with the slides class.

Even without the color feature, the slides class is useful for producing black-and-white viewgraphs with the special fonts and the other advantages that it offers. Most of the formatting commands of regular LaTeX may be used, except for the page breaking and sectioning commands. The special features that are unique to slides are described in the next sections.

The color layer commands of LaTeX 2.09 are so obsolete that they will not be described here at all.

8.10.2 The slide environments

The source file for producing slides is structured much the same as that for a normal document, except that it is the slides class that is invoked as

```
2ε \documentclass{slides}
      preamble text
   \begin{document}
      slide text
   \end{document}
```

If true colors are wanted (not color layers in black and white), the `color` package can be added in the preamble as for example

[2ε] `\usepackage[dvips]{color}`

Since this package is not specific to the `slides` class, it is described in Section 6.3 and not here.

The preamble may contain global specifications, for changing the paper size or selecting a page style, as usual. It may not contain any printable material.

Any text that appears after `\begin{document}` but before the special slide environments described below is output to an unnumbered *leading page* that comes before any slides. This may serve as a cover page for the slides.

There are three environments available for organizing different parts of a slide: the main *slide* itself, possible *overlays*, and additional *notes* to the slide.

Slides

A slide or viewgraph is created by means of the environment

[2ε] `\begin{slide}`   *text and commands*   `\end{slide}`

The contents of a `slide` environment may be any text that one pleases. Note that the `slides` class makes use of its own set of character fonts. The standard font in the normal size is roughly equivalent to the LaTeX sans serif font `\sffamily` in the size `\LARGE`. All the regular font commands and declarations of Section 4.1 are available, along with the other display and list environments of Chapter 4. However, there can be no page breaks within the environment, for the entire text is expected to fit on one page (slide or viewgraph). The successive `slide` environments will be numbered consecutively. If the page should overflow, a warning message is printed.

Any color commands from the `color` package may be employed, if that package has been loaded. The colors are then output directly, and not as layers as for the older SLiTEX.

Overlays

An *overlay* is an addition to a slide that can be laid on top of it to fill in certain gaps. The idea is to create suspense during a presentation, by filling in some key words a few minutes later, or to be able to replace some text by alternatives.

The `slides` class generates overlays with the `overlay` environment, which functions exactly the same as the `slide` environment, except that the numbering is done as a sub-number of the last slide. Thus the overlays following slide 6 are numbered 6-a, 6-b, and so on.

[2ε] \begin{overlay}   *text and commands*   \end{overlay}

The slide environment for a viewgraph that is to have overlays must come before the overlay environments that go with it. Both the slide and overlay should contain identical texts, except that certain parts are printed in 'invisible' ink by means of the two declarations

\invisible   and   \visible

These two commands function just like font declarations. They may be set within curly braces {} to limit their scope, or used to apply to the whole environment. Normally, in the slide, a few words might be made invisible, while in the corresponding overlay, \invisible is declared at the start and only those words that were blanked out in the slide are made visible. This may be seen in the sample on page 254.

One application for overlays is to present alternative text. One might, for example, show a table of cost estimates, with the figures on an overlay. By exchanging this overlay for another, new numbers that can be achieved by certain procedural changes may be fitted into the same table.

Notes

During a presentation, it is often necessary to refer to a list of keywords or other notes between the actual projected slides. The slides class has the note environment for producing such reminders for the speaker.

\begin{note}   *text*   \end{note}

Like the overlays, notes are sub-numbered with respect to the previous slide, but with numbers instead of letters. Following slide 4, notes are numbered 4-1, 4-2, and so on.

8.10.3 Further features

Page styles

The LATEX command \pagestyle{*style*} may also be used with the slides class. The following styles are available:

plain   All slides, overlays, and notes have their number at the lower right corner.

headings

The same as plain except that if the clock option is selected (see below), a time marker is printed in the lower left corner of the notes. This is the default page style.

empty   The sheets are printed with no page numbers.

In normal LaTeX, the command \pagestyle is issued in the preamble and is valid for the whole document. Individual pages may be given a different style by means of the \thispagestyle command which remains in effect for only one page. This command should not be employed within one of the environments, but rather a new \pagestyle command may be given *outside* of the slide, overlay, and note environments.

Timing notes

The option clock can be selected with \documentclass; it activates two additional commands:

$\boxed{2_\varepsilon}$ \settime{*secs*}
$\boxed{2_\varepsilon}$ \addtime{*secs*}

and prints the time in minutes at the bottom of each note. In this way, the notes contain a reminder of how far advanced the presentation should be at that point. The value of the time marker is set and incremented with the above two commands, which take a number in seconds as argument. The initial value of the timer is 0.

Selective slide processing

If the slide file possesses very many slide environments, it is possible that the user may want to alter only a small number of them, without having to output the complete set once again. This may be accomplished with the command

\onlyslides{*slide_list*}

in the preamble. The *slide_list* stands for a list of slide numbers in ascending order, for example, 2,5,9–12,15, specifying the slides, or range of slides, that are to be processed.

Non-existent slide numbers may also appear in the *slide_list*. If the slide file contains, say, 20 slides, the command \onlyslides{1,18–999} will allow only slides 1 and 18–20 to be processed. Any overlays that belong to these slides will also be output.

Finally, the command

\onlynotes{*note_list*}

in the preamble limits the notes that are output to those listed in *note_list*. Supposing that slide 5 has three note environments associated with it, \onlynotes{5} arranges that only the note pages 5-1, 5-2, and 5-3 are printed.

The \onlyslides and \onlynotes commands function similarly to the command \includeonly described in Section 8.1.2. An interactive means of selecting slides and notes to process may be set up exactly as illustrated in Section 8.1.3 with the \typein command:

```
\typein[\slides]{Which slides to do?}
\onlyslides{\slides}
```

During processing, the message 'Which slides to do?' is typed to the screen, and the user replies with the desired slide numbers. A similar scheme may be set up for \onlynotes.

A sample slide file

An example of input text with the slides class, with a leading page, overlay, and note, and using the clock option, is given here. The four-page result is shown on the next page.

```
\documentclass[a4paper,clock]{slides}
\begin{document}
\begin{center}\Large\bfseries
Sample Viewgraphs
\end{center}

\begin{slide}
\begin{center}\large Advantages of \texttt{slides}\end{center}

\begin{itemize}
\item Uses special fonts
\item Forces key words instead of long text
\invisible
\item Supports color layers
\visible
\end{itemize}

Color commands may be used {\invisible as color layers}

\end{slide}

\begin{overlay}
\invisible
\begin{center}\large Advantages of \texttt{slides}\end{center}

\begin{itemize}
\item Uses special fonts
\item Forces key words instead of long text
\visible
\item Supports color layers
\invisible
```

The Leading Page

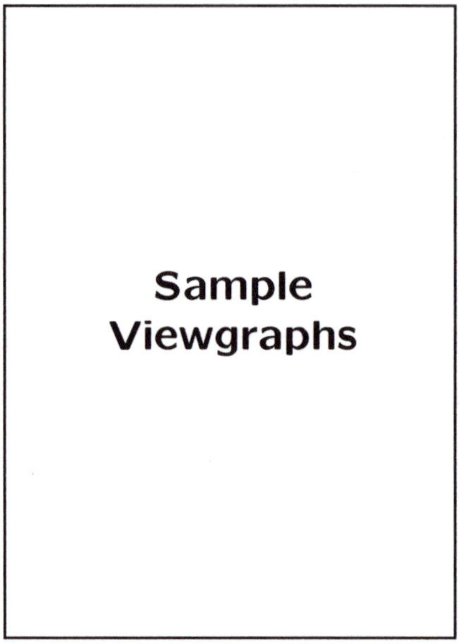

Sample
Viewgraphs

The Slide Page

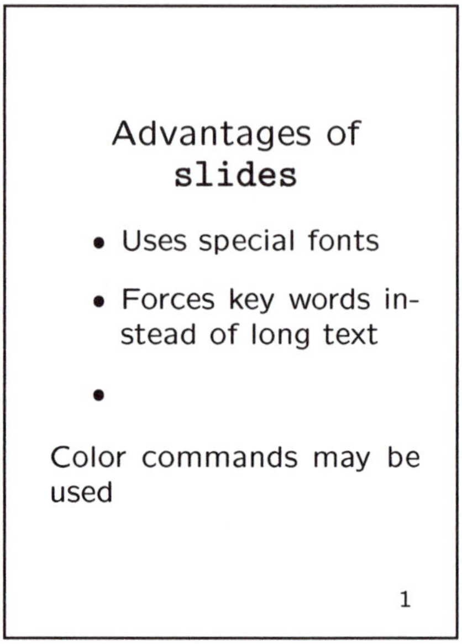

Advantages of
`slides`

- Uses special fonts

- Forces key words in-
  stead of long text

-

Color commands may be
used

1

The Overlay Page

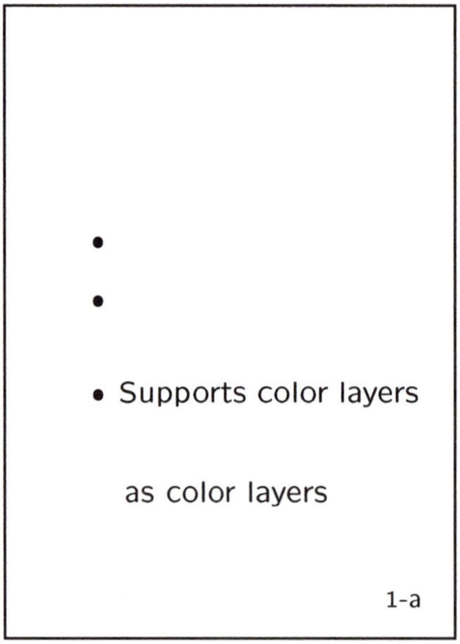

-

-

- Supports color layers

  as color layers

1-a

The Note Page

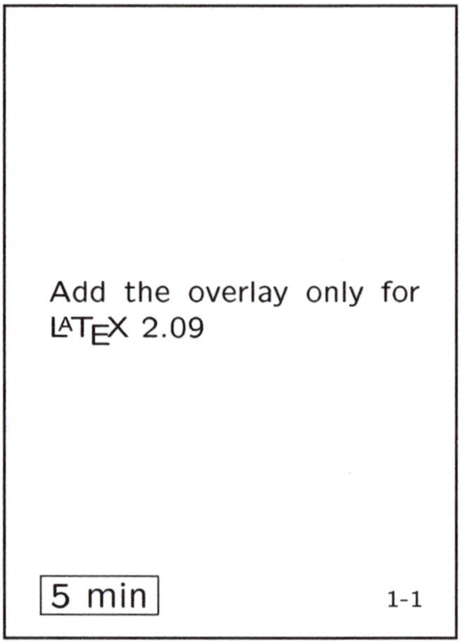

Add the overlay only for
LaTeX 2.09

5 min   1-1

```
\end{itemize}

Color commands may be used {\visible as color layers}
\end{overlay}

\addtime{300}
\begin{note}
Add the overlay only for \LaTeX~2.09
\end{note}

\end{document}
```

9 Error Messages

Errors are bound to be made at times during the preparation of a long LaTeX document. The mistakes can be of various kinds, from simple typing errors for command names to forgetting that some commands must be paired or giving an incorrect syntax for a complicated command.

Errors during the LaTeX processing produce a list of messages on the monitor screen, which appear totally incomprehensible to the beginner. Even the advanced user can have difficulty figuring out a particular error message. However, these messages contain information about fundamental structures that can aid an experienced TeX programmer to see much deeper into the heart of the problem.

In addition, the error messages contain useful information even for the beginner. It is the purpose of this chapter to explain some of these messages that can be of assistance to non-programmers.

9.1 Basic structure of error messages

Error messages have two sources: those from LaTeX and those from the underlying TeX program. The LaTeX messages are often followed by TeX messages since LaTeX operates at a higher level.

9.1.1 TeX error messages

We start with a simple error as an example.

```
\documentclass{article}
\begin{document}
The last words appear in \txetbf{bold face}.
\end{document}
```

Here the command \textbf has been mistakenly typed as \txetbf. During the processing, LaTeX assumes that the user wants to invoke a TeX command \txetbf. Since LaTeX does not know the TeX commands, it passes this text on to TEX, which then decides that there is no such command in its repertoire. The following error message is written to the screen:

```
! Undefined control sequence.
1.3 The last words appear in \txetbf
                                    {bold face}.
?
```

The program stops at this point and waits for a response from the user. This message can be understood even by beginners. It consists of an error indicator that starts with an exclamation point !. Here the indicator is: ! Undefined control sequence, meaning that an *unknown command name* (control sequence) was the cause of the error. Next comes a *pair of text lines*, the *first* of which is prefixed with 1.3, meaning that the error occurred in 'line 3' of the input text. The error itself was encountered at the last symbol printed in this *upper* line. The *lower* line shows the continuation of the input line being processed when the error was found, here the words {bold face}. Before continuing, TEX waits for a reaction from the user, as indicated by the question mark in the last line of the message.

Entering another ? and carriage return as response produces the following information:

```
Type <return> to proceed, S to scroll future error messages,
R to run without stopping, Q to run quietly,
I to insert something, E to edit your file,
1 or ... or 9 to ignore next 1 to 9 tokens of input,
H for help, X to quit
?
```

This is a list of the possible user responses:

1. ⟨*return*⟩: simply typing the return key tells TEX to continue processing after making an attempt to handle the error according to some preprogrammed rules. In the case of an unknown command name, the error treatment is to ignore it.

2. S *scroll mode*: TEX continues the processing, writing further error messages to the screen as they are encountered, but without stopping for a user response. This is as though ⟨*return*⟩ were pressed after all subsequent errors.

3. R *run mode*: TEX continues processing as with S, but does not even stop as it would in *scroll mode* if the file named in an \input or \include command is missing.

4. Q *quiet mode*: the same as with R except that no further error messages are printed to the screen. They are, however, written to the .log file.

5. I *insert*: the mistake can be corrected by inserting the proper text. TEX puts the line of text entered from the keyboard into the processing in place of the error and then continues. Such a correction applies only to the processing: the original text in the .tex file is unchanged and must be altered with an editor. Typing in I\stop brings the program to a halt with the current page in the .dvi file.

6. 1 ...: entering a number less than 100 will delete that many characters and commands from the subsequent text. The program then stops to await further response from the user.

7. H *help*: an extended account of the problem is printed to the screen, which is more informative than the brief error indicator and which may contain tips for relieving the error.

8. X *exit*: The TEX processing is halted at this point. The current page does *not* appear in the .dvi file.

9. E *edit*: The process is halted as with X and a message is printed saying on which line of which file the error occurred. For some implementations, the editor may actually be called automatically, going directly to the faulty line.

The above response letters may be typed in either in capitals or in lower case. The response does not take effect until after ⟨*return*⟩ is pressed.

Typing H or h for *help* on the previous sample error message produces the text:

```
The control sequence at the end of the top line
of your error message was never \def'ed. If you have
misspelled it (e.g., '\hobx'), type 'I' and the correct
spelling (e.g., 'I\hbox'). Otherwise just continue
and I'll forget about whatever was undefined.

?
```

The error is described in more detail: the command name at the end of the upper line is unknown; if this is simply a typing error, enter the proper text with the I response, in this case I\textbf. Otherwise, press ⟨*return*⟩ and the faulty command will be ignored. The line of text will be processed as though it had been The last word appears in bold face. Of course, no bold face can actually appear.

The basic structure of a TEX error message is summarized as:
Every error message begins with an error indicator, marked by a ! at the beginning of the first line. The indicator text is a brief description of the problem. Following are one or more lines of input text. The last symbol of the first of these lines is the one that has caused TEX to stop and print the error message. The last line contains the text or commands

that are the next to be processed. TEX waits for a response from the user. If that response is the typing of an H for help, a more detailed description, with possible tips, is printed to the screen, and TEX waits for a further response.

9.1.2   LaTeX error messages

One of the major differences between LaTeX 2.09 and LaTeX 2ε is the appearance of their error messages. While the older version prints many lines of incomprehensible text in order to show the deep, internal coding that underlies the error, the newer one prints only that line of text where the actual error presumably occurred. In the next section we will show samples of error messages from LaTeX 2.09, but otherwise all examples are from LaTeX 2ε.

As an example of a text with a LaTeX error, we take

```
\documentclass{article}
\begin{document}
\begin{qoute}\slshape
  Text indented at both ends
\end{quote}
\end{document}
```

Here the call \begin{quote} was incorrectly typed as qoute. The LaTeX processing writes the error message:

```
! LaTeX Error: Environment qoute undefined.

See the LaTeX manual or LaTeX Companion for explanation.
Type  H <return>  for immediate help.
 ...

l.3 \begin{qoute}
                 \slshape
?
```

The first line of this message states that LaTeX itself discovered this error, with a brief error indicator, in this case Environment qoute undefined. All LaTeX error messages begin with a line like this, followed by a reference to the LaTeX manuals for detailed explanation (which is also to be found in Section 9.3 of this book). The third line of text is a reminder that the response H ⟨*return*⟩ will also type additional clarification.

The line of three dots ... indicates that there are more lines of internal coding that have been suppressed. In LaTeX 2.09, these internal lines are actually printed, and even in LaTeX 2ε, it is possible to choose to have them appear. See Section 9.1.3.

The next pair of lines shows just where the processing was halted when the error was detected. As for TEX messages, the current input line of text is broken at the error, with the processed part on the first line, here \begin{qoute}, and the remainder on the next line. Here this remainder is \slshape. The line indicator 1.3 (small letter L, not number one) shows on which line (3) of the input file the error was found.

LaTeX now awaits a response from the user. Typing H ⟨*return*⟩ yields the additional information:

```
Your command was ignored.
Type I <command> <return> to replace it with another command,
or  <return>  to continue without it.
?
```

This means the last command in the upper line of the pair beginning with 1.3 has not been read into the processing. The mistake may thus be corrected *for the current process run only* with I\begin{quote} ⟨*return*⟩. The misspelling will still be in the text file, though, and needs to be removed by an editor run later.

However, if the processing is continued by simply pressing ⟨*return*⟩, then \begin{qoute} is ignored, as though this text were not in the source file. This leads directly to another error when the command \end{quote} is encountered, since now the matching \begin{quote} command is missing. The actual text of this second error message is:

```
! LaTeX Error: \begin{document} ended by \end{quote}.

See the LaTeX manual or LaTeX Companion for explanation.
Type  H <return>  for immediate help.
 ...

1.5 \end{quote}

?
```

The message again contains the standard LaTeX announcement, with error indicator, in the first line, followed by a reference to the manuals, and an invitation to type H for help. The error indicator in this case is

```
    \begin{document} ended by \end{qoute}.
```

This arises because when LaTeX encountered \end{quote}, it checked the name of the current environment. Since \begin{quote} was missing, the matching \begin command is the \begin{document} statement, producing a mismatched \begin...\end pair.

The last pair of lines again shows just how far the processing had occurred before the error was detected. In this case, the whole of the current line has already been taken in, so the remainder line is blank.

The response H ⟨*return*⟩ at this point leads to the same message as for the first error. A correction with an I entry will not do any good since the missing \begin{quote} can no longer be inserted ahead of the environment text. Entering I\begin{quote} will replace the \end{quote} with \begin{quote}, but that does not solve the real problem. The best response now is to type ⟨return⟩ so that the command \end{quote} is ignored and the processing continues.

Now both the faulty \begin{qoute} and correct \end{quote} commands are removed and the processing takes place as if the quote environment had never been in the input text at this point.

If the typing error qoute had been made in the \end instead of in the \begin command, the error message would have been:

```
! LaTeX Error: \begin{quote} on input line 3 ended by \end{qoute}

See the LaTeX manual or LaTeX Companion for explanation.
Type  H <return>  for immediate help.
 ...

l.5 \end{qoute}

?
```

The previous explanations should be enough for the reader to understand what is now being said. The error indicator here is

```
\begin{quote} on input line 3 ended by \end{qoute}
```

and the last pair of lines indicate that the problem is in line 5 and that the troublesome command is \end{qoute}. The obvious thing to do now, and the H message encourages this conclusion, is to make a correction with I\end{quote}. However, now a new message appears on the screen:

```
! Extra \endgroup.
<recently read> \endgroup
l.5 \end{qoute}

?
```

With the exception of the last pair beginning with l.5, this makes no sense at all. The error indicator ! Extra \endgroup is seemingly meaningless. That this is a TeX error message and not a LaTeX one is hardly any compensation.

The frustrated user is not really to blame. The response was perfectly reasonable even if it was wrong. Only experience could say that the best action at that point was to have just pressed ⟨return⟩. The quote environment is closed off anyway, although any special actions associated with \end{quote} would be left off.

The help message at this point is

```
Things are pretty mixed up, but I think the worst is over.
```

This is at least encouraging, and the reader should not lose heart. The best thing is to just keep on pressing ⟨*return*⟩ to get through the run.

We give here two recommendations for choosing a response, one specific and one general. The specific one is:

> If a false environment name appears in a \begin command, the proper correction of the error is
>
> I\begin{*right_name*}
>
> If the misspelling has occurred in the \end command, the best way to handle this mistake is just to press ⟨*return*⟩. The environment will be closed and any local declarations or definitions will be terminated. However, if there are any commands to be executed or text to be printed by the \end command, these will be missing.

The general recommendation is

> If the user knows how to correct the error with the help of the error message, this should be done with
>
> I *correction*
>
> Otherwise, one may give ⟨*return*⟩ and wait and see what happens. Even if more peculiar (TEX) error messages appear, one can keep on pressing ⟨*return*⟩ until the processing is finally at an end. The following print-out should indicate where the mistake was.

Instead of continually pressing ⟨*return*⟩, one may also enter S, R, or Q and ⟨*return*⟩ to speed up the error treatment (Section 9.1.1). Here, as with the simple ⟨*return*⟩ entry, the faulty commands are not just ignored. TEX attempts to treat them by making assumptions about what the user wanted to do at this spot. Only when this is no longer possible will TEX ignore the command completely. For example, if the indicator reads

\begin{*environment*} ended by \end{*environment*}

there was at least a \begin command with a valid environment name earlier. It could be assumed that the name of the environment in the \end command is wrong. LATEX thus tries to execute this command using the name of the current environment.

9.1.3 Error messages from LATEX 2.09

For comparison, we show the equivalent error messages from LATEX 2.09, after replacing \documentclass with \documentstyle in the sample input on page 260. The first message reads

```
LaTeX error.   See LaTeX manual for explanation.
             Type  H <return>  for immediate help.
! Environment qoute undefined.
\@latexerr ...diate help.}\errmessage {#1}

1.3 \begin{qoute}

?
```

The major difference here is that the error indicator appears in the third line by itself, and not as part of the LaTeX error announcement in the first line. The other difference is the strange text in the fourth line that begins with \@latexerr .... This is actually part of the internal command that produced the message, and should be simply ignored.

The second error message, which follows when ⟨return⟩ is pressed without any corrective text, is more radically different from that for LaTeX 2ε:

```
LaTeX error.   See LaTeX manual for explanation.
             Type  H <return>  for immediate help.
! \begin{document} ended by \end{quote}.
\@latexerr ...diate help.}\errmessage {#1}

\@checkend ...urrenvir \else \@badend {#1}
                                          \fi
\end ...me end#1\endcsname \@checkend {#1}
                                       \expandafter ...
1.5 \end{quote}

?
```

The extra pairs of line between the \@latexerr... and 1.5... show the internal decoding of the \end command, something that is of no interest or use to the regular LaTeX user. They only serve to confuse. If you are still using LaTeX 2.09 and are presented with such error messages, you should just ignore these lines altogether. Otherwise, the error indicators, the input lines beginning with 1.n, the help messages, and the responses are to be interpreted and employed as for LaTeX 2ε.

Sometimes it is desirable to look deeper behind the error message, especially for practiced TeXperts who know how to interpret them. In this case, LaTeX 2ε can be made to output the missing lines of coding by setting

> \setcounter{errorcontextlines}{*num*}

where *num* is the number of levels a macro will be decoded on an error. By default, *num* = −1 for LaTeX 2ε, while for LaTeX 2.09 it is equal to 5. The knowledgeable user can set it to 5 or more to gain this extra insight.

9.1.4 Error messages from TeX macros

The majority of TeX commands and practically all the LaTeX commands are so-called TeX macros. These are combinations of primitive commands grouped together under a new command name, by which they may be called as a unit. TeX macros are structures similar to those that may be defined by the LaTeX command \newcommand. Up to nine arguments may be passed to them, just as for LaTeX commands. However, the corresponding TeX commands for generating macros are more general than \newcommand.

In fact, of the approximately 900 Plain TeX commands available, only 300 are primitives, or fundamental commands. The remaining 600 are macros. If an error occurs within a macro, the other commands within it may also be affected.

Here is an example for clarification. The command \centerline is a macro defined as

```
\def\centerline#1{\@@line{\hss#1\hss}}
```

in which \@@line is itself a macro while \hss is a TeX primitive, a rubber length that can be infinitely stretched or shrunk. To avoid leading the reader into the murky depths of TeX commands, we will simply point out that the above macro definition is more or less equivalent to the LaTeX command sequence

```
\newcommand{\centerline}[1]{\makebox[\textwidth][c]{#1}}
```

Now take the following sample input text

```
\documentclass{article}
\begin{document}
\centerline{This is an \invalid command}
\end{document}
```

in which a \ has been placed before the word invalid, making a false command named \invalid. During the LaTeX processing, the following TeX error message appears:

```
! Undefined control sequence.
<argument> This is an \invalid
                                command
l.3 \centerline{This is an \invalid command}

?
```

This TeX message is now a bit more understandable. The error indicator is the same as for the example in Section 9.1.1:

```
! Undefined control sequence.
```

The following pair of lines claim that the error was recognized after the 'command' \invalid was processed, and that the next piece of text to be read in is the word command. At the same time, the <argument> at the beginning of the upper line says that this text is the argument of some other command.

The next pair of lines are familiar: the error occurred in line 3 of the input text, after the entire line, command and argument, had been read in for processing.

9.2   Some sample errors

9.2.1   Error propagation

The example with the incorrect \begin{qoute} environment showed that a simple response with the ⟨return⟩ key led to a second error message in spite of the fact that the \end{quote} command that produced it was correct. That one wrongly corrected error leads to further errors is more often the rule than the exception.

Let us look at the input text

```
\documentclass{article}
\begin{document}
\begin{itemie}
  \item This is the first point in the list
  \item And here comes the second
\end{itemize}
\end{document}
```

The only mistake in this text is the misspelled environment name itemie in place of itemize. The LaTeX processing first produces the same error message as for the incorrect quote environment earlier:

```
! LaTeX Error: Environment itemie undefined.

See the LaTeX manual or LaTeX Companion for explanation.
Type  H <return>  for immediate help.
 ...

1.3 \begin{itemie}

?
```

At this point, entering ⟨return⟩ leads to a new error messages:

```
! LaTeX Error: Lonely \item--perhaps a missing list environment.

See the LaTeX manual or LaTeX Companion for explanation.
Type  H <return>  for immediate help.
```

```
   ...

1.4    \item T
            his is the first point in the list
?
```

What has happened here is that without the \begin command, the \item has been issued outside of a list environment, where it is rather meaning-less. (Actually, it does have a meaning here: it prints an error message!) It is now too late to try to insert the missing start of the itemize envi-ronment. Typing H ⟨return⟩ for help yields

```
Try typing  <return>  to proceed.
If that doesn't work, type  X <return>  to quit.
?
```

Following this advice and pressing ⟨return⟩, one obtains the same er-ror message once again, but this time on line 5, for the second \item command. Persevering further and pressing ⟨return⟩ once again, one is presented with

```
! LaTeX Error: \begin{document} ended by \end{itemize}.

See the LaTeX manual or LaTeX Companion for explanation.
Type  H <return>  for immediate help.
   ...

1.6 \end{itemize}

?
```

The itemize environment has now come to its end, but since it was never properly started, LaTeX complains about a mismatch of \begin and \end commands. Now at last, typing another ⟨return⟩ leads to a proper continuation of the processing. Of course the itemized text will be incorrectly formatted, but the rest of the document will not be affected.

In this example, a single mistake in the input text generated three further error messages. This is by no means unusual. Some LaTeX errors can produce hundreds of such successive errors. It is even possible that the chain of errors never ceases and that the processing never advances. In this case, there is nothing else to do but to terminate the program. This should be done with the entry I\stop after the next error message. It may be necessary to give it several times before it takes effect. If this does not work, that is, if the same message appears every time, then the response X ⟨return⟩ will halt the program immediately.

It is better to stop the program with I\stop than with X, since then the output will include the last page being processed. This can be of assistance in trying to deduce what the source of the error was.

The final lesson of this section is simply: *even when faced with a host of error messages, don't panic! Continue to press the return key to advance the processing.*

Pressing S ⟨*return*⟩ instead will produce the same set of error messages on the screen but without waiting for a user response in between (Section 9.1.1).

9.2.2   Typical serious errors

Occasionally it may happen that a user forgets to include one of the commands \documentclass or \begin{document}, or the entire preamble. The last can easily occur when a LaTeX file is to be read in with \input or \include, but is mistakenly entered directly with the LaTeX program call. If a file with the text

```
This file has no preamble.
```

is put directly into the LaTeX processing, the following message appears on the screen:

```
! LaTeX Error: Missing \begin{document}.

See the LaTeX manual or LaTeX Companion for explanation.
Type  H <return>  for immediate help.
 ...

1.1 T
    his file has no preamble.
?
```

with the accompanying help message after typing H ⟨*return*⟩:

```
You're in trouble here.  Try typing  <return>  to proceed.
If that doesn't work, type  X <return>  to quit.
?
```

One sees from the error message that LaTeX has discovered a mistake on reading the very first letter of the text. The help message is also not very encouraging. There is no point trying to continue the processing with ⟨*return*⟩. Rather one should stop it with X or E right away, since nothing useful can be achieved here.

Even when the above example file contains the environment

```
\begin{document}
    This file has no preamble.
\end{document}
```

it is not possible to carry out a proper processing. The TeX error message now printed is

```
! LaTeX Error: The font size command \normalsize is not defined:
            there is probably something wrong with the class file.

See the LaTeX manual or LaTeX Companion for explanation.
Type  H <return>  for immediate help.
 ...

1.1 \begin{document}

?
```

The error indicator is rather peculiar since the command \normalsize never appears in the input. Only the last two lines of this message are easily understood. They state that the error was recognized on line 1 after \begin{document} was read in. The fact that this is the text of the first line is enough to show that no meaningful processing can take place, since the mandatory \documentclass command that must precede it is missing.

Here again there is no choice but to halt the program with X or E, since a continuation of the processing would be meaningless.

(The reason for the strange error indicator is that many formatting parameters are initiated by the \begin{document} statement, including the standard font. It is here that \normalfont is called internally, and that command must be defined in the class file, which is missing. Other initiating commands are defined in the LATEX format itself, and so they do not cause this error.)

If the name of the document class is wrongly typed, say as

```
\documentclass{fred}
```

the following message appears:

```
! LaTeX Error: File 'fred.cls' not found.

Type X to quit or <RETURN> to proceed,
or enter new name. (Default extension: cls)

Enter file name:
```

This message should be fairly clear: the program is looking for a file with the name fred.cls but cannot find it, so would the user be so kind as to enter an alternative name. If the new name has the same extension as that requested (.cls), it need not be explicitly given. In this case, the name of any standard LATEX class, such as article, report, book, or letter, may be typed in, as well as that of any additional classes that you might have access to. Of course, it should be the class for which the document was written.

This same error message is printed whenever a file is to be read in that cannot be found on the system. The issuing command may be \input,

\include, or \usepackage. If the file sought really does exist, perhaps it is not located where TeX is looking for files. Usually there is a system parameter set by the installation to point TeX to its files. If your file is somewhere else, you can trying entering its full name, with directory or path designations.

If LaTeX is asking for a file that you know does not exist, or which is totally unknown to you, you will want to halt the processing right away or tell LaTeX to skip the silly file. At this point, you have a problem. LaTeX insists on a valid file name. Typing X ⟨*return*⟩ only produces the message once again, this time stating that it cannot find the file x.*ext*. Some installations provide a dummy file named null.tex, while one of the extension packages available (Section D.3.3) offers a set of files named after the standard response letters x.tex, e.tex, r.tex, h.tex, s.tex to emulate the responses to normal error messages.

Emergency stop: sometimes one reaches a situation where the program cannot be halted after an error, even with I\stop or X. In such a case, one has to apply the operating system's program interrupt. Details from the computer center or PC manuals.

9.2.3 Mathematical errors

Surprisingly few errors occur as a result of incorrect application of mathematical formula commands themselves, even for a user with only a little experience. More often, the errors in mathematical formulas are fiddly ones such as forgetting a closing brace } or neglecting to switch back to text mode. Another type of common error is to use a symbol in text mode that may only appear in math mode. We will point out a few typical examples here.

One wants to produce the text: 'The price is $3.50 and the order number is type_sample', and types the input text:

```
The price is $3.50 and the order number is type_sample.
```

This text contains two errors, such that the first cancels the second one: the $ sign is the math mode switch for generating formulas within text (Section 5.1). The proper way to write a true dollar sign in text is as \$. Instead, in the sample text, the $ alone switches to math mode, setting everything that follows as a formula. However, the closing switch-back $ is missing, something that TeX will first notice at the end of the current paragraph.

```
! Missing $ inserted.
<inserted text>
                $
l.4

?
```

If ⟨*return*⟩ is simply pressed as a response, TEX inserts a $ at this point. For our sample text, this would be at the end of the text before the blank line. This means that the text from the first $ to the end of the paragraph will be set as a text formula:

The price is $3.50 and the order number is type_sample.$

This output should show the user right away what went wrong. Near the end of the line, there is a letter s printed as an index. This is the second error in the example. The underbar character _ is only permitted in math mode and should have been typed here as _. However, after the first $ had switched (incorrectly) into math mode, the _ sign became allowed, with the result that the following letter s was lowered.

If we now correct the text by replacing the $ sign with \$, then the _ symbol is no longer in math mode, producing the error message:

```
! Missing $ inserted.
<inserted text>
                $
l.5 ...ice is \$3.50 and the order number is type_
                                                    sample.
?
```

Pressing ⟨*return*⟩ here tells TEX to recover from the error by inserting the apparently missing $ sign at this point before the math command _. The processing continues, and at some place before the end of the current paragraph, TEX will notice again that the closing $ is not present. The same error message is printed as in the first case. The processing goes on with another ⟨*return*⟩ response, with the result:

The price is $3.50 and the order number is type_sample.$

In all three cases, asking for more help with H would type to the screen:

```
I've inserted a begin-math/end-math symbol since I think
you left one out. Proceed, with fingers crossed.
```

The last sentence above is just what we recommend for mathematical error messages: *keep on pressing ⟨return⟩ or S to get through to the end of the processing, and then look at the printed output to find the mistake.*

9.2.4   Errors from multi-file texts

If the document text is split over many files that are to be read in with \input or \include commands, the line number in the error message refers to the file currently being read. A response with E ⟨*return*⟩ will call the editor program and open that file, going to the indicated line where the error was recognized. With the other responses, the editor has to be called separately, with the faulty file explicitly named.

It is possible to determine which file was being processed at the time of the error by examining the processing messages or the .log file. As the files are opened for reading, TEX writes an opening parenthesis (and the name of the file to the monitor screen and to the .log file. When the file is closed, a closing parenthesis) is printed. Output page numbers are similarly written to both the screen and .log file in square brackets. For example, if the screen contains the processing messages

```
..(myfile.tex [1] [2] [3] (part1.tex [4] [5]) (part2.tex [6] [7]
! Undefined control sequence
1.999 \finish
?
```

this can be interpreted as follows: a file myfile.tex was being read, and after pages 1, 2, and 3 had been output, the file part1.tex was read in with an \input or \include command inside the file myfile.tex. Pages 4 and 5 were output and part1.tex was closed, after which another file part2.tex was opened for input. An error was discovered in line 999 of this file. If this error is corrected from the keyboard, or if the processing is otherwise continued, the screen messages proceed as

```
[8] [9]) [10]
! Too many }'s
1.217 \em sample}
```

The closing parenthesis after page 9 indicates that the file part2.tex has been closed. The next error is on page 10 in the main file myfile.tex, since there is no closing) for it. The error was discovered on line 217 of this file.

9.3   List of LATEX error messages

The LATEX error messages are listed here, divided into general, package, or font error messages. Within each group, they are ordered alphabetically according to the error indicators. A description of the possible causes and solutions is also given.

Many messages are new to LATEX 2ε, while others have a slightly altered text from the 2.09 version. All messages in LATEX 2ε are preceded by the words ! LaTeX Error:, while in 2.09 the error identifier is prefixed only by the exclamation point and appears on the third line. If these are the only differences between the two texts, the 2.09 version is not shown.

The usual icons are used to indicate messages that are exclusive to one version or the other.

9.3.1   General LaTeX error messages

The following error messages are those not involving font selection or definition, nor any of the special features for LaTeX programming class and package files (Appendix C).

`! LaTeX Error: ... undefined.`

The argument of a `\renewcommand` or `\renewenvironment` has not been previously defined. The corresponding `\new...` command should be used instead.

`! LaTeX Error: \< in mid line.`

The command `\<` in a `tabbing` environment occurred in the middle of a line. This command may only appear at the beginning of a line (Section 4.6.3).

`! LaTeX Error: Bad \line or \vector argument.`

The first argument of a `\line` or `\vector` command specifies the angle of the line or arrow. This message states that the selected angle entries are invalid. See pages 157–159.

`! Bad use of \\.`   (LaTeX 2.09)

2.09 (This message has been replaced by `There's no line here to end` in LaTeX 2ε.)

`! LaTeX Error: Bad math environment delimiter.`

LaTeX has encountered a math switch command in the wrong mode: either a `\[` or `\(` command in math mode or the corresponding `\]` or `\)` in normal text mode. Either the math switches have been improperly paired or some braces {...} are incorrect.

`! LaTeX Error: \begin{...} on input line ... ended by \end{...}.`
`! \begin{...} ended by \end{...}.`   (LaTeX 2.09)

LaTeX has encountered an `\end` command without a corresponding `\begin` of the same name. This may be due to a typing mistake in the name of the environment, or to the omission of a previous `\end` command. A good way to avoid this error is always to enter the `\end` command immediately after the `\begin`, inserting the actual environment text with the editor in between the two. This is especially useful for long, nested environments. It also reduces the risk of typing the environment name incorrectly in the `\end` command.

`! LaTeX Error: Can be used only in preamble.`

Many commands may only be called within the preamble. These include

\documentclass, \usepackage, \nofiles, \includeonly, \makeindex, \makeglossary, and several others. Certain commands that only have meaning within class or package files, such as \ProvidesClass and \ProvidesPackage (Section C.2.1), as well as many \Declare.. and \Set.. commands are also only allowed in the preamble. If one of these commands is issued after \begin{document}, this message is printed.

! LaTeX Error: Command ... invalid in math mode.

[2ε] A command has been issued in math mode that only makes sense in text mode, such as \item or \circle. The LaTeX 2ε font commands \itshape, \bfseries, and so on, also produce this error in math mode, since the math alphabet commands \mathit, \mathbf should be used instead.

! LaTeX Error: Command ... already defined.
! Command name ... already used.   (LaTeX 2.09)

The user has tried to redefine an existing structure with one of \newenvironment, \newcommand, \newtheorem, \newsavebox, \newfont, \newlength, \newcounter, or \DeclareMathAlphabet. Either a different name must be selected or, in the case of commands and environments, the \renew... version must be employed. (Note that when an environment named sample is defined, the commands \sample and \endsample are also created.)

! LaTeX Error: Command ... undefined in encoding ....

[2ε] The specified command has been defined with \DeclareTextCommand for a certain NFSS encoding (say OT1) but was executed while another encoding (for example T1) for which there is no definition was active.

! LaTeX Error: Counter too large.

A counter that is to be printed as a letter contains a value greater than 26.

! LaTeX Error: Environment ... undefined.

LaTeX has encountered a \begin command with an unknown environment name. This is probably due to a typing mistake. It may be corrected during the processing with the response I followed by the correct name. (This does not alter the source file, where the error remains.)

! LaTeX Error: File '...' not found
Type X to quit or <RETURN> to proceed,
or enter new name. (Default extension: ...)
Enter file name:.

[2ε] A file is to be loaded with one of the inputting commands, but it cannot be found. You may type a new file name, or quit, or proceed without it.

If the file name has been given correctly, and it does exist, maybe it is not located where LaTeX looks for files. Make sure that it is in one of the right directories. The user is offered a chance to type in an alternative, or correctly typed, name via the keyboard. Type in the file name, with optional extension, and press ⟨return⟩. LaTeX proposes a default extension that is the same as that requested. There is no default for the main part of the file name.

`! LaTeX Error: Float(s) lost.`

A `figure` or `table` environment or a `\marginpar` command was given within a vertical box (`\parbox` or `minipage` environment), or these structures were issued within a LaTeX command that uses vertical boxes internally, such as a footnote. This error is first recognized by LaTeX when a page is output so that the actual cause could be many lines earlier in the text. A number of tables, figures, or marginal notes may have been lost as a result, but certainly not the one that triggered the problem.

`! LaTeX Error: Illegal character in array arg.`

A `tabular` or `array` environment contains an unknown column formatting character (see Section 4.8.1) or the formatting entry in the second argument of a `\multicolumn` command is wrong.

`! LaTeX Error: \include cannot be nested.`

2ε An attempt has been made to call `\include` from a file that has already been read in by an `\include` command. All `\include` commands must be issued in the main file (Section 8.1.2).

`! LaTeX Error: LaTeX2e command ... in LaTeX 2.09 document.`

2ε This error occurs if a LaTeX 2_ε command is used in compatibility mode, that is, if the document uses `\documentstyle` instead of `\documentclass`. Some of these are `\LaTeXe`, `\usepackage`, `\ensuremath`, the `lrbox` environment, as well as the new syntax for adding an optional argument with `\newcommand` and `\newenvironment`. The reason that these are forbidden in compatibility mode is so that the author can be sure that the document conforms to LaTeX 2.09 if it is to be sent to other installations that only have the older version.

`! LaTeX Error: Lonely \item--perhaps a missing list environment.`

An `\item` command has been given outside of a list environment (Section 4.3). Either the environment name has been misspelled in the `\begin` command and not corrected by keyboard entry, or the `\begin` command has been forgotten.

! LaTeX Error: `Missing @-exp in array arg.`

The column formatting argument of a `tabular` or `array` environment contains an @ symbol without the necessary following text in curly braces { } that must go with it (Section 4.8.1 for @-expressions), or the same thing occurs in the second argument of a `\multicolumn` command.

! LaTeX Error: `Missing begin{document}.`

Either the `\begin{document}` command has been forgotten or there is printable text inside the preamble of the document. In the latter case, there could be a declaration with incorrect syntax, such as a command argument without curly braces { } or a command name without the backslash \.

! LaTeX Error: `Missing p-arg in array arg.`

The column formatting argument of a `tabular` or `array` environment contains a p symbol without the necessary width specification that must go with it (Section 4.8.1), or the same thing occurs in the second argument of a `\multicolumn` command.

! LaTeX Error: `No counter '...' defined.`
! `No such counter.`   (LaTeX 2.09)

A `\setcounter` or `\addtocounter` command was called that referred to a counter name that does not exist. Most likely the name was typed incorrectly. If this error occurs while an .aux file is being read and if the name of the counter is indeed correct, the defining `\newcounter` command was probably given outside of the preamble. Therefore it is highly recommended always to give the `\newcounter` commands inside the preamble. (If the other LaTeX counter commands are used with an undefined counter name, a long list of funny TeX error messages appears.)

! `No theorem environment '...' defined.`   (LaTeX 2.09)

2.09 (In LaTeX 2_ε, this message is replaced by `No counter defined.`)
A `\newtheorem` command has been given with an optional argument naming another theorem-like declaration that shares the same counter, but this other theorem does not exist (Section 4.5). Either the name of the other theorem has been misspelled or it has not been previously defined.

! LaTeX Error: `No \title given.`

2_ε The `\maketitle` command has been given before `\title` has been declared.

! LaTeX Error: `Not in outer par mode.`

A `figure` or `table` environment or a `\marginpar` command was given within math mode or inside a vertical box (\parbox or minipage environment). In the first case, the math switch-back command was probably forgotten.

! LaTeX Error: Page height already too large.

[2ε] The command \enlargethispage is trying to extend the vertical size of the page, which LaTeX already considers too large.

! LaTeX Error: \pushtabs and \poptabs don't match.

The number of \poptabs commands in a tabbing environment does not agree with the number of previous \pushtabs given (Section 4.6.4).

! LaTeX Error: Something's wrong--perhaps a missing \item.

The most likely cause of this error is that the text in a list environment (list, itemize, enumerate, or description) does not begin with an \item command. It will also occur if thebibliography environment is given without the argument {*sample_label*} (Section 4.3.6).

! LaTeX Error: Suggested extra height (...) dangerously large.

[2ε] The command \enlargethispage is trying to extend the vertical size of the page more than LaTeX considers reasonable.

! LaTeX Error: Tab overflow.

The last \= command in a tabbing environment exceeded the maximum number of tab stops allowed by LaTeX.

! LaTeX Error: There's no line here to end.
! Bad use of \\.   (LaTeX 2.09)

The command \newline or \\ has been issued after a \par or blank line where it makes no sense. If additional vertical space is to be inserted here, this should be done with a \vspace command.

! LaTeX Error: This may be a LaTeX bug.

This message states that LaTeX is fully confused. This could be the result of a previous error after which the user response was to press ⟨*return*⟩. In this case, the processing should be brought to a halt with I\stop or X or E and the earlier error corrected. It is also possible, although unlikely, that there is a bug in the LaTeX program itself. If this is the first error message in the processing, and the text seems otherwise to be satisfactory, the file should be saved and submitted to the computing center for further investigation.

```
! LaTeX Error: Too deeply nested.
```

Too many list environments (`description`, `itemize`, `enumerate`, or `list`) have been nested inside one another. The maximum depth of such nesting is dependent on the installation but should always be at least four.

```
! LaTeX Error: Too many columns in eqnarray environment.
```

$\boxed{2\varepsilon}$ The `eqnarray` environment may only have three columns per line. You may have forgotten to start a new row with \\, or have placed an extra & in the line.

```
! LaTeX Error: Too many unprocessed floats.
```

This error may occur if there are too many \marginpar commands on one page. However, it is more likely that LaTeX is retaining more `figure` and `table` floats than it can hold. This happens if too many such float objects have been given before they can be output (Section 6.4). In this case, the last figure or table should be added later in the text. Another cause could be that the figure or table cannot be located on a normal text page, but rather on a special float page at the end of the text or after a \clearpage or \cleardoublepage command. Since the output sequence of figures and tables will be the same as that with which they were input, one such float can block the entire queue. A \clearpage or \cleardoublepage command can free the blockage.

```
! LaTeX Error: Undefined tab position.
```

A \>, \+, \-, or \< command in a `tabbing` environment has tried to move to a tabulator stop that does not exist (Section 4.6).

```
! LaTeX Error: \verb ended by end of line.
```

$\boxed{2\varepsilon}$ The text of an in-line verbatim command \verb+...+ extends over more than one line of text. This is permissible in LaTeX 2.09, but in LaTeX 2_ε, it is forbidden in order to catch a common error: the missing terminating character. Make sure that the entire text between the initial and terminating characters is all on one input line.

```
! LaTeX Error: \verb illegal in command argument.
```

$\boxed{2\varepsilon}$ The \verb command may not be used in the argument of any other command, except \index and \glossary. It may not be used within section titles or footnotes, for example.

9.3.2   LaTeX package errors $\boxed{2\varepsilon}$

The special programming commands for handling class and package files (Appendix C) have their own set of error messages. If one of these features issues a serious error, there is often little the user can do about it other than report the problem to the author of the file. On the other hand, some errors are due to the improper use of the file or of the options that it provides.

Often classes and packages contain their own error or warning messages with text and meanings that are peculiar to them. These are indicated as such, along with the name of the class or package that issued them. For example, the package `mypack` could print the error

```
Package mypack Error: cannot mix options 'good'
(mypack)                and 'bad'.
```

A help message should also be available when H is typed. Obviously, such error (and warning) messages cannot be explained here, since they depend entirely on the package in question.

! LaTeX Error: \LoadClass in package file.

A package file has called `\LoadClass`, which it is not allowed to do. A class file can only be loaded from another class file.

! LaTeX Error: Option clash for package ....

The specified package has been requested a second time with a different set of options. A package file will only be loaded once and a second attempt will be ignored. Thus, if two `\usepackage` or `\RequirePackage` commands load the same package with different options, there is a conflict. Try to arrange for a consistent set. Typing H for help after this message will print out the two sets of options.

! LaTeX Error: \RequirePackage or \LoadClass in Options Section.

These two commands may not appear in the definition of a class or package option made with the `\DeclareOption` command. Instead, the option should set some flag or other indicator that is later tested before calling the command in question.

! LaTeX Error: This file needs format '...' but this is '...'.

The `\NeedsTeXFormat` command specifies a different format from the one being used. The *format* is the prestored set of instructions that determine what type of TeX is being run. For LaTeX2_ε, the format name specified must be `LaTeX2e`. This file cannot be processed at all with the current format in use.

! LaTeX Error: Two \documentclass or \documentstyle commands.

A document may contain only one `\documentclass` or `\documentstyle` command. If you can only see one in the main document file, check that other files being loaded do not contain an offending second one.

! LaTeX Error: Two \LoadClass commands.

The class file contains more than one \LoadClass command, something that is not allowed. The class file has been improperly written.

! LaTeX Error: Unknown option '...' for package '...'.

An option has been specified with \usepackage but it is not defined for that package. Check the instructions for the package or look for a misprint.

! LaTeX Error: \usepackage before \documentclass.

The \usepackage command may not appear before \documentclass. A class file must be loaded before any packages.

9.3.3   LaTeX font errors [2ε]

The following error messages occur when defining or selecting fonts with the New Font Selection Scheme (Sections 8.5 and C.5). Some of these messages indicate that the fonts have not been properly set up, in which case the font description files may be corrupted or contain mistakes. In either case, the system manager must see that a proper set of files are installed.

! LaTeX Error: ... allowed only in math mode.

A math alphabet command such as \mathbf has been used in text mode. Perhaps a $ has been forgotten.

! LaTeX Error: Command '...' not defined as a math alphabet.

The name of a non-existent math alphabet has been specified in one of the \Set.. or \Declare.. commands that take a math alphabet name as argument. The math alphabet name must be declared with the command \DeclareMathVersion before it may be used.

! LaTeX Error: Command ... not provided in base LaTeX2e.

A number of symbols that were part of the basic LaTeX 2.09, but not in TeX, are no longer automatically included in LaTeX 2ε. Add one of the packages latexsym or amsfonts to include them.

! LaTeX Error: \DeclareTextComposite used on inappropriate
                command ....

\DeclareTextComposite (Section C.5.7) is used to redefine an existing accenting command to print a single character for a certain encoding. If the accenting command does not already exist, this message is issued.

`! LaTeX Error: Encoding scheme '...' unknown.`

A font declaration or selection command that refers to a non-existent encoding scheme has been issued. There is probably a typing error involved here.

`! LaTeX Error: Font family '...+...' unknown.`

A `\DeclareFontShape` command has been issued for a font encoding and family combination that has not been previously declared by means of a `\DeclareFontFamily` command (Section C.5.6).

`! LaTeX Error: Font ... not found.`

The font with the specified attributes could not be found, nor could an adequate substitution be made. The font defined by `\DeclareErrorFont` is used instead.

`! LaTeX Error: Math alphabet identifier ... is undefined`
`              in math version '...'.`

A math alphabet (Section C.5.3) that has not been defined for the current math version has been invoked. This means that the alphabet has been created by `\DeclareMathAlphabet` with an empty shape attribute, but no `\SetMathAlphabet` declaration has been given to define it for the selected math version.

`! LaTeX Error: Math version '...' is not defined.`

The name of a non-existent math version has been specified in one of the `\Set...` or `\Declare...` commands that take a math version name as argument. The version name must be declared with `\DeclareMathVersion` before it may be used.

`! LaTeX Error: *** NFSS release 1 command ... found`
`              *** Recovery not possible. Use ....`

A command from the first version of NFSS has been used, which is no longer valid. Use the suggested replacement command instead.

`! LaTeX Error: Not a command name: '...'.`

The first argument of `\DeclareMathAccent` must be a command name, like `\acute`. Most likely the backslash `\` character has been forgotten.

`! LaTeX Error: Symbol font '...' not defined.`

The name of a non-existent symbol font has been specified in one of the `\Set..` or `\Declare..` commands that take a symbol font name as argument. The font name must be declared with `\DeclareSymbolFont` before it may be used as a symbol font.

> ! LaTeX Error: The font size command \normalsize is not defined:
>   there is probably something wrong with class file.

It is necessary for class files to define \normalsize, the basic size of standard text in the document. This is a change from the practice in LaTeX 2.09. If the class file does not do this, if it only defines \@normalsize, then it must be repaired. This message also appears if \documentclass is missing, for an empty class file is clearly a faulty one.

> ! LaTeX Error: This NFSS system isn't set up properly.

Something is wrong with the font descriptions in the .fd files, or there is no valid font declared by \DeclareErrorFont. This is a serious problem, to be reported to the system manager.

> ! LaTeX Error: Too many math alphabets used in version ....

There is a limit of 16 math alphabets possible, a number set by TeX itself. Any additional math alphabet definitions are ignored.

> ! LaTeX Error: Unknown symbol font '...'.

The name of a non-existent symbol font has been used as argument to the command \DeclareSymbolFontAlphabet. The font name must first be declared with \DeclareSymbolFont before it may be used.

9.4   TeX error messages

This section contains a list of some of the most common TeX error messages, ordered alphabetically according to the error indicator, together with a brief description of the possible cause. Each one begins with an exclamation point only. Since these messages are generated by TeX, there is no difference between LaTeX 2.09 and LaTeX 2_ε.

> ! Counter too large.

As a TeX error, this message refers to a footnote marked either with letters or symbols in which the counter has exceeded a value of 26 or 9. It may also occur if there are too many \thanks commands on a title page.

> ! Double subscript.

A mathematical formula contains two subscripts for the same variable, for example, x_2_3 or x_{2}_{3}. The proper way to produce x_{2_3} is with x_{2_3} or x_{2_{3}} (Section 5.2.2).

> ! Double superscript.

A mathematical formula contains two superscripts for the same variable, for example, x^2^3 or x^{2}^{3}. The proper way to produce x^{2^3} is with x^{2^3} or x^{2^{3}} (Section 5.2.2).

! Extra alignment tab has been changed to \cr.

A line in a tabular or array environment contains more & commands than there are columns defined. The error is probably due to a forgotten \\ at the end of the previous line.

! Extra }, or forgotten $.

In math mode, either an opening brace { has been left off or an extra closing brace } has been included by mistake. Another possibility is that a math mode switch command such as $, \[, or \(has been forgotten.

! Font ... not loaded: Not enough room left.

The text processing requires more character fonts to be loaded than TₑX can handle owing to memory limitations. If certain parts of the document need different fonts, you can try to split it up and to process the parts separately.

! I can't find file '...'.

This message has actually been replaced by the LATₑX 2ₑ equivalent File ... not found, and functions much the same way. The specified file is to be read in but it cannot be found in the directories being searched. The user has an opportunity to type in an alternative (or the correctly spelled) name. After printing the error indicator, TₑX asks for the name of another file to read:

Please type another input file name:

and then waits for you to enter the new name and to press ⟨return⟩. Contrary to LATₑX 2ₑ, the default extension here is always .tex and not the extension requested.

! Illegal parameter number in definition of ....

This error message is probably due to a \newcommand, \newenvironment, \renewcommand, or \renewenvironment command in which the substitution character # has been applied incorrectly. This character may appear within the defining text only in the form #n, where n is a number between 1 and the number of arguments specified in the command. Otherwise the character # may only appear in the definition as \#. This error may also arise if the substitution character is applied within the last argument {end_def} (Section 7.4).

! Illegal unit of measure (pt inserted).

If this error message comes right after another error with the message

! Missing number, treated as zero.

the problem lies with this previous error (see below). Otherwise, the mistake is that TeX expects a length specification at this point but has only been given a number without a length unit. This occurs most often when a length is to be set to zero and 0 has been typed in instead of 0mm or 0pt. If this is the case, then responding with ⟨*return*⟩ produces the right result since for a zero value any unit specification is all right. This error may also occur if a length specification has been completely forgotten.

! Misplaced alignment tab character &.

The single character command & has been given in normal text outside of the tabular and array environments. Possibly the intention was to print &, in which case \& should be typed. This may still be achieved during the processing by responding with I\& to the error message.

! Missing control sequence inserted.

This error is most likely caused by a \newcommand, \renewcommand, \newlength, or \newsavebox command in which the backslash \ is missing from the first argument. Pressing ⟨*return*⟩ as response will complete the processing correctly since TeX assumes a backslash is missing and inserts it.

! Missing number, treated as zero.

This error is most likely due to a LaTeX command that expects a number or length as argument, which is missing. It may also occur when a command that takes an optional argument is followed by text beginning with [. Finally another cause can be a \protect command preceding a length or \value command.

! Missing { inserted.
! Missing } inserted.

TeX is totally confused when either of these error indicators is written. The line number printed will probably not be where the source of the error is to be found, which is a missing opening or closing brace. If the error is not obvious, continue the processing by pressing ⟨*return*⟩ and try to deduce where it might be from the printed output.

! Missing $ inserted.

Most likely a symbol or command that may only appear in math mode was used in normal text. Recall that all those commands described in Chapter 5 are only allowed in math mode unless otherwise stated. If an \mbox command is inserted within a math formula, its argument has temporarily exited from math to normal text mode. The message may also occur if a blank line appears within a math formula, signalling a new paragraph and the end of the formula without the necessary $ sign.

`! Not a letter.`

The word list in a `\hyphenation` command contains a character that is not recognized as a letter, such as an accent command like `\'{e}`. Such words can only be hyphenated by explicitly inserting the hyphenation possibilities (Sections 3.6.1 and 3.6.2).

`! Paragraph ended before ... was complete.`

The argument of a command contains a blank line or a `\par` command, something that is not allowed. Probably a closing brace } has been omitted.

`! \scriptfont ... is undefined (character ...).`
`! \scriptscriptfont ... is undefined (character ...).`
`! \textfont ... is undefined (character ...).`

These messages arise in LaTeX 2.09 when a math formula makes use of symbols in a font which was not designed for math mode, such as `\sf` (sans serif). Such character fonts must be made available with the command `\load{`*size*`}{`*style*`}`, for example `\load{\normalsize}{\sf}`. If such a formula is to appear in a footnote, it is necessary to give `\load {\footnotesize}{\sf}` beforehand. In LaTeX 2_ε, this problem does not arise with the math alphabet commands such as `\mathsf`.

`! TeX capacity exceeded, sorry [...].`

TeX sets up various storage buffers in the computer memory to carry out the text processing. This message appears when one of these buffers is full and can no longer be used. The name of the buffer and its maximum size are printed in the square brackets of the error indicator. With this message, the TeX processing is terminated. The source of this problem is hardly ever due to insufficient memory, no matter how much complicated text is being processed, but rather to an error in the text itself. The methods described in Section 9.6 may be applied to try to detect the true error.

The following descriptions of the various buffers should help to decide whether the storage capacity allotted to TeX really is too small and to explain what one might do to correct this.

`buffer size` The problem here can be that the text in the argument of a sectioning, `\caption`, `\addcontentsline`, or `\addtocontents` command is too long. The message then normally appears when the `\end{document}` command is processed, but may also arise at one of `\tableofcontents`, `\listoffigures`, or `\listoftables`. The way to avoid this is to use the optional argument for the short form of the heading text (Sections 3.3.3 and 6.4.4). Indeed, such a long entry in the table of contents is a nuisance anyway and should be shortened. After the correction has been made in the input text, the previous LaTeX `.aux` file must be deleted before reprocessing.

This problem can occur on a PC if a word processing program has been used to generate the input text instead of a text editor. Some of these programs put an entire paragraph into a single line even though the text on the screen is broken up into lines.

`exception dictionary` The list of hyphenation exceptions that have been entered with the \hyphenation commands has become too large. Words used less frequently should be removed and their possible word divisions indicated explicitly with the \- command.

`hash size` The source file contains too many command definitions or uses too many cross-reference markers. This does not mean that the input text really needs all these commands, for it may be that the user has developed a large collection of private commands that are stored in a single file and read into every document, whether they are all applicable or not.

`input stack size` An overflow of this buffer is probably due to a mistake in a command definition. For example, the command defined with

```
\newcommand{\com}{One more \com}
```

produces One more {One more {...One more \com}...}} going on for ever, since it continually calls itself. Actually, it does not go on for ever, but only until this buffer is full.

`main memory size` This buffer contains the text for the page currently being processed. It also overflows if a recursively defined command has been called. However, the more usual reasons are: (1) a large number of very complicated commands have been defined on one page; (2) there are too many \index or \glossary commands on one page; (3) the page itself is too complex to fit within the allotted buffer space.

The solution to the first two situations is clear: reduce the number of command definitions and/or \index and \glossary commands on that page. In the third case, the cause might be a long tabbing, tabular, array, or picture environment, or a stuck float (figure or table) waiting for an output command.

To find out whether the memory overflow is really due to an overly complex page, add the command \clearpage just before the spot where the overflow occurs. If the error message no longer appears on the next processing run, then this page was indeed too complicated for TeX. However, if the overflow still persists, the error is probably a mistake in the input text. If necessary, it may have to be located with the method explained in Section 9.6.

If the page really is too complex for the TeX processing, it must be simplified. However, first recall that the entire last paragraph is processed before that page is output, even if the page break occurs near the beginning of the paragraph. Introducing a \newpage may solve the problem and should be tried out before attempting a tedious restructuring of the text. If the error is due to a stuck figure or table, it might be alleviated by moving the float object to later in the text, or by changing the positioning argument

(Section 6.4.1). If the whole text is not yet complete, one can try giving
\clearpage for now in order to clear the blocked floats and then decide
on a better ordering when the text is finalized.

pool size Most likely there are too many command definitions and/or labels,
or their names are too long. Try shortening the names. This error may
also occur if a closing right brace } has been forgotten in the argument
of a counter command such as \setcounter or in a \newenvironment or
\newtheorem command.

save size This buffer overflows when commands, environments, and the scope
of declarations are nested too deeply. For example, if the argument of a
\multiput command contains a picture environment, which in turn possesses a \footnotesize declaration with another \multiput command,
and so on. Such a nesting must be simplified, unless the real problem
is a forgotten closing brace } that merely makes the structure appear so
complex.

! Text line contains an invalid character.

The input text contains a strange symbol that TEX does not recognize.
This could be a problem with the editor itself, that it is inserting extra
characters. If an examination of the source file does not reveal the strange
symbol, consult the computing center for help.

! Undefined control sequence.

Every TEX user encounters this error message at some point. It is usually
the result of an incorrectly typed command name. It may be amended
during the processing by responding with I and the proper name of the
command, plus ⟨return⟩. This does not alter the source file, which must
be corrected separately after the LATEX run. If the command name has
been entered correctly in a LATEX command, it may be that it was issued
in an improper environment where it is not allowed (that is, not defined
there).

! Use of ... doesn't match its definition.

If '...' is the name of a LATEX command, it is likely that one of the picture
commands from Sections 6.1.3 and 6.1.4 has been called with the wrong
syntax for its arguments. If the name is \@array, there is a faulty @-
expression (Section 4.8.1) in a tabular or array environment. Possibly
a fragile command was given in the @-expression without the \protect
command preceding it.

! You can't use 'macro parameter #' in ... mode.

The special symbol # has been used in normal text. Probably there should
have been a \# in order to print # itself. This can be corrected during the
processing with the response I\# and ⟨return⟩.

9.5 Warnings

TeX and LaTeX errors both bring the processing run to a temporary stop and wait for a reaction from the user, or they may halt the program completely. Warnings, on the other hand, merely inform the user that the processed output may contain some faults that he or she might want to correct. Warnings appear on the screen along with the page number where they occur, without the program coming to a stop. They are also written to the .log file where they may be examined after the LaTeX processing and possible printing. Warnings may be issued either by LaTeX or by TeX itself.

9.5.1 General LaTeX warnings

LaTeX warnings are indicated by the words 'LaTeX Warning:' at their start, followed by the warning message itself.

```
LaTeX Warning: Citation '...' on page ... undefined on
                input line ....
```

The keyword in a \cite command has not yet been defined with a corresponding \bibitem command (Sections 4.3.6 and 8.3.3).

```
LaTeX Warning: Citation '... ' undefined on input line ....
```

$\boxed{2_\varepsilon}$ The keyword in a \nocite command has not yet been defined.

```
LaTeX Warning: Command ... has changed.
                Check if current package is valid.
```

$\boxed{2_\varepsilon}$ The \CheckCommand statement is used to test if a given command has a certain definition. This warning is issued if the test fails. This is used to be sure that a given package is doing what the programmer thinks it is doing.

```
LaTeX Warning: Float too large for page by ..pt on input line...
```

$\boxed{2_\varepsilon}$ A float (figure or table environment) is too big to fit on the page. It will be printed anyway, but will extend beyond the normal page margins.

```
LaTeX Warning: 'h' float specifier changed to 'ht'.
```

$\boxed{2_\varepsilon}$ The float placement specifier 'h' for 'here' (Section 6.4.1) does not really mean exactly at the location of the float environment. It will be placed 'here' only if there is enough room remaining on the page for it, otherwise it appears at the top of the next available page. This message is a reminder of this fact.

```
LaTeX Warning: inputting '...' instead of obsolete '...'.
```

[2ε] Some packages written for LaTeX 2.09 explicitly input certain files, such as `article.sty`, that have a new equivalent with a different name (in this case `article.cls`). Dummy files with the old names that issue this message and input the right file are provided.

```
LaTeX Warning: Label '...' multiply defined.
```

Two `\label` or `\bibitem` commands have defined markers with the same name (Sections 8.3.1 and 8.3.3). Even after the correction has been made, this message will be printed once more to the screen since the information that led to it is to be found in the .aux file from the last processing run. On the run after that, it should be gone.

```
LaTeX Warning: Label(s) may have changed.
               Rerun to get cross-references right.
```

The printed outputs from the `\ref`, `\pageref`, and `\cite` commands may be incorrect since their values have been altered during the processing. LaTeX must be run once more so that the right values are used.

```
LaTeX Warning: Marginpar on page ... moved.
```

A marginal note on the given page has been shifted downwards to prevent it from being too close to another one. This means that it will not be beside the line of text where the `\marginpar` command was actually given.

```
LaTeX Warning: No \author given.
```

[2ε] The `\maketitle` command has been issued without a previous `\author` command. Unlike a missing `\title` command, this is not an error, but merely peculiar. The warning is printed just in case you did mean to specify an author.

```
LaTeX Warning: Optional argument of \twocolumn too tall on page.
```

[2ε] The `\twocolumn` command (Section 3.2.5) starts a new page and switches to two-column formatting. The text in the optional argument is printed in one wide column above the double column text. If that text is too large to fit onto one page, this warning is printed.

```
LaTeX Warning: Oval too small on input line ....
```

The size specified for an `\oval` command is too small for LaTeX to find an appropriate quarter circle for it. (As of December 1, 1997, this warning has been replaced by the following one.)

```
LaTeX Warning: \oval, \circle, or \line size unavailable on
                input line ....
```

[2ε] The size specified for an \oval, \circle, or slanted \line command in a picture environment is too small for LaTeX to print. (First included as of December 1, 1997.)

```
LaTeX Warning: Reference '...' on page ... undefined on
                input line ....
```

The marker name in a \ref or \pageref command has not been defined by a \label command in the previous processing run (Section 8.3.1). If this message does not disappear on the next run, the corresponding \label command is missing.

```
LaTeX Warning: Text page ... contains only floats.
```

[2ε] This message points out that the float style parameters (Section 6.4.3) have excluded any regular text from appearing on the specified page. This is not necessarily bad, but you might want to check that page visually.

```
LaTeX Warning: There were multiply-defined labels.
```

[2ε] This message is printed at the end of the LaTeX run if there were any \label or \bibitem commands that used the same marker name more than once. A warning is also printed near the beginning of the run for each repeated marker name.

```
LaTeX Warning: There were undefined references.
```

[2ε] This message is printed at the end of the LaTeX run if there were any \ref or \pageref commands whose markers had not been defined during a previous run. A warning is also printed earlier for each undefined marker used.

9.5.2   LaTeX package warnings [2ε]

The class and package warnings are those that check the names and versions of the loaded files against those requested, or the use of options, or whether the filecontents environment has written some text to a file. The commands involved are described in Section C.2.9.

Classes and packages can also issue their own warnings which are peculiar to them. These cannot be listed here, since they depend entirely on the class or package itself.

```
LaTeX Warning: File '...' already exists on the system.
```

```
                    Not generating it from this source.
```

The `filecontents` environment is *not* extracting text from the main file because it has discovered a file of the same name already on the system.

```
LaTeX Warning: Unused global option(s):.
```

Options specified in the `\documentclass` statement are *global*, meaning they can apply to the class and/or to any following packages. However, if no class or package recognizes any of these options, this warning is printed, followed by a list of the unused options.

```
LaTeX Warning: Writing file '...'.
```

The `filecontents` environment (Section C.2.9) is extracting text out of the main file and writing it to a file of the given name.

```
LaTeX Warning: You have requested class/package '...',
               but the class/package provides '...'.
```

The name of the class or package as given by its internal identifying command `\ProvidesClass` or `\ProvidesPackage` does not agree with the file that was read in with `\usepackage` or `\RequirePackage`.

```
LaTeX Warning: You have requested, on input line ..., version
               '...' of class/package ...,
               but only version '...' is available.
```

The date of a class or package file, as given by its internal identifying command `\ProvidesClass` or `\ProvidesPackage`, is earlier than that asked for by the `\usepackage` or `\RequirePackage` command. The class or package may not have all the features that the inputting file expects.

```
LaTeX Warning: You have requested release '...' of LaTeX,
               but only release '...' is available.
```

The date of your LaTeX version is earlier than that specified for some input file in a `\NeedsTeXFormat` command (Section C.2.1). Your LaTeX may not provide all the features needed by that file.

9.5.3   LaTeX font warnings $\boxed{2\varepsilon}$

Font warnings are those involving the NFSS commands (Sections 8.5 and C.5). They are indicated by the text: `LaTeX Font Warning:` plus the warning text.

```
LaTeX Font Warning: Command ... invalid in math mode.
```

A command that should only appear in text mode has been given in math mode. The command is simply ignored. The commands `\mathversion`,

\boldmath, \unboldmath, and \em lead to this message. There are other commands that produce an *error* message with the same text.

```
LaTeX Font Warning: Command \tracingfonts not provided.
(Font)              Use the 'tracefnt' package.
(Font)              Command found: on input line ....
```

The font tracing diagnostic tool \tracingfonts may only be used if the tracefnt package (page 242) has been loaded. Otherwise, this command is ignored.

```
LaTeX Font Warning: Encoding '...' has changed to '...' for
(Font)              symbol font '...' in the math version '...'.
```

To make use of the specified symbol font in the given math version, it was necessary to change the font encoding temporarily.

```
LaTeX Font Warning: Font shape '...'in size <...> not available
(Font)              size <...> substituted.
```

No font has been defined for the size and shape requested, so a substitute size is used instead.

```
LaTeX Font Warning: Font shape '...' undefined
(Font)              using '...' instead.
```

The shape attribute that has been requested is unknown, or has not been defined, so a substitute shape will be used instead.

```
LaTeX Font Warning: *** NFSS release 1 command ... found
(Font)              *** Update by using release 2 command ....
(Font)              *** Recovery is probably possible.
```

A command from the first version of NFSS has been used, which is no longer valid. However, LaTeX will try to insert the newer equivalent and proceed with it.

9.5.4   TeX warnings

A TeX warning is recognized by the fact that it is not an error message (not prefixed with !) and that the processing is not halted. The most common TeX warnings are:

```
Overfull \hbox ....
```

TeX could not break this line in a reasonable way so part of it will extend into the right margin. The rest of the information in the message can be of assistance.

For example, with the complete warning message

!

```
Overfull \hbox (17.2122pt too wide) in paragraph at lines 4--6
[]\OT1/cmr/m/n/10 If T[]X can-not find an ap-pro-pri-ate
 spot to di-vide a word at the end of the line, as right
 here aaaaaaaaaaaaaaaaaaaaaaaaaaa
```

one knows that the line is about 17.2 pt (6 mm) too long and extends this amount into the right-hand margin. The line is part of the paragraph in lines 4 to 6. The font used is designated by its attributes \OT1\cmr/m/n/10. The text of the problem line is If ... ... right here aaaaaaaaaaaaaaaaaaaaaaaaaaa. Possible word divisions are shown with hyphens, such as di-vide. The last word cannot be divided, which is the cause of the problem.

A way around this is to include some suggested hyphenations with the \-command, such as aaaaaaaaa\-aaaaaaaaaaaaa. Similarly, a \linebreak command before the problem word, or setting the whole paragraph in a sloppypar environment, will get rid of this warning message.

If a badly broken line extends only a tiny bit into the right margin, say 1 pt or less, in most cases this will hardly be noticed and may be left as it is. A sample output should be printed just to verify the appearance.

```
Overfull \vbox ....
```

This warning occurs very rarely. TeX could not break the page properly so that the text extends beyond the bottom of the page. More often TeX sets less text on a page than too much. Thus this warning arises only when the page contains a very large vertical box, higher than the value of \textheight, such as a long table.

```
Underfull \hbox ....
```

This is the opposite of the Overfull \hbox warning. It appears when TeX has filled a line right and left justified, but with so much interword spacing that it considers the appearance to be undesirable. This is often the result of a sloppypar environment, a \sloppy declaration, or a \linebreak command. It may also come about after an inappropriate application of a \\ or \newline command, such as two \\ commands one after the other. The additional information in the warning message contains the text of the badly formatted line plus an evaluation of the 'badness' of the word spacing.

If in the example of the Overfull \hbox a \linebreak is included in the text at the spot ..., as right\linebreak here, then the warning becomes

```
Underfull \hbox (badness 5504) in paragraph at lines 4--6
[]\OT1/cmr/m/n/10 If T[]X can-not find an ap-pro-pri-ate
 spot to di-vide a word at the end of the line, as right
```

which states that the paragraph in lines 4 to 6 contains an output line in which the interword spacing may be unacceptably wide. The text of this line reads If ... ..., as right and is set in the font \OT1/cmr/m/n/10. The evaluation badness 5504 is TeX's estimate of how unacceptable the spacing is: *the smaller this number the better.*

In order to appreciate the magnitude of this 'badness', it is necessary to know how it is calculated. All interword spacings have a basic size and optimal amounts by which they may be stretched or shrunk. For each output line, the sum of the actual word spacings as well as those of the ideal stretches and shrinkages are found. The value of the 'badness' is then

badness = $100 \times$ (actual shrinkage/optimal shrinkage)^3 or
badness = $100 \times$ (actual stretch/optimal stretch)^3

where 'actual shrinkage' and 'actual stretch' are the sums of the additional or subtracted spacings to make the line justified at both ends.

Normally TEX tolerates an over-stretching of the word spacings up to a value of badness = 200, which corresponds to a factor of 1.26 between the true and ideal stretch values. Within the sloppypar environment or after a \sloppy declaration, lines may be infinitely stretched. An Underfull \hbox warning appears when the badness exceeds a value of 1000, which means the stretching has reached a value 2.15 times its ideal. The badness value is limited to 10 000 if the above formula yields a number greater than this.

In practice, a value of badness < 2000 can be tolerated even though it will be noticeable that the spacing is not ideal. A sample printing should be examined to see how bad it looks.

Underfull \vbox ....

The page has been broken with head and foot justified, but TEX judges the amount of inter-paragraph spacing to be possibly unacceptable. The badness number here corresponds to the quantity with the same name from the Underfull \hbox warning.

9.6   Search for subtle errors

At some point or another you will encounter an error message for which you cannot identify any cause, try as you may. For such devious errors, we recommend the following search strategy:

1. Copy the file twice into a *previous* and a *working* copy (in addition to the original, which remains untouched during this search).

2. In the working copy, find the outermost environment where the error occurred and remove one or more inner environments. If there are no inner environments, shorten the remaining text. Process the file with LaTeX once more.

3. If the error still occurs, copy the shortened working copy to the previous copy and repeat step 2. If the outer environment in step 2 is \begin{document} ... \end{document}, the shortening may be carried out by simply inserting \end{document} at some earlier point.

4. If the error is no longer in the shortened working copy, copy the previous copy back to the working copy. The error is still present in this version. Remove less of the text than last time and repeat steps 2 to 4.

5. If the error is found to be in the next innermost environment, repeat the procedure for this environment with steps 2 to 4.

With this strategy, the error may be localized to one command or to the innermost environment with only a small remaining structure. If the error still cannot be identified in spite of its being precisely localized, seek help from a more experienced colleague or from the computing center. However, it is normally possible to recognize the mistake once the position of the error has been found.

It does happen that even though the error has been corrected, the same error message appears on the next LaTeX process run. This is because of the internal transfer in information through the LaTeX auxiliary files, which are always one run behind on the current situation. For example, if there is a mistake in one of the sectioning commands, then after it has been eliminated, the faulty entry still exists in the `.toc` file. If the document contains the `\tableofcontents` command, LaTeX will read in that `.toc` file on the next run, and issue the error message once more, since a new `.toc` file is only created after a successful processing.

In this case, the `.toc` file should also be edited and the error removed. If that is not possible, the file should be erased and the corrected source file processed *twice* with LaTeX. If the error was in `\caption`, `\addcontentsline`, or `\addtocontents`, the same applies to the corresponding `.lof` or `.lot` file.

Occasionally the `.aux` file itself must be erased to prevent an error in it from being repeated even after the `.tex` file has been repaired. Here one must be careful that the command `\nofiles` has not been given in the preamble, for then a corrected `.aux` file will not be generated after a further LaTeX process run.

A ╎ Letters

In addition to the three standard LATEX document classes, there is a fourth one named `letter` for formatting correspondence. As provided, the file `letter.cls` is intended for private letters without any frills such as letterheads or business reference codes. However, this can be changed by the local LATEX expert.

We first present the standard LATEX `letter` class, and then demonstrate how a house style can be written using our own institute style as an example. As always, it is highly recommended that each of these modifications be given its own file name so that `letter.cls` refers only to the provided LATEX version.

A.1 The LATEX `letter` class

The `letter` document class is meant for writing letters. A single input file may contain the text for more than one letter and addressee, all from the same sender. Address labels may also be printed automatically if one wishes. Most of the normal LATEX commands function as usual within the `letter` class. One exception, however, is the sectioning commands, which will lead to the error message `! Undefined control sequence` when issued. It actually makes little sense to try to divide a letter up into chapters, sections, etc. On the other hand, there are a number of special commands that may be applied only within this style.

The input text for a letter begins as for every LATEX document with

> \documentclass[*options*]{letter}

in which all the options listed in Section 3.1 may be given for *options* except for `twocolumn` and `titlepage` which hardly make any sense within a letter. The option `twoside` is allowed in LATEX 2_ε, but not in LATEX 2.09.

Every letter must contain the name and address of the sender, which are set for all the letters in one file by including the commands

```
\address{sender_address}
\signature{sender_name}  or  \name{sender_name}
```

in the preamble. The *sender_address* normally consists of several lines, separated by \\ commands, as in the example

```
\address{Max-Planck-Institut f\"ur Aeronomie\\
         Max-Planck-Str.\ 2\\
         D--37191 Katlenburg--Lindau\\Germany}
```

The entry in the \name command will be used in the return address in the letterhead, if such has been programmed. The entry in the \signature command will be printed at the end of the letter below the space left blank for the writer's signature. If \signature has not been specified, the \name entry is inserted here instead. This allows a more formal \name to be used for the return address and a different form, perhaps with multiple lines, for the signature, as in

```
\name{Prof.\ M.\ Ostmann}
\signature{Martin Ostmann\\Project Leader}
```

When the above commands are issued in the preamble, they remain valid for all the letters in the document, except for those letters that contain new versions of these commands. Thus one letter might have a different \signature from the others. The scope of these entries extends only to the end of the environment in which they were called (see Section 7.5.4).

Two other sender entries are possible in standard LaTeX letter class. They are intended to be employed in local modifications for a house style. The idea is that if \address is *not* called, the preprogrammed company letterhead which might also contain the sender's room and/or telephone number is generated. Thus the commands

```
\location{room_number}  and  \telephone{tel_number}
```

are provided. As letter.cls is delivered, the entries *room_number* and *tel_number* are printed at the bottom of the first page *only if* \address *is not issued*.

The preamble may also contain the \pagestyle command with the usual entries plain, empty, or headings. The first is the default, putting the page number centered at the bottom of all pages after the first. The headings page style adds the recipient's name, the date, and page number in a line at the top of all pages after the first.

After the preamble commands, the actual text begins as in all LaTeX files with the command \begin{document}. The text consists of one or more letters, each enclosed in a letter environment with the syntax:

```
\begin{letter}{recipient} text of letter \end{letter}
```

where *recipient* stands for the name *and* address of the addressee, divided into several lines separated by \\ commands.

```
\begin{letter}{Mr Donald J. Burns\\
            Ontario Institute of Physics\\
            41 Adelaide St.\\
            London, Ontario\\Canada N4R 3X5}
```

The *text of letter* normally begins with the command \opening and ends with \closing, between which the body of the letter appears, mixed with whatever LaTeX commands are desired. The syntax of these two commands is

\opening{*dear*}
\closing{*regards*}

where *dear* is the salutation, such as Dear Mr Tibs, and *regards* stands for the terminating text, for example Yours sincerely,. The \opening command could also contain other text, for example a subject entry line, with the true salutation as part of the following body text.

LaTeX places the *sender's* name and address in the upper right corner of the first page with the current date set right justified below it. Then the *recipient's* name and address appear at the left margin, followed by the *salutation* and the body of the letter. The letter ends with the *terminating text* and the sender's *name* or *signature* left justified from the center of the line with sufficient vertical space between them for the handwritten signature.

After the \closing, a number of other commands may appear as part of the letter. One is \cc to produce a copy distribution list:

\cc{*name1* \\ *name2* \\ ... }

The text 'cc:' (or more correctly, the text defined in \ccname) is printed at the left margin, followed by an indented list of names of the copy recipients.

The second additional command is \encl for making a list of enclosures:

\encl{*enclosure1* \\ *enclosure2* \\ ... }

The word 'encl:' (actually the text in \enclname) is printed at the left margin and then the list of enclosures.

Finally, the command \ps may be used to add a postscript after the signature. The command itself does not generate any text, nor does it take an argument. The postscript text is everything located between the \ps and \end{letter} commands.

Normally the letter is dated automatically with the current date. However, if it is desired that the letter be back-dated, or that the date be otherwise fixed in the text, it may be set with

\date{*date_text*}

A sample letter produced with the standard `letter` class.

Max–Planck–Institut für Aeronomie
Max–Planck-Str. 2
D-37191 Katlenburg–Lindau
Germany

September 8, 1998

TEXproof Ltd
P. O. Box 123
9876 Wordtown
Textland

Dear Sir;

We are most pleased to be able to answer your request for information about the use of LATEX for general text processing at a scientific institute.

1. After some initial trepidation on the part of the secretarial staff, which was mainly due to the first experience with a computer in any form, the system is now fully accepted and appreciated.

2. Much to the surprise of many secretaries, they find that they are able to set the most complicated mathematical formulas in a reasonably short time without difficulties. The same applies to the production of detailed tables.

3. Creating cross-references and keyword indices no longer causes horror, even when the boss is well known for demanding constant changes.

4. Finally, the high quality appearance of the output has assisted in winning acceptance for LATEX in our house.

An additional positive note is the ability to write business letters readily, making use of the `letter` class provided with LATEX. In our institute, we have designed a special version to print our own letterhead, saving the need to have special letter paper printed.

Yours truly,

Patrick W. Daly

encl: Listing of our `mpletter.cls`
      Sample output

cc: H. Kopka
    B. Wand

in which case *date_text* appears where the current date would be placed.

A letter file may contain any number of letter environments, one per letter. As stated already, when \address, \name, and \signature have been specified in the preamble, they remain in force for all the letters in the file. It is possible to alter any one or more of these sender entries by reissuing the command within one of the letters, *before* the \opening command, but then this change is valid solely for that letter. If both \name and \signature have been declared, it is the latter that is printed below the signature space.

The first page contains no page number. Subsequent pages have either a centered page number at the bottom (default) or a heading line with recipient, date, and page number at the top (page style headings).

The sample letter on the facing page was generated with the following input text.

```
\documentclass[a4paper,11pt]{letter}
\name{Dr P. W. Daly}
\address{Max--Planck--Institut f\"ur Aeronomie\\
        Max--Planck--Str.\ 2\\
        D-37191 Katlenburg--Lindau\\Germany}
\signature{Patrick W. Daly}
\date{September 8, 1998}
\begin{document}
\begin{letter}{\TeX proof Ltd\\P.\,O. Box 123\\
              9876 Wordtown\\Textland}
\opening{Dear Sir;}
We are most pleased to be able to answer your request for
information about the use of \LaTeX{} for general text
processing at a scientific institute.
\begin{enumerate}
  \item
  After some initial trepidation on the part of the secretarial
  . . . . . . . . . . . . . . . . . . . . . . . . . . . . . .
  in winning acceptance for \LaTeX{} in our house.
\end{enumerate}
An additional positive note is the ability to write business
. . . . . . . . . . . . . . . . . . . . . . . . . . . . . .
special paper printed.
\closing{Yours truly,}
\encl{Listing of our \texttt{mpletter.cls}\\Sample output}
\cc{H. Kopka\\B. Wand}
\end{letter}
\end{document}
```

By adding the command \makelabels in the preamble, the user can print out address stickers after all the letters have been output. The entries for the addresses are taken from the recipients' names and addresses in the argument to the letter environment. The standard LATEX letter class is designed for a page of labels $4\frac{1}{4} \times 2$ inches, ordered in two columns. This could be altered for other formats. An address sticker without a corresponding letter may be printed by including an *empty* letter environment of the form

> \begin{letter}{*recipient*}\end{letter}

A.2   A house letter style

The sample letter above demonstrates the possibilities of the standard LaTeX letter class. The height and width of the text may easily be altered with appropriate declarations in the preamble. The use of explicit English words and date style is no problem for other languages since they are all contained in special commands that can be redefined. (This actually only applies to LaTeX versions after December 1, 1991, for before then the English words were explicitly used.)

The letter class has been designed so that if the \address command is omitted, a company letterhead will be printed instead. This presupposes that the file has been reprogrammed at the local installation for this purpose. Each employee using this house letter style will have certain personal entries to make, such as his or her room and/or telephone number. These commands have been provided for in the standard letter class.

We have such a house letter style at our institute which we will illustrate here. However, it was found necessary to add some more personal entries, such as 'Our Ref.', 'Your Ref.', and email addresses. In addition, a command to print out 'Subject:' has also been added.

To distinguish our local house style from that of standard LaTeX, we have named it mpletter. It contains most of the features of letter, since it in fact reads in that class file and then makes its alterations and additions. The \address command is not necessary, since all letters are printed with the institute letterhead, including its address. The recipient's name and address are taken from the argument of the letter environment and are vertically centered in the space provided in the letterhead.

The writer's name and telephone extension are entered with

> \name{*author*}   and   \telephone{*ext_number*}

which, if given in the preamble, apply to all the letters in the file. If different letters have other authors, these commands must be made in the appropriate letter environment *before* the \opening command. New entry commands specific to mpletter are

> \yref{*your_code*}
> \ymail{*your_date*}
> \myref{*our_code*}
> \subject{*subj_text*}

MAX–PLANCK–INSTITUT FÜR AERONOMIE
Max–Planck–Str. 2
Katlenburg–Lindau
GERMANY

MPI für Aeronomie, D–37191 Katlenburg–Lindau

Mr George Murphy
35 Waterville Rd.
Centertown, Middlesex
United Kingdom

Dr P. W. Daly
Tel.: 05556-401-279
Email:
daly @ linmpi.mpg.de

May 10, 1998

Your Ref.: GFM/sdf   *Your letter from*: April 28, 1998   *Our Ref.*: PWD/sib

Subject: LATEX information

Dear George,

Thank you for your inquiry about the latest version of the LATEX installation and additional packages. As you may know, since June 1, 1994, LATEX 2$_\varepsilon$ is the official version, replacing the now obsolete 2.09 version. Updates have been issued every six months, in June and December, with absolutely urgent fixes coming out as patches when necessary.

I am sending you a copy of the current version of the entire LATEX package on a CD-ROM, as you requested. In a separate directory named `bibtex` you will find the special bibliography formatting package files mentioned in 'A Guide to LATEX'. I hope you will find these of use.

Do not hesitate to get in touch with me again if you have any further questions about the installation or running of the package.

                                                                Regards,

                                                                Patrick W. Daly

encl: 1 CD-ROM with LATEX package

cc: H. Kopka

Telephone	05556–401– 1	Bank		Train Station
Telefax	05556–401– 240	Kreis–Sparkasse Northeim		Northeim
Telex	9 65 527 aerli	41 104 449 (BLZ 262 500 01)		(Han.)

which produce the words

Your Ref.:, Your letter of:, Our Ref.:, Subject:

properly positioned under the letterhead together with the corresponding text argument. If any of these commands are missing, their text will not appear in the letter.

Since ours is a German institute, there is an option `german` that translates all these words into their German equivalents. The entry commands have the same names, however.

As in the standard `letter` class, the current date is printed automatically, but may be changed to any desired text with the command

\date{*date_text*}

if one wishes to back-date a letter or to fix the dating within the letter file itself. This is often handy if one only keeps electronic copies of the letters for the record. Otherwise, when a hard copy is run off months later, it will appear with the new current date and not with the original one.

The entry *author* that is given as the argument to the \name command appears in the letterhead as the name of the sender. It will also appear below the space left for the signature, unless it is overridden by the \signature command that specifies an alternative form of the name for this purpose.

The sample letter on the previous page has been generated with our house style, using the following input text.

```
\documentclass[12pt]{mpletter}
\name{Dr P. W. Daly} \signature{Patrick W. Daly}
\myref{PWD/sib}
\date{May 10, 1998}
\subject{\LaTeX{} information}
\telephone{279} \internet{daly}
\ymail{April 28, 1998} \yref{GFM/sdf}
\begin{document}
\begin{letter}{%
Mr George Murphy\\35 Waterville Rd.\\
Centertown, Middlesex\\United Kingdom}
\opening{Dear George,}
Thank you for your inquiry about the latest version of the
\LaTeX{} installation and additional packages.  As you may know,
. . . . . . . . . . . . . . . . . . . . . . . . . . . . . . . .
Do not hesitate to get in touch with me again if you have any
further questions about the installation or running of the
package.

\closing{Regards,}
\encl{1 CD-ROM with \LaTeX{} package}
\cc{H. Kopka}
```

```
\end{letter}
\end{document}
```

The first page of our institute letter appears as shown, without a page number. If the text must continue beyond one page, the next page will have the heading

MAX–PLANCK–INSTITUT FÜR AERONOMIE

To Mr George Murphy   *May 10, 1998*   *Page 2*

The addressee's name that appears in this heading is taken from the first line of the *recipient* argument in the \begin{letter}. This argument is split up by the LATEX processing so that the first line is contained in the command \toname, and the rest of the lines in \toaddress. The words 'To' and 'Page' are in the standard commands \headtoname and \pagename, and may be changed by an appropriate language adaptation option as shown in Section C.4.2.

A.3   A model letter customization

Adapting the letter.cls class file to the requirements of a company style should present few difficulties to an experienced LATEX programmer. Even a normal user may be able to make the necessary changes with the help of the example in this section.

We present here the class file mpletter.cls that was used to produce the sample letter on page 303. It makes heavy use of the LATEX programming features described in Appendix C. In order to understand it, one should be familiar with Section C.2. It will not be necessary to make any changes to the file letter.cls itself, since all modifications are in a separate class file that inputs the original.

The new class file is to be called mpletter.cls. It begins by specifying the TEX format that it requires, and by identifying itself.

```
\NeedsTeXFormat{LaTeX2e}
\ProvidesClass{mpletter}
```

It will be necessary to execute conditionals, so we will need the ifthen package described in Section 7.3.5. We will want a flag to decide if the letter is to be in German or not, determined by an option. Create the flag and define the option german to set it.

```
\RequirePackage{ifthen}
\newboolean{@german}
\setboolean{@german}{false}
\DeclareOption{german}{\setboolean{@german}{true}}
```

All other options that are valid in the standard letter class will also be accepted here, so simply pass them on to that class with the default option. Then process all options before loading letter itself with the a4paper option. We only have A4 paper in our institute.

```
\DeclareOption*{\PassOptionsToClass{\CurrentOption}{letter}}
\ProcessOptions
\LoadClass[a4paper]{letter}
```

This completes the preliminaries. So far, we have read in the standard letter class, along with the package ifthen, and defined a new option that is not present in the original class. Otherwise, all options and functions are unchanged, so far.

We now define the new 'name' commands to contain language-sensitive text, such as '*Subject*', which are not provided for in the basic class. The actual definitions will be executed by two commands \englishnames and \germannames.

```
\newcommand{\englishnames}{%
 \newcommand{\yrefname}{\textsl{Your Ref.}}
 \newcommand{\ymailname}{\textsl{Your letter from}}
 \newcommand{\myrefname}{\textsl{Our Ref.}}
 \newcommand{\subjectname}{\textsl{Subject}}
 \newcommand{\telephonename}{Telephone}
 \newcommand{\stationname}{Train Station}
 \newcommand{\germanname}{GERMANY}
 \newcommand{\telcode}{[49]-5556-401}
 \newcommand{\postcode}{D--37191}
}
```

```
\newcommand{\germannames}{%
. . . . . . . . . . . . . . . . . . . . . .
 \newcommand{\telcode}{(05556) 401}
 \newcommand{\postcode}{37191}
}
```

```
\ifthenelse{\boolean{@german}}
   {\RequirePackage{german}\germannames}{\englishnames}
```

The last lines test whether the flag @german has been set (by the option german), and if so, the package german is loaded, and the German names are defined, otherwise the English names. The package german already translates the standard names commands \toname, \headtoname, and \pagename, so they are not included in \germannames.

Having settled the language problem, we now attack those commands for entering extra information in the header. Each of these stores its text argument in an internal command for future use. First the internal storage commands must be created.

```
\newcommand{\@yref}{}     \newcommand{\@ymail}{}
\newcommand{\@myref}{}    \newcommand{\@subject}{}
\newcommand{\@internet}{}
```

```
\newcommand{\yref}[1]{\renewcommand{\@yref}{\yrefname: #1}}
\newcommand{\ymail}[1]{\renewcommand{\@ymail}{\ymailname: #1}}
\newcommand{\myref}[1]{\renewcommand{\@myref}{\myrefname: #1}}
\newcommand{\subject}[1]{\renewcommand{\@subject}
                         {\subjectname: #1}}
\newcommand{\internet}[1]{\renewcommand{\@internet}{#1}}
\newcommand{\INTERNET}{@linmpi.mpg.de}
```

Set the dimensions of the text on the page and its margins. These numbers are appropriate for A4 paper (which does make the a4paper option superfluous).

```
\setlength{\textheight}{215mm}   \setlength{\textwidth}{160mm}
\setlength{\oddsidemargin}{0pt}  \setlength{\evensidemargin}{0pt}
\setlength{\topmargin}{-20pt}    \setlength{\headheight}{12pt}
\setlength{\headsep}{35pt}
```

The next step is to define some fixed fonts that are needed for the letterhead. We refer to the fonts with their explicit NFSS attributes; in this case they are all Computer Modern sans serif fonts in various sizes. These fonts will not change if totally different families are used for the body of the letter.

```
\DeclareFixedFont{\xviisf}{OT1}{cmss}{m}{n}{17}
\DeclareFixedFont{\xsf}{OT1}{cmss}{m}{n}{10}
\DeclareFixedFont{\viiisf}{OT1}{cmss}{m}{n}{8}
```

The letterhead itself is divided into two fields, the left one containing the institute name in large letters, the right one the address in a smaller font. Below the first horizontal line, the left field displays the name and address of the recipient, positioned to fit in the window of an envelope, and the right one has the personal data of the letter writer, name, extension, computer address. The widths of these fields are established.

```
\newlength{\leftfield}    \setlength{\leftfield}{117mm}
\newlength{\rightfield}   \setlength{\rightfield}{43mm}
```

The total width of these two fields equals 160 mm, which is the same as \textwidth.

Next, we place the institute name and address in several saved boxes.

```
\newsavebox{\FIRM}        \newsavebox{\firmaddress}
\newsavebox{\firm}        \newsavebox{\firmreturn}

\sbox{\FIRM}
  {\parbox[t]{\leftfield}
      {\xviisf MAX--PLANCK--INSTITUT F\"UR AERONOMIE}}

\sbox{\firm}
  {\xsf MAX--PLANCK--INSTITUT F\"UR AERONOMIE}
```

```
\sbox{\firmreturn}
  {\viiisf\underline{MPI f\"ur Aeronomie,
                    \postcode{} Katlenburg--Lindau}}

\sbox{\firmaddress}
  {\parbox[t]{\rightfield}{\viiisf\baselineskip10pt
    Max--Planck--Stra{\ss}e 2\\
    \postcode{} Katlenburg--Lindau\\\germanname}}
```

Using these boxes as building blocks, we put together the actual head and foot of the letterhead page in two further save boxes.

```
\newsavebox{\firmhead}    \newsavebox{\firmfoot}

\sbox{\firmhead}
  {\parbox{\textwidth}{\usebox{\FIRM}\raisebox{6pt}
    {\usebox{\firmaddress}}\\[3pt] \rule{\textwidth}{1pt}}}

\sbox{\firmfoot}
  {\parbox{\textwidth}{\rule{\textwidth}{0.6pt}\\[5pt]
    \viiisf\setlength{\baselineskip}{12pt}%
    \begin{tabular}[t]{@{}ll}
      \underline{\telephonename} & \telcode-1\\
      \underline{Telefax}        & \telcode-240\\
      \underline{Telex}          & 9\,65\,527 aerli
    \end{tabular}\hfill
    \begin{tabular}[t]{l}
      \underline{Bank}\\
      Kreis--Sparkasse Northeim\\
      41\,104\,449 (BLZ 262\,500\,01)
    \end{tabular}\hfill
    \begin{tabular}[t]{l@{}}
      \underline{\stationname}\\
      Northeim\\
      (Han.)
    \end{tabular} }}
```

The box \firmhead is fairly clear: it is a \parbox of width \textwidth containing the boxes \FIRM and \firmaddress side by side, with a line below. The foot \firmfoot is also a \parbox of the same width, but containing three columns of general institute information, set in tabular environments.

It now remains to have the head and foot boxes placed on the first page. In the letter class, there is a special page style named firstpage that is always invoked for the first page of a letter. This must be redefined.

```
\renewcommand{\ps@firstpage}
  {\setlength{\headheight}{41pt}\setlength{\headsep}{25pt}%
   \renewcommand{\@oddhead}{\usebox{\firmhead}}%
```

```
\renewcommand{\@oddfoot}{\raisebox{-20pt}[0pt]
                        {\usebox{\firmfoot}}}%
\renewcommand{\@evenhead}{}\renewcommand{\@evenfoot}{}}
```

This page style must (re)define the commands \@oddhead and \@oddfoot, which are always inserted at the top and bottom of odd pages, to place our special \firmhead and \firmfoot. The even pages are unimportant, since the first page is always odd. It is necessary to enlarge \headheight and \headsep to allow the big boxes to fit in.

Subsequent pages are set with the headings or plain page styles. We want to modify the former to include the firm address once more.

```
\renewcommand{\ps@headings}
    {\setlength{\headheight}{41pt}%
     \renewcommand{\@oddhead}
        {\parbox{\textwidth}{\usebox{\firm}\\[5pt]
            \slshape \headtoname{} \toname\hfill\@date\hfill
                                    \pagename{} \thepage\\
            \rule[3pt]{\textwidth}{1pt}}}
     \renewcommand{\@oddfoot}{}
     \renewcommand{\@evenhead}{\@oddhead}
     \renewcommand{\@evenfoot}{\@oddfoot}}
```

One small problem remains: the first time one of these page style commands is executed, the head and foot commands may not yet exist, causing \renewcommand to complain. We ensure that they are there at the start by predefining them with \providecommand (Section 7.3.1).

```
\providecommand{\@evenhead}{}\providecommand{\@oddhead}{}
\providecommand{\@evenfoot}{}\providecommand{\@oddfoot}{}
```

Now make headings the active page style.

```
\pagestyle{headings}
```

There is only one last thing to do, and that is to redefine the opening command that prints the recipient's address and the salution. We add a bit more, including the personal data of the sender, as well as the reference information. The address goes in the left field, the personal data to the right. The references go in a line below the rule, followed by the subject line. These entries are tested first, and are only included if they are not blank. Several stored entry commands used here are part of the standard letter class, such as \toname and \toaddress. The \@date entry is either \today or whatever text was stored with \date.

```
\renewcommand{\opening}[1]{\thispagestyle{firstpage}%
    \parbox[t]{\leftfield}
        {\usebox{\firmreturn}\\
         \parbox[b][3.5cm][c]{\leftfield}{\toname\\\toaddress}}%
    \parbox[t]{\rightfield}
        {\fromname
```

```
\ifthenelse{\equal{\telephonenum}{}}
    {}{\\ Tel.: \telcode-\telephonenum}
\ifthenelse{\equal{\@internet}{}}
    {}{\\\{\viiisf Email: \@internet\INTERNET}}
    \\[5mm] \@date}
\par
\rule{\textwidth}{0.6pt}
\makebox[\leftfield][l]
    {\ifthenelse{\equal{\@yref}{}}
        {\@ymail}{\@yref\hfill\@ymail\hfill}}
\@myref\par
\ifthenelse{\equal{\@subject}{}}
    {}{\@subject\par}
\vspace{2\parskip} #1 \par\nobreak}
```

The result of this formatting can be seen on page 303. It should be possible to modify the coding as needed for other organizations without too much difficulty.

The field for the recipient's name and address has been positioned so that it will appear in the address window of an envelope when properly folded. The smaller return address will also be visible through this window. The printing of extra address stickers is thus superfluous.

As an exercise, the user should design a document class pletter *for writing personal letters. With the tips for formatting company letters, it should not be difficult to add a letterhead of the form*

Sheila Joan McDonald

Tel.: 234-9871
31 Maple Drive
Willowtown

Hint: The name has been printed with the font cmdunh10 scaled \magstep4, *declared with* \newfont. *Such a* pletter *class must remain in the personal directories of the user at a computer center. It would be rather embarrassing if everyone made use of the same personal letterhead.*

B Bibliographic Databases

In academic publications, a bibliography or list of references is standard procedure. We have already described in Sections 4.3.6 and 8.3.3 how a bibliography can be formatted with the thebibliography environment and how its entries are referred to within the text. Often authors find that they are constantly referring to the same publications in most of their own papers. Similarly, researchers working in the same field will frequently be referring to the same set of papers. This means that many of the entries in the thebibliography environment will be repeated from one work to the next, or even among co-workers at one institute.

It would be very useful if the bibliographic entries could be stored in one database file once and for all to be available for all documents with a list of references in that field. Such a database system is possible with the BibTeX program supplied with the LaTeX package. The information about the various publications is stored in one or more files with the extension .bib. For each publication there is a *key* that identifies it, and which may be used in the text document to refer to it. Such a file is called a *bibliographic database*.

B.1 The BibTeX program

BibTeX is an auxiliary program to LaTeX that automatically constructs a bibliography for a LaTeX document by searching one or more databases. To this end, the LaTeX file must contain the command

> \bibliography{*database1, database2, ...* }

at that point in the text where the bibliography is to appear. Here the argument *database1, database2, ...* is a list of root names, separated by commas, of the bibliographic database files that are to be searched. The extension .bib is not explicitly written.

Reference can be made to a publication in one of the databases at any place in the text with the command

> \cite{*key*}

as is explained in Section 8.3.3. The *key* is the database identifier for that publication, which of course the user must know beforehand. After the first LATEX processing, the BIBTEX program must be run. How this is called depends on the computer installation (see the *Local Guide*), but it is often with the operating command bibtex plus the root name of the LATEX file. Supposing this root name is comets, the call

> bibtex comets

produces a new file with the name comets.bbl, containing the extracted information for those publications for which there was a \cite reference, packaged in a thebibliography environment to be input into the document on the next LATEX run.

Occasionally the bibliography is to include publications that were *not* referenced in the text. These may be added with the command

> \nocite{*key*}

given anywhere within the main document. It produces no text at all but simply informs BIBTEX that this reference is also to be put into the bibliography. With \nocite{*}, *every* entry in all the databases will be included, something that is useful when producing a list of all entries and their keys. (**Does not work for BIBTEX version 0.98 and earlier.**)

After running BIBTEX to make up the .bbl file, it is necessary to process LATEX *at least twice* to establish both the bibliography and the in-text reference labels. The bibliography will be printed where the \bibliography command is issued; it in fact inputs the .bbl file.

The style of the bibliography may be selected with the declaration

> \bibliographystyle{*style*}

which may be issued anywhere after the preamble. The *style* argument can take on one of the following values:

plain   The entries in the bibliography are ordered alphabetically; each is assigned a running number in square brackets as the in-text reference marker, printed where the \cite commands are issued.

unsrt   The entries are ordered according to their first references by the \cite and \nocite commands. The entry for the first \cite receives the number 1, that of the next \cite with a different key the number 2, and so on. The markings and listings are otherwise the same as for plain.

alpha   The ordering in the bibliography is the same as for plain but the markers are an abbreviation of the author's name plus year of publication. A reference to Smith (1987) would appear as [Smi87] instead.

abbrv   The ordering and marking are the same as for plain, but the bibliographic listing is shortened by abbreviating first names, months, and journal names.

There may be other styles available on your computer. In particular, those for author–year citations, as described in Section B.3.1, deviate a great deal from those of the standard LATEX package. Bibliography styles are contained in files with the extension .bst.

The next section describes the bibliographic databases for use with the program BIBTEX version 0.99. There are a number of new features over version 0.98, which are indicated with a double asterisk **. Databases (.bib files) that worked under the older version will still function under 0.99. However, the bibliography style files (.bst) operate with one version or the other and may not be mixed. To find out what version you have, look at the first line of output from the BIBTEX run, either on the screen or in the .blg protocol file.

B.2   Creating a bibliographic database

Creating a bibliographic database might seem like more work than typing up a list of references with the thebibliography environment; the great advantage is that the entries need to be included in the database once and only once and are then available for all future publications. Even if a different bibliography style is demanded in later works, all the information is already on hand in the database for BIBTEX to write a new thebibliography environment in another format. In fact, we feel that even for a single document, it is simpler to make an entry into the database than to adhere to the very precise and fiddly requirements of a literature list, especially regarding punctuation and positioning of the authors' initials. The data bank entry proceeds very quickly and easily if one has a generalized template, as illustrated in Section B.2.6.

The entries in a bibliographic database are of the form

```
@BOOK{knuth:86a,
   AUTHOR = "Donald E. Knuth",
   TITLE = {The \TeX{}book},
   EDITION = "third",
   PUBLISHER = "Addison--Wesley",
   ADDRESS = {Reading, Massachusetts},
   YEAR = 1986   }
```

The first word, prefixed with @, determines the *entry_type*, as explained in the next section. The *entry_type* is followed by the reference information for that entry enclosed in curly braces { }. The very first entry is the *key* for the whole reference by which it is referred to in the \cite command.

In the above example this is knuth:86a. The *key* may be any combination of letters, numerals, and symbols, except commas. The actual reference information is then entered in various *fields*, separated from one another by commas. Each *field* consists of a *field_name*, an = sign, with optional spaces on either side, and the *field text*. The *field_names* shown above are AUTHOR, TITLE, PUBLISHER, ADDRESS, and YEAR. The *field text* must be enclosed either in curly braces or in double quotation marks. However, if the text consists solely of a number, as for YEAR above, the braces or quotation marks may be left off.

For each entry type, certain fields are *required*, others are *optional*, and the rest are *ignored*. These are listed with the descriptions of the various entry types below. If a required field is omitted, an error message will occur during the BibTeX run. Optional fields will have their information included in the bibliography if they are present, but they need not be there. Ignored fields are useful for including extra information in the database that will not be output, such as a comment or an abstract of the paper. Ignored fields might also be ones that are used by other database programs.

The general syntax for entries in the bibliographic database reads

> @*entry_type*{*key*,
>   *field_name* = {*field text*},
>   . . . .
>   *field_name* = {*field text*} }

The names of the *entry_types* as well as the *field_names* may be written in capitals or lower case letters, or in a combination of both. Thus @BOOK, @book, and @bOOk are all acceptable variations.

The outermost pair of braces for the entire entry may be either curly braces { }, as illustrated, or parentheses ( ). In the latter case, the general syntax reads

> @*entry_type*(*key*, ...)

However, the *field text* may only be enclosed within curly braces {...} or double quotation marks "..." as shown in the example above.

B.2.1   The entry types

The following is a list of the standard entry types in alphabetical order, with a brief description of the types of works for which they are applicable, together with the required and optional fields that they take. The meanings of the fields are explained in the next section.

@article Entry for an article from a journal or magazine.
       required fields   author, title, journal, year.
       optional fields   volume, number, pages, month, note.

@book Entry for a book with a definite publisher.
required fields author or editor, title, publisher, year.
optional fields volume or number, series, address, edition, month, note.

@booklet Entry for a printed and bound work without the name of a publisher or sponsoring organization.
required fields title.
optional fields author, howpublished, address, month, year, note.

@conference Is the same as @inproceedings below.

@inbook Entry for a part (chapter, section, certain pages) of a book.
required fields author or editor, title, chapter and/or pages, publisher, year.
optional fields volume or number, series, type, address, edition, month, note.

@incollection Entry for part of a book that has its own title.
required fields author, title, booktitle, publisher, year.
optional fields editor, volume or number, series, type, chapter, pages, address, edition, month, note.

@inproceedings Entry for an article in conference proceedings.
required fields author, title, booktitle, year.
optional fields editor, volume or number, series, pages, address, month, organization, publisher, note.

@manual Entry for technical documentation.
required fields title.
optional fields author, organization, address, edition, month, year, note.

@mastersthesis Entry for a Master's thesis.
required fields author, title, school, year.
optional fields type, address, month, note.

@misc Entry for a work that does not fit under any of the others.
required fields none.
optional fields author, title, howpublished, month, year, note.

@phdthesis Entry for a PhD thesis.
required fields author, title, school, year.
optional fields type, address, month, note.

@proceedings Entry for conference proceedings.
> *required fields*   title, year.
> *optional fields*   editor, volume or number, series, address, month, organization, publisher, note.

@techreport Entry for a report published by a school or other institution, usually as part of a series.
> *required fields*   author, title, institution, year.
> *optional fields*   type, number, address, month, note.

@unpublished Entry for an unpublished work with an author and title.
> *required fields*   author, title, note.
> *optional fields*   month, year.

Each entry type may also have the optional fields key and crossref. The former provides additional information for alphabetizing the entries, especially when the author information is missing. The author information is normally found in the author field but may also be in the editor or even organization fields. This key field has nothing to do with the *key* for identifying the entry in the \cite command. The crossref field gives the *key* for another entry in the database that shares many of the information fields, as illustrated in Section B.2.3.

B.2.2   The fields

The fields that may be used within a bibliographic entry are listed below together with their meanings. They are always given in the form:

> *field_name* = { *field text* }   or
> *field_name* = " *field text* "

address
> The address of the publisher or other institution. For major publishing houses, it is sufficient to give just the city. For smaller publishers, giving the full address is recommended.

annote An annotation that may be used by non-standard bibliography styles to produce an annotated bibliography. The standard BibTeX styles ignore it.

author The name(s) of the author(s) as described in Section B.2.4.

booktitle
> The title of a book when only part of it is being cited. See Section B.2.4 for special considerations on capitalization.

chapter
> A chapter or section number.

`crossref`
> The key of another entry in the database that shares many of the same field entries. See Section B.2.3.

`edition`
> The edition of a book, usually written in full and capitalized, as 'Second'. The standard styles will change it to lower case as necessary.

`editor` The name(s) of the editor(s) in the form described in Section B.2.4. If there is also an `author` field, then this gives the editor of the book or collection in which the citation appears.

`howpublished`
> States anything unusual about the method of publishing. Should be capitalized. Example: 'Privately published'.

`institution`
> The name of the sponsoring institution for a technical report.

`journal`
> The name of a journal or magazine. Abbreviations are provided for the most common ones (see Section B.2.5).

`key` An addition for alphabetizing purposes when the author information is missing. This is not the same as the *key* for identifying the entry to a `\cite` command.

`month` The month in which the work was published or, if unpublished, when it was written. Abbreviations exist in the form of the first three letters of the (English) names.

`note` Any additional information that should be added. Capitalize the first letter.

`number` The number of a journal, magazine, technical report, or work in a series. Journals are usually identified by volume and number; technical reports are issued a number by the institution; books in series sometimes have a number given to them.

`organization`
> The sponsoring organization for a conference or a manual.

`pages` A page number or a range of pages, in the form 32,41,58 or 87--101 or 68+. The last form indicates page 68 and following pages. A single hyphen given for a range will be converted by the standard styles to the double hyphen to produce a dash, as '87-101'.

`publisher`
> The publisher's name.

`school`   The name of the academic institution where a thesis was written.

`series`   The name of a series or set of books. When citing a book from a series, the `title` field gives the name of the book itself while the optional `series` specifies the title of whole set.

`title`   The title of the work, obeying the capitalization rules listed in Section B.2.4.

`type`   The type of technical report, for example, 'Research Note'.

`url`   The *universal resource locator*, or Internet address, for online documents; this is not standard but supplied by more modern bibliography styles.

`volume`   The volume number of a journal or multi-volume book.

`year`   The year in which the work was published or, if unpublished, in which it was written. It should normally consist of four numerals, such as 1993.

Additional field names may be included, which BiBTeX will simply ignore. For example, to add the abstract of an article in the database,

> `abstract` = {*text of an abstract*}

This can be of use in other applications as well as for database management.

B.2.3   Cross-referencing fields

This section applies only to BiBTeX version 0.99 and later.
If many entries in the bibliographic database share a common set of field information, such as for a number of works all appearing in the same conference proceedings, it is possible to refer to another entry containing that common set with the `crossref` field. For example,

```
@INPROCEEDINGS{xyz-1,
   crossref = {xyz-proceedings},
   author = {J. S. Jones},
   title = {The First Results from the {Appleville Experiment}},
   pages = {34--38}   }
 .  .  .  .  .  .  .  .  .  .  .  .
@PROCEEDINGS{xyz-proceedings,
   editor = {C. H. Kelvin},
   title = {Proceedings of the First Conference on the
              {Appleville Experiment}},
   booktitle = {Proceedings of the First Conference on the
                 {Appleville Experiment}}
   year = 1991     }
```

The first entry, with key `xyz-1`, is to obtain all its missing fields from the second entry, by means of the `crossref` field referring to the entry with key `xyz-proceedings`. The missing fields are `editor`, `booktitle`, and `year`, those that are common to all articles in the conference proceedings. Note that `booktitle` is an ignored field for `@PROCEEDINGS` but needs to be included here since it is required for `@INPROCEEDINGS`.

If an entry is referred to by two or more other entries, then it too will be included in the bibliography, even though its key never appeared as the argument of a `\cite` or `\nocite` command.

In order for this system to function properly, the entry that is referred to must appear in the database(s) *after* all those that refer to it. It is therefore recommended to put all such referenced entries at the end of the database. Cross-references may not be nested.

B.2.4   Special field formats

There are some special rules for entering the texts to the fields `author`, `editor`, `title`, and `booktitle`. BIBTEX will process names, putting surnames first, abbreviating given names with initials, and so on, according to the instructions in the style file. Thus it is important that the program knows what is a given name and what a surname. Similarly for titles, capitalization may change depending on style and/or entry type, so BIBTEX must know what words are always capitalized.

Names

The names in the `author` and `editor` fields may be typed in either in the form {*Given Names Surname*} or as {*Surname, Given Names*}. That is, BIBTEX assumes that if there is no comma the last capitalized name is the surname, or family name; otherwise what comes before the comma is taken to be the surname. Thus the name texts `"John George Harrison"` and `"Harrison, John George"` are equivalent for Mr J. G. Harrison. However, if a person has a double surname, without a separating hyphen, the second form must be employed, or the double name must be enclosed in braces, as

    `"San Martino, Maria"`   or   `"Maria {San Martino}"`

for Ms M. San Martino.

Auxiliary words to a surname that are not capitalized, such as *von* or *de*, may be entered in either form:

    `"Richard von Mannheim"`   or   `"von Mannheim, Richard"`
    `"Walter de la Maire"`   or   `"de la Maire, Walter"`

Anything enclosed in braces will be treated as a single item, something that is used in ambiguous cases, or when the name contains a comma or the word *and*. An example is

```
"{Harvey and Sons, Ltd}"
```

If the name contains a *Junior* or some other addition, it must be entered {*Surname, Junior, Given Name*}, for example as

```
"Ford, Jr, Henry"  or  "Ford, III, Henry"
```

However, if there is to be no comma, then the *Jr* must be treated as part of a double surname, *something that is not recommended at all*:

```
"{Filmore Jr}, Charles"  or  "Charles {Filmore Jr}"
```

BⁱᵦTEX 0.99 and later. Accents within a name formed with a backslash command should be enclosed in braces *with the backslash as the first character after the opening brace.* In this way, the alphabetization and the formation of labels with the alpha bibliography style will function properly. For example,

```
author = "Kurt G{\"o}del",
year = 1931
```

will produce the label [Göd31] as desired. The accent text must not be enclosed any deeper than shown here. (In all versions of BⁱᵦTEX, accents may be put into the name text for printing in the bibliography. What is new here is the treatment of labels and alphabetization.)

BⁱᵦTEX 0.99 and later. Hyphenated first names are properly abbreviated. Thus "Jean-Paul Sartre" becomes 'J.-P. Sartre'. (In earlier versions, the result is 'J. Sartre'.)

If an author or editor field is to contain more than one name, the names are separated by the word and. For example,

```
author = "Helmut Kopka and Daly, Patrick William"  or
AUTHOR = {Peter C. Barnes and Tolman, Paul and Mary Smith}
```

If the and is actually part of the name, the whole name must be enclosed in braces, as pointed out above.

If the author list is too long to type in all the names, it may be terminated with and others. This will be converted to the form of *et al.* prescribed in the style file.

Titles

The capitalization of the title depends on the bibliography style: usually book titles are capitalized while article titles are not. The text in the fields title and booktitle should be written in the capitalized form so that BⁱᵦTEX can change to lower case as required.

The general rules for capitalizing titles in English state that the first word of the title, the first word after a colon, and all other words are capitalized except articles and unstressed prepositions and conjunctions. For example:

```
title="The Right Way to Learn: A Short-Cut to a Successful Life"
```

Words that are always to be capitalized, such as proper nouns, must be enclosed in braces. It is sufficient to enclose only the letter that must be capitalized. The two following examples are equivalent:

```
title = "The {Giotto} Mission to Comet {Halley}"  or
TITLE = {The {G}iotto Mission to Comet {H}alley}
```

B.2.5   Abbreviations

It is always possible to use an abbreviation in the field text in place of actual text. Some abbreviations, such as the names of the months and some standard journal names, are already available, and the user is free to define new ones for his or her personal use.

The name of an abbreviation consists of any combination of letters, numbers, and symbols except

```
" # % ' ( ) , = { }
```

An abbreviation is defined with the command

```
@string{abbrev_name = {text}}      or
@string(abbrev_name = {text})
```

where *abbrev_name* stands for the name of the abbreviation and *text* is the replacement text. For example, if the abbreviation

```
@string{JGR = {Journal of Geophysical Research}}
```

has been defined, the following two field statements are identical:

```
journal = JGR
journal = {Journal of Geophysical Research}
```

The name of the abbreviation is *not* enclosed in braces or double quotes, otherwise the name itself will be interpreted literally as the field text. In both the @string command and the name of the abbreviation, the case of the letters is unimportant. The above abbreviation could just as well be defined as

```
    @STRING{jgr = {Journal of Geophysical Research}}
 or @StrinG{jGr = {Journal of Geophysical Research}}
```

and it may be referred to in the fields as JGR, JGr, JgR, Jgr, jGR, jGr, jgR or jgr.

BIBTEX 0.99 and later. Abbreviations may be concatenated with the symbol # between them. Thus after giving

```
@string{yrbk = {Institute Yearbook}}
```

this abbreviation may be combined with other abbreviations and text, as in

```
title = "Max-Planck˜" # yrbk # 1993
```

to produce 'Max-Planck Institute Yearbook 1993'.

Standard abbreviations exist for the months of the year, the names consisting of the first three letters of the name, as jan, feb, etc. Further predefined abbreviations are available for some standard journal names, but the user will have to inquire about these at the computing center, or consult the *Local Guide*, since these will be dependent on the installation.

The @string commands may be issued anywhere in the database between two bibliographic entries, but must already be defined before they are used. Therefore, it makes most sense to insert all the abbreviation definitions at the beginning of the database file.

B.2.6   Using a template

Typing in the text for bibliographic entries into a database might seem to be rather complicated since there are so many things to remember, such as which fields are required and which are optional, what their formats are, and so on. One way to simplify this task is to make up *templates* for the most common entry types that you use. A template is basically a complete entry with all the fields left blank. It is stored in a separate file on its own, to be inserted into the database file whenever a new entry of that type is to be made. Then the field texts are written in.

An appropriate template for the entry type @article could be

```
@ARTICLE{<key>,
    AUTHOR  = {},
    TITLE   = {},
    YEAR    = {},
    JOURNAL = {},
        VOLUME  = {},
        NUMBER  = {},
        MONTH   = {},
        PAGES   = {}
}
```

in which the <key> is a reminder that the keyword is to go here, the required fields come first and are indented relative to the entry type name, and the optional fields are further indented. The field note is not included since it is less commonly used and since it may be included in all entry types.

Finally, we mention that BibTEX has been written by Oren Patashnik, Stanford University, in close association with Leslie Lamport. The BibTEX installation should contain the two files `btxdoc.tex` and `btxhak.tex` for additional information about BibTeX. In particular, `btxhak.tex` contains instructions for writing bibliographic style files. Both these files are to be processed with LaTeX and output in the usual way for document files.

B.3   BibTeX extensions

B.3.1   Author–year bibliography styles

The bibliography style used in this book, as well as in many of the journals of the natural sciences, is completely different from that provided by LaTeX. In this style, reference is made to other published works by citing the author's name and year of publication. Furthermore, the citation may be either parenthetical [*Jones et al.*, 1990] or textual, as shown by *Jones et al.* [1990]. There may be some variations within this style: for example, in this book parentheses are used in place of square brackets and the name is left in normal type.

Standard LaTeX produces a citation as a running number within square brackets [10], and will permit a comment to be added [10, page 188], but does not recognize textual citations. The numbers are considered to be labels which are then listed in the bibliography itself beside the corresponding literature references. The author–year bibliography contains no labels since the correspondence between citation and reference is clear from the list of authors and year.

The standard bibliographic styles cannot support author–year citations because they do not transfer the necessary information from the BibTeX database to the `.bbl` file. However, there are a number of non-standard `.bst` files that do transfer this information but they can only be used with appropriate package files that interpret it properly. The packages must also redefine the `thebibliography` environment to suppress the label, as a minimum.

The simplest of these, by Oren Patashnik, the originator of BibTeX, are called `apalike.bst` and `apalike.sty`. This style gives the `\bibitem` command an optional argument containing the author and year, as

```
\bibitem[Jones et al., 1990]{jone90}
```

The `\cite` command operates as normal, which means that only parenthetical citations are possible.

More refined packages offering both parenthetical and textual citations are `newapa.sty`, `chicago.sty`, `astronomy.sty`, and `harvard.sty`. Each of these has its own set of `.bst` files which cannot be used with any of the other packages. And each package has its own modified citation syntax.

In an attempt to unify this situation, Patrick W. Daly has written a citation package named `natbib.sty`, which, although having its own

method of transferring author–year information to LaTeX, does accept
.bst files intended for all the other packages listed above too. One can
even choose numerical or author–year citation mode from a single one
bibliographic style file. Furthermore, the standard .bst files can also be
used with natbib, although only for numerical citations.

The natbib package has a flexible citation syntax permitting parenthetical (\citep) and textual (\citet) citations, with the abbreviated or
full author list, with optional notes both before and after the citations.
For example,

\citet{jon90}	⇒	Jones et al. (1990)
\citet[pg.~22]{jon90}	⇒	Jones et al. (1990, pg. 22)
\citep{jon90}	⇒	(Jones et al., 1990)
\citep[pg.~22]{jon90}	⇒	(Jones et al., 1990, pg. 22)
\citep[e.g.][]{jon90}	⇒	(e.g. Jones et al., 1990)
\citep[e.g.][pg.~22]{jon90}	⇒	(e.g. Jones et al., 1990, pg. 22)
\citet*{jon90}	⇒	Jones, Baker, and Toms (1990)
\citep*{jon90}	⇒	(Jones, Baker, and Toms, 1990)

The citation punctuation (type of brackets, points between multiple citations, dates, and so on) can be selected by options when loaded, specified
by declarations, or programmed to be associated with the bibliography
style.

Multiple citations are available as usual, with compression of repeated
authors:

\citet{jon90,jam91}	⇒	Jones et al. (1990); James et al. (1991)
\citep{jon90,jam91}	⇒	(Jones et al., 1990; James et al. 1991)
\citep{jon90,jon91}	⇒	(Jones et al., 1990, 1991)
\citep{jon90a,jon90b}	⇒	(Jones et al., 1990a,b)

Additional options allow multiple citations to be sorted according to their
order in the reference list (usually alphabetical then by date), or to give
the full author list automatically on the first citation.

In numerical citation mode, the command \citep behaves like the
standard \cite command, printing the reference's number in brackets,
while \citet gives the authors' names followed by the reference's number.
This is textual citation in numerical mode, allowing the author to use the
same text in both modes without any rewriting.

\citet{jon90}	⇒	Jones et al. [21]
\citet[pg.~22]{jon90}	⇒	Jones et al. [21, pg. 22]
\citep{jon90}	⇒	[21]
\citep[pg.~22]{jon90}	⇒	[21, pg. 22]
\citep[e.g.][]{jon90}	⇒	[e.g. 21]
\citep[e.g.][pg.~22]{jon90}	⇒	[e.g. 21, pg. 22]
\citep{jon90a,jon90b}	⇒	[21, 32]

The `natbib` package is available from the CTAN file servers listed in Section D.6, as are all the other packages mentioned here.

B.3.2 Customizing bibliography styles

It seems that every journal and publishing house has its own set of rules for formatting bibliographies. The differences are really minor, such as whether to use commas or colons here, or to put volume numbers in bold or italic typeface. Nevertheless, in spite of this triviality, the rules are rigid for each house.

Standard LaTeX provides only four bibliography styles, and their differences have more to do with sorting and labeling than with such fussy details. It is possible to design one's own `.bst` file, by modifying existing ones; however, this requires a knowledge of the peculiar BIBTEX programming language, something that puzzles even experienced LaTeX users. There are some 50 `.bst` files in the BIBTEX directories of the TeX file servers, and probably many more in contributed LaTeX directories. Each is designed for a specific journal and there is no guarantee that it would be accepted by another one. What is a user to do if his or her publisher demands a certain format that just is not to be found? How can one seek a given format anyway among those offered?

Some help can be found in the `custom-bib` package written by Patrick W. Daly, available in the contributed LaTeX directories of the file servers. The heart of this package is a generic (or master) bibliography style `merlin.mbs` containing alternative coding for a multitude of different bibliographic features or options. It is processed with the DocStrip program which is capable of including alternative coding according to selected options (see Section D.3.1).

Because of the large number of options offered (well over 100), a menu-guided interface is provided, called `makebst.tex`. When processed under TeX or LaTeX, this 'program' first asks which `.mbs` file is to be read, and then interactively constructs a DocStrip batch job to produce the selected bibliographic features according to menu information contained in the `.mbs` file itself. This means that there could be a number of `.mbs` files to choose from, each one explaining its set of options to the user via `makebst`.

Support for other languages is provided by means of additional `.mbs` files, one for each language, which contain the translations for such words as *volume*, *editor*, *edition*, and so on.

Some of the features provided by the supplied generic bibliographic style file are:

- numerical or author-year citations; in the latter case, one may also choose which \bibitem style is to be used;

- order of references: by citation order, alphabetical by all authors, alphabetic by author–year label;

- format of authors' names: first name plus surname, initials plus surname, surname plus initials, reversed initials only for first author, and more;

- number of author names to include before giving *et al.*;

- typeface to be used for the author names;

- position of date, parentheses or brackets around the year;

- format of volume, number, and pages for journals;

- capitalization of article titles as sentence or title style;

- whether to use the word *and* or an ampersand *&*;

- placement of commas with the word *and*;

- whether to abbreviate *editor, volume, chapter*, etc.;

- first and last page numbers or only first;

- optional addition of shorthand designations of common journal names;

- and much more.

LaTeX Programming

One of the major improvements in LaTeX 2_ε is the support for class and package writers, or more simply, for LaTeX programmers. Most users will not really appreciate the other changes from LaTeX 2.09, seeing only some extra font commands, a \documentclass declaration at the start, and the \usepackage command to load what used to be *option files*. There are some additional extras here and there, but generally, they will feel that the new version has not brought them anything truly new. However, the LaTeX programmer who has written those option files, or who has tried to install new font schemes, certainly will appreciate the NFSS and the new class and package control features. By now these contributed extensions to the basic LaTeX installation have become so dependent on LaTeX 2_ε that any user who wants to take advantage of them has no choice but to update his or her system.

In this appendix, we present the special commands that were designed for class and package files, with several examples of useful packages. We also explain how fonts are installed under the New Font Selection Scheme, especially those with encodings that differ from those of the standard CM fonts.

C.1   Class and package files

C.1.1   Style files in LaTeX 2.09

The concept of class and package files is new to LaTeX 2_ε, replacing those of *main style* and *style option* files of LaTeX 2.09. In that system, one had to select the main style with

2.09 \documentstyle[*option_list*]{*style*}

This loaded a file named *style.*sty that defined the overall format for the type of document wanted. The major styles were article, report, and book, just like the most important classes in LaTeX 2_ε. Somewhere in the style file there would be the command \@options which processes the options in the *option_list* as follows:

1. if a command \ds@*option* exists, it is executed; otherwise, the option name is put into a second list;

2. after the *option_list* is exhausted, the second list is taken and for each entry the file *option.*sty is input.

This procedure allows options to be internally defined in the style file, or to be stored in a separate style option file with the same name as that option, and the extension .sty. The idea is that the coding for some options is independent of the main style, so it could be stored external to the style files.

That may have been the original philosophy behind the concept of option files; what developed out of it is a multitude of additional 'options' written by hosts of contributors. LATEX is programmable, so anyone could add on valuable extra features, store them in a file, and read them into other documents. With network servers, these additions became widely distributed and used. Since many of them tinker with internal LATEX commands, they could not be simply read in by an \input command in the preamble, but rather needed to be given the .sty extension so that they could be included as a pseudo-option under point 2 above. They were not really options at all, but extended coding or extra features.

Another difficulty arose in writing new main styles. If one needed an article-like style for a certain journal, or a book-like style for a certain publisher, one could take the existing article.sty or book.sty files as a basis and make new ones with the necessary modifications, but these would not be immune to future updates to the originals. Sometimes these updates were vital to be consistent with changes to LATEX itself. Alternatively, one could simply write an 'option' to contain the modifications, or even a new main style that input the original. However, these would not be true main styles with their own option lists.

C.1.2   The LATEX 2$_\varepsilon$ concept

The wealth of contributed LATEX programming was probably never anticipated by Leslie Lamport when he released LATEX. It is now a fact of life, and indeed one of the great strengths of the system. LATEX 2$_\varepsilon$ not only accommodates such 'foreign' extensions, it actually supports and encourages them, as witnessed by the copious presentations of such packages in *The LATEX Companion* (Goossens *et al.*, 1994).

And this is the way it should be. The extensions have been written by people who needed them, who realized that LATEX was missing something vital for them. On the other hand, to add all of them to the basic LATEX installation would overload it with features that 90% of the users would never require. The philosophy now is that LaTeX provides a fundamental core, or kernel, which is extended first by the standard class files, and then by the myriad of contributed packages and other classes.

It is the role of the LaTeX Team to establish guidelines for programming, to ensure that packages do not clash needlessly with the kernel, or with each other, and to provide a basis of stability so that useful packages continue to operate through further updates to the kernel and the standard classes. This security factor was missing in LaTeX 2.09, where programmers were really left on their own to 'hack' as they felt best, and not to do true programming. The new features of LaTeX 2_ε for class and package control, together with a set of programming tools, should offer an enhanced degree of reliability and durability, both among packages and against future updates of the kernel.

C.1.3 Levels of commands

There are a number of levels of commands with varying degrees of security for the future:

user commands (highest level) described in this and the other manuals, consisting of lower case letters, such as \texttt, are part of the LaTeX external definition to be supported forever;

class and package commands with longish names of mixed upper and lower cases (like \NeedsTeXFormat) are intended mainly for programmers, and are also guaranteed; most are preamble-only commands, but there is otherwise no real restriction to class and package files;

internal LaTeX commands containing the character @ in their names can only be used in class and package files; they are not guaranteed forever, although many of them are indispensable for special effects; a programmer makes use of them at the risk that some day his or her package may become obsolete;

low-level TeX commands also have names with lower case letters, and no @; they should be safe against future evolution of LaTeX, but even this is not absolutely certain; they should be avoided where possible, as explained below;

internal private commands are those used within a contributed class or package file; it is recommended that they all be prefixed with some upper case letters representing the package name and @ in order to avoid clashes with other packages; for example, \SK@cite, from the showkeys package.

A question that confronts LᴬTEX programmers is to what extent the internal LᴬTEX commands may be used in class and package files. There is always a danger that such commands may vanish in later versions, since they have never been documented in the official books (Lamport, 1985, 1994; Goossens *et al.*, 1994). Like the TEX commands discussed in the next section, their use cannot be forbidden, but one must be aware that a certain degree of risk will accompany them.

The guidelines issued by the LᴬTEX Team strongly recommend employing the high-level LᴬTEX commands whenever possible.

- Use \newcommand and \renewcommand instead of \def; if one of the TEX defining commands must be used, because a template is required, or because it must be \gdef or \xdef, issue a dummy \newcommand beforehand to test for a name clash. If it is unimportant whether the command name already exists, issue a dummy \providecommand followed by \renewcommand. The ability to define commands with one optional argument at the high level removes one reason for wanting to reach down to the lower ones.

- Use \newenvironment and \renewenvironment instead of defining *myenv* and \end*myenv*.

- Assign values to lengths and glues (rubber lengths) with the \setlength command, rather than by simply equating.

- Avoid the TEX box commands \setbox, \hbox, and \vbox; use instead \sbox, \mbox, \parbox and the like. With the extra optional arguments in LᴬTEX 2$_\varepsilon$, the need for the TEX equivalents is greatly diminished, and the LᴬTEX versions are far more transparent. Moreover, the LᴬTEX boxes will function properly with the color package, while the others are unpredictable.

- Issue error and warning messages with \PackageError and \PackageWarning rather than with \@latexerr or \@warning; the former also inform the user of the source of the message, instead of labeling them all as LᴬTEX messages.

- There is no suggestion that one should exclusively use the \ifthenelse command from the ifthen package (Section 7.3.5) in place of the TEX conditionals. It seems that this package is offered to simplify employing conditionals, in a manner more consistent with LᴬTEX syntax. Although most of the examples in this book use it rather than the TEX versions, we never employ it ourselves in our own programming.

Adhering to these and similar rules will help ensure that a package will remain fit through future extensions of the LᴬTEX kernel.

C.1.4   TEX commands

Why should primitive TEX commands be shunned? To define commands with \def rather than \newcommand must be just as good, and is often unavoidable. Is there really a chance that it might be removed from a future LATEX3? The primitives are the building blocks on which all flavors of LATEX are constructed. Surely they must remain!

This is not really the point. The primitive TEX commands form the bedrock of any format, and anything defined with them will always do exactly what the programmer expected. However, the equivalent LATEX tools could actually do more as time goes on. The \newcommand checks for name clashes with existing commands, for example. It might even be possible that a debugging device that keeps track of all redefinitions will be added later; any commands defined with \def would be excluded from such a scheme. Even now in LATEX 2$_\varepsilon$, there is something like this to keep track of all files input with middle and high-level commands.

Another example of how low-level programming can go astray is the case of robust commands in LATEX 2$_\varepsilon$. Many commands are intrinsically fragile, meaning that they are prematurely interpreted when used as arguments of other commands, but they may be made robust by prefixing them with \protect. In LATEX 2.09, several fragile commands were defined in a robust manner by including the \protect in the definition, as for the LATEX logo command:

```
\def\LaTeX{\protect\p@LaTeX}
\def\p@LaTeX{...}
```

The true definition is in the internal \p@LaTeX, not the external \LaTeX. Since that definition for the logo actually possessed many flaws, several packages have included an improved version. Those that simply redefined \LaTeX itself made the command fragile; those that were cleverer only redefined \p@LaTeX, with the result that they are totally left behind in LATEX 2$_\varepsilon$, where commands are made robust in a completely different (and much better) manner. (Incidentally, the internal definition of the LATEX logo has also been greatly improved.)

In spite of the desirability of employing only official LATEX commands, there are many occasions when either the internal LATEX commands or the TEX primitives just must be used. The risk of future incompatibility must be taken in order to have a workable package now. However, one should not take this risk lightly, where a high-level equivalent is available.

C.2   LATEX programming commands

All the commands described in this section are new to LATEX 2$_\varepsilon$. They are not essential to class and package files, but they do extend their usefulness and guarantee that they are employed properly.

C.2.1   File identification

Three commands test that the external environment in which the class or package has been inserted is correct. The first of these is

\NeedsTeXFormat{*format*}[*version*]

The first statement in a class or package should be the declaration of the TEX format needed. Although there are existing formats with other names, only the one named LaTeX2e actually recognizes this statement. All others will immediately issue the error message

```
! Undefined control sequence.
1.1 \NeedsTeXFormat
                   {LaTeX2e}
```

which all by itself is fairly informative.

What is perhaps of more use within LATEX 2$_\varepsilon$ is the optional *version* argument, which must contain the date of issue in the form yyyy/mm/dd. If a package makes use of features that were introduced in a certain version, its date should be given, so that if it is used with an earlier version of LATEX 2$_\varepsilon$, a warning is printed. For example, the command \DeclareRobustCommand did not exist in the preliminary test release of LATEX 2$_\varepsilon$, but was first introduced with the official release of June 1, 1994. Thus any package containing this command should begin with

\NeedsTeXFormat{LaTeX2e}[1994/06/01]

The form of the date is important, including the zeros and slashes.

This declaration is not limited to class and package files: it may also be issued at the start of the document itself to ensure that it is processed with the right LATEX. It must, however, be given in the preamble.

The next two commands identify the class or package file itself:

\ProvidesClass{*class*}[*version*]
\ProvidesPackage{*package*}[*version*]

In both cases, the *version* consists of three parts: date, version number, and additional information. The date is in the same format as above, while the version number can be any designation without blanks, and the additional information is text with or without blanks. An example is

\ProvidesPackage{shortpag}[1995/03/24 v1.4 (F. Barnes)]

Only the date part is actually checked by LATEX against the date specified in the calling \usepackage command. The version number and additional information are printed out if \listfiles has been requested. However, the above format is necessary for the \GetFileInfo command in the doc package (Section D.3.2).

Both the \documentclass and \usepackage commands (as well as \LoadClass and \RequirePackage) may take an optional argument to specify the earliest acceptable release date for the class/package. For example, with

> \documentclass[12pt]{article}[1995/01/01]

if the article class file loaded contains

> \ProvidesClass{article}[1994/07/13 v1.2u
>   Standard LaTeX document class]

a warning message is printed. The same procedure applies to the commands \usepackage and \ProvidesPackage.

This system of version checking allows a document to insist that suitable versions of the class and package files are loaded. It assumes, however, that all later versions are fully compatible with earlier ones.

There is a further identifying command for general files, those to be loaded with \input.

> \ProvidesFile{*file_name*}[*version*]

There is no checking of the name or version in this case, but both pieces of information will be printed by \listfiles.

C.2.2   Loading further classes and packages

In the main document file, classes are read in by means of the initializing \documentclass command, and packages with \usepackage. Within class and package files, the commands

> \LoadClass[*options*]{*class*}[*version*]
> \RequirePackage[*options*]{*package*}[*version*]
> \LoadClassWithOptions{*class*}[*version*]
> \RequirePackageWithOptions{*package*}[*version*]

must be used instead. The first allows one class file to load another, with selected options, if desired; the second permits class and package files to load other packages. Only one \LoadClass command may appear within any class file; it may not be called from a package file. Neither command may be invoked in the document file. The *packages* argument may be a list of several package names, separated by commas.

The ...WithOptions variants load the class or package with all those options that were specified for the current one, something that is often required. (These are only available as of December 1995.)

How the optional *version* arguments interact with the corresponding \Provides.. command has been explained in the previous section; how the *options* argument is treated is described below.

C.2.3   Processing options

Both classes and packages may take on options which are defined with

> \DeclareOption{*option*}{*code*}

where *option* is the name of the option and *code* is the set of instructions that it is to execute. Internally, a command named \ds@*option* is created. Often the code does nothing more than set flags, or input an option file. (\RequirePackage may *not* be used within the option code!) Two examples from article.cls are

> \DeclareOption{fleqn}{\input{fleqn.clo}}
> \DeclareOption{openbib}{\setboolean{\@openbib}{true}}

A default option is defined with \DeclareOption*, which takes no option name, specifying the code to be executed for all requested options that are undefined.

There are two special commands that may be used only within the *code* of the default option definition:

\CurrentOption contains the name of the option being processed;

\OptionNotUsed declares \CurrentOption to be unprocessed.

For example, to have a class file emulate LᴬTᴇX 2.09 behavior where all undefined options load a .sty file of the same name, define

> \DeclareOption*{\InputIfFileExists{\CurrentOption.sty}%
>   {}{\OptionNotUsed}}

which first checks if there is a .sty file of the requested name, and if not declares the option to be unused. Requested options that have not been used (processed) are listed in a warning message.

The options are then processed with the commands

> \ExecuteOptions{*option_list*}
> \ProcessOptions
> \ProcessOptions*

where \ExecuteOptions calls the \ds@*option* commands for those options in *option_list*. This is normally done to establish certain options as being present by default. \ProcessOptions executes all the requested options *in the order in which they were defined* and then erases them. Options are therefore executed only once by this command. The *-version is similar, except that the options are executed *in the order requested*. For compatibility with LᴬTᴇX 2.09 styles, the command \@options is maintained as another name for \ProcessOptions*.

It is also possible to specify options for a class or package file with

> \PassOptionsToClass{*options*}{*class_name*}
> \PassOptionsToPackage{*options*}{*package_name*}

where *options* is a list of valid options recognized by the specified class or package file. These commands may be used within the definition of other options. The class or package named must later be loaded with \LoadClass or \RequirePackage.

If the default options of class and package files have not been altered by \DeclareOption*, the standard procedure for handling options that have been requested but are undefined is

- all options requested in the \documentclass statement are designated *global*; they are considered to apply to all subsequent packages, but not to classes loaded with \LoadClass; if they are not defined in the main class, no error or warning is issued;

- all options requested with other commands, including \LoadClass and the \PassOptionsTo.., are *local*; if they are not defined in that class or package, an error is issued;

- if there global options that are defined in neither the class nor any of the packages, a warning is issued;

- options, global and local, are executed in the order in which they are defined in the class and packages, unless \ProcessOptions* has been called in which case they are executed in the order in which they are invoked.

C.2.4 Deferred processing

Sometimes, to achieve certain special effects or to avoid possible conflicts with other packages, it is desirable to have some commands executed at the end of the package or class, or at the beginning or end of the document. This can be accomplished with

\AtEndOfClass{*cmds*}
\AtEndOfPackage{*cmds*}
\AtBeginDocument{*cmds*}
\AtEndDocument{*cmds*}

The first two store away *cmds* to be carried out at the end of the class or package file. They can be used by local configuration files that are read in at the beginning, but contain modifications that should be made at the end so that they are not overwritten by the defaults. The last two declarations store away the *cmds* to be executed with \begin{document} and \end{document} respectively. All of these may be issued more than once, in which case the *cmds* are processed in the order in which they were issued.

The *cmds* stored with \AtBeginDocument are inserted into the processing stream effectively within the preamble, but after the command

\begin{document} has done almost everything else that it does. Thus the *cmds* may be considered to be part of the main body but preamble-only commands are also allowed.

C.2.5   Robust commands

In Section 7.3 we showed how new commands may be defined with \newcommand, redefined with \renewcommand, and provisionally created with \providecommand. Two other defining commands, with exactly the same syntax, also exist.

\DeclareRobustCommand{\com_name}[narg][opt]{def}

is used to define or redefine a command named \com_name such that it is *robust*, meaning it may be used in the argument of another command without a \protect before it. If the command already exists, a message is written to the transcript file, and the old definition is overwritten.

The other command checks the current definition of \com_name.

\CheckCommand{\com_name}[narg][opt]{def}

issues a warning message if the definition is not the same as *def*, with the same number of arguments, and so on. This is used to ensure that the state of the system is as one expects, and that no previously loaded packages have altered some important definition.

Both \DeclareRobustCommand and \CheckCommand may be called at any point in the document.

C.2.6   Commands with 'short' arguments

Normally, the arguments to user-defined commands are allowed to contain new paragraphs, with the \par command or with a blank line. In TᴇX jargon, these commands are said to be 'long'. This is not the standard behavior for commands created with the TᴇX \def command, where the arguments must be short in order to act as a test for forgotten closing braces.

As of December 1, 1994, LᴬTᴇX 2_ε provides *-forms of all the defining commands:

```
\newcommand*            \renewcommand*
\newenvironment*        \renewenvironment*
\providecommand*
\DeclareRobustCommand*   \CheckCommand*
```

which create user-defined commands with 'short' arguments in the same way as does \def.

It is recommended that one should almost always take the *-version of these defining commands, unless there is some very good reason to expect that the possible arguments may be 'long', that is, contain new paragraphs. Long arguments should be the exception, not the rule.

C.2.7 Issuing errors and warnings

Classes and packages may be programmed to issue their own error messages and warnings. This is useful to indicate which file is responsible for the message.

Error messages are generated with

> \ClassError{*class_name*}{*error_text*}{*help*}
> \PackageError{*package_name*}{*error_text*}{*help*}

where *error_text* is the message printed to the monitor and to the transcript file, and *help* is additional text printed after the user responds with H. If the texts contain command names that are to be printed literally, they must be preceded by \protect; spaces are generated with \space, and new lines with \MessageBreak. For example,

```
\PackageError{ghost}{%
   The \protect\textwidth\space is too large\MessageBreak
   for the paper you have selected}
   {Use a smaller width.}
```

produces the error message

```
! Package ghost Error: The \textwidth is too large
(ghost)                     for the paper you have selected.

See the ghost package documentation for explanation.
Type  H <return>  for immediate help.
```

Typing H ⟨*return*⟩ produces

```
Use a smaller width.
```

after which LATEX halts again to wait for a response as described in Section 9.1.

Warnings may also be issued from classes and packages in a similar way. The difference is that there is no *help* text, and the processing does not stop for a response. The line number of the input file where the warning occurred may be optionally included.

> \ClassWarning{*class_name*}{*warning_text*}
> \ClassWarningNoLine{*class_name*}{*warning_text*}
> \PackageWarning{*package_name*}{*warning_text*}
> \PackageWarningNoLine{*package_name*}{*warning_text*}

For example, with the warning

```
\PackageWarning{ghost}
   {This text is haunted}
```

one obtains the message

`Package ghost Warning: This text is haunted on input line 20.`

and the processing continues. Warnings may be split into several lines with the \MessageBreak command, just like error texts.

Two last commands of this type are

\ClassInfo{*class_name*}{*info_text*}
\PackageInfo{*package_name*}{*info_text*}

which write their texts only to the transcript file, and not to the monitor. They are otherwise just like the corresponding NoLine warnings.

C.2.8   Inputting files

Files other than classes and packages may also be input, in which case it is often desirable to make sure that they exist beforehand. Or, alternative actions might be taken depending on the existence of a certain file. These goals are met with

\IfFileExists{*file_name*}{*true*}{*false*}
\InputIfFileExists{*file_name*}{*true*}{*false*}

Both these commands test for the presence of the specified *file_name* in the area that LATEX is looking for files, and execute *true* if it is found, otherwise *false*. In addition, \InputIfFileExists reads in that file after executing *true*.

These commands are not restricted to the preamble, nor to class or package files. In fact, the regular \input command is defined in terms of them.

Many special classes make use of these commands to read in a local configuration file. For example, the class ltxdoc contains

```
\InputIfFileExists{ltxdoc.cfg}
    {\typeout{Local config file ltxdoc.cfg used}}
    {}
```

just before \ProcessOptions is called. This allows one to have a local configuration that might specify

\PassOptionsToClass{a4paper}{article}

for a European installation, without altering the files that are processed with the ltxdoc class.

C.2.9   Checking files

Although not really part of programming, two LATEX 2_ε features to keep track of files used are described here. The first of these, already mentioned in Section 8.1.1, is the command

```
\listfiles
```

which may be given in the preamble, even before \documentclass. It causes a list of all input files to be printed at the end of the processing, along with their version and release data. In this way, one has a record of just which files were included, something that may be of use when deciding to send a document file to another installation for processing there. Since any non-standard files may also have to be included, these may be more readily identified from such a listing.

For example, the simple document file

```
\documentclass{article}
\usepackage{ifthen}
\listfiles
\begin{document}
  \input{mymacros}
    This is \te.
\end{document}
```

produces the listing

```
*File List*
article.cls 1998/05/05 v1.3y Standard LaTeX document class
  size10.clo 1998/05/05 v1.3y Standard LaTeX file (size option)
   ifthen.sty 1997/11/02 v1.0n Standard LaTeX ifthen package (DPC)
mymacros.tex
  ***********
```

In this case, the local file mymacros.tex contains no version information because it is missing a \ProvidesFile command.

What should one do if a local file, such as mymacros.tex above, is needed for the processing of a document file that is to be sent elsewhere? One could send it along with the main file, but that requires giving the recipient more instructions on what to do. Or, its contents could simply be included in the main file, for shipping purposes. For a package file, this is not so easy, since internal commands containing the @ sign would cause trouble, and the options would not be handled properly. LATEX 2_ε provides the environment

```
\begin{filecontents}{file_name}
    file contents
\end{filecontents}
```

which may only appear at the very beginning of the document, before the \documentclass command. It tests to see if there is a file on the system with the name *file_name*, and if not, it writes its contents literally to a file of that name. This may be a package file that is subsequently input with \usepackage. In this way, the missing non-standard files can be ported together with the main document file.

If we extend the above simple example by including at the very start

```
\begin{filecontents}{mymacros}
\newcommand{\te}{the end}
\end{filecontents}
```

the newly written file `mymacros.tex` contains

```
%% LaTeX2e file 'mymacros'
%% generated by the 'filecontents' environment
%% from source 'mydoc' on 1998/03/27.
%%
\newcommand{\te}{the end}
```

Note that the `filecontents` environment adds some comment lines to explain where the new file came from. If this is undesirable, the `filecontents*` environment may be used instead.

C.2.10 Compatibility mode

A *compatibility* mode is retained in order to process LᴬTEX 2.09 documents under LᴬTEX 2ε. This is initiated with the `\documentstyle` command in place of `\documentclass`. The entire document file should then conform to the older standard, avoiding all of the new features of LᴬTEX 2ε.

However, compatibility mode will still load the newer class files, rather than the older style files, which may no longer be on the system. It looks first for a file with extension `.cls`, and only if that does not exist will it try to load the `.sty` file. This is to accommodate older, non-standard, style files that can still function as classes.

Since the standard classes, like `article`, exist as `.cls` files, they will be loaded even by `\documentstyle`, but must function like their obsolete `.sty` counterparts. There are very many differences, such as how the margins and text dimensions are set. Nevertheless, the output in compatibility mode must be the same as if it had been processed by LᴬTEX 2.09 with `article.sty`. To accomplish this, compatibility mode sets the boolean switch (Section 7.3.5) `@compatibility` to ⟨*true*⟩, so that any class or package may test for the mode. The form of the test is

$$\ifthenelse{\boolean{@compatibility}}{true}{false}$$

where *true* represents the commands that are to apply only in compatibility mode, and *false* those in the normal, LᴬTEX 2ε native mode.

!   For all the standard class files, dummy `.sty` files, such as `article.sty`, also exist, which do nothing more than issue a warning and load the class file. This is for compatibility with older packages which loaded these files explicitly.

C.2.11 Useful internal commands

!   Notwithstanding the guidelines in Section C.1.3 that recommend avoiding internal LᴬTEX commands, there are a number that are fairly fundamental, and indeed form many of the building blocks of the LATEX kernel and many standard packages. Since they are still internal commands, they are not guaranteed for all future updates. However, if they were to vanish, many of the interesting extension packages provided by the LATEX Team itself would have to be drastically overhauled. We merely present them briefly here for the sake of the bolder user. They all exist in both LATEX 2ε and 2.09.

> \@namedef{*cmd*}{*def*}
> \@nameuse{*cmd*}

define and execute a command named *cmd*, where the backslash is not included in the command name. This name may contain any characters, even those normally forbidden in command names.

> \@ifundefined{*cmd*}{*true*}{*false*}

executes *true* if the command *cmd* does not exist, else *false*. Again, the backslash is not included in *cmd*, and any characters may appear in the command name. This test is often used to define commands conditionally, a task that has been taken over by \providecommand. It may also be employed to determine whether the main class is article-like or not: \@ifundefined{chapter}{..}{..} tests for the existence of the \chapter command.

> \@ifnextchar*char*{*true*}{*false*}

tests if the next character is *char*, and if so, executes *true*, else *false*. This command is traditionally used to define commands with optional arguments, where *char* is [. The new extended syntax of \newcommand offers a high-level means of achieving this.

> \@ifstar{*true*}{*false*}

tests if the next character is a star *, and if so, executes *true*, else *false*. It is used to define *-forms of commands and environments, something that still cannot be done at the high level.

> \@for *obj* := *list* \do {*cmds*}

where *list* is a command that is defined to be a list of elements separated by commas, and *obj* is successively set equal to each of these elements while the code *cmds* is executed once for each element. For example,

> \newcommand{\set}{start,middle,end}
> \@for \xx:=\set \do {This is the \xx. }

prints 'This is the start. This is the middle. This is the end.'

C.2.12 Useful TEX commands

Many of the most sophisticated features of LATEX and its packages can only be programmed with the help of Plain TEX commands. These are described not only in *The TEXbook* (Knuth, 1984), but also in the excellent reference manual by Eijkhout (1992), *TEX by Topic, a TEXnician's Reference.*

We do not intend this book to a manual for TEX; nevertheless, there are a few common TEX commands that appear in many packages, and even in the examples to follow. A brief description of what they do will aid the understanding of these codings. A true TEXpert or TEXnician can skip this section altogether.

> \def*cmd*#1#2..{*definition*}

is the standard defining command in TEX. It is the equivalent of \newcommand* except that there is no check for name clashes, and the arguments are specified differently. For example, the command \Exp to write scientific notation can be defined as

> `\def\Exp#1#2{\ensuremath{#1\times10^{#2}}}`
> or as   `\newcommand*{\Exp}[2]{\ensuremath{#1\times10^{#2}}}`

In both cases, \Exp{1.1}{4} produces 1.1×10^4. However, \def can go further: it can put the arguments in a template, such as

> `\def\Exp#1(#2){\ensuremath{#1\times10^{#2}}}`

to allow the more convenient notation \Exp1.1(4), something that cannot be produced with \newcommand. The \def command is often used when a command is to be defined without knowing (or caring) whether its name already exists, or when a template is needed. Incidently, the TEX equivalent of the unstarred \newcommand is written \long\def.

> \gdef   \edef   \xdef

are variations on \def: the first makes a *global* definition, valid even outside the current environment or {..} bracketing; the second is an *expanded* definition, such that any commands in it have their meanings and not the command itself inserted in the definition; the last is a combination of the other two, expanded and global.

> \noexpand   \expandafter

control the expansion of commands in definitions and execution. Any commands in the definition part of \edef are expanded (their meanings inserted) unless they have \noexpand before them. The opposite is achieved with \expandafter, which jumps over the following command, expands the next one, and then executes the one skipped. This is very deep TEXnology, and is best illustrated by an example with the \Exp command defined above.

> `\newcommand*{\mynums}{1.1(4)} \expandafter\Exp\mynums`

is identical to \Exp1.1(4), whereas \Exp\mynums is not; \mynums is expanded to 1.1(4) *before* \Exp is executed.

> \let*cmd_a* = *cmd_b*   or   \let*cmd_a**cmd_b*

makes *cmd_a* take on the *current* meaning of *cmd_b*. This is often employed to save the current meaning of a command before redefining it, possibly using the older meaning too.

```
\relax
```

does absolutely nothing, but it is often inserted in places where something should be but nothing is wanted.

```
\ifcond true_code \else false_code \fi
```

is the form of a TeX conditional. There are too many variations on the condition *cond* to explain here, but one common application is the equivalent of the LaTeX boolean switch commands:

```
\newif\ifflag        = \newboolean{flag}
\flagtrue            = \setboolean{flag}{true}
\flagfalse           = \setboolean{flag}{false}
\ifflag ..\else..\fi = \ifthenelse{\boolean{flag}}{..}{..}
```

For those who are used to it, the TeX form is more compact, but does not conform to the general LaTeX style of doing things.

```
\ifcase num text_0 \or text_1 \or ... \fi
```

executes one of the *text_num* according to the value of *num*.

```
\endinput
```

terminates the current file being input. This is not really necessary, but it is considered good programming to end all files this way. The main document file does not need it since `\end{document}` has the same effect.

C.3   Sample packages

We present here some demonstration packages to illustrate the programming commands of the previous section. These packages are not trivial but are all useful in their own right.

C.3.1   Modifying the text size

The standard LaTeX classes set the text size parameters `\textwidth` and `\textheight` according to the size option specified in `\documentclass`, such as a4paper or legalpaper. They also adjust the margins so that the text is centered both horizontally and vertically. The value of `\textwidth` is restricted, however, so that there are at most 60–70 characters per line, this being considered optimal by the rules of typography.

Often one wants to ignore this limitation and use the paper to its maximum extent. Only with trial and error can one find a good combination of parameter values that fits a certain paper size. It would be nice to have a package to do this for us.

The `fullpage` package presented here uses the values of `\paperwidth` and `\paperheight`, which have been set according to the paper size option, to produce a margin of one inch on all sides. It even considers what the page style is so that room for head and footlines may be taken into account. Optionally, it may set an even narrower margin of 1.5 cm.

Before proceeding, it might be useful to review the page format parameters shown in the figures on pages 35 and 555.

The package file begins by stating that it needs LATEX 2_ε and by identifying itself. The package information contains the date in prescribed form, the version number, and additionally the initials of the author. The ifthen package is then loaded since conditionals will be required.

```
\NeedsTeXFormat{LaTeX2e}
\ProvidesPackage{fullpage}[1994/02/15 1.0 (PWD)]
\RequirePackage{ifthen}
```

Next the options are prepared. These are to be in and cm for margins of 1 in and 1.5 cm respectively (1 cm is really too narrow). A special length is used to store the margin value. This is a private, internal command, conforming to the convention mentioned on page 329.

```
\newlength{\FP@margin}
\DeclareOption{in}{\setlength{\FP@margin}{1in}}
\DeclareOption{cm}{\setlength{\FP@margin}{1.5cm}}
```

Furthermore, the four standard page styles are also to be option names. They will set two internal boolean switches and the page style.

```
\newboolean{FP@plain}
\newboolean{FP@empty}
\DeclareOption{plain}{\setboolean{FP@plain}{true}
                      \setboolean{FP@empty}{false}
                      \pagestyle{plain}}
\DeclareOption{empty}{\setboolean{FP@plain}{true}
                      \setboolean{FP@empty}{true}
                      \pagestyle{empty}}
\DeclareOption{headings}{\setboolean{FP@plain}{false}
                         \setboolean{FP@empty}{false}
                         \pagestyle{headings}}
\DeclareOption{myheadings}{\setboolean{FP@plain}{false}
                           \setboolean{FP@empty}{false}
                           \pagestyle{myheadings}}
```

Finally, the default set of options is executed, and then the selected options are processed, in this case, in the order specified. The reason for this is that if more than one page style has been given, the last one should dominate.

```
\ExecuteOptions{in,plain}
\ProcessOptions*
```

Now the calculation can begin. First, for plain and empty, there are no headlines, so set the relevant parameters to zero. This is when FP@plain

is ⟨*true*⟩. Then set the space reserved for the footline to zero for the
empty page style (FP@empty is ⟨*true*⟩). For headings and myheadings,
this space is maintained since these styles normally have a plain page
(with footline) on the first page.

```
\ifthenelse{\boolean{FP@plain}}
   {\setlength{\headheight}{0pt}
    \setlength{\headsep}{0pt}}{}
\ifthenelse{\boolean{FP@empty}}
   {\setlength{\footskip}{0pt}}{}
```

With all the margins and headline and footline spacings set, the actual
calculation can be made. Recall that the driver program leaves a 1 inch
margin at the left and above, which must be subtracted from \topmargin
and \oddsidemargin.

```
\setlength{\textwidth}{\paperwidth}
\addtolength{\textwidth}{-2\FP@margin}
\setlength{\oddsidemargin}{\FP@margin}
\addtolength{\oddsidemargin}{-1in}
\setlength{\evensidemargin}{\oddsidemargin}

\setlength{\textheight}{\paperheight}
\addtolength{\textheight}{-\headheight}
\addtolength{\textheight}{-\headsep}
\addtolength{\textheight}{-\footskip}
\addtolength{\textheight}{-2\FP@margin}
\setlength{\topmargin}{\FP@margin}
\addtolength{\topmargin}{-1in}
```

Remember that \paperheight and \paperwidth are set to the full dimensions of the paper by the paper size option. If this is wrongly specified,
of course, the final result will be incorrect.

This completes the coding for the package file fullpage.sty. An
example of how it might be invoked is

```
\documentclass[a4paper,12pt]{article}
\usepackage[headings]{fullpage}
. . . . . .
```

C.3.2   Redesigning the head and footlines

Something that is often demanded by LaTeX users is the possibility of
altering the head and footlines that appear on each page. Standard LaTeX
does provide a limited number of *page styles* (Section 3.2) but these are
not always sufficient. The most flexible of them is the myheadings page
style which allows the author to determine the text of the running heads
with \markright and \markboth. However, the general format, including
font style and placement of the page number, is still fixed by LaTeX.

There is an excellent package by Piet van Oostrum called `fancyhdr` (replacing his earlier `fancyheadings` package) which allows very flexible head and footlines. Here we show an example of how you can do it yourself.

We wish to modify the headline for documentation of large projects requiring particular information in the running head. Apart from a short version of the title and the section and page numbers, they might demand the document serial number, date, and version number. It would be useful to have a generalized documentation page style that allows these entries to be specified outside of the page style definition itself.

Since this coding is only feasible for the `article` class, we will create a new class, called `dochead`, that inputs `article.cls` and then defines a new page style. Actually this class could contain many other special features of which our new page style is only one.

Again we start by declaring the TEX version that we need (release 4 from December 1, 1995), and by identifying the class file.

```
\NeedsTeXFormat{LaTeX2e}[1995/12/01]
\ProvidesClass{dochead}[1998/04/27 1.1 (PWD)]
```

This class has no options of its own, passing all specified ones on to the underlying `article` class by means of the `\LoadClassWithOptions` command, which only became available with release 4.

```
\LoadClassWithOptions{article}
```

Now come the commands that are used to enter the four pieces of information for the headline: short title, date, serial number, and version. Each of these commands stores its argument in an internal `\DH@` command, each of which is initially set to be empty. Finally, the entering commands are declared to be allowed only in the preamble, since it would be disastrous if they were called after the main text had started.

```
\newcommand*{\DocTitle}[1]   {\renewcommand*{\DH@title}{#1}}
\newcommand*{\DocDate}[1]    {\renewcommand*{\DH@date}{#1}}
\newcommand*{\DocNumber}[1]
          {\renewcommand*{\DH@number}{\MakeUppercase{#1}}}
\newcommand*{\DocNumber}[1] {\renewcommand*{\DH@number}{#1}}
\newcommand*{\DocVersion}[1]{\renewcommand*{\DH@version}{#1}}
\newcommand*{\DH@title}{}    \newcommand*{\DH@date}{}
\newcommand*{\DH@number}{}   \newcommand*{\DH@version}{}
\@onlypreamble{\DocTitle}    \@onlypreamble{\DocDate}
\@onlypreamble{\DocNumber}   \@onlypreamble{\DocVersion}
```

The `\@onlypreamble` command is an internal LATEX one.

Note: the commands `\MakeUppercase` (and `\MakeLowercase`, not used here), which change the case of their arguments, are preferable to the corresponding TEX commands used in LATEX 2.09. The former is used here to ensure that the document number always appears in upper case letters.

We now define the new page style, dochead, which means we must create a command \ps@dochead, to be executed by \pagestyle{dochead}. What this command must do is redefine the four internals \@oddhead, \@evenhead, \@oddfoot, and \@evenfoot. We make the footlines empty, and the even and odd headlines identical. That headline is to be a minipage, the full width of the page, containing a table with the relevant document information.

```
\newcommand*{\ps@dochead}{%
 \renewcommand*{\@oddhead}{%
   \begin{minipage}{\textwidth}\normalfont
     \begin{tabular*}{\textwidth}{@{}l@{\extracolsep{\fill}}%
       l@{\extracolsep{0pt}:~}l@{}}%
     \DH@number      & Version & \DH@version   \\
     \DH@title       & Section & \thesection   \\
     Date:~\DH@date  & Page    & \thepage
     \end{tabular*}\vspace{0.5ex} \rule{\textwidth}{0.6pt}%
   \end{minipage}}
 \renewcommand*{\@evenhead}{\@oddhead}
 \renewcommand*{\@oddfoot}{{}}
 \renewcommand*{\@evenfoot}{{}}
}
\pagestyle{dochead}
```

The last line activates the new page style.

It is also necessary to increase the size of \headheight and \headsep because our headline is much higher than normal. We pick a height of 3.5 times a regular line.

```
\setlength{\headheight}{3.5\baselineskip}
\setlength{\headsep}{3em}
```

We now add a further refinement. Such documentation normally wants the pages to be numbered within the sections. Therefore, we want to redefine the command \thepage which prints the page number from the page counter (Section 7.1.4), and we need to modify the \section command to start a new page and to reset the page counter. This is done by storing the current definition of \section and then redefining it.

```
\let\DH@section=\section
\renewcommand*{\thepage}{\thesection-\arabic{page}}
\renewcommand*{\section}{\newpage\setcounter{page}{1}\DH@section}
```

Notice that the new \section command calls the old one stored in the internal command \DH@section.

If the above is stored in a file named dochead.cls, the following main text could be used to invoke it:

```
\documentclass[12pt,a4paper]{dochead}
\DocTitle{Spacecraft Cleanliness}
\DocNumber{Esa--xy--123}
```

```
\DocDate{1998 Feb 26}
\DocVersion{4.2}
\begin{document}
```

. . . .

The headline on the fourth page of Section 3 would then appear as below.

ESA–XY-123	Version: 4.2
Spacecraft Cleanliness	Section: 3
Date: 1998 Feb 26	Page : 3-4

With this example as a model, it should not be difficult to create similar headlines for other applications with different entries.

C.3.3   Reprogramming the sectioning commands

Another aspect of LATEX formatting that one might want to alter is the appearance of the sectioning titles. The fonts used, as well as their sizes and the spacing around them, are all rigidly set in the class files. We know users who are so frustrated by the predefined sectioning that they type in their own by hand, including the numbers, something that completely violates the basic idea of LATEX.

In fact, all the sectioning commands from \section to \subparagraph are defined by means of a generalized internal section command with the following syntax:

\@startsection{*sec-name*}{*level*}{*indent*}{*pre-skip*} {*post-skip*} {*style*}*[*short title*]{*title*}

where both the * and [*short title*] are optional. A command like \section is defined to be \@startsection with all the arguments up to and including *style*; the remaining optional *, *short title*, and *title* are given as arguments to \section itself (Section 3.3.3). The meanings of these arguments are:

sec-name: the name of the section level, for example subsubsection; this name is used to select the appropriate counter and to enter the *title* correctly in the table of contents and page headlines;

level: is the number in the sectioning hierarchy as described on page 40, 1 for \section, 2 for \subsection ... ;

indent: is the amount of indentation from the left margin;

pre-skip: is a rubber length, the absolute value of which is the space inserted above the title; if this is negative, the first paragraph following the section title is *not* indented;

post-skip: is a rubber length; if positive, it is the space inserted below the title; if negative, its absolute value is the space between the title and the subsequent run-in text;

style: font declarations issued when printing the section title, such as \bfseries and \Large;

*: optional; if present, the section counter is not incremented, the section is not numbered, no entry is made in the table of contents;

short title: optional alternative text for the table of contents and page headline; may only be given if the * is absent;

title: the text printed as the section title; if *short title* is missing, this text is also entered in the table of contents and page headline.

The font style of the title is determined by the declarations in *style*; these must be declarations, like \bfseries, and not commands with arguments like \textbf. However, as of June 1, 1996, the last item in the *style* list may indeed be a command with an argument. This permits something like \bfseries\MakeUppercase to force the title to be in capital letters.

Rubber lengths with generous stretch and shrink should be given for *pre-skip* and *post-skip* to allow LATEX more freedom in avoiding bad page breaks.

The section number is written with the command \@seccntformat that takes one argument, the name of the sectioning counter (*sec-name* above) that is to be printed. This may be redefined for special effects, as illustrated in the example below.

Since both \@startsection and \@seccntformat contain @, they are internal commands that can only be used in a class or package file. We give here a brief package, mysects.sty, which redefines all the sectioning commands. The values of *pre-skip* and *post-skip* are those in the standard article.cls.

```
\NeedsTeXFormat{LaTeX2e}[1996/06/01]
\ProvidesPackage{mysects}[1998/03/21 v1.0 (PWD)]
```

The date of the fifth LATEX release is specified because we want to use a command in the font styles. We start with the \section command which is to be centered, \Large and upper case.

```
\renewcommand{\section}{\@startsection {section}{1}{0pt}%
            {-3.5ex plus -1ex minus -.2ex}%
            {2.3ex plus.2ex}%
            {\centering\normalfont\Large\MakeUppercase}}
```

The subsections and subsubsection titles are in small caps and bold slanted fonts respectively.

```
\renewcommand{\subsection}{\@startsection{subsection}{2}{20pt}%
        {-3.25ex plus -1ex minus -.2ex}%
        {1.5ex plus .2ex}%
        {\normalfont\large\scshape}}
\renewcommand{\subsubsection}{\@startsection{subsubsection}{3}%
        {10pt}%
        {-3.25ex plus -1ex minus -.2ex}%
        {1.5ex plus .2ex}%
        {\normalfont\normalsize\bfseries\slshape}}
```

The paragraph and subparagraph titles are to be run-in types (negative *post-skip*).

```
\renewcommand{\paragraph}{\@startsection{paragraph}{4}{0pt}%
        {3.25ex plus 1ex minus.2ex}%
        {-1em}%
        {\normalfont\normalsize\underline}}
\renewcommand{\subparagraph}{\@startsection{subparagraph}{5}%
        {\parindent}%
        {3.25ex plus 1ex minus .2ex}%
        {-1em}%
        {\normalfont\normalsize\itshape}}
```

Finally, redefine the section numbers to have a period after them.

```
\renewcommand{\@seccntformat}[1]{\@nameuse{the#1}.\quad}
```

As an example of output with mysects, consider the short document:

```
\documentclass{article}
\usepackage{mysects}
\setcounter{secnumdepth}{3}
\begin{document}
\section{Historical Outline}
\subsection{Medieval Life}
\subsubsection{The Role of the Priest}
\paragraph{In the Church}
The priest fulfills several functions.
\subparagraph{At the Altar}
Here the priest was master of ceremonies.
\end{document}
```

which produces output:

1.   HISTORICAL OUTLINE

1.1.   MEDIEVAL LIFE

1.1.1.   The Role of the Priest

<u>In the Church</u>   The priest fulfills several functions.

At the Altar   Here the priest was master of ceremonies.

Because the counter `secnumdepth` was set to 3, numbering only goes to the third level.

C.4 Changing preprogrammed text

C.4.1 Caption names

There are a number of titles that are preprogrammed into LaTeX, such as 'Contents', 'Bibliography', 'Chapter'. It might be desirable under certain circumstances to alter these, for example if the work is written in a language other than English. However, even an English writer may prefer to have the word 'Summary' instead of 'Abstract'.

With the LaTeX 2.09 release of December 1, 1991, all such explicit words were replaced by commands that can be easily redefined. This change was motivated by a parallel International LaTeX (Section D.1) which was then incorporated into the official version.

These name commands are not defined in the basic LaTeX format itself, but rather in the various class files, as they are needed. Thus \chaptername exists only in classes book and report, but not in article. This means, if a package is to redefine the word 'Chapter' for all classes, it must do so with something like

```
\providecommand*{\chaptername}{}
\renewcommand*{\chaptername}{Chapitre}
```

The standard set of name commands and their initial values are:

(defined in book, report, and article classes)

\contentsname	{Contents}
\listfigurename	{List of Figures}
\listtablename	{List of Tables}
\indexname	{Index}
\figurename	{Figure}
\tablename	{Table}
\partname	{Part}
\appendixname	{Appendix}

(defined in book and report classes)

\chaptername	{Chapter}
\bibname	{Bibliography}

(defined in article class)

\abstractname	{Abstract}
\refname	{References}

(defined in letter class)

\ccname	{cc}

```
\enclname          {encl}
\pagename          {Page}
\headtoname        {To}
```

(defined in makeidx package)

```
\seename           {see}
```

(defined only in certain non-standard packages)

```
\alsoname          {see also}
\prefacename       {Preface}
\subjectname       {Subject}
\notesname         {Notes}
```

The redefinition of these name commands is a fundamental part of the language packages described in Section D.1. The commands are not redefined directly but rather by means of certain \captions*language* such as \captionsgerman and \captionsenglish that allow for convenient switching back and forth between different languages. For example,

```
\newcommand*{\captionsgerman}{%
  \renewcommand*{\contentsname}{Inhaltsverzeichnis}
  ...
  \renewcommand*{alsoname}{siehe auch}}
```

C.4.2   The date

The \today command for the current date is another one that outputs explicit English words. Its standard definition conforms to the American style of giving dates, that is 'November 15, 1998'. If one wants to redefine this, either for the British style (15th November 1998), or for another language altogether, the best method is to follow the example of the other names commands: do not redefine \today directly, but create commands that allow one to switch back and forth.

For example, we can define \dateUSenglish and \dateenglish making use of the internal TeX counters \year, \month, \day and the TeX \ifcase command (page 343).

```
\newcommand*{\dateUSenglish}{\renewcommand*{\today}{%
  \ifcase\month \or
  January\or February\or March\or April\or May\or June\or
  July\or August\or September\or October\or November\or
  December\fi \space\number\day, \number\year}}
```

```
\newcommand*{\dateenglish}{\renewcommand*{\today}{%
  \number\day \ifcase\day \or
  st\or nd\or rd\or th\or th\or th\or th\or th\or th\or th\or
  th\or th\or th\or th\or th\or th\or th\or th\or th\or th\or
  st\or nd\or rd\or th\or th\or th\or th\or th\or th\or th\or
```

```
st\fi\space \ifcase\month \or
January\or February\or March\or April\or May\or June\or
July\or August\or September\or October\or November\or
December\fi \space\number\year}}
```

Definitions for other languages can be modeled after these examples.

C.5   Installing fonts with NFSS $\boxed{2\varepsilon}$

In Section 8.5 we explained the system of font attributes involved in the New Font Selection Scheme (NFSS). These form the basis of the high-level user commands, such as `\textbf` or `\sffamily`, by which fonts are selected by attributes rather than by explicit names. In this section, we wish to elaborate on the internal workings of NFSS, showing how a set of attributes are associated with a particular font and how special symbols are assigned to their proper positions within various encoding schemes.

C.5.1   Default attribute values

We implied in Section 8.5.2 that the font attribute declarations like `\itshape` are defined as `\fontshape{it}\selectfont`, whereas in fact they make use of certain default attributes. Thus the true definition of `\itshape` is

> `\fontshape{\itdefault}\selectfont`

The default commands available are

Family:	`\rmdefault`	`\sfdefault`	`\ttdefault`
Series:	`\mddefault`	`\bfdefault`	
Shape:	`\updefault`	`\itdefault`	`\sldefault` `\scdefault`

It is `\itdefault` that is defined to be the attribute `it`.

It is also necessary to define the standard attributes chosen when the command `\normalfont` is issued. These are contained in the four defaults

> `\encodingdefault` `\familydefault` `\seriesdefault` `\shapedefault`

All this may sound like a complicated route linking the high-level commands to a particular font. However, it does provide flexibility and modularity. The author only needs to know that three families, two series, and four shapes are available, and does not care what they really are. A programmer defines these with the defaults at a lower level.

An example of how all the standard fonts may be replaced by PostScript ones, by simply redefining the three family defaults, is shown on page 360.

Such redefinitions are much simpler than trying to alter the font declarations themselves, including \normalfont. Those definitions are in fact much more complex than implied here, whereas the default commands really are as simple as indicated.

C.5.2   Defining font commands

!

There are a number of commands available for defining new font declarations and commands. These are intended mainly for LaTeX package programmers but may be used in a normal document as well.

> \DeclareFixedFont{\cmd}{code}{family}{series}{shape}{size}

defines \cmd to be a declaration that selects the font of the specified attributes. It is rigidly fixed in all attributes. This is equivalent to \newfont except that the font is determined by attributes and not by name.

> \DeclareTextFontCommand{\cmd}{font_specs}

defines \cmd to be a font *command* that sets its argument according to *font_specs*. This command is used internally to define all the font commands like \textbf, which is defined with *font_specs* as \bfseries.

> \DeclareOldFontCommand{\cmd}{text_specs}{math_specs}

defines \cmd to be a font *declaration* that may be used in math mode in the manner of LaTeX 2.09: as a declaration, not a command. It is useful for defining commands to be compatible with the old version, but should be avoided. For example, \it is defined with

> \DeclareOldFontCommand{\it}{\normalfont\itshape}{\mathit}

C.5.3   Mathematical alphabets

The font that has been activated for text processing does *not* influence the characters and their fonts in math mode, since special mathematical symbol fonts are used for this purpose. If a formula is to appear in bold face, the command \boldmath (Section 5.4.9) must be issued, which remains in effect until it is countermanded by \unboldmath. Both of these declarations must be made outside of math mode.

These commands may still be employed in the same way under NFSS. However, the internal math font selection command is

> \mathversion{vers_name}

in which the argument *vers_name* currently takes on values normal and bold. The declarations \unboldmath and \boldmath are defined in terms of this command. It is planned that special package files will become available to allow additional sets of math symbols.

On the other hand, mathematical alphabet commands may be issued within math mode itself, to set letters as symbols in particular fonts (Section 5.4.2):

```
\mathrm \mathcal \mathnormal \mathbf \mathsf \mathit \mathtt
```

These are all *commands* operating on arguments rather than *declarations*, unlike the LATEX 2.09 equivalents.

New math font alphabets may be defined by the user. For example, to define a slanted math font \maths1, give

```
\DeclareMathAlphabet{\maths1}{OT1}{cmr}{m}{s1}
```

which means that the new math font command \maths1 selects that font in family cmr with weight m and shape s1, which, with the normal font definitions, is font cms1 in the appropriate size. However, this font will be selected in *all* math versions, whereas it would be more suitable if a bold font were selected when \mathversion{bold} is in effect. This is accomplished by adding to the definition of \maths1

```
\SetMathAlphabet{\maths1}{bold}{OT1}{cmr}{bx}{s1}
```

which redefines \maths1 exceptionally for math version bold to be a font with weight bx. For the normal font definitions, this is cmbxs1 in the current size.

New math versions can be created with

```
\DeclareMathVersion{vers_name}
```

and the fonts belonging to it are determined by issuing \Set... commands for each math alphabet or symbol font.

C.5.4 Mathematical symbol fonts

Mathematical symbols must be defined in a totally different manner from text characters: they bear a command name (like \alpha), may come from various fonts, behave differently depending on type, and can appear in different sizes. Under LATEX 2.09, the symbol names were fixed to the Computer Modern math fonts, but NFSS offers more flexibility for additional (or replacement) symbol fonts.

A symbol font name is declared with the command

```
\DeclareSymbolFont{sym_fnt_name}{code}{family}{series}{shape}
```

which associates the name *sym_fnt_name* with the given set of attributes. This name is not a command, but rather an internal designation for use in defining symbols. The selected font is valid for all versions, but if a different font is to be associated with the same name under other versions,

```
\SetSymbolFont{sym_fnt_name}{version}{code}{family}{series}{shape}
```

may be used to redefine *sym_fnt_name* for that one version.

The standard LATEX setup declares

```
\DeclareSymbolFont{operators}{OT1}{cmr}{m}{n}
\DeclareSymbolFont{letters}{OML}{cmm}{m}{it}
\DeclareSymbolFont{symbols}{OMS}{cmsy}{m}{n}
\DeclareSymbolFont{largesymbols}{OMX}{cmex}{m}{n}
```

the sequence of which is important for reasons that are built deeply into TEX itself.

Once the symbol font names have been defined, they may be used to construct math alphabets and various types of symbols.

`\DeclareSymbolFontAlphabet{`*\math_alph*`}{`*sym_fnt_name*`}`

defines *math_alph* to be a math alphabet based on the font with the internal name *sym_fnt_name*. This command is to be preferred over `\DeclareMathAlphabet` if a symbol font exists with the proper attributes for the math alphabet.

The primary command for defining symbols is

`\DeclareMathSymbol{`*\symbol*`}{`*type*`}{`*sym_fnt_name*`}{`*pos*`}`

which makes *symbol* print the symbol in position *pos* of font *sym_fnt_name*. The *pos* is a number, in decimal (10), octal ('12) or hexadecimal ("0A) representation. The *type* specifies the functionality of the symbol and is one of

`\mathord`	an ordinary symbol
`\mathop`	a large operator like $\sum$
`\mathbin`	a binary operator like $\times$
`\mathrel`	a relational operator like $\geq$
`\mathopen`	an opening bracket like {
`\mathclose`	a closing bracket like }
`\mathpunct`	punctuation
`\mathalpha`	an alphabetic character

Math alphabet commands operate only on the symbols of type `\mathalpha`; other types always produce the same symbol, for a given math version, within all math alphabets.

Similarly, math accent commands are established with

`\DeclareMathAccent{`*\cmd*`}{`*type*`}{`*sym_fnt_name*`}{`*pos*`}`

where *type* is either `\mathord` or `\mathalpha`; in the latter case, the symbol changes with the math alphabet.

Math delimiters and radicals can appear in two different sizes. They are set up with

`\DeclareMathDelimiter{`*\cmd*`}{`*type*`}{`*sym_fnt1*`}{`*pos1*`}{`*sym_fnt2*`}{`*pos2*`}`
`\DeclareMathRadical     {`*\cmd*`}          {`*sym_fnt1*`}{`*pos1*`}{`*sym_fnt2*`}{`*pos2*`}`

which define *cmd* to print the smaller variant from position *pos1* of font *sym_fnt1* and the larger from position *pos2* of *sym_fnt2*.

The sizes have not been specified in any of the above math declarations. This is because there are normally four different sizes available, depending on the math style, as explained in Section 5.5.2. However, these sizes must somehow be specified. This is done with

`\DeclareMathSizes{`*text*`}{`*math_text*`}{`*script*`}{`*sscript*`}`

where the four arguments are numbers giving a point size. When the normal text font is size *text* pt, `\textstyle` will be in *math_text* size, `\scriptstyle` in *script*, and `\scriptscriptstyle` in *sscript* size. For example,

`\DeclareMathSizes{10}{10}{7}{5}`

All the `\Declare...` and `\Set...` commands may only be called in the preamble.

C.5.5   Addressing the attribute values

!

In some programming situations, it is necessary to make use of the current values of the attributes without knowing what they are. These are stored in the internal commands

\f@encoding	\f@shape	\tf@size
\f@family	\f@size	\sf@size
\f@series	\f@baselineskip	\ssf@size

The values of these commands should never be changed directly. However, they may be used to test if they possess a certain value. Since they all contain the character @ in their names, they may only be used in class or package files, and not in the main document file.

C.5.6   Defining fonts under NFSS

Under NFSS, fonts are specified within a document by giving the attributes required and then calling \selectfont. How is the set of font attributes then associated with a particular external font name such as those found in Appendix F? This is done by means of *font definition commands* which are usually stored in files with extensions .def and .fd.

First the declaration

\DeclareFontEncoding{*code*}{*text_set*}{*math_set*}

sets up a new encoding attribute named *code*; whenever a text font of this encoding is selected, *text_set* is executed in order to redefine accent commands or other things that are coding dependent; similarly *math_set* is called for every math alphabet of this encoding. It is possible to define default *text_set* and *math_set* with

\DeclareFontEncodingDefaults{*text_set*}{*math_set*}

This allows general text and math mode settings to be declared with this command, while more specialized ones, which are executed afterwards, appear in \DeclareFontEncoding.

If no font can be found for the specified attributes,

\DeclareFontSubstitution{*code*}{*family*}{*series*}{*shape*}

declares the values of the attributes that should be substituted; substitutions are made in order of *shape*, *series*, then *family*; the encoding is never substituted. If even this fails to find a valid font, then

\DeclareErrorFont{*code*}{*family*}{*series*}{*shape*}{*size*}

determines which font is to be used as a last resort.

A new family with a specified encoding scheme is established with

\DeclareFontFamily{*code*}{*family*}{*option*}

where *option* is a set of commands that may be executed every time a font of this family and encoding is selected.

The main font-defining declaration that associates external font names with font attributes is

> \DeclareFontShape{*code*}{*family*}{*series*}{*shape*}
> {*font_def*}{*option*}

where *option* is additional commands that are executed when one of these fonts is selected.

The *font_def* contains a series of size/font associations, each consisting of a size part, a function, an optional argument, and a font argument. For example,

```
\DeclareFontShape{OT1}{cmr}{m}{n}
    {   <5> <6> <7> <8> <9> <10> <12> gen * cmr
        <10.95> cmr10
        <14.4>   cmr12
        <17.28> <20.74> <24.88> cmr17}{}
```

states that the medium series, normal shape members of the cmr family are to be represented by external fonts cmr5 ... cmr12 for sizes 5–12 pt, by cmr10 scaled to 10.95 pt for sizes near 11 pt, and so on. If the specified size is not present, the nearest size within certain limits is taken instead.

The size part consists of numbers in angle brackets, representing point sizes. The brackets may also contain ranges, as <-10> means all sizes up to, but excluding, 10 pt, <10-14> means 10 pt to less than 14 pt, and <24-> indicates 24 pt and higher. The possible functions are:

(empty) loads the named font scaled to the requested point size; if an optional argument in square brackets precedes the font name, it acts as an additional scaling factor;
>   <11> [.95] cmr10   would load cmr10 scaled to 95% of 11 pt;

gen * generates the font name by appending the point size to the font argument;
>   <12> gen * cmr   loads cmr12;

genb * generates the font name by appending the point size times 100 to the font argument (for DC and EC fonts, page 448);
>   <14.4> genb * ecss   loads ecss1440;

sub * substitutes a different font whose attributes are given in the font argument as *fam/ser/shp*;
>   <-> sub * cmtt/m/n   this is best used when there is no font of the required attributes; a message is output to the monitor and to the transcript file;

subf * is like the empty function, but issues a warning that an explicit substitute font has been loaded;

fixed * loads the specified font at its normal size, ignoring the size part; if an optional argument is given, it is the point size to which the font is scaled, as in

    `<10> fixed * [11] cmr12`   which loads `cmr12` at 11 pt when 10 pt is requested.

All the above functions may be preceded by an `s` to suppress messages to the monitor. Thus silent `sub` * is `ssub` *, and silent empty is `s` *.

As another example, consider the definition of bold, italic, typewriter, for which there is no font in the Computer Modern collection:

```
\DeclareFontShape{OT1}{cmtt}{bx}{it}{
    <-> ssub * cmtt/m/it }{}
```

This substitutes (silently) for all sizes (`<->`) the medium italic typewriter attributes. Which fonts those are is determined by a `\DeclareFontShape` command with those attributes.

The font definition commands may be issued in a package file, or even in the document itself. However, the normal procedure is to store each of the `\DeclareFontEncoding` commands in a file named *code*`enc.def` (for example, `ot1enc.def` for the OT1 encoding), and to place the commands `\DeclareFontFamily` and `\DeclareFontShape` in a file whose name consists of the encoding and family designations plus the extension `.fd`. For example, the shape specifications for encoding OT1 and family `cmr` are to be found in `ot1cmr.fd`. When a coding and family combination that is not already defined is selected, LATEX tries to find the appropriate `.fd` file for input. It is therefore not necessary to input such files explicitly, for they are loaded automatically as required.

However, it is important that the encoding be declared beforehand. If it is not already known in the current format, `\DeclareFontEncoding` must be issued, either explicitly, or by loading the *code*`enc.def` file. One way to do this is to invoke the provided package `fontenc`, as for example

    `\usepackage[OT2,T1]{fontenc}`

where the desired codings are listed as options in square brackets, the last of which is made current.

The normal user will never need to worry about such problems. However, LATEX programmers will find things considerably easier for them. For example, to install PostScript fonts under NFSS is almost trivial. The common PostScript fonts are already defined as separate families in `.fd` files of their own; for example, `ot1ptm.fd` associates PostScript Times fonts to a family named `ptm`. A package to activate these fonts is supplied under the name `times.sty` containing essentially the lines

```
\renewcommand{\rmdefault}{ptm}
\renewcommand{\sfdefault}{phv}
\renewcommand{\ttdefault}{pcr}
\renewcommand{\bfdefault}{b}
```

This makes Times ptm the default Roman family, invoked with the command \rmfamily, Helvetica phv the default sans serif family (called by \sffamily), and Courier pcr the default typewriter family (activated by \ttfamily). The default attribute for bold face is defined to be b instead of the regular bx, since bold extended is not provided by these fonts.

Another two examples of the usefulness of NFSS are the Cyrillic fonts of the University of Washington (Section E.5.2) and the extended EC fonts with the Cork encoding (Section F.4.3). Both of these may be activated within a document simply by selecting another encoding, OT2 for Cyrillic, T1 for the EC fonts. (These encodings must first be declared, for example with the fontenc package as illustrated on the previous page.)

C.5.7 Encoding commands

!

In LaTeX, special characters and accents are addressed by means of commands, such as \O to print the Scandinavian letter Ø. The position of this character in the font tables depends on the encoding (character 31 in OT1 and 216 in T1), so that it is necessary to redefine all such symbol commands when the encoding is altered. This is carried out with the help of certain *encoding commands*, which normally appear in the *code*enc.def file along with the \DeclareFontEncoding command.

To define a command that functions differently in the various encodings,

```
\ProvideTextCommand{\cmd}{code}[narg][opt]{def}
\DeclareTextCommand{\cmd}{code}[narg][opt]{def}
```

are available and behave just like \providecommand and \newcommand except \cmd has the definition *def* only when the encoding *code* is active. Thus \cmd may have different definitions for each encoding.

```
\DeclareTextSymbol{\cmd}{code}{pos}
```

defines \cmd to print the character in the font position *pos* when encoding *code* is active.

```
\DeclareTextAccent{\cmd}{code}{pos}
```

defines \cmd to be an accent command, using the character in font position *pos* as the accent symbol, when encoding *code* is active.

```
\DeclareTextComposite{\cmd}{code}{letter}{pos}
\DeclareTextCompositeCommand{\cmd}{code}{letter}{def}
```

define the action of command \cmd followed by the single *letter* either to print the character in font position *pos* or to execute the definition *def*. These declarations are most useful with the T1 encoding, where many accented letters are separate symbols on their own (Section F.4.3). Thus \'{e} in OT1 prints an acute accent (character 19) over the letter *e*, while in T1, it prints the single character in position 233. This behavior is achieved with

```
\DeclareTextAccent{\'}{OT1}{19}
\DeclareTextComposite{\'}{T1}{e}{233}
```

The command must already have been defined for the encoding, either with \DeclareTextAccent or with \DeclareTextCommand; in the latter case, it must be defined to take a single argument.

All of the above definition commands create new commands for a specific encoding. If the defined commands are invoked in some other encoding, an error message is issued. Default definitions may be provided for all unspecified encodings with

```
\DeclareTextCommandDefault{\cmd}[narg][opt]{def}
\ProvideTextCommandDefault{\cmd}[narg][opt]{def}
\DeclareTextAccentDefault{\cmd}{code}
\DeclareTextSymbolDefault{\cmd}{code}
```

where the first two create a default definition that applies to all unspecified encodings, while the second two declare which encoding is to be taken as the default.

Note: \DeclareTextCompositeCommand, the \Provide... and default commands were added in the LaTeX 2_ε version from December 1, 1994. Any file that uses them should therefore contain \NeedsTeXFormat{LaTeX2e}[1994/12/01].

LATEX Extensions

D

One of the reasons for the great popularity of LATEX is that it can be reprogrammed: if some feature is missing from the basic installation (the kernel) it is possible for a knowledgeable user to add it by writing a package. Under the LATEX 2.09 regime, hundreds of packages (then called *style options*) were made public by means of computer file servers. Many of these are still compatible with LATEX 2ε, while others have been converted to take advantage of the new features, and many new ones have been written for LATEX 2ε only.

It is not the intention of this book to go much beyond the standard LATEX packages; a description of *some* of the many contributed packages is to be found in *The LATEX Companion* by Goossens *et al.* (1994). Instead, we outline only some few extensions for multilingual LATEX and PostScript fonts. We then describe some advanced features of the basic installation, like integrated documentation of packages, some semi-official extensions, and how LATEX 2ε is to be installed. Finally we describe how LATEX documents can be put online in the World Wide Web, as HTML or PDF files.

All LATEX packages described in this appendix may be obtained from the sources listed in Section D.6.

D.1 International LATEX

In the original version of LATEX, several English words such as 'Figure' and 'Bibliography' were included explicitly in certain commands. This in fact violates the rules of good programming which forbid doing anything explicitly. In Europe, LATEX users quickly made customizations for their particular languages. The adaptation for German, german.sty, is a set of macros from many contributors collected together by H. Partl of the Technical University of Vienna, and has become the standard for the German-speaking TEX Users Group (DANTE). It contains some facilities for other languages, including French and English. More details are given in the next section.

The key to `german.sty` is the fact that the standard LaTeX document styles (classes) have been altered so that explicit English words in the output are replaced by command names. These names had become standardized among European users for application to all languages. The modified LaTeX version by J. Schrod from Darmstadt was known as ILaTeX, for International LaTeX. These names and their standard English values are to be found on page 351.

As of December 1, 1991, these naming features became part of the LaTeX standard. They represent good programming practice, since they allow even English-speaking users to change certain titles easily, for example 'Abstract' into 'Summary' or 'Contents' into 'Table of Contents'.

Another important tool for multilingual usage is the TeX \language counter which allows more than one set of hyphenation patterns to be stored in the format with `initex` (Section D.5.2). Different patterns are activated by setting \language to the appropriate number. This is only possible with TeX version 3.0 or greater.

D.1.1   The german package

At the 6th Meeting of German TeX Users in October 1987, a standard set of TeX and LaTeX commands was agreed upon for application to the German language. The package `german.sty` was set up by H. Partl of the Technical University of Vienna, and extended by many other workers. The current version is maintained by Bernd Raichle of the University of Stuttgart. It is meant to function not only with LaTeX 2_ε, but also with Plain TeX and LaTeX 2.09, with the standard font encoding as well as that of Cork (T1 encoding, Section F.4.2).

It is not our purpose to present and explain the entire `german` package, but rather only to illustrate the features it contains specific for German. In the next section, we do the same for the `french` package.

Since `german.sty` was the inspiration for International LaTeX and for the `babel` multilingual system, it does deserve this extra attention at this point.

The original, minimal adaptation for German before `german.sty` existed consists of three commands:

```
\catcode'\"=\active \def"{\"} \def\3{\ss}
```

which make the double quote " a command (active), define it to be the umlaut accent (\"), and define \3 to be the German double *s*, or *eszet*, ß. These were meant only to simplify the typing of German texts.

The double quote command symbol

The modern `german.sty` elaborates the definition of the double quote command so that:

- the umlaut in Computer Modern fonts is positioned closer to the letter than it is in normal LATEX; compare ö ("o) with ö (\"o);

- the double quote may be used in \verb commands and verbatim environments;

- hyphenations are allowed in the whole word in which the umlaut appears;

- the *eszet* may be printed with "s; the older form \3 is kept for consistency with earlier texts, but is not encouraged.

The double quote symbol is also employed for other special functions in German, in particular for word division:

"ck The German rules of hyphenation require that when *ck* is divided, it becomes *k-k*. This is achieved by typing "ck, for example Dru"cker.

"mm Another rule states that when two words are joined together, there may only be at most two doubled letters (for example, *Schwimm + Meister = Schwimmeister*), but when they are divided again, the missing third letter must be reinserted (*Schwimm-meister*). This is entered in the text as Schwi"mmeister. Other double-triple letter possibilities also exist.

"- Suggested hyphenations are typed in with "-, which permits further automatic hyphenations within the word, something that the normal TEX command \- does not allow.

"| Finally, ligatures may be broken with "| where they are not appropriate, such as between joined words (*Auf + Lage = Auflage* auf"|lage and not *Auflage*).

Note: in the controversial reformed German spelling that became official in August 1998, and which has been taught in the schools since two years earlier, triple consonants are retained and ck is never divided. A revised package ngerman implements these new features.

Quotation marks

Instead of the English 'inverted commas', German uses „Gänsefüßchen", or ‚geese-feet', for double and single quotation marks. The commands to produce these are \glqq for German left quote quote, the opening double quote („), \grqq ("), \glq (‚), and \grq ('). The two double quote commands may be abbreviated as "' and "'.

Provision is made for the French «guillemets» which are typed in with commands \flqq and \frqq. These may be abbreviated to "< and ">. Single ‹guillemets› may also be generated with commands \flq and \frq. Both of these often appear in German texts.

Captions and dates

The name commands of International LATEX are activated for German in the manner described in Section C.4, that is, with the command \captionsgerman. They may be switched back with \captionsenglish; a command \captionsfrench is also available in german.sty, although a better French adaptation is achieved with french.sty (Section D.1.2).

The \today command to print out the current date may be redefined with the declarations \dategerman, \dateaustrian, \dateUSenglish, \dateenglish, and \datefrench. The definitions of the two English variants are the same as those in Section C.4.2. Depending on the language chosen, \today produces 'November 15, 1998', '15th November 1998', „15. November 1998", or «15 novembre 1998». The Austrian date differs from the German one only in the name of the first month of the year: *Jänner* instead of *Januar*.

Language selection

With german.sty it is possible to select the French and English languages by means of a \selectlanguage command, which takes as its argument one of german, french, english, USenglish, or austrian. This command activates the hyphenation patterns by setting \language to the corresponding value and sets the corresponding caption names.

D.1.2   The french package

The French language package french.sty was originally initiated independently of german.sty by a number of people writing documents in French. It was then enlarged and secured by Bernard Gaulle who offered his work to the French-speaking TEX Users Group (GUTenberg).

We present here an overview of *some* of the special commands available in this French package for both TEX and LATEX. As for the German example, the purpose is to show the nature of the customizations necessary and the complexity of the problem.

Whereas the main concerns of german.sty are to simplify the umlaut accent and to accommodate the German hyphenation rules, french.sty has no special accent commands but does consider French punctuation and abbreviations. The alterations may be grouped into five categories, each of which has a pair of commands to activate and deactivate the whole group. The declaration \french calls all five activating commands.

Hyphenations

With the command \frenchhyphenation, TEX switches over to the French word division patterns which were loaded into the format by initex. This means that words are divided according to French rules. With the counter-manding \nofrenchhyphenation, it switches back to the standard set, language 0, which is normally English.

Captions and date

The command \frenchtranslation calls \captionsfrench to redefine all the name commands in International LATEX. It also sets \today to be the French version, shown above. There are a number of additional name commands, especially for letters that are not present in standard LATEX. There is also a \resume command parallel to \abstract since many French papers require a summary (a *résumé*) in both French and English.

The command \nofrenchtranslation deactivates the French trans-lations.

Punctuation

The rules of French typography require that extra spacing be inserted before the 'double punctuation marks ! ? ; :'. This spacing is less than the usual interword spacing; the input text should contain a full space before these characters (but see \untypedspaces below), which is reduced automatically when the command \frenchtypography has been given: this paragraph is being set with this command in effect! Is this not obvious? The TEX command \frenchspacing is also activated so that the spacing after the end of a sentence remains the same as that between words.

The only quotation marks allowed in French are the *guillemets* which are printed here as « guillemets » as they appear with TEX and now as « guillemets » with LATEX. They are produced with << and >>, and are some-what different from those generated by \flqq and \frqq in german.sty («German»). They automatically include extra spacing around the en-closed text. Quoted text may also be put into the environment guillemets. The opening guillemets are added at the start of every paragraph within the quotation.

There are a number of sub-options also available, some of which are:

\untypedspaces Inserts the extra space before ! ? ; : even if it has been omitted in the input text (for the benefit of non-French typists).

\noTeXdots Changes the \ldots and \dots commands to print the ellip-sis in the French manner (...) as three closely spaced periods instead of in the English way (...).

\todayguillemets Paragraphs within a second-level quotation, or in the guillemets environment, have an opening « preceding them, the same as the first-level paragraphs: this is the modern style and is the default.

\ancientguillemets   The paragraphs within a second-level quotation are started with closing (») instead of opening («) guillemets; this is an old-fashioned practice in French.

\noenglishquote Turns the single quote characters ' and ' into grave and acute accents ` and ´.

\noenglishdoublequotes Turns the double quote characters '' and '' into the « guillemets ».

\idotless Produces a dotless *i* for accents *î* and *ï*.

The command \nofrenchtypography switches the special punctuation and any sub-options off.

Layout

With the command \frenchlayout, several changes are made in the LᴬTᴇX formats. These are

- All paragraphs are indented, even the first one in a section, which LᴬTᴇX normally does not indent.

- The \item labels in the itemize environment are made into dashes at every level, as illustrated here.

- The sectioning counters are reset after a new part has been started.

- A new environment order is defined like enumerate but numbers its items as 1º 2º 3º ... .

These format changes may be removed with the \nofrenchlayout command.

Abbreviations

A number of new commands are defined when \frenchmacros is issued to help in printing certain common French abbreviations.

\ier To print the masculine 1er with 1\ier.

\iere To print the feminine 1re with 1\iere.

\ieme To print other ordinal numbers, such as 3e with 3\ieme.

\at To print the « a enroulé » (@).

\bv To print the « barre verticale » with the \tt font (|).

\chap To print the « chapeau » or « circonflexe » (ˆ).

\boi To print the « barre oblique inversée » (backslash) with the \tt font (\\).

\tilde To print the tilde (˜).

\Numero To print Nº.

\degres To print the degree sign º.

\fup{x} To raise and reduce x, as M\fup{me} for Mme.

\primo \secundo \tertio \quatro To print 1º, 2º, 3º, and 4º. When followed by a closing parenthesis, for example \tertio), they print 3°).

\quando=n To print any number in the above form, for example: \quando={10} for 10º.

\minMAJ{oe} To print œ or Œ following the automatic case selection in the table of contents and page head. Other arguments are possible, such as ae and i.

These special commands may be turned off with the \nofrenchmacros declaration.

Language selection

The routines in french.sty also provide the possibility to switch back to English, with the command \english. The companion file english.sty contains the caption names in English as well as the relevant switching commands.

Extra languages that make use of an existing set of hyphenation patterns may be defined with \NouveauLangage[num]{lang}. This command then creates a new command \lang which switches to *lang* by setting the counter \language to *num*. A command \langTeX is also executed, which must be separately defined, to activate any other special features of this language.

D.1.3 Multilingual LaTeX—the babel system

One problem with both german.sty and french.sty is that, although they allow further languages to be added, they do so in incompatible ways. It is therefore not possible to make use of both these style files within one document.

The babel system has been developed by Johannes Braams at the Dutch PTT Research Laboratories. Its main purpose is to provide a standard means of switching languages with a flexible method for loading the hyphenation patterns. It has been written to work with both TeX (versions 2.x and 3.x) and LaTeX (international and English only).

The three main requirements for a single language adaptation are: translation of the explicit English words, special commands to simplify certain features of that language, and selection of the appropriate hyphenation patterns. To these babel adds: removal of the special commands and testing for the current language. The language-specific files must conform to the babel selection command structure. The concept of 'dialects' is also included, that is, two 'languages' that share a common set of hyphenation patterns. Examples are German and Austrian as well as British and US English, which only differ in the form of the dates (see page 366).

At present, the babel package contains definition files for the following languages:

> Afrikaans, Bahasa, Breton, Catalan, Croatian, Czech, Danish, Dutch, English, Esperanto, Estonian, Finnish, French, Galician, German, Greek, Hungarian, Irish, Italian, Norwegian, Polish, Portuguese, Romanian, Russian, Scottish, Spanish, Slovakian, Slovenian, Sorbian (Upper and Lower), Swedish, Turkish, Welsh

The necessary definition files are delivered with the installation. The hyphenation patterns for the various languages must be obtained from another source, however. For most of these files, there are no special commands for accents or punctuation: they only redefine the naming commands for International LaTeX.

The german.sty file has been a great inspiration for babel. However, since the command \selectlanguage is employed slightly differently, it was necessary to create a new germanb.sty for babel applications, which contains the same set of special commands, but with a few babel additions. Similarly, the babel French file had to be named francais.sty to distinguish it from french.sty.

The babel files

The babel installation provides the following files for its implementation:

babel.def contains some of the basic macros for running babel, and must be loaded by the first language file read in; the remaining macros are either already in the format itself (loaded from hyphen.cfg) or read in from switch.def; babel.def determines which is the case and reacts accordingly;

switch.def contains the additional babel macros, to be read in if these macros are not already stored in the LaTeX format;

hyphen.cfg contains the same macro definitions as in switch.def, plus some more that are to be run by initex; if this file is available during the creation of the LaTeX 2ε format (Section D.5.2), these macros are built into that format and file `switch.def` is not necessary at run time;

`babel.sty` is a master package that loads the language files specified as options;

`language.dat` contains a list of languages and the file names of their hyphenation patterns; this file is read in by `hyphen.cfg` during the initex run; it is the only file that may (must!) be edited for the particular installation (Section D.1.4);

`esperant.ldf...` the language definition files.

`esperant.sty...` compatibility files that read in the `.ldf` files.

Invoking babel

Under LATEX 2_ε, the `babel` package is loaded with the desired languages as options:

```
\usepackage[english,esperanto]{babel}
```

Alternatively, the language names may be used as global options, something that is recommended if there are other packages that take the languages as options, as

```
\documentclass[english,esperanto]{article}
\usepackage{babel,varioref}
```

In both cases, the last named language is the one that is immediately active.

The recognized language names that may be used as options are:

afrikaans	dutch	germanb	romanian
american	english	greek	russian
austrian	UKenglish	hungarian	scottish
bahasa	USenglish	irish	slovak
brazil	esperanto	italian	slovene
brazilian	estonian	lowersorbian	spanish
breton	finnish	magyar	swedish
british	francais	norsk	turkish
catalan	french	nynorsk	uppersorbian
croatian	frenchb	polish	welsh
czech	galician	portuges	
danish	german	portuguese	

Many of these options are synonyms for other language files; for example, both `hungarian` and `magyar` load the file `magyar.ldf`.

If the `babel` package is loaded, all it does is read in the specified language packages; they in turn read in `babel.def` and possibly `switch.def` (if its macros are not already in the format).

Language switching commands

The normal user needs to know very little about the `babel` internal operations. The new high-level user commands are:

```
\selectlanguage{language}
\begin{otherlanguage}{language} text \end{otherlanguage}
\begin{otherlanguage*}{language} text \end{otherlanguage*}
\foreignlanguage{language}{text}
\iflanguage{language}{yes_text}{no_text}
```

The `\selectlanguage` command and the `otherlanguage` environment are two ways to switch to another language, with all its features and translations. The `\foreignlanguage` command and the `otherlanguage*` environment also switch to *language* but without the translations or date changes; they are meant for setting short sections of text in a language different from the regular one.

The `\iflanguage` executes *yes_text* if *language* is current otherwise *no_text*.

In addition, `\languagename` contains the name of the currently selected language.

Of course, each language definition file can also have its own special commands to simplify typing, as illustrated in Sections D.1.1 and D.1.2 for German and French.

Contents of a language definition file

Although one does not normally need to know anything about how the switching mechanism functions, we will outline it here for the interested user.

Two additional internal `babel` commands are:

```
\addlanguage{lang_num}    and
\adddialect{lang_num_1}{lang_num_2}
```

where *lang_num* is an internal language number for specifying the set of hyphenation patterns. The command `\addlanguage` sets its argument equal to the next available language number. The form of *lang_num* used in `babel` is `\l@`*language*; for example, `\l@english`. This command is executed on those language names listed in `language.dat` during the initex run. The command `\adddialect` sets the first argument equal to the second so that the two languages make use of the same set of hyphenation patterns. For example, in `english.ldf` there is the command `\adddialect{\l@american}{\l@english}`.

A language definition file must provide four commands:

\captions*lang* to redefine the naming commands like \tablename;
\date*lang* to define \today;
\extras*lang* to define any language-specific commands;
\noextras*lang* to remove the language-specific commands.

If the language *lang* is not one of those with prestored hyphenation patterns in the current format file (that is, if \l@*lang* is undefined), it is set to be a 'dialect' of language number 0.

Finally, the language definition file calls \selectlanguage{*lang*} to activate that language. This is accomplished by

- calling \language\l@*lang* to select the set of hyphenation patterns,

- invoking \originalTeX in order to remove any existing language-specific commands,

- activating \captions*lang*, \date*lang*, and \extras*lang* to invoke the language-specific names, date, and commands,

- redefining \originalTeX to be \noextras*lang* so that the next language switch will remove those features specific to *lang*.

D.1.4 Contents of the `language.dat` file

As has been mentioned before, with TeX version 3.0 and later it is possible to store more than one set of hyphenation patterns in the format file. The counter \language is used to switch between them by setting it to a different number.

However, there is no standard to dictate which languages belong to which numbers. If a package were to assume a certain ordering, it would most certainly function incorrectly at installations other than the one for which it was designed. The babel system invokes a much more reliable procedure, which is also used by the french package.

During the production of the format by initex (Section D.5.2), the file hyphen.cfg is input, which in turn reads in language.dat, the only file to be tailored to the local installation. This file contains a list of languages to add as well as the name of the file with the hyphenation patterns and the name of any additional file to be included. It also indicates names of dialects that use the same hyphenations by prefixing the name with an equals sign. For example, if language.dat contains

```
english    hyphen.tex
=UKenglish
=USenglish
germanb    ghyphen.tex
french     fhyphen.tex   french.exc
```

the hyphenation patterns stored in files hyphen.tex, ghyphen.tex, and fhyphen.tex are loaded under languages 0, 1, and 2 respectively, and \l@english, \l@germanb, and \l@francais are defined as numbers 0, 1, and 2 for use with the \selectlanguage command. Languages \l@UKenglish and \l@USenglish are equated to the last loaded language, in this case, 0. A list of French exceptions in french.exc is loaded as part of language 2.

D.2 LATEX and PostScript

The introduction of PostScript as a printing and plotting language has established a new standard for ready-to-print files that is very flexible and still portable by electronic transfer to other institutions. A number of DVI driver programs exist to translate the .dvi output of TEX into this versatile output format. The most popular of these is called dvips, a non-commercial program by Tomas Rokicki, available on the CTAN file servers (Section D.6).

PostScript can offer many additional features to LATEX, such as the importation of external graphics, color, scaling, and rotating. Originally, each driver program had its own set of packages for implementing these features, each with different syntax. Now LATEX 2$_\varepsilon$ has standardized the high-level user commands for all drivers with the official extension packages graphics and color. Since these packages are considered to be standard, albeit advanced, features, not necessarily restricted only to PostScript, they are not explained here but rather in Chapter 6, Sections 6.2 and 6.3.

D.2.1 Invoking PostScript fonts

Using a PostScript driver for the output of a LATEX document means that the many PostScript Type 1 fonts may also be exploited.

A collection of packages can be obtained to simplify the selection of these fonts. They are to be found in the directory psnfss on the CTAN servers (see the diagram on page 401). They contain the package files as well as the necessary font definition .fd files for the 35 standard PostScript fonts, plus a number of other commercially available fonts. The packages are actually quite simple; for example, the times.sty file contains only four lines, which are listed on page 360. All these packages do is redefine the three font families, \rmdefault, \sfdefault, \ttdefault and the bold face default attribute \bfdefault.

The packages for the 35 standard fonts, and the three assignments that they make, are listed in Table D.1 opposite. (The 35 fonts include italic and bold variants of these.)

Table D.1: The psnfss packages and their fonts

Package	\rmfamily	\sffamily	\ttfamily
times.sty	Times-Roman	Helvetica	Courier
palatino.sty	Palatino	Helvetica	Courier
newcent.sty	NewCenturySchlbk	AvantGarde	Courier
bookman.sty	Bookman	AvantGarde	Courier
avant.sty		AvantGarde	
helvet.sty		Helvetica	

What is missing from the psnfss collection are the necessary font metric .tfm files. These are located on CTAN in the directory font/psfonts, with a subdirectory for each font. For example, the Times-Roman font metrics are in subdirectory Adobe/times.

The .tfm files are not sufficient to install these fonts; one also requires the virtual font .vf files, the NFSS font definition .fd files, and possibly the .map mapping files to translate external font names to their true internal ones. These are available in the metric directory in various subdirectories.

See Section F.5 for more information on PostScript fonts and how LATEX handles them.

D.3 Further LATEX accessories

D.3.1 The DocStrip utility

Since the TEX program is a programming as well as a text formatting language, it can be exploited to provide a number of utility 'programs' to manipulate files. Such programs are immediately portable: if you have TEX, you can run them. One example of such a program is the DocStrip utility which removes comment lines from files. This program was originally written by Frank Mittelbach and further developed by Johannes Braams, Denys Duchier, Marcin Woliński, and Mark Wooding.

At first sight one might ask why anyone would want to do such a thing, since comments are ignored anyway. In the original LATEX release, Leslie Lamport provided all the support files with copious comments, which are many times longer than the coding itself. In those days, storage space and speed were at a premium, so for working purposes he provided the same files without the comments. The commented versions could be stored off-line, for reference. Mittelbach's idea was to create a program to simplify the decommenting of files for compact storage and fast loading.

Out of this simple concept came two additional features that extended the application of DocStrip: alternative lines of coding could be selectively suppressed or included depending on options chosen at processing time; and multiple files, or modules, could be input to form a single output file. This means that one source file can be the home of several different LᴬTEX packages, and that a single package file can be constructed out of many different components. In fact, the main LᴬTEX 2ₑ generation file `latex.ltx` is built up out of over 30 individual files, while the standard class files, with much common coding, along with their option files, are all extracted from the one file `classes.dtx`.

It is with the integrated documentation system of Section D.3.2 that DocStrip finds one of its most important applications today. However, it has other uses as well, such as with the generic bibliographic style files presented in Section B.3.2. It has become something like the C preprocessor and `makefile` utility of Unix, but running under TEX to make it fully portable to all systems.

Running interactively

The simplest way of running DocStrip on a file is to invoke it with TEX (or LᴬTEX, but TEX is faster), as

```
tex docstrip
```

which produces the response

```
***************************************************************
* This program converts documented macro-files into fast *
* loadable files by stripping off (nearly) all comments! *
***************************************************************
****************************************************************
* First type the extension of your input file(s):  *
\infileext=
****************************************************************
```

One replies by entering the input extension (usually .dtx); then one is asked in turn for the output extension, the options wanted, and finally, for the root name of the file(s) to be processed. Execution follows.

There are some limitations to this method, since the input and output files have the same root name, differing only in the extension, and initial and final comments added to the output are fixed. The more flexible means of running the utility is with a batch job.

Running as a batch job

A DocStrip batch job is a file containing instructions for the utility. For example, suppose the `fullpage` package illustrated in Section C.3.1 is put into a documented source file called `fullpage.dtx`, and the true package file `fullpage.sty` is to be extracted with the option `package`, while a documentation driver file `fullpage.drv` is obtained with option `driver`; the batch file, named `fullpage.ins`, could look like

```
\input docstrip

\preamble
This is a stripped version of the original file.
\endpreamble

\postamble
This is the end of the stripped file.
\endpostamble

\declarepreamble\predriver
This is a documentation driver file.
\endpreamble

\declarepostamble\postdriver
End of documentation driver file.
\endpostamble

\keepsilent
\askforoverwritefalse

\generate{\file{fullpage.sty}{\from{fullpage.dtx}{package}}
         \file{fullpage.drv}{\usepreamble\predriver
                             \usepostamble\postdriver
                             \from{fullpage.dtx}{driver}}
        }
\endbatchfile
```

The first and last lines are vital: first the file docstrip.tex is loaded, defining all the special DocStrip commands; then after the instructions have been executed, \endbatchfile ensures an orderly termination.

The commands \preamble and \postamble allow one to insert explanatory comments at the beginning and end of the extracted file. The preamble is often a copyright notice and/or a caveat that the extracted file should never be distributed without the source. Additional labeled pre- and postambles may be declared with \declarepreamble and \declarepostamble, and activated with declarations \usepreamble and \usepostamble.

The instructions \keepsilent and \askforoverwritefalse are optional. The former suppresses processing information during the run while the latter turns off the warning that an existing file may be overwritten.

The main command is \generate which specifies the files to be created with a series of \file commands, each taking two arguments: the name of the new file and a list of instructions for its production. These instructions can be declarations like \keepsilent or \usepreamble, but the main one is the command \from. This again takes two arguments: the name of the input file and a list of the options to be applied. In the above example, each generated file has only a single input source, but multiple input files are possible with a series of \from commands.

The created fullpage.sty file now contains

```
%%
%% This is file 'fullpage.sty',
%% generated with the docstrip utility.
%%
%% The original source files were:
%%
%% fullpage.dtx   (with options: 'package')
%% This is a stripped version of the original file.
%% Copyright (C) 1994 Patrick W. Daly
\NeedsTeXFormat{LaTeX2e}
. . . . . . . . . . . . . . . . . . . . . . . .
\addtolength{\topmargin}{-1in}
%% This is the end of the stripped file.
%%
%% End of file 'fullpage.sty'.
```

It is possible to have a master batch job to process individual ones, in which case the master must input them with \batchinput, and not with \input.

The DocStrip syntax has undergone considerable changes over the years. The above example illustrates the usage as of December 1996. Prior to that, it was necessary to have

> \def\batchfile{*name*}

as the very first line, where *name* is the name of the batch file. This sometimes caused problems if the file were renamed without a corresponding change to this line. The \endbatchfile command did not exist at this time.

Furthermore, before December 1995, it was not possible to generate multiple files with a single pass through the input file(s). The \generate and \file commands did not exist; instead one called \generateFile for each output file.

Rules for removing lines

The \generate and \file commands cause the input file(s) to be transferred to the output file, line by line, according to the following rules:

1. any lines beginning with a *single* % sign are removed;

2. any lines beginning with two % signs are retained;

3. any line beginning with %<opt> (or %<+opt>) will have the rest of its text transferred if opt is one of the selected options; otherwise it is removed;

4. any line beginning with %<!opt> (or %<-opt>) will have the rest of its text transferred if opt is *not* one of the selected options; otherwise it is removed;

5. all lines between %<*opt> and %</opt> are retained or removed depending on whether opt is a selected option or not.

Options can be combined logically, negated, and grouped:

```
a&b        (a and b);
a|b        (a or b);
!a         (not a);
(a|b)&c    (c and one of a or b)
```

For more information on DocStrip, see the documentation which can be obtained by processing docstrip.dtx with LATEX.

D.3.2 Documenting LATEX coding

The documenting of software products is extremely important: it provides on the one hand a manual for the user, and on the other details about the coding for the programmer. The original LATEX styles, as well as the basic latex.tex file, were heavily commented by Leslie Lamport, but only with straightforward normal text. It was Frank Mittelbach's doc package that first allowed sophisticated, integrated documentation of the source codes.

By *integrated* documentation, we mean that the descriptions and the coding are to be found merged together in a single source file. Thus two processes are necessary, one to extract the actual coding on its own (the DocStrip utility of Section D.3.1) and another to print the documentation (the doc package). The basic idea behind this package is that the comments are in fact regular LATEX text, with some extra features to allow automatic indexing and to record the program's evolution. The coding itself appears in a special type of verbatim environment.

The documentation is produced by running a special driver file through LATEX. Such a driver for a source file named fullpage.dtx would be, in its simplest form,

```
\documentclass{article}
\usepackage{doc}
\begin{document}
  \DocInput{fullpage.dtx}
\end{document}
```

The \DocInput command reads in the specified file, but it first alters the function of the % character from 'comment' to 'do nothing'. This means all comment lines in fullpage.dtx become real text to be processed!

Originally, the source files were given the extension .doc, and they were nothing more than a .sty file with the special doc package commands. One could rename the file to .sty and use it as the package file itself. With the introduction of the LATEX 2$_\varepsilon$ standard, both doc.sty and DocStrip have become part of the basic installation, no longer optional extensions. Now the source files are labeled .dtx and they behave as the documentation driver file. This means that one only has to process them under LATEX to obtain the documentation. The package .sty file must be extracted with DocStrip before it can be used.

We describe some of the extra features that the doc package makes available for the 'comments' in the .dtx files, illustrated by an example, the source file fullpage.dtx for the fullpage package presented in Section C.3.1. As usual, a more complete manual can be acquired by processing doc.dtx.

The description part

The documentation consists of two parts: the *description*, which is a manual for the end user, and the *coding*, a detailed explanation of how the software works, including the lines of code themselves. It is possible to suppress the coding part and to print only the description by issuing

    \OnlyDescription

in the preamble.

Special commands for the description part are:

    \DescribeMacro{\macro_name}
    \DescribeEnvironment{env_name}

which are placed at the start of the text that illustrates a new high-level command (macro) or environment. These commands do two things: they place the macro or environment name in a marginal note at that location (for easy reference when reading) and they insert an entry in the index.

Because documentation needs to use the \verb command frequently for printing input text, some abbreviations are provided with

    \MakeShortVerb{\c}   and   \DeleteShortVerb{\c}

which first turn the character *c* into shorthand for \verb*c*, and then restore its normal use. For example, after \MakeShortVerb{\|}, |\mycom| prints \mycom. (These commands can be made available for any document by loading the package shortvrb, which is actually extracted from the doc package.)

If the comment character % has been deactivated, how can one put comments into the documentation text? One way is to make use of the TEX conditional \iffalse to form a block, or *meta* comment, as

```
% \iffalse
%     These lines are ignored even when the
%     percent character is inactive
% \fi
```

The other method is to use ˆˆA in place of %, a special doc feature.

The description part is terminated with

\StopEventually{*final text*}

where *final text* is to appear at the very end of the article; if only the description part is printed, *final text* is printed immediately and the documentation is ended.

The coding part

The coding part should normally contain the more specialized material that is of no interest to the everyday user. The special commands that may be used here are

\begin{macro}{*macro_name*} *text and code* \end{macro}
\begin{environment}{*env_name*} *text and code* \end{environment}

both of which again insert a marginal note and make an entry in the index. They also organize any \changes commands, as explained below.

The most important environment in the coding part is macrocode, which prints its contents as in verbatim, optionally with a code line number. The form of this environment is somewhat special:

␣␣␣␣\begin{macrocode}
lines of code
␣␣␣␣\end{macrocode}

The four spaces before the \end{macrocode} are obligatory; those before the \begin are not necessary, but it is good practice to insert them for symmetry. This environment also counts all the backslashes within it for a checksum test, and makes an index entry for every command name that it finds. So it is something more than a mere verbatim environment!

The coding part is brought to an end with

\Finale

which carries out the checksum test and prints the stored *final text* from \StopEventually. There may actually be more text following it, which is only printed when both description and coding parts are output.

Index of macros and record of changes

The doc package makes automatic entries into an index by means of the two \Describe*xxx* commands and the two environments presented above. As well, all commands that appear in the coding are indexed. However, the indexing is turned on only if one of

> \CodelineIndex   or   \PageIndex

is given in the preamble. The first references the indexed commands to the number of the code line where they appear, the second to the page number. In the second case, the code lines are not numbered, unless the declaration \CodelineNumbered is also issued.

Since not all commands in the coding really need to be indexed, especially those that are part of standard LATEX and TEX, the command

> \DoNotIndex{*list of command names*}

is given, often repeatedly, near the beginning, to exempt the listed commands from being indexed.

The automatic indexing of all the commands in the code slows down the processing considerably. Once the index .idx file has been produced, future runs do not need to repeat this effort (unless there have been changes to the code). The command

> \DisableCrossrefs

will suppress this indexing, but it may be countermanded by an earlier \EnableCrossrefs which neutralizes the disabling command.

The text of the index is generated from the .idx data file by the MakeIndex program (Section 8.4), which must be run with the special indexing style gind.ist as

> makeindex -s gind.ist *filename*

The indexing style file gind.ist can be extracted from doc.dtx.

To print the index, the command

> \PrintIndex

is placed where it should appear. Often this is part of the *final text* in \StopEventually. An up-to-date index can only appear after MakeIndex has been run between two LATEX processings.

A record of changes to the software can be made by inserting

> \changes{*version*}{*date*}{*text*}

throughout the documentation, in both the description and coding parts. To form the change history list, the command

> \RecordChanges

must be placed in the preamble. This enables the change entries to be placed in a glossary file, which is then processed by MakeIndex as

> makeindex -s gglo.ist -o *filename*.gls *filename*.glo

(The indexing style file `gglo.ist` is also extracted from `doc.dtx`.) The change history is then printed in the documentation where

> `\PrintChanges`

is located, again often as part of *final text*. The texts of the \changes commands are ordered, first by version number, then by the name of the macro or environment in which they appear.

Integrity tests

If the source file is to be sent over electronic networks, there is a danger that it might be corrupted or truncated. Two tests are possible to check for this. By placing

> `\CheckSum{`*num*`}`

near the start of the documentation (before the coding anyway), all the backslashes in the `macrocode` environments will be added up, and the total compared with the number *num* by the `\Finale` command. If *num*=0, the true total will be printed on the monitor; otherwise, if the sum does not agree with *num*, an error is printed, with the two values.

The other test checks that the character set has not been corrupted by passing through computer systems with different character codes.

```
\CharacterTable
{Upper-case      \A\B\C\D\E\F\G\H. . .
 . . . . .
 . . . . .   Tilde     \~}
```

The argument must agree exactly with that expected by `doc` (except for inactive % signs and multiple spaces). It should be copied from the `doc.sty` or `doc.dtx` files.

Obtaining the file information

A new feature in the LATEX 2$_\varepsilon$ version of doc is the command

> `\GetFileInfo{`*filename*`}`

which defines \filename, \filedate, \fileversion, and \fileinfo from the optional release information to be found in the \Provides*xxx* command identifying the specified file (Section C.2.1). The idea is that the .dtx file should contain this information once, and only once, but it needs to be known to print it in the title of the article. These \file*xxx* commands may be used for this purpose. The release information must conform to the sequence *date*, blank, *version*, blank, *text*.

The ltxdoc class

A special class called ltxdoc is provided to assist running the doc package. It invokes the article class with the doc package, and then issues commands

```
\AtBeginDocument{\MakeShortVerb{\|}}
\CodelineNumbered
\DisableCrossrefs
```

and defines a number of other useful commands for aiding the documentation. See the description by processing ltxdoc.dtx. It also provides for local configuration: if ltxdoc.cfg exists, it is read in. This can pass paper size or other formatting options to article, issue \OnlyDescription by means of \AtBeginDocument, and so on.

A sample .dtx file

To illustrate these features, we show part of the source file fullpage.dtx. The initial lines ensure that all the files that can be extracted from it (the package .sty and the documentation driver .drv) receive their proper identifying commands. The date and version information appears only once, but is transferred to both extracted files.

```
% \iffalse        (This is a meta-comment)
%% Copyright (C) 1994 Patrick W. Daly
\NeedsTeXFormat{LaTeX2e}
%<*dtx>
\ProvidesFile        {fullpage.dtx}
%</dtx>
%<package>\ProvidesPackage{fullpage}
%<driver>\ProvidesFile{fullpage.drv}
% \fi
%\ProvidesFile{fullpage}
                [1994/02/15 1.0 (PWD)]
```

It is the last \ProvidesFile{fullpage} that enables the proper functioning of \GetFileInfo; the first one is only a dummy to absorb the information line when the .dtx file is read directly.

Next, the driver part is given. This is what LaTeX sees when it processes the file directly.

```
%\iffalse
%<*driver>
\documentclass[a4paper,11pt]{article}
\usepackage{doc}
\EnableCrossrefs
\RecordChanges
\CodelineIndex
\begin{document}
```

```
    \DocInput{fullpage.dtx}
\end{document}
%</driver>
%\fi
```

Now the checksum and list of commands that are not to be indexed are given, followed by the start of the article.

```
% \CheckSum{73}
% \DoNotIndex{\addtolength,\boolean,\ExecuteOptions,\ifthenelse}
% \DoNotIndex{\newboolean,\newlength,\ProcessOptions}
% \DoNotIndex{\RequirePackage,\setboolean,\setlength}
% \changes{1.0}{1994 Feb 15}{Initial version}

% \GetFileInfo{fullpage}
% \title{\bfseries A Package to Set Margins to Full Page}
% \author{Patrick W. Daly}
% \date{This paper describes package \texttt{\filename}\\
%       version \fileversion, from \filedate}
% \maketitle
% \MakeShortVerb{\|}
%
% \section{Purpose}
% To set a uniform margin of one inch or 1.5~cm on all four
% . . . . . .
```

We advance to the end of the description and start of the coding part.

```
% \StopEventually{\PrintIndex\PrintChanges}
%
% \section{The Coding}
% The first thing is to read in the \texttt{ifthen} package,
% if it is not already there.
%     \begin{macrocode}
%<*package>
\RequirePackage{ifthen}
%     \end{macrocode}
%
% \begin{macro}{\DeclareOption}
% \begin{macro}{\FP@margin}
% Define the options with help of the length |\FP@margin|. The
% options |in| and |cm| select the actual margin size.
%     \begin{macrocode}
\newlength{\FP@margin}
\DeclareOption{in}{\setlength{\FP@margin}{1in}}
\DeclareOption{cm}{\setlength{\FP@margin}{1.5cm}}
%     \end{macrocode}
% \end{macro}
```

The end of the file finishes the coding and calls \Finale.

```
. . . . . .
\setlength{\topmargin}{\FP@margin}
\addtolength{\topmargin}{-1in}
%</package>
%    \end{macrocode}
% \end{macro}\end{macro}
% \Finale
```

D.3.3   Further useful packages

The members of the LATEX3 Team, as individuals, have written a number of packages that they collectively make available in a special directory called tools (see the diagram on page 401). Many of these originate from the days of LATEX 2.09 and have now been updated for LATEX 2_ε.

We describe here briefly what some of these packages do; a detailed explanation of each one can be obtained by processing the corresponding .dtx file with LATEX.

afterpage permits commands to be saved and executed at the end of the current page. This is useful for forcing stubborn floats to be output right away, with \afterpage{\clearpage}. If \clearpage were issued directly, a new page would be inserted at that point. With \afterpage, the command is reserved until the end of the page.

array (Section 4.8.4 on page 104) is a re-implementation of the tabular and array environments with several additional features.

bm permits individual math symbols to be printed in bold face with command \bm{\sym}. Thus $\alpha + \bm{\beta}$ produces $\alpha + \boldsymbol{\beta}$ without recourse to the \boldmath command which would set the whole formula in bold face. There is also a command \DeclareBoldMathCommand{\name}{\sym} that defines \name to be \bm{\sym}.

calc re-implements the counter and length commands to allow 'normal' arithmetic with them, such as

```
\setlength{\mylen}{3cm + \textwidth}
\setcounter{page}{\value{section} * 3}
\parbox{\linewidth / \real{1.6}}{...}
```

Note that the arithmetical expressions can be used not only to set values but also as arguments to commands and environments.

dcolumn (Section 4.8.4) requires the array package; it allows decimal point alignment in the tabular tables.

delarray (Section 4.8.4) requires the array package; it permits large bracketing symbols to be put around array environments, for making matrices of various sorts.

enumerate (Section 4.3.5, page 70) re-implements the enumerate environment so that an optional argument determines the numeration style.

fileerr.dtx when unpacked, produces a set of small files that may be used to reply to TEX when a file is not found; this makes the responses similar to those for error messages; the files are named h.tex, e.tex, s.tex, and x.tex; thus, replying x⟨*return*⟩ loads the last one, which terminates the input.

fontsmpl is a package to output a sample text, accents, and special characters for a given font; with the accompanying file fontsmpl.tex, a family such as cmr can be printed with all attributes and encodings.

ftnright puts footnotes in the right-hand column in two-column mode.

hhline requires the array package; allows more flexibility in putting horizontal rules in tables.

indentfirst indents the first paragraph of sections, which normally are not indented.

layout defines the command \layout which draws a diagram of the current page format, with the values of the parameters printed.

longtable (Section 4.8.4 on page 106) makes tables that extend over several pages, with breaks occurring automatically.

multicol introduces the environment multicols which switches to multi-column output without a page break either before or after. It is called with

> \begin{multicols}{*num_cols*}[*header text*][*pre_space*]
> *Text set in* num_cols *columns*
> \end{multicols}

The optional *header text* is printed in a single column across the top of the multi-column part. A \newpage is issued only if the remaining space on the current page is less than a certain value, stored in the length \premulticols; the optional length argument *pre_space* overrides this value if given. Similarly the length \postmulticols determines whether a new page is inserted at the end of the multi-column text. Column separation and possible rule are set by the standard LATEX lengths \columnsep and \columnseprule as for the \twocolumn command. The text in the multi-column format is balanced; that is, the columns are all equally long.

showkeys prints out all the cross-reference keys defined by \label and \bibitem and used by \ref, \pageref, and \cite; these are marginal and interlinear notes for a draft only, to check the cross-referencing.

somedefs allows only selected commands to be defined in a package, depending on options in \usepackage.

tabularx (Section 4.8.4) defines a tabularx environment which is like tabular*, a table of desired width, except that it is the column widths that expand and not the intercolumn spacing.

theorem provides extensions to the \newtheorem command for more flexible 'theorem-like' environments; it is very similar to the $\mathcal{AMS}$ amsthm package (Section E.4.1).

varioref (Section 8.3.2) defines \vref and \vpageref commands, analogous to \ref and \pageref, which check if the referenced object is only one page away, and if so, print text like 'on the previous page'; the actual text printed can alternate between two variants.

verbatim (Section 4.9.1) is a re-implementation of the verbatim environment that prevents memory overflow for long texts; it also defines a command \verbatiminput to input a file and to print its contents literally, as well as a comment environment for a block comment in the input text.

xr (Section 8.3.2) is a package to allow \ref to cross-reference \label commands from other documents.

xspace contains a device to fix the problem of command names swallowing up the blank that follows them; thus the command \PS defined as

```
\newcommand{\PS}{PostScript\xspace}
```

may be used as \PS file without having to terminate the \PS command with \␣ or {}.

D.4   LaTeX and the World Wide Web

Today it is no longer sufficient to produce professional-looking printed output on paper; one has to be able to get it online as well, that is, make it available in electronic form of some kind. To some extent, PostScript output fulfills this requirement in that it may be considered electronic paper. However, true electronic publishing is something quite different.

The Internet used to be a nice, quiet neighborhood where academics could exchange simple, plain text emails or obtain known files by FTP, until it was invaded by the rowdy *World Wide Web* with its corporate identities, slick advertizing, energetic yuppies, anarchists, and revolutionaries. In other words, it went cosmopolitan. It is here in this glittering marketplace that electronic documents have to compete for attention.

An electronic document is not something that is simply sent to a printer or leisurely viewed on a monitor: it is interactive, in that it guides the reader to other parts of itself or to other documents located anywhere else on the Web. With these *links*, the viewer can jump to other relevant passages with a mere mouse click. Colored illustrations are naturally part of such documents, but so are sounds and movies, as well as means of sending feedback to the author. This is a totally new medium for information exchange, more radical than the Gutenberg revolution from handwritten manuscripts to printed books.

D.4.1 Hypertext Markup Language (HTML)

Text containing links to other sections or to other documents is referred to as *hypertext*, and so-called Web pages are normally written in a language named HTML (*Hypertext Markup Language*). Structurally, this format has many similarities with LATEX which is also partially a markup language. However, 'markup' means that only the logical structure, such as sectioning, is indicated in the original file, not the actual formatting. For HTML, it is the viewer/reader who determines the font type, size, and line widths of the output, not the originator.

Once an author has invested considerable time and effort in producing his or her LATEX source files for an important paper, report, or even a book, he or she may feel the need to 'put it on the Web', that is, to make it available for anyone with an Internet connection to view, copy, and print it. For a complicated document, it could be a ponderous task to recast the entire text, plus tables, figures, and math formulas, into HTML. Here a number of additional programs can be applied to simplify the task. More complete instructions are delivered with them, or can be found in *The LATEX Web Companion* (Goossens and Rahtz, 1999).

The LATEX2HTML program

The LATEX2HTML translator, written by Nikos Drakos with additions by a large number of contributors, is a huge Perl script that converts a LATEX source file into HTML with the help of several other programs, notably LATEX itself, dvips, Ghostscript, and the netpbm library of graphics utilities.

When LATEX2HTML processes a LATEX file, it creates a new subdirectory with the same name as that file, to which it writes the resulting HTML output and any generated images as .gif files. The program interprets the LᴬTᴇX input text in the same way as LᴬTᴇX itself does, but instead of producing typesetting instructions in the .dvi file, it writes appropriate HTML code to an .html file. For example, \section{Introduction} is interpreted as <H1>Introduction</H1> for the output.

This means that LᴬTᴇX2HTML is essentially duplicating the LᴬTᴇX processing, an enormous undertaking. Since the document classes contain varying commands or have common ones behaving differently, there must be Perl scripts for each one (article.perl, and so on) to program these commands properly. Additional packages loaded with \usepackage can also define new commands, or alter the functionality of existing ones; the translator must be informed about these by means of corresponding Perl scripts. For example, the natbib package described in Section B.3.1 defines citation commands \citet and \citep, which need to be made available to LᴬTᴇX2HTML in a file natbib.perl. Most of the tools packages of Section D.3.3 are included as Perl scripts, as are many other popular contributed packages. In other words, the entire LᴬTᴇX2HTML installation must mirror the LᴬTᴇX one.

This process is somewhat simplified by the fact that both formats are markup languages written in pure typewriter text. However, this similarity soon reaches its limits with the many LᴬTᴇX features not available in HTML, such as complex math, included figures, cross-referencing. In the other direction, HTML exhibits hyperlinks, both internally and to external documents, something not provided by normal LᴬTᴇX.

LᴬTᴇX2HTML attempts to reproduce math with the limited HTML possibilities, but failing that, formulas are handled the same as figures and other environments that cannot be directly rendered: they are converted to GIF image files along the route .tex → .dvi → .ps → .pbm → .gif, which are then included in the HTML file as in-line images.

Cross-references and citations are automatically provided with hyperlinks to their targets; the keyword index also links back to the text. Other features of HTML can be included by means of special commands defined in the html.sty package. For example:

- explicit hyperlinks, internal and external

- conditional text for the HTML or LᴬTᴇX versions only

- raw HTML code

- segmentation of the HTML output

- finer image control

- ...

This package thus provides the LᴬTᴇX writer with a more convenient interface for producing hypertext documents; furthermore, the same document can still be processed with LᴬTᴇX to generate the 'paper' version at any time. The conditional text mentioned above allows the electronic and paper versions to exhibit some differences.

LaTeX2HTML is available on the CTAN servers (Figure D.4 on page 401) under the `support` directory. It can also be downloaded from the LaTeX2HTML home page `http://cbl.leeds.ac.uk/nikos/tex2html/doc/latex2html/latex2html.html`, where more information as well as a manual can be obtained.

The TeX4ht program

An alternative program is that by Eitan M. Gurari, called TeX4ht. It produces much the same results as LaTeX2HTML, but by a different route. Rather than trying to interpret the LaTeX input text itself, it processes instead the `.dvi` file to extract the HTML output. It is therefore applicable both to LaTeX and Plain TeX, as its name implies.

In reality, it is not as simple as that. Since the DVI output is normally only a list of typesetting instructions, containing no markup information at all, it is necessary to process the LaTeX input with the matching `TeX4ht.sty` package that redefines all the regular LaTeX commands to write appropriate code to the `.dvi` file. This is accomplished by means of the `\special` command, a basic TeX command for writing instructions to a particular DVI driver. In this case the driver is the TeX4ht program; no other driver will be able to process this `.dvi` file.

The TeX4ht processor writes the HTML output according to various options specified in the LaTeX input. For example, the HTML output can be segmented by chapters or sections, tables of contents may be added to each segment, links are automatically established for cross-references. As with LaTeX2HTML, manual links to external documents or internal points can be made, raw HTML coding included, explicit images included. It is also possible to turn off the HTML conversion in order to produce a normal `.dvi` file; conditional input for whether HTML is on or off is also available.

Any symbols, tables, equations, or imported figures that cannot be directly rendered in HTML are written to a new DVI file, with extension `.idv`, one item per page. Finally, a log file is written containing conversion instructions for each such object; this file is so constructed as to act as a script, or batch job, for creating GIF images. Just how this conversion is realized and how the instructions are formatted depend on the local installation and can be configured by the user. However, the usual route is something like `.idv` → `.ps` (dvips) → `.pcx` (Ghostscript) → `.gif` (some other image conversion program). An example of TeX4ht output with a GIF representation of a math equation can be seen in Figure D.1 on page 393.

Since TeX4ht needs to redefine all the LaTeX markup commands, one could argue that it too is duplicating LaTeX just like LaTeX2HTML. To some extent this is true, although there is a very important difference: it is only the markup, not the formatting commands, that needs to be massaged. Most classes and additional packages will need no extra configuring to be processed by TeX4ht. A package like `natbib` which adds new citation commands `\citet` and `\citep` will totally confuse LaTeX2HTML without the additional Perl script `natbib.perl`, whereas under TeX4ht these commands will produce the correct output text in the HTML file without any additional definitions. (However, the automatic links between the citations and list of references will be missing, since this is a markup feature, but the printed text will be all right.)

TeX4ht is also available on CTAN, in the `support` directory, or from `http://www.cis.ohio-state.edu/~gurari/TeX4ht/mn.html`.

LaTeX2HTML and TeX4ht are only two examples of procedures for creating HTML output from LaTeX. In both cases, one should not try to convert arbitrary LaTeX files, but should view them as means of producing normal LaTeX and HTML output from a common source file. This source file must be constructed accordingly, possibly with conditional texts for the two outputs, with manual hyperlinks, formatting options for HTML, and so on.

The use of GIF bitmap images for rendering items not readily available in HTML, while producing the visual features on screen, does not allow these objects to be searched for or otherwise further processed. Such images also do not participate in automatic font size changes and other Web manipulations.

The techexplorer Hypermedia Browser

A completely different approach to Web viewing of TeX and LaTeX documents is offered by the IBM `techexplorer` Hypermedia Browser, a plug-in for Netscape Navigator and Microsoft Internet Explorer under Windows 95 or Windows NT. When properly installed, it is launched automatically within the parent browser window when one opens a file with one of the extensions `.tex`, `.bbl`, `.ltx`, `.latex`, or `.tcx`.

The `techexplorer` interprets the TeX and LaTeX commands directly and displays the results on the monitor. The original page height and line width are ignored, the paragraphs being fitted to the size of the window as is normal for any browser. Fonts, tables, environments, but most importantly mathematics are reproduced very well. Figure D.1 on the opposite page demonstrates a single formula represented by both TeX4ht and `techexplorer`. Even user commands defined with `\def` and `\newcommand` are recognized, provided they occur within the same file. However, no external packages may be loaded, nor are the class files accepted. (Both `\usepackage` and `\documentclass` are simply ignored.)

Hypertext features can be included with additional TeX-like commands that establish internal and external links, pop-up menus, multiple-file documents (segmentation) with automatic links between them, and in-

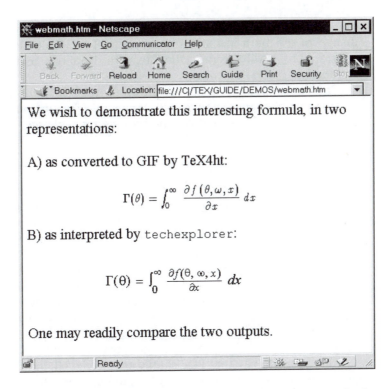

Figure D.1: Example of a math formula represented by TEX4ht (as a GIF image of the original LATEX output) and by `techexplorer` rendering.

clusion of images. In other words, a `techexplorer` document can have the full range of hypertext features of HTML. Here one must make the same comment as for LATEX2HTML and TEX4ht: it is not primarily intended for viewing arbitrary LATEX documents, but rather for preparing hypertext Web pages by means of TEX or LATEX. The only problem here is that most Internet users will probably not have `techexplorer` installed, so that they will only be presented with the input text, and not its rendering.

One additional application of `techexplorer` which is not available to the other programs is to embed TEX commands, especially mathematics, inside a regular HTML file. If the commands are stored in a file `formula.tex`, they could be embedded with

```
<EMBED SRC="formula.tex" WIDTH=300 HEIGHT=70>
```

or they can be entered directly as

```
<EMBED TYPE="application/x-techexplorer"
    TEXDATA="$\Gamma(\theta)=\int_0^\infty\;
    \frac{\partial f\,(\theta,\omega,x)}
    {\partial x}\; dx$" WIDTH=300 HEIGHT=70>
```

The result, shown in the lower part of Figure D.1, is a very acceptable display of the math formula in the middle of an HTML file. (This file was generated by TeX4ht from a LaTeX file, with the above `techexplorer` text included as raw HTML code.)

The `techexplorer` is obtainable from `http://www.software.ibm.com/enetwork/techexplorer`. It comes in a free introductory version, as well as in a professional version for a nominal charge.

D.4.2   Portable Document Format (PDF)

In 1993, Adobe Systems Inc. published the *Portable Document Format*, or PDF, as a page-oriented format for electronic documents. Based very strongly on PostScript, which is also an Adobe creation, it combines fixed textual contents with hypertext elements, compression methods, and security measures. Whereas HTML determines only the document markup and not the format (font size, line width, page breaks), with PDF the document layout is finalized, as it is with PostScript. However, PDF with its interactive hypertext features corresponds much more to electronic publishing with rigid layout than does HTML with its malleable form. If PostScript is electronic paper, then HTML is electronic rubber, while PDF is a true electronic document with hardcopy possibilities.

This is not to say that PDF should replace HTML in all instances. For presenting volatile information, or for guiding a reader around a Web site, or for publicity purposes, HTML is far more suitable. PDF is appropriate for documents with more permanency, that are to be cited. One wants to be sure that line 5 on page 21 will be the same for all other readers.

Producing PDF output

Adobe has made the PDF specification public and freely available so that anyone is entitled to write software to produce and to read files in this format. They have also made available a free reader program, *Acrobat Reader*, so it is hardly worthwhile writing another one, at least not commercially. (Ghostscript can also interpret PDF as well as PostScript.)

Where Adobe makes its money is in the sales of programs to produce and manipulate PDF files. Of these, the main ones are:

- Acrobat Distiller to convert PostScript into PDF, and

- Acrobat Exchange to modify existing PDF files.

In the latter case, 'modification' means adding links, changing colors, including navigational aids, and inserting notes; it does not mean altering the actual textual content in any way.

Today most graphical and text processing programs are capable of generating PDF output. For the LaTeX user, there are two routes that may be taken to produce PDF output. The obvious one is .tex → .dvi → .ps → .pdf using a DVI driver to generate the PostScript and then Distiller to create the PDF file. This is a simple, effective procedure, but the result is little more than compressed PostScript with none of the hypertext or electronic features of a true PDF file. However, some of these features can be added automatically with the help of the hyperref package described below in Section D.4.3.

Another disadvantage of this procedure is that the bitmap Computer Modern fonts appear very badly when displayed. One should take care to make use of PostScript (Type 1) fonts only. The CM fonts are themselves freely available as Type 1 fonts, and these should be used whenever PDF is to be the final output.

pdfTEX

The more direct method is to process the LATEX file with pdfTEX, a variation on the TEX program developed by Hàn Thế Thành. This is distributed with some TEX installations although it is still (at the time of writing) a test version. Feedback from the users directs the further development. The following points can be made:

- pdfTEX behaves just like the regular TEX program unless the declaration \pdftouput=1 is issued before the first page is output; in this case, it generates PDF instead of DVI output and activates many new TEX-like commands for including PDF features.

- Since pdfTEX combines the functionality of both the TEX program and a DVI driver, it must be able to deal with fonts in the way that a driver program would. Originally this meant that it could only handle PostScript Type 1 fonts, but now it can also cope with METAFONT bitmap fonts (Type 3) as well as TrueType fonts.

- When started, the program reads a configuration file pdftex.cfg which can set many parameters, as such page size, offsets, PDF compression level, DVI or PDF output, and can specify the names of font mapping files.

- The font mapping files are similar to those used by dvips (page 458); they relate the TEX name for the font to various characteristics, encoding scheme, true name, and specify the file containing the character drawing instructions.

- New TEX-like commands exist to include PDF features, like compression level, page attributes, document information, opening setup, forms, annotations, links, bookmarks, and article threads.

- Graphics can be included in PNG, TIFF, or JPEG formats; PDF files may be inserted provided they contain only a single page without any fonts or bitmaps. The graphics packages (Section 6.2) support pdfTeX by means of a `pdftex` graphics option.

pdfTeX is a very exciting prospect for electronic publishing of LaTeX documents. One generates a new LaTeX format (Section D.5.2) using pdfTeX instead of TeX, with the `-ini` option, names this `pdflatex.fmt`, and then invokes it with

```
pdftex &pdflatex
```

Normally the installation includes a command `pdflatex` that translates to the above.

To simplify exploiting PDF features with LaTeX-style commands, and to create internal links automatically, the `hyperref` package is indispensable.

D.4.3   The hyperref package

Sebastian Rahtz has written a LaTeX package `hyperref.sty` to add automatic hypertext links to LaTeX documents that are intended to become HTML or PDF files. Not only do all internal cross-references link to their reference points, citations are also linked to the list of references, table of contents to the section headings, and index listings to the original text. Additional links to external documents are possible by means of a single syntax for all types of output.

It is especially for PDF output, whether by means of PostScript plus Distiller or directly with pdfTeX, that this package is extremely useful. It provides LaTeX-like commands for the PDF features, most of which are invoked by options to the package. A typical LaTeX file for pdfTeX might include

```
\usepackage[pdftex,
   pdftitle={Graphics and Color with LaTeX},
   pdfauthor={Patrick W Daly},
   pdfsubject={Importing eps files and use of color in LaTeX},
   pdfkeywords={LaTeX, graphics, color},
   pdfpagemode={UseOutlines},
   bookmarks,bookmarksopen,
   pdfstartview={FitH},
   colorlinks,linkcolor={blue},citecolor={blue},
   urlcolor={red},
]
{hyperref}
```

The first option above, `pdftex`, specifies the type of output; for PostScript, `dvips` or one of the other supported drivers must be given. The next four

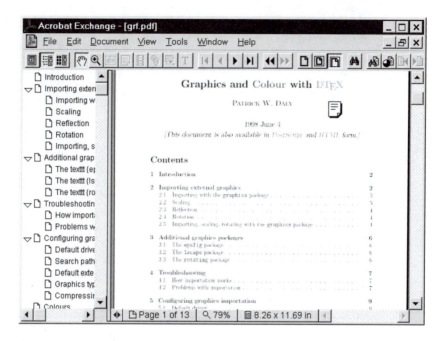

Figure D.2: Output produced by pdfTeX with the `hyperref` package for automatic links, with an outline list at the left.

options are entries for the PDF document information stored in the file but not displayed in the text. The remaining options determine how the file is to be viewed when opened and to establish the colors of the links. Sample output can be seen in Figure D.2.

Many more options are available. The package can also produce HTML and PDF forms, set explicit links, and create buttons to execute PDF viewer commands, such as 'Next Page'.

Like pdfTeX itself, `hyperref` is still being developed. However, even in its present form it is a powerful tool for interfacing to the print medium of the future. It can be found on CTAN in the `macros/latex/contrib/supported` directory (Figure D.4 on page 401).

D.5   Installing LaTeX

D.5.1   TeX implementations

One must have the TeX program and its auxiliaries (METAFONT, font files) before LaTeX can be set up on top of it. Installing TeX is a somewhat daunting experience, but thankfully it need not be done very often and there are many ready-to-run implementations available for practically

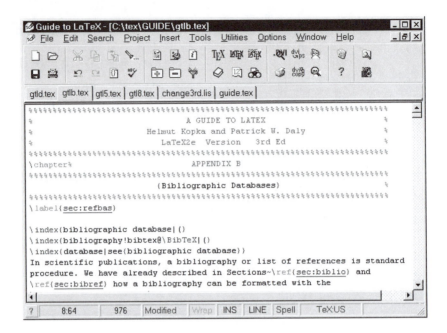

Figure D.3: Sample display with the WinEdt editor for interfacing to LATEX.

every computer operating system. These can be obtained from CTAN under the directory `systems` (Figure D.4 on page 401) or from the TEX Users Group (Section D.6.1).

A widely-used system developed by Karl Berry converts all the TEX source programs into the C language for subsequent compilation. This is named *Web2c* since the original programs were written in Donald Knuth's Web language (no relation to the World Wide Web). The teTeX implementation, by Thomas Esser, applies Web2c to many Unix machines, while Fabrice Popineau has extended it to Windows 95 and NT (`fpTeX`). Another excellent TEX implementation for Windows is called MikTEX by Christian Schenk.

Commercial packages for TEX also exist, offering graphics display of the output while you type: for Windows (Y&Y Inc., 45 Walden St., Concord, MA 01742 USA, `http://www.YandY.com`) and *Textures* for Macintosh (Blue Sky Research, 317 SW Alder, Portland, OR 97204 USA, `http://www.bluesky.com`). A true 'what-you-see-is-what-you-get' TEX system is *Scientific Workplace*, which can be purchased from TCI Software Research (Brooks/Cole Publishing Company, 511 Forest Lodge Road, Pacific Grove, CA 93950 USA, `http://www.tcisoft.com/tcisoft.html`).

While not a TEX implementation at all, the Windows text editor *WinEdt* by Aleksander Simonic (`winedt@istar.ca`) is an excellent interface to all TEX programs. Fully configurable, it makes use of key combinations familiar from other Windows applications, invokes LATEX, BIBTEX, MakeIndex, previewers, and printer drivers with a mouse click, hot key, or pull-down menu. As delivered, it is set up to enhance editor functions for LATEX files by highlighting commands and comments, for example. Figure D.3 demonstrates the appearance of the editing area.

D.5.2   Making the LATEX format

Once a viable TEX implementation is running, one can install LATEX following the provided instruction files. These are both general and specific to certain operating systems. The latter bear the extension .txt and the name of the TEX supplier, such as decustex.txt or oztex.txt, while the general instructions are in a file install.txt. These instructions should be read carefully before proceeding.

The files needed come in either packed or unpacked form. The packed files consist of a large number of .dtx files, the integrated source and documentation files. These are unpacked by running initex on the batch job file unpack.ins. This generates the necessary .cls, .sty, .clo, and other files. It also constructs the fundamental file latex.ltx which is needed to produce the format file.

If your set of files contains latex.ltx then they are already unpacked. This is convenient because the unpacking can take from a few minutes to some hours, depending on the computer. On the other hand, it takes much less time to transfer the packed files over networks.

The next step should be to run initex on the latex.ltx file to produce the LATEX format latex. However, before doing that, there are a number of configuration aspects to be considered. At several points during the processing, certain files are read in, but if a file of the same name but extension .cfg exists, it is loaded instead. It is this device that permits one to configure the final format for local conditions or desires. The possible configuration files are:

texsys.cfg offers the possibility of adding some very machine-specific adjustments for older versions of TEX or for some peculiarities of the TEX installation; information is to be found in the specific .txt file, or in ltdirchk.dtx;

fonttext.cfg is loaded in place of fonttext.ltx if it is present; this defines the fonts that are to be available for processing text; details can be found in fontdef.dtx;

fontmath.cfg is loaded in place of fontmath.ltx if it is present; this is the equivalent of fonttext.cfg for math fonts;

preload.cfg is loaded in place of preload.ltx if it is present; this determines which fonts are preloaded into the format; a number of other `preload` files may be extracted from `preload.dtx`, any one of which may be renamed to `.cfg` to be implemented;

`hyphen.cfg` is loaded in place of `hyphen.ltx` if it is present; this specifies the hyphenation patterns and their assignments; by default, the patterns in `hyphen.tex` are loaded into language 0; the `babel` system provides such a configuration file (Section D.1.4).

Once any configuration files have been set up and located where TEX can read them, initex may be run on `latex.ltx` to generate the format file `latex.fmt`. This file must be placed where format files are read, the other class, option, and package files where TEX reads its files. Finally the LATEX processing command must be defined to call TEX with the `latex` format, for example, as

```
tex &latex
```

Note: the LATEX 2.09 format is called `lplain`; thus the name of the format itself reveals the version of LATEX being activated.

D.5.3   The LATEX 2ε releases

Since its first release in June, 1994, LATEX 2ε has been regularly updated every six months in June and December, nominally on the first of these months but in fact often up to a month delayed. Not only are the kernel files for the basic LATEX improved and extended, so are the support programs such as **DocStrip** (Section D.3.1), the packages in the **tools** collection (Section D.3.3), and even the extended fonts (Section F.4.3).

A new release requires that the file `latex.ltx` be remade from the source files, as described in the previous section, and then that a new format file be created from it. Any urgent changes that are to be made in between these regular releases take place by means of patches in an `ltpatch.ltx` file, which is read in by `latex.ltx`. In this way, one only needs to obtain the patch file and to regenerate the format file from the existing `latex.ltx`.

The first few upgrades saw a fair number of changes as the new system was tried out by actual users in real practice; the more recent updates have seen changes that are mostly internal, increasing efficiency and speed while reducing size.

A document, class, or package that employs any of the newer features should include that release date as an optional argument to the identifying `\NeedsTeXFormat` (page 332). We indicate these dates when describing the features, and in the command summary of Appendix G.

D.6   Sources for LATEX files

The easiest way to obtain the extended LATEX software is via one of the network servers supported by various educational institutions. They also provide the most up-to-date standard LATEX installation, TEX extensions, programs for running TEX (and thus LATEX) on various PCs, collections of drivers for different printers, hyphenation patterns for other languages, etc. Such a server is a treasure trove for the dedicated TEX and LATEX user.

There are three main network servers for TEX and LATEX located in the United States, Great Britain, and Germany, forming the Comprehensive TEX Archive Network, CTAN. They are not only kept up to date but also regularly exchange contributions among themselves. They are all reachable on the Internet.

Country	Internet address
USA	`ftp://ctan.tug.org/tex-archive/`
UK	`ftp://ftp.tex.ac.uk/tex-archive/`
Germany	`ftp://ftp.dante.de/tex-archive/`

In addition, there are many full and partial mirrors of these sites. See the TUG home page, `http://www.tug.org`, for details.

The directory structure is the same for all three servers. A sketch of the main parts of the directory tree is presented in Figure D.4.

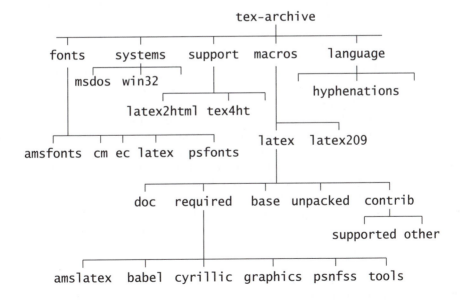

Figure D.4: Partial directory tree of CTAN servers

The latest version of the LaTeX package may be found under the sub-directories `base` or `unpacked`, where in the latter the files are already unpacked and `latex.ltx` is ready. The LaTeX extensions and general style files that have been contributed by other users are located in the subdirectory `contrib/supported`, with each contribution in a subdirectory of its own. Extensions offered by the LaTeX3 Team are found in the `required` subdirectory, which include `babel`, `graphics`, `tools`, and so on.

Rather than using an Internet browser to obtain files from CTAN, one can also make use of the more powerful *file transfer protocol* (FTP). With this, one logs on as user `anonymous`, giving the email address as password. One major advantage of this system is that the CTAN servers are set up to allow the transfer of whole directories via FTP as a single compressed archive file. One simply moves to the parent directory of the one sought, and fetches the file with the desired directory name plus extension `.zip` or `.tar.gz`, depending on which compression scheme one can handle.

A sample FTP session could look as follows:

`ftp ftp.dante.de`	*Connect to a CTAN server*
`anonymous`	*Logon as user* `anonymous`
`daly@linmpi.mpg.de`	*Email as password*
`quote site index verbatim.dtx`	*Search for file*
`cd /tex-archive/macros/latex`	*Change to* `latex` *directory*
`cd required/tools`	*Move further down to* `tools`
`pwd`	*Print current directory name*
`dir`	*List directory contents*
`cd ..`	*Move back one directory level*
`binary`	*Set to binary transfer*
`get tools.zip`	*Transfer* `tools` *directory*
`quit`	*End session*

Note that the command `binary` must be issued since the compressed file is in binary code; for pure text files, the command `ascii` is given, which is also the default at the start.

Help for finding a sought package is provided by the `quote site index` command above. Additionally, the CTAN Web sites offer search engines for locating files. A comprehensive list of CTAN contents is maintained by Graham Williams (`ftp://ctan.tug.org/tex-archive/help/Catalogue/catalogue.html`, with mirrors at the other sites).

For users without Internet connections, the TeX organizations listed below can provide most of the CTAN contents on CD-ROM.

D.6.1   TeX organizations

There are a number of organizations dedicated to disseminating the TeX and LaTeX ideas and programs. These user groups will also provide the necessary files for those who are not on a network.

The following list of mail and Internet addresses is based on those appearing in the TUG Web site, as of mid-1998, with the friendly help of the TUG staff for the latest updates. The addresses are, of course, subject to change without notice.

TUG (International Users Group)
Mimi Jett, President
TEX Users Group
1466 NW Front Avenue, Suite 3141
Portland, OR 97209-2820, USA

Tel: [+1] 503 223-9994
Fax: [+1] 503 223-3960
Email: `tug@mail.tug.org`
URL: `http://www.tug.org/`

AsTEX (French-speaking)
Michel Lavaud, President
Association pour la diffusion de
logiciels scientifiques liés TeX
Association AsTEX
BP 6532, 45066 Orléans CEDEX 2,
France
Tel: [+33] 2 38 64 09 94
Email: `astex@univ-orleans.fr`
Discussion list:
`astex-admin@univ-orleans.fr`
URL: `http://www.univ-orleans.fr/EXT/ASTEX/`

CervanTEX (Grupo de Usuarios de TEX Hispanoparlantes)
Julio Sanchez
GMV SA
Isaac Newton 11
PTM Tres Cantos
E-28760 Madrid, Spain
Email: `jsanchez@gmv.es`
Public mailing list:
`spanish-tex@eunet.es`
(Send subscription requests to
`listserv@eunet.es`.)
URL: `http://gordo.us.es/Actividades/GUTH`

CsTUG (Czech and Slovak Republics)
Petr Sojka, President
Československé sdružení uživatelů TEXu
CsTUG, C/o FI MU
Botanická 68a
CZ-602 00 Brno, Czech Republic
Tel: [+420] 5-41512352
Fax: [+420] 5-41512568
Email: `cstug@cstug.cz`
ftp: `ftp.cstug.cz/pub/tex`
URL: `http://www.cstug.cz/`

CyrTUG (Russia)
Eugenii V. Pankratiev, President
Irina Makhovaya, Executive Director
Associaciia Pol'zovatelei Kirillicheskogo
TeX'a
Mir Publishers
2, Pervyi Rizhskii Pereulok
Moscow 129820, Russia
Tel: [+7] 95 286 0622
Fax: [+7] 95 288 9522
Email: `cyrtug@cemi.rssi.ru`
URL:
`http://www.cemi.rssi.ru/cyrtug`

DANTE (German-speaking)
Thomas Koch, President
Deutschsprachige Anwendervereinigung
TeX e.V.
Postfach 101840
D-69008 Heidelberg, Germany
Tel: [+49] 6221 29766
Fax: [+49] 6221 167906
Email: `dante@dante.de`
URL: `http://www.dante.de`

Estonian User Group
Enn Saar, Tartu
Astrophysical Observatory, Toravere
EE 2444 Estonia
Email: `saar@aai.ee`

ΕϕΤ The Greek TEX Friends Group
Apostolos Syropoulos, President
366, 28th October Str.
GR-671 00 Xanthi, Greece
Tel: [+30] 541 28704
Email: `apostolo@platon.ee.duth.gr`
URL: `http://obelix.ee.duth.gr/eft`

GUST (Poland)
Tomasz Przechlewski, President
Polska Grupa Użytkowników Systemu TEX

Instytut Matematyki Uniwersytetu
Gdańskiego
ul. Wita Stwosza 57
80-952 Gdansk, Poland
Email: ekotp@univ.gda.pl
URL:
http://www.gust.org.pl/gust.html

GUTenberg (French-speaking)
Michel Goossens, President
Groupe francophone des Utilisateurs de
TEX
Association GUTenberg
C/o IRISA, Campus universitaire de
Beaulieu
F-35042 Rennes CEDEX, France
Tel: [+33] 1 44 32 37 96
Fax: [+33] 1 44 32 20 80
Email: gut@irisa.fr
URL: http://www.ens.fr/gut/

ITALIC (Irish)
No formal user group yet.
Public mailing list:
ITALIC-L@irlearn.ucd.ie
(send subscription requests to
listserv@irlearn.ucd.ie.)
Peter Flynn
Computer Centre
University College
Cork, Ireland
Email: pflynn@www.ucc.ie

JTUG (Japan)
Nobuo Saitoh, Chairman
Japan TEX Users' Group
Faculty of Environmental Information
Keio University
5322 Endo, Fujisawa-shi
JP-252 Japan
Tel: [+81] 466 47 5111
Email: ns@keio.ac.jp

Lithuanian TEX **Group**
Vytas Statulevièius, Chair
Akademijos 4
LT-2600 Vilnius, Lithuania
Tel: [+370] 22 729 609
Email: vytass@mii.lt
URL: http://www.vtex.mii.lt/tex/

Nordic TEX **Group** (Scandinavian
countries)
Dag Langmyhr, Chair
Department of Informatics

PO Box 1080 Blindern
University of Oslo
N-0316 Oslo, Norway
Tel: [+47] 22 85 24 50
Fax: [+47] 22 85 24 01
Email: dag@ifi.uio.no
URL:
http://www.ifi.uio.no/~dag/ntug/

NTG (Dutch-speaking)
Erik Frambach, Chair
Nederlandstalige TEX Gebruikersgroep
Postbus 394
NL-1740 AJ Schagen
The Netherlands
Email: ntg@nic.surfnet.nl
URL: http://www.ntg.nl/

TEXCeH (Slovenian TEX User Group)
Vladimir Batagelj
Jadranska 19
SI-61111 Ljubljana, Slovenia
Email: Tex-Ceh@fmf.uni-lj.si
URL: http:
//vlado.mat.uni-lj.si/texceh/
texceh.html

TUGIndia (Indian TEX Users Group)
Prof. (Dr.) K.S.S. Nambooripad (Chairman)
Kripa, TC 24/548
Sastha Gardens
Thycaud, Trivandrum 695014, India
Tel: [+91] 471 324341
Fax: [+91] 471 333186
Email: tugindia@mailexcite.com

TUG-Philippines (Philippines TEX Users
Group)
Dr. Severino V. Gervacio, President
De La Salle University
2401 Taft Avenue, Manila, Philippines
Tel: [+63] 632 5360270
Fax: [+63] 632 5360270
Email: gervacio@skyinet.net

UK TUG (United Kingdom)
Philip Taylor, Chairman
For information:
Peter Abbott
1 Eymore Close
Selly Oak, Birmingham B29 4LB, England
Email: uktug-enquiries@tex.ac.uk
URL: http:
//www.tex.ac.uk/UKTUG/home.html

E Math Extensions with $\mathcal{A_MS}$-LATEX

The American Mathematical Society, $\mathcal{A_MS}$, has supported the development and usage of TEX since its first release. It in fact owns the TEX logo as a registered trademark.

Shortly after TEX 82 became available, the $\mathcal{A_MS}$ produced a macro package for generating a special format `amstex`, described in *The Joy of TEX* by Spivak (1986). The macros in this $\mathcal{A_MS}$-TEX complement the mathematical typesetting features of Plain TEX by adding additional ones and simplifying others.

However, $\mathcal{A_MS}$-TEX is not a documentation preparation system like LATEX, describing the logical layout of a document by means of markup commands, but rather is simply an extension to Plain TEX.

The great popularity of TEX as a text formatting program is due primarily to the availability of LATEX as a user-friendly interface to the underlying TEX machinery. Many authors therefore asked the $\mathcal{A_MS}$ to provide LATEX with the same mathematical features as in $\mathcal{A_MS}$-TEX. The $\mathcal{A_MS}$-LATEX project was thus launched in 1987, with version 1.0 completed three years later by Frank Mittelbach and Rainer Schöpf, together with Michael Downes of the $\mathcal{A_MS}$. $\mathcal{A_MS}$-LATEX has been fully converted to LATEX 2_ε with version 1.2 in 1996.

The $\mathcal{A_MS}$-LATEX collection consists of three parts: packages for extending mathematical typesetting, extra classes for articles and books published by the $\mathcal{A_MS}$, and supplemental fonts for additional symbols, math alphabets, and Cyrillic fonts.

In the next sections, we give an overview of the math extensions available in the package `amsmath`; a more detailed user's manual is delivered with the collection, in the file `amsldoc.tex`. How to invoke the additional fonts is described below in Section E.5, or in the $\mathcal{A_MS}$-supplied manual found in the file `amsfndoc.tex`. The extra classes will not be described in this book; $\mathcal{A_MS}$ authors should refer to the instructions in the document `instr-l.tex` provided with $\mathcal{A_MS}$-LATEX.

E.1 Obtaining the $\mathcal{A}_{\mathcal{M}}$S packages

The $\mathcal{A}_{\mathcal{M}}$S packages are available on the CTAN servers (Section D.6) just like all the other LATEX extensions. Referring to Figure D.4 on page 401, the $\mathcal{A}_{\mathcal{M}}$S-LATEX files are to be found in `tex-archive/macros/latex/required/amslatex` while the extra fonts and their support packages are located in `tex-archive/fonts/amsfonts`.

The files are distributed in various subdirectories. For `amslatex`, all the files that the user needs are in the subdirectory `input`. These must be placed somewhere where LATEX looks for files during processing. In modern TEX installations, this would be something like `texmf/tex/latex/amslatex`, or its equivalent on other operating systems. The files in the other subdirectories, `classes` and `math`, are documented source files and may be stored with other such files. LATEX never needs them during processing.

For `amsfonts`, the files required for processing are located in the subdirectory `latex`. The font metric .tfm files are in `tfm` while the METAFONT .mf source files are found under `sources` in subdirectories `euler`, `extracm`, `cyrillic`, and `symbols`. These too must be placed in appropriate places in the user's TEX directory tree. This could be `texmf/tex/latex/amsfonts` for the files in subdirectory `latex`, `texmf/fonts/tfm/ams` for the .tfm files, and `texmf/fonts/source/ams` for the .mf files. The files in the subdirectory `plaintex` are intended for use with Plain TEX; if they are wanted, they may be placed in `texmf/tex/plain/amsfonts`.

The pixel .pk files are also provided on CTAN in various sizes. However, it is better to create them with METAFONT (Section F.3) for the local printer type. Most viewers and drivers today call this program automatically as the pixel files are needed, so it is not necessary to generate them explicitly.

E.2 Invoking $\mathcal{A}_{\mathcal{M}}$S-LATEX

If the \documentclass statement at the beginning of the LATEX document selects one of the $\mathcal{A}_{\mathcal{M}}$S classes `amsbook`, `amsart`, or `amsproc`, then most of the $\mathcal{A}_{\mathcal{M}}$S-LATEX features are loaded automatically. These features may still be employed with other classes by including the main extension package `amsmath` by means of

    \usepackage[*options*]{amsmath}

in the document's preamble. The list of allowable options is described below in Section E.3.8.

The amsmath package defines many of the new math typesetting features itself, but it also loads a number of other packages from the $\mathcal{A}_{\mathcal{M}}$S-LATEX collection that contain further extensions, such as `amsopn`, `amstext`, and `amsbsy`. These packages could be loaded separately without amsmath if only their limited features are wanted. On the other hand, the packages amscd and amsthm are not included in amsmath and must be loaded explicitly if their features are desired. Simply add their names to the list of packages in \usepackage.

The following Section E.3 describes the new commands and environments made available with amsmath and its associated packages. We call these the standard features of $\mathcal{A}_{\mathcal{M}}S$-LATEX. Further extensions added by other packages are explained afterwards.

!

The examples in this chapter employ the user-defined commands \mi, \me, and \dif defined in Section 5.4.10 for printing upright i, e, and d in math mode.

E.3   Standard features of $\mathcal{A}_{\mathcal{M}}S$-LATEX

This relatively long section presents those standard features of $\mathcal{A}_{\mathcal{M}}S$-LATEX that are activated by loading the amsmath package, or by selecting one of the $\mathcal{A}_{\mathcal{M}}S$ classes amsart, amsbook, or amsproc.

E.3.1   Additional font switching commands

Standard LATEX provides the math alphabet commands \mathcal, \mathrm, \mathbf, \mathsf, \mathit, \mathtt for changing fonts within math mode (Section 5.4.2). With $\mathcal{A}_{\mathcal{M}}S$-LATEX, one may also use the commands

   \boldsymbol{*symbol*}   and   \pmb{*symbol*}

to print *symbol* in a bold face, provided there is an appropriate bold font for it. Whereas the command \mathbf sets only Latin letters, numbers, and Greek upper case letters in bold, these commands also affect math symbols and Greek lower case letters. Compare the result of $\mathbf{\nabla\times V\,d\sigma}$ ($\nabla \times \mathbf{V}d\sigma$) with that of $\boldsymbol{\nabla\times V\,d\sigma}$ ($\boldsymbol{\nabla \times V\,d\sigma}$). With the standard LATEX command \mathbf, only the letters V and d appear bold, while the other characters remain in normal weight. Not only that, these letters are upright, and not italic as required by international standards. With \boldsymbol, all symbols are bold and italic.

Those symbols for which no bold face font exists will remain in normal weight with the \boldsymbol command. Such symbols are, for example, those that come in two sizes (Section 5.3.7) like $\sum$, $\int$, $\bigcup$, and so on. The command \pmb (poor man's bold) simulates a bold face even for these symbols by printing them several times slightly displaced. The result of $\pmb{\sum\;\int\;\bigcup}$ is $\pmb{\sum \int \bigcup}$.

These commands are defined in the package amsbsy, which may be loaded separately without amsmath.

A short piece of normal text can be given within a formula with

   \text{*short_text*}

In contrast to \mbox{*short_text*} for standard LATEX, the \text command switches font size correctly when used as superscripts and subscripts. Thus ..._{\text{Word}} sets $_{\text{Word}}$ lower and changes to \scriptstyle font size. To achieve the same effect in standard LATEX, the font size must be given explicitly, as ..._{\mbox{\scriptstyle Word}}. Furthermore, the name \text is more precise than \mbox.

This command is defined in the package amstext. Like amsbsy, it may be loaded on its own without amsmath if none of the other $\mathcal{A}_{\mathcal{M}}$S-LATEX features are wanted.

Another command for inserting normal text within a displayed equation is

\intertext{*insert_text*}

The *insert_text* is inserted as a left-justified line of text between those of the formula. The alignment of the formula lines remains unaffected, something that is not guaranteed if one closes the displayed equation, inserts text and reopens the equation. For example:

$$(x + iy)(x - iy) = x^2 - ixy + ixy - i^2 y^2$$
$$= y^2 + y^2 \quad \text{since} \quad i^2 = -1 \quad \text{is true.}$$

On the other hand

$$(x + iy)^2 = x^2 + 2ixy + i^2 y^2 = x^2 + 2ixy - y^2$$
$$(x - iy)^2 = x^2 - 2ixy + i^2 y^2 = x^2 - 2ixy - y^2$$

```
\begin{align*}
  (x+\mi y)(x-\mi y) & = x^2 - \mi xy + \mi xy - \mi ^2y^2\\
              & = y^2 + y^2\quad\text{since}\quad \mi ^2=-1
              \quad\text{is true.}\\
  \intertext{On the other hand}
  (x +\mi y)^2 & = x^2 + 2\mi xy +\mi ^2y^2 = x^2 + 2\mi xy-y^2\\
  (x -\mi y)^2 & = x^2 - 2\mi xy +\mi ^2y^2 = x^2 - 2\mi xy-y^2
\end{align*}
```

The alignment of the lines on the first equals sign is maintained through the interruption with the line of text 'On the other hand'. The environment align (Section E.3.6) is one of several new ones provided by $\mathcal{A}_{\mathcal{M}}$S-LATEX to replace the standard LATEX environment eqnarray. Note that \intertext may only be issued immediately after the \\ for starting a new line.

E.3.2 Multiple mathematical symbols

Mathematical formulas often require the same symbol to appear several times, such as multiple integral signs, or symbols and arrows to be stacked above or below a mathematical expression. Usually the distances between these symbols depends on the mathematical meaning, something that LATEX cannot automatically recognize. $\mathcal{A}_{\mathcal{M}}$S-LATEX provides a number of structures to help the user find the right spacing without tedious trial and error.

Multiple integrals

With $\mathcal{A}_{\mathcal{M}}$S-LATEX, the commands \iint, \iiint, \iiiint, and \idotsint output multiple integrals, with upper and lower limits being added in the usual manner. In text formulas, they are printed as $\iint$, $\iiint$, $\iiiint$, $\int \cdots \int$, while in displayed formulas, they appear as

$$\iint\limits_S f(x,y)\,\mathrm{d}S$$ `\[ \iint\limits_S f(x,y)\,\dif S \]`

$$\iiint\limits_V f(x,y,z)\,\mathrm{d}V$$ `\[ \iiint\limits_V f(x,y,z)\,\dif V\]`

$$\iiiint\limits_G f(x,y,z,t)\,\mathrm{d}G$$ `\[ \iiiint\limits_G f(x,y,z,t)`
`                \,\dif G \]`

$$\int \cdots \int\limits_U f(x_1,\ldots,x_k)\,\mathrm{d}U$$ `\[ \idotsint\limits_U`
`    f(x_1,\ldots,x_k)\,\dif U \]`

whereby the command \limits may be left off if the option intlimits has been specified when amsmath was loaded (see Section E.3.8).

Multiline limits

One often needs multiline limits or indices for summations, as in the examples below.

$$\Delta_{\substack{p_1p_2\cdots p_{n-k}\\q_1q_2\cdots q_{n-k}}} \qquad \sum_{\substack{k_0,k_1,\ldots\geq 0\\k_0+k_1+\cdots=0}} a_{0k_0}a_{1k_1}\cdots$$

$\mathcal{A}_{\mathcal{M}}$S-LATEX provides a command \substack for this purpose, with syntax

    \substack{*1st line**2nd line*\\...\\ *last line*}

where the command must immediately follow the ^ or _ shifting commands, and be entirely enclosed in curly braces { }. The index for the left-hand example above was generated with

    \Delta_{\substack{p_1p_2\cdots p_{n-k}\\q_1q_2\cdots q_{n-k}}}

The lines of text printed by \substack are centered horizontally, as is apparent in the right-hand example above. This was produced with

```
\[ \sum_{\substack{k_0,k_1,\ldots\ge0\\ k_0+\k_1+
   \cdots=0}}a_{0k_0} a_{1k_1}\cdots \]
```

On the other hand, the environment subarray offers more control over
the horizontal alignment:

```
\begin{subarray}{pos} 1st line\\2nd line\\...\\ last line
\end{subarray}
```

The argument *pos* may be c for centered, or l for left-justified lines.

$$\sum_{\substack{i\in\Lambda\\i<j<n}} P(i,j)$$

```
\[ \sum_{\begin{subarray}{l} i\in\Lambda\\
   i<j<n \end{subarray}} P(i,j) \]
```

Special limits

To add a differential prime sign to a summation symbol without a lower
limit, like $\sum' E_n$, in a displayed formula, one can easily give

```
\[ \sum\nolimits' E_n \]
```

$$\sum{}' E_n$$

However, if there to be an upper limit as well, the prime can only be
added with much fiddling. The \sideset command simplifies this task
considerably.

$$\sum_{n=0}^{\infty}{}'(2n+1)E_{2n+1}$$

```
\[ \sideset{}{'}\sum_{n=0}^\infty
   (2n+1) E_{2n+1} \]
```

The complete syntax for this command is:

```
\sideset{pre}{post}\symbol
```

where *pre* and *post* are the superscripts and subscript commands to be
added before and after the \symbol, respectively. They must contain
raising and lowering operators ^ and _.

The product symbol $\prod$ is given daggers above and asterisks
below with \sideset{_\dag^*}{_\dag^*}\prod, as shown
at the right.

$${}^*_\dagger\prod{}^*_\dagger$$

Additional shifting commands are

```
\overset{char}{\symbol}   and   \underset{char}{\symbol}
```

which place the arbitrary *char* above or below the \symbol in the size
appropriate for superscripts and subscripts. Thus $\overset{*}{X}$
produces $\overset{*}{X}$ and $\underset{*}{X}$ yields $\underset{*}{X}$.

Extended arrows

The amsmath package provides a number of commands to produce extra long arrows for combining with mathematical expressions. The commands

```
\overleftarrow{expr}          \underleftarrow{expr}
\overrightarrow{expr}         \underrightarrow{expr}
\overleftrightarrow{expr}     \underleftrightarrow{expr}
```

produce lengthened arrows pointing left and right, as well as double arrows above and below the mathematical expression *expr*.

$$\overrightarrow{ABCD} = \underrightarrow{AB} + \underrightarrow{BC} + \underrightarrow{CD}$$
$$\overleftarrow{ABCD} = \underleftarrow{DC} + \underleftarrow{CB} + \underleftarrow{BA}$$
$$\overleftarrow{ABCD} = \underleftarrow{DCAB}$$

```
\begin{eqnarray*}
\overrightarrow{ABCD} & = &
   \underrightarrow{AB} +
   \underrightarrow{BC} +
   \underrightarrow{CD} \\
\overleftarrow{ABCD} & = &
   \underleftarrow{DC} ...
\end{eqnarray*}
```

The lower arrows are actually too close to the expression, as is seen on the right-hand side of the last example. This should also be the case for all the other examples with a lower arrow, except that we have added a strut to push them down somewhat:

```
\underrightarrow{\rule[-2pt]{0pt}{2pt}AB} ...  and
\underleftarrow{\rule[-2pt]{0pt}{2pt}DC} ...
```

A smaller font size will be used for the arrows when they appear in exponents, indices, superscripts, and subscripts:

$$\int_{\overrightarrow{0.2\pi}} r\, d\varphi = 2\pi r$$

```
\int_{\overrightarrow{0.2\pi}}
   r\,d\varphi = 2\pi r
```

There are two more commands for horizontal arrows of variable length:

```
\xleftarrow[below]{above}    \xrightarrow[below]{above}
```

which place the mandatory *above* in superscript size over the arrow, and the optional *below* in subscript size beneath it.

$$A \xleftarrow{n+\mu-1} B \xrightarrow[T]{n\pm i-1} C$$

```
\[ A \xleftarrow{n+\mu-1} B
   \xrightarrow[T]{n\pm i-1} C \]
```

The package amscd (Section E.4.2) offers further possibilities for combining arrows and text.

Stacked accents

Attempting to place multiple accents over a character with standard LATEX, say with `$\hat{\hat{A}}$`, results in a misplacement: $\hat{\hat{A}}$. The same applies to the other mathematical accents commands \check, \breve, \acute, \grave, \tilde, \bar, \vec, \dot, and \ddot (Section 5.3.9). The amsmath package provides a set of math accent commands with the same names but capitalized,

```
\Hat    \Breve  \Grave  \Bar  \Dot
\Check  \Acute  \Tilde  \Vec  \Ddot
```

which can be safely combined with one another for the expected results: $\Hat{\Hat{A}}$, $\Breve{\Bar{B}}$ and $\Tilde{\Tilde{C}}$ produce $\hat{\hat{A}}$, $\breve{\bar{B}}$, and $\tilde{\tilde{C}}$.

Three or four dots in a row over a symbol are often used for time derivatives of third and fourth order. They can be placed with the commands

    \dddot{*sym*}   and   \ddddot{*sym*}

In this way $\dddot{u}$ and $\ddddot{u}$ are produced with $\dddot{u}$ and $\ddddot{u}$.

Continuation dots

In standard LATEX, one has the commands \ldots and \cdots for printing three continuation dots, either on the baseline or raised to the center of the line. $\mathcal{A}_{\mathcal{M}}\mathcal{S}$-LATEX offers a number of additional possibilities. The most general of these is the \dots command which adjusts the vertical height according to the symbol that follows it. If this is an equals sign or binary operator, such as + or −, the dots are raised, as with \cdots, otherwise they are on the baseline, as with \ldots.

```
$a_0+a_1+\dots+a_n$    ⇒    a_0 + a_1 + ⋯ + a_n
$a_0,a_1,\dots,a_n$    ⇒    a_0, a_1, …, a_n
```

If the continuation dots come at the end of a formula, there is no following symbol to determine the height of the dots. In this case, one must manually indicate the height by means of one of the commands \dotsc (comma), \dotsb (binary), \dotsm (multiplication), or \dotsi (integral).

The \dotsc is to be used with commas, so $A_1,A_2,\dotsc$ produces $A_1, A_2, \ldots$; the \dotsb sets them for a binary operator, thus $A_1+A_2+\dotsb$ yields $A_1 + A_2 + \cdots$; the command \dotsm is actually identical to \dotsb, but it is logically meant to be applied to multiplication, $A_1A_2\dotsm$ makes $A_1 A_2 \cdots$; finally \dotsi places the dots at the mean height of an adjacent integral sign.

$$\int_{A_1}\int_{A_2}\dotsi$$ $\int_{A_1}\int_{A_2}\cdots$

E.3.3   Fractions

The TEX fraction commands \atop, \choose, and others may be allowed in standard LATEX, but not in $\mathcal{A}_{\mathcal{M}}$S-LATEX. With the amsmath package, only those fraction commands described here may be used.

Basic fraction commands

In addition to the regular LATEX command \frac{*over*}{*under*}, $\mathcal{A}_{\mathcal{M}}$S-LATEX provides the commands \tfrac and \dfrac with the same syntax. These are effectively the same as \frac but with the font set to \textstyle or \displaystyle, respectively (Section 5.5.2). We demonstrate their effects with some examples from the $\mathcal{A}_{\mathcal{M}}$S-LATEX user's manual.

```
\[ \frac{1}{k}\log_2 c(f)\qquad
   \tfrac{1}{k}\log_2 c(f)\qquad
   \sqrt{\frac{1}{k}\log_2 c(f)}\qquad
   \sqrt{\dfrac{1}{k}\log_2 c(f)}   \]
```

$$\frac{1}{k}\log_2 c(f) \qquad \tfrac{1}{k}\log_2 c(f) \qquad \sqrt{\frac{1}{k}\log_2 c(f)} \qquad \sqrt{\dfrac{1}{k}\log_2 c(f)}$$

Binomial expressions

A binomial expression looks something like a fraction, but is enclosed in round parentheses and is missing the horizontal rule. The basic command in the amsmath package is

$\qquad$ \binom{*over*}{*under*}

which functions in the same way as \frac and the other fraction commands.

```
\[ \binom{n+1}{k} = \binom{n}{k}
   + \binom{n}{k-1 \]
```
$$\binom{n+1}{k} = \binom{n}{k} + \binom{n}{k-1}$$

Similarly there are the commands \tbinom and \dbinom analogous to \tfrac and \dfrac.

User-defined fractions

The amsmath package provides a powerful tool for defining fraction-like structures:

$\qquad$ \genfrac{*left_brk*}{*right_brk*}{*thickness*}{*mathsize*}{*over*}{*under*} where *left_brk* and *right_brk* are the parenthesis characters on the left and right, *thickness* is the thickness of the horizontal line, and *{mathsize}* is a number 0–3 representing the math sizes \displaystyle, \textstyle, \scriptstyle, and \scriptscriptstyle, respectively. The last two arguments, *over* and *under*, are the texts in the two parts of the fraction, the same arguments as in \frac and \binom.

If the *thickness* is left blank, the standard thickness for LATEX fractions is used. If *mathsize* is empty, the font size is determined automatically by the normal rules in Section 5.5.2.

Rather than repeating \genfrac with the same first four arguments time and again, one should define new fraction commands with it. For example, the following definitions are given in amsmath.sty:

```
\newcommand{\frac}[2]{\genfrac{}{}{}{}{#1}{#2}}
\newcommand{\dfrac}[2]{\genfrac{}{}{}{0}{#1}{#2}}
\newcommand{\tfrac}[2]{\genfrac{}{}{}{1}{#1}{#2}}
\newcommand{\binom}[2]{\genfrac{(}{)}{0pt}{}{#1}{#2}
```

As a further example, consider the redefinition of the command \frac

```
\renewcommand{\frac}[3][]{\genfrac{}{}{#1}{}{#2}{#3}}
```

in which the line thickness is now an optional first argument; without this optional argument, the command behaves as normal. Thus

```
\[ \binom{n}{m} =
   \frac[2pt]{n!}{M!(n-m)!} \]
```

yields

$$\binom{n}{m} = \frac{n!}{M!(n-m)!}$$

Continued fractions

Continued fractions can be made in $\mathcal{A}_{\mathcal{M}}\mathcal{S}$-LATEX with the command

```
\cfrac[pos]{over}{under}
```

whereby the denominator *under* may contain further \cfrac commands.

```
\[ a_0 + \cfrac{1}{a_1 + \cfrac{1}{a_2 +
        \cfrac{1}{a_3 + \cfrac{1}{a_4 +
        \dotsb }}}} \]
```

produces

$$a_0 + \cfrac{1}{a_1 + \cfrac{1}{a_2 + \cfrac{1}{a_3 + \cfrac{1}{a_4 + \cdots}}}}$$

If the optional argument *pos* is missing, the numerator *over* is centered on the horizontal rule; otherwise it may take values of l or r to left or right justify the numerator.

E.3.4   Matrices

Standard LATEX possesses the environment `array` for producing arrays and matrices. The `amsmath` package provides the additional environments `pmatrix`, `bmatrix`, `Bmatrix`, `vmatrix`, and `Vmatrix`, which automatically add the enclosing braces (), [], {}, | | and ‖ ‖ around the array, and in the right size. For completeness, there is also a `matrix` environment with no braces.

In contrast to the standard `array` environment (Section 4.8.1), these matrix environments do not require an explicit column specification as argument. By default, up to 10 centered columns may be used without any argument. (This maximum number may be changed by giving the special counter `MaxMatrixCols` a new value with either `\setcounter` or `\addtocounter`.) Otherwise the matrix environments are used in the same way as the `array` environment.

The following example is taken from Section 5.4.3 on page 133 and is recast here using $\mathcal{A}_{\mathcal{M}}S$-LATEX constructs.

$$\sum_{p_1<p_2<\dots<p_{n-k}}^{(1,2,\dots,n)} \Delta_{\substack{p_1p_2\dots p_{n-k}\\p_1p_2\dots p_{n-k}}} \sum_{q_1<q_2<\dots<q_k} \begin{vmatrix} a_{q_1q_1} & a_{q_1q_2} & \cdots & a_{q_1q_k} \\ a_{q_2q_1} & a_{q_2q_2} & \cdots & a_{q_2q_k} \\ \cdots\cdots\cdots\cdots\cdots\cdots \\ a_{q_kq_1} & a_{q_kq_2} & \cdots & a_{q_kq_k} \end{vmatrix}$$

```
\[ \sum_{p_1<p_2<\dots<p_{n-k}}^{(1,2,\dots,n)}
   \Delta_{\substack{p_1p_2\dots p_{n-k}\\p_1p_2
   \dots p_{n-k}}}
   \sum_{q_1<q_2<\dots<q_k}
   \begin{vmatrix}
     a_{q_1q_1} & a_{q_1q_2} & \dots & a_{q_1q_k}\\
     a_{q_2q_1} & a_{q_2q_2} & \dots & a_{q_2q_k}\\
     \hdotsfor[2.0]{4}\\
     a_{q_kq_1} & a_{q_kq_2} & \dots & a_{q_kq_k}
   \end{vmatrix}                   \]
```

Comparing this input text with that on page 133, one sees that it is simpler and easier to follow. The only new command used here is `\hdotsfor` which has the syntax

   `\hdotsfor[`*stretch*`]{`*n*`}`

and which prints a continuous line of dots through *n* columns. The optional argument *stretch* is a multiplicative number to increase the dot density, being 1.0 by default.

```
\[ \begin{matrix} a & b & c & d & e\\
                  x & \hdotsfor{3} & z
   \end{matrix}  \]
```

$$\begin{matrix} a & b & c & d & e \\ x & \cdots\cdots\cdots & z \end{matrix}$$

Compare the standard dot spacing above with that from `\hdotsfor[2.0]` in the previous example.

The initial letter of each of the xmatrix environments indicates the type of braces that enclose it: pmatrix for (round) parentheses, bmatrix for (square) brackets, Bmatrix for (curly) braces, vmatrix for vertical lines, and Vmatrix for double vertical lines. They appear as

$$
\begin{matrix} r & s & t \\ u & v & w \\ x & y & z \end{matrix} \quad
\begin{pmatrix} r & s & t \\ u & v & w \\ x & y & z \end{pmatrix} \quad
\begin{bmatrix} r & s & t \\ u & v & w \\ x & y & z \end{bmatrix}
$$

$$
\begin{Bmatrix} r & s & t \\ u & v & w \\ x & y & z \end{Bmatrix} \quad
\begin{vmatrix} r & s & t \\ u & v & w \\ x & y & z \end{vmatrix} \quad
\begin{Vmatrix} r & s & t \\ u & v & w \\ x & y & z \end{Vmatrix}
$$

where each matrix has been produced with

```
\[ \begin{xmatrix} r & s & t\\ u & v & w\\ x & y & z
   \end{xmatrix}  \]
```

where xmatrix is set to matrix, pmatrix, bmatrix, Bmatrix, vmatrix, and Vmatrix one after the other.

To generate a small array within a text formula, one can apply the smallmatrix environment. In this way $\left(\begin{smallmatrix} a & b & c \\ e & m & r \end{smallmatrix}\right)$ can be made with

```
$ \bigl( \begin{smallmatrix} a & b & c\\ e & m & r
   \end{smallmatrix} \bigr) $
```

E.3.5   User extensions and fine adjustments

Function names

Standard LATEX recognizes a number of preprogrammed function names (Section 5.3.8) that are printed in math made by placing a backslash in front of that name: arccos, arcsin, arctan, arg, cos, cosh, cot, coth, csc, deg, det, dim, exp, gcd, hom, inf, ker, lg, lim, liminf, limsup, ln, log, max, min, Pr, sec, sin, sinh, sup, tan, tanh. Not only do these names appear in an upright font, as is required for function names, but the spacing with adjacent parts of the mathematical expression is adjusted automatically.

$\mathcal{A}_{\mathcal{M}}$S-LATEX provides some more function names, as variations on the standard \lim name:

\varlimsup	$\varlimsup$	\varinjlim	$\varinjlim$
\varliminf	$\varliminf$	\varprojlim	$\varprojlim$

These functions may take on limits with the raising and lowering operators ^ and _; for example \varliminf_{n\to\infty} for $\varliminf_{n\to\infty}$.

It is also possible to define new function names with the same font and spacing properties as the predefined ones. The command

```
\DeclareMathOperator{\cmd}{name}
```

which may only be issued in the preamble, before \begin{document}, defines a command \cmd that prints the function name *name*. For example, to define a function name \doit, give

        \DeclareMathOperator{\doit}{doit}

and then $A=3\doit^2(B)$ yields $A = 3\,\text{doit}^2(B)$. Note that superscripts and subscripts are printed beside the operator; if they are to be printed as limits, that is, above and below the operator in displayed math mode, use the *-form to define them. For example:

        \DeclareMathOperator*{\Lim}{lim} $\lim\limits_{n\to-\infty}^{n\to+\infty}$
        \[\Lim_{n\to-\infty}^{n\to+\infty}\]

The *name* text need not be identical to the command name. In particular, it may contain special characters not allowed in command names.

Modulo expressions are printed in standard LATEX with \bmod and \pmod commands, and are complemented in $\mathcal{A}_{\mathcal{M}}S$-LATEX by \mod and \pod. The possibilities are:

$z \equiv x + y \bmod n^2$	z \equiv x+y \bmod{n^2}
$z \equiv x + y \pmod{n^2}$	z \equiv x+y \pmod{n^2}
$z \equiv x + y \mod n^2$	z \equiv x+y \mod{n^2}
$z \equiv x + y \pod{n^2}$	z \equiv x+y \pod{n^2}

The automatic parentheses are missing with \mod, while with \pod the name 'mod' is omitted. Furthermore, \pmod is redefined for text formulas to reduce the preceding space: $y\pmod{a+b}$: $y \pmod{a + b}$.

The \DeclareMathOperator command and additional function names are defined in the amsopn package, which may be loaded on its own without amsmath.

Fine-tuning roots

The positioning of an index to a root sign is not always ideal under standard LATEX. In $\sqrt[\beta]{k}$, for example, the β could be somewhat higher and shifted slightly to the right. The $\mathcal{A}_{\mathcal{M}}S$-LATEX commands

        \leftroot{*shift*}   \uproot{*shift*}

cause such manual displacements, where *shift* is a number specifying the size in small, internal units. Negative numbers represent a shift in the opposite direction. Compare the above standard result with that of $\sqrt[\leftroot{-1}\uproot{3}\beta]{k}$: $\sqrt[\beta]{k}$.

The size of the root sign depends on its contents. If they hang below the baseline, the root sign extends lower down than for contents that have no depth. Note the differences between $\sqrt{x}$, $\sqrt{y}$, and $\sqrt{z}$. Some publishers want all root signs be at the same height, as $\sqrt{x} + \sqrt{y} + \sqrt{z}$. This is accomplished with the TeX command \smash which places its argument in a box of zero height and depth. The $\mathcal{A}_{\mathcal{M}}S$-LaTeX version of this command allows an optional argument b or t to zero only the depth or height, respectively. The above example is produced with $\sqrt{\smash[b]{y}}$. The option b is taken because we want only the depth to be ignored, not the height of the letter y.

Spacing adjustment

With standard LaTeX, there are a number of commands to fine-tune the spacing in a math formula (Section 5.5.1). These are \, \: \; \quad and \qquad for increasing amounts of positive spacing, and \! for negative spacing. With the amsmath package, the first three still exist, but may also be called with the more obvious names \thinspace, \medspace, and \thickspace. There is also \negthinspace as an alias for \!.

The complete set of spacing commands are summarized in the table below taken from the $\mathcal{A}_{\mathcal{M}}S$-LaTeX manual.

Short form	Command name	Demo	Short form	Command name	Demo
\,	\thinspace	⅃L	\!	\negthinspace	⊥
\:	\medspace	⅃L		\negmedspace	⊥
\;	\thickspace	⅃L		\negthickspace	⊥
	\quad	⅃ L			
	\qquad	⅃ L			

The general math spacing command is

\mspace{*mu*}

which inserts space in mathematical spacing units 'mu' (=1/18 em). For example, \mspace{-9mu} puts in negative spacing of 1/2 em.

Vertical bars

In standard LaTeX, the commands | and \| are used for single and double vertical bars, | and ‖. However, these symbols are often used as delimiters (that is, like braces) in which case different spacing requirements are needed. In particular, a distinction must be made between the left and right delimiter in expressions like $|a|$ and $\|v\|$. The LaTeX commands are only appropriate for single appearances, like $p|q$ or $f(t, x)|_{t=0}$.

The amsmath package defines the delimiter commands \lvert, \rvert for a single bar, and \lVert, \rVert for a double bar. They are useful for defining commands that set their arguments in such delimiters, as

\newcommand{\abs}[1]{\lvert#1\rvert}
\newcommand{\norm}[1]{\lVert#1\rVert}

Now $\abs{a}$ produces $|a|$, and $\norm{v}$ $\|v\|$.

A similar recommendation can be made for the standard commands `\langle` and `\rangle`. By defining

```
\newcommand{\mean}[1]{\langle#1\rangle}
```

one gives $\mean{x}$ to generate $\langle x \rangle$, rather than $<x>$ which produces $< x >$.

Boxed formulas

A formula may be placed in a box with the command

```
\boxed{formula}
```

For example,

```
\[ \boxed{\int_0^\infty f(x)\,\dif x \approx
     \sum_{i=1}^n w_i \me^{x_i} f(x_i)}   \]
```

produces

$$\boxed{\int_0^\infty f(x)\,\mathrm{d}x \approx \sum_{i=1}^n w_i \mathrm{e}^{x_i} f(x_i)}$$

E.3.6   Multiline equations

Equations consisting of several lines which are horizontally aligned at set points, such as the equals sign, can be generated in standard LATEX with the eqnarray and eqnarray* environments (Section 5.4.7). Many authors consider these to be far too limited for publications with complicated multiline equations. $\mathcal{A}_{\mathcal{M}}\mathcal{S}$-LATEX therefore provides a range of further alignment environments for formulas extending over a single line:

align   gather   falign   multline   alignat   split

With the exception of split, all exist in a standard and a *-form. As for eqnarray, the standard form adds an automatic equation number to each line, while the *-form does not. The standard LATEX equation environment for single line formulas is also available in $\mathcal{A}_{\mathcal{M}}\mathcal{S}$-LATEX in a *-form. It may be used in combination with multiline environments.

Common features of alignment environments

All the alignment, or multiline, environments switch to math mode at the start and back to text mode at the end, except for split which must be called in math mode. A new line is forced in the formula with the

\\ command, as usual; an optional argument \\[*len*] can be added to increase the line spacing by *len*, again as usual.

The automatic numbering with the standard forms can be suppressed for single lines by adding \notag before the \\ line break. Alternatively, the line can be given a desired marker with \tag{*mark*}. For example, with \tag{$\dag$}, the marker is (†). Using the *-form instead, the marker text is printed without the parentheses.

The vertical position of the equation number or marker can be adjusted with the command

> \raisetag{*len*}

which moves the marker upwards by *len* for that line only. A negative value moves it downwards.

The multline environment

The multline environment is a variant of the equation environment for *single* formulas that are too long for one line. The line breaks occur where the user forces them with the \\ command. The first line is left justified, the last right justified, and lines in between are centered. However, if the option fleqn has been given, all the lines appear left justified.

The equation number, if present, appears at the right of the *last* line by default or if the option reqno has been selected; if the option leqno has been chosen, the number is placed at the left of the *first* line. (See Section E.3.8 for the amsmath options.)

It is possible to shift individual lines fully to the left or right with the commands \shoveleft{*formula*} and \shoveright{*formula*}. The entire formula text for that line, except the terminating \\, is placed in their arguments.

The left and right margins for the formula are set by the length parameter \multlinegap which is initially 10 pt. This may be altered by the user with the \setlength or \addtolength commands.

An equation with five lines could be broken to look as follows:

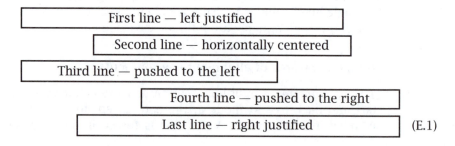

First line — left justified

Second line — horizontally centered

Third line — pushed to the left

Fourth line — pushed to the right

Last line — right justified   (E.1)

```
\begin{multline}
  \framebox[.75\columnwidth]{First line --- left justified}\\
  \framebox[.6\columnwidth]{Second line --- horizontally
      centered}\\
  \shoveleft{\framebox[.6\columnwidth]{Third line --- pushed
      to the left}}\\
  \shoveright{\framebox[.6\columnwidth]{Fourth line --- pushed
      to the right}}\\
  \framebox[.75\columnwidth]{Last line --- right justified}
\end{multline}
```

A real equation would contain mathematical expressions and not the `\framebox` commands in the above demonstration.

The split environment

Like `multline`, the `split` environment is meant for a single equation that does not fit on one line. Line breaks are again forced with the \\ command; the difference is that in each line there is an alignment marker & such that the lines are horizontally positioned to line up the markers.

The `split` environment does not switch into math mode, nor does it produce an equation number. It is intended to be applied within another math environment, such as `equation` or `gather`. This is why there is an `equation`* environment in $\mathcal{A}_{\mathcal{M}}\mathcal{S}$-LATEX.

The equation number, if present, is provided by the outer environment. It is applied to the entire multiline formula, which by default, or with the option `centertags`, is centered on the group of lines. With the option `tbtags`, it is placed either at the left of the last line, or at the right of the first line, depending on the further options `leqno` and `reqno`, respectively. (See Section E.3.8.)

$$H_c = \frac{1}{2n} \sum_{l=0}^{n} (-1)^l (k-l)^{p-2} \sum_{l_1+\cdots+l_p=l} \prod_{i=1}^{p} \binom{n_i}{l_i}$$

$$\times [(k-l) - (k_i - l_i)]^{k_i-l_i} \times \left[(k-l)^2 - \sum_{j=1}^{p} (k_i - l_i)^2 \right] \tag{E.2}$$

```
\begin{equation}\begin{split}
  H_c&=\frac{1}{2n}\sum_{l=0}^n (-1)^l (k-l)^{p-2}
      \sum_{l_1+\dots+l_p=l} \prod_{i=1}^p \binom{n_i}{l_i}\\
      &\quad\times[(k-l) - (k_i-l_i)]^{k_i-l_i}\times
      \Bigl[(k-l)^2 - \sum_{j=1}^p (k_i-l_i)^2\Bigr]
\end{split}\end{equation}
```

The alignment has been chosen to be just before the equals sign. The second line begins with the alignment marker & so that the equation continues below the = in the first line. A \quad has been inserted so that the $\times$ is not immediately below the equals sign. Alternatively, one could place the & after the = and dispense with \quad in the second line. However, the above is more suitable when there are continuation lines beginning with =. Note the centered equation number at the right.

The gather environment

The gather environment switches to math mode, centering each of its formula lines without any alignment. The formula lines are separated by \\ commands. Each line receives an equation number, unless the *-form has been used, or \notag has been issued in that line.

$$\frac{1}{2} + \left(\frac{2}{3}\right)^4 + \left(\frac{3}{4}\right)^9 + \cdots + \left(\frac{n}{n+1}\right)^{n^2} + \cdots = \sum_{n=1}^{\infty} \left(\frac{n}{n+1}\right)^{n^2} \tag{E.3}$$

$$\text{converges since} \quad \lim_{n\to\infty} \sqrt[n]{\left(\frac{n}{n+1}\right)^{n^2}} = \lim_{n\to\infty} \left(\frac{1}{1 + \dfrac{1}{n}}\right)^n = \frac{1}{e} < 1$$

root condition

$$2 + \frac{3}{4} + \frac{4}{9} + \cdots + \frac{n+1}{n^2} + \cdots = \sum_{n=1}^{\infty} \frac{n+1}{n^2} \tag{E.4}$$

$$\text{diverges since} \quad \int_c^\infty \frac{x+1}{x^2}\,dx = \left[\ln x - \frac{1}{x}\right]_c^\infty = \infty \quad \text{(integral condition)}$$

```
\begin{gather}
\frac{1}{2} + \left(\frac{2}{3}\right)^4 + \left(\frac{3}{4}
\right)^9 + \dots + \left(\frac{n}{n+1}\right)^{n^2} + \dotsb
= \sum_{n=1}^\infty \left(\frac{n}{n+1}\right)^{n^2} \\
\text{converges since}\quad\lim_{n\to\infty}
\sqrt[n]{\left(\frac{n}{n+1}\right)^{n^2}} = \lim_{n\to\infty}
\left(\frac{1}{1 + \dfrac{1}{n}}\right)^n = \frac{1}{\me} < 1
\tag*{root condition}\\
2 + \frac{3}{4} + \frac{4}{9} + \dots + \frac{n+1}{n^2} +
\dotsb = \sum_{n=1}^\infty \frac{n+1}{n^2}\\
\text{diverges since}\quad\int_c^\infty \frac{x+1}{x^2}\,
\dif x = \left[ \ln x -\frac{1}{x}\right]_c^\infty = \infty
\tag{integral condition}
\end{gather}
```

Note the use of \tag and \notag (page 420) to add text as markers to the second and fourth lines.

The align environment

The align environment is intended for multiple equations with horizontal alignment, usually on an equals sign or equivalent. New lines are indicated with \\ as usual. Each line is split into aligned columns such that the first column is right justified against the & character, the second left justified; the third column is right justified against the *third* &, the fourth column left justified again, and so on. This is the same as an array environment with column specification {rl rl rl ... }.

$$(x^n)' = nx^{n-1} \qquad (e^x)' = e^x \qquad (\sin x)' = \cos x$$

$$\left(\frac{1}{x^n}\right)' = -\frac{n}{x^{n+1}} \qquad (a^x)' = a^x \ln a \qquad (\cos x)' = -\sin x$$

$$\left(\sqrt[n]{x}\right)' = \frac{1}{n\sqrt[n]{x^n - 1}} \qquad (\ln x)' = \frac{1}{x} \qquad (\tan x)' = \frac{1}{\cos^2 x}$$

$$(\log_a x)' = \frac{1}{x \ln a} \qquad (\cot x)' = -\frac{1}{\sin^2 x}$$

```
\begin{align*}
  \left(x^n\right)' &= nx^{n-1} & \left(\me^x\right)' &= \me^x &
  (\sin x)'         &= \cos x   \\
  \left(\frac{1}{x^n}\right)' &= -\frac{n}{x^{n+1}} &
  \left(a^x\right)' &= a^x\ln a & (\cos x)'          &= -\sin x\\
  \left(\sqrt[n]{x}\right)'   &= \frac{1}{n\sqrt[n]{x^n -1}} &
  (\ln x)'          &= \frac{1}{x} & (\tan x)'
                    &= \frac{1}{\cos^2 x}\\
  & &   (\log_a x)' &= \frac{1}{x\ln a} & (\cot x)'
                    &= -\frac{1}{\sin^2 x}
\end{align*}
```

The align* environment is used here to prevent the lines from being numbered, something that is not appropriate for such a collection of formulas. The alignment within each of the three column pairs is on the equals sign. The input for the last line begins with a double && to produce an empty column pair.

Occasionally a set of formulas is to be aligned on several equals signs in one line, as in the equations below for the volume V, inertial moment I_z, and mass M of an arbitrary body, in Cartesian, cylindrical, and spherical coordinates. In this case, the second and third parts are separated by a double && so that the left-hand sides of these column pairs are empty: the equals signs are always on an odd-numbered alignment marker.

$$V = \int\limits_V d\nu \quad = \iiint dx\, dy\, dz \qquad = \iiint \rho\, dx\, d\rho\, d\phi$$

$$= \iiint r^2 \sin\theta\, dr\, d\theta\, d\phi \qquad \text{(E.5)}$$

$$I_z = \int\limits_V \rho^2\, d\nu = \iiint (x^2 + y^2)\, dx\, dy\, dz = \iiint \rho^3\, dz\, d\rho\, d\phi$$

$$= \iiint r^4 \sin^3\theta\, dr\, d\theta\, d\phi \qquad \text{(E.6)}$$

$$M = \int\limits_V \delta\, d\nu \quad = \iiint \delta\, dx\, dy\, dz \qquad = \iiint \delta\rho\, dz\, d\rho\, d\phi$$

$$= \iiint \delta r^2 \sin\theta\, dr\, d\theta\, d\phi \qquad \text{(E.7)}$$

```
\begin{align}
V    &= \int\limits_V\dif v        &&= \iiint\dif x\,\dif y\,\dif z
     &&= \iiint \rho\,\dif x\,\dif \rho\,\dif \phi \notag \\
&&&&= \iiint r^2 \sin\theta\,\dif r\,\dif\theta\,\dif\phi\\
I_z &= \int\limits_V \rho^2\,\dif v &&=\iiint(x^2 + y^2)
                                   \,\dif x\,\dif y\,\dif z
     &&= \iiint \rho^3\,\dif z\,\dif \rho\,\dif \phi \notag \\
&&&&= \iiint r^4\sin^3\theta\,\dif r\,\dif\theta\,\dif\phi\\
M    &= \int\limits_V \delta\,\dif v &&= \iiint \delta
                                   \,\dif x\,\dif y\,\dif z
     &&= \iiint \delta\rho\,\dif z\,\dif\rho\,\dif\phi\notag \\
&&&&=\iiint\delta r^2\sin\theta\,\dif r\,\dif\theta\,\dif\phi
\end{align}
```

There are two variations on the `align` environment, `falign` and `alignat`. The first has exactly the same syntax as `align` but it inserts so much spacing between the column pairs that the entire line is filled out. The `alignat` environment is just the opposite: no spacing is inserted automatically between the column pairs. It must take the number of column pairs as a mandatory argument, but otherwise the syntax of the contents is the same as that for `align`. The example above with the volume, inertial moment, and mass of a body could just as well have been given with

```
\begin{alignat}{3}  formula_text  \end{alignat}
```

(In fact, this is precisely what was done in order to fit it within the line width of this book.)

If one or more columns are empty, as in this example, it is possible to control their widths precisely in the `alignat` environment by adding explicit spacing between the two & characters in one of the lines. See Section 5.5.1 for spacing in math mode.

In summary, for the align environment and its variants, the first, third, fifth, etc. & characters are alignment markers, while the second, fourth, sixth, etc. are column pair separators.

Nested alignment environments

We have already pointed out on page 421 how the split environment is to be placed inside an equation environment. The same is true for the environments aligned and gathered, which may be used as building blocks within formulas. Their contents and behavior are otherwise the same as their related environments.

Both of these environments take an optional argument *pos*

```
\begin{aligned}[pos]    lines  \end{aligned}
\begin{gathered}[pos]   lines  \end{gathered}
```

which takes values of t or b to determine the vertical alignment (top or bottom) when they appear beside other elements. When no *pos* is given, they are centered. In this way

$$
\begin{aligned}
\alpha &= aa \\
\beta &= bbbbb \\
\gamma &= g
\end{aligned}
\qquad \text{versus} \qquad
\begin{aligned}
\delta &= dd \\
\eta &= eeeeee \\
\varphi &= f
\end{aligned}
\qquad \text{versus} \qquad
\begin{gathered}
s = x + y \\
d = u - v - w \\
p = x \circ y
\end{gathered}
$$

is produced with

```
\begin{equation*}
  \begin{aligned} \alpha&=aa\\ \beta&=bbbbb\\ \gamma&=g
  \end{aligned}
  \qquad\text{versus}\qquad
  \begin{aligned}[t] \delta&=dd\\ \eta&=eeeeee\\ \varphi&=f
  \end{aligned}
  \qquad\text{versus}\qquad
  \begin{gathered}[b] s= x+y\\ d= u - v - w\\ p = x\circ y
  \end{gathered}
\end{equation*}
```

The cases environment

Although it is possible with standard LATEX to produce structures of the form

$$
P_{r-j} = \begin{cases} 0 & \text{if } r - j \text{ is odd,} \\ r!\,(-1)^{(r-j)/2} & \text{if } r - j \text{ is even.} \end{cases}
\tag{E.8}
$$

as demonstrated by a similar example in Section 5.4.1 on page 130, the $\mathcal{A}_{\mathcal{M}}$S-LATEX cases environment allows a simpler input:

```
\begin{equation}
P_{r-j}=\begin{cases} 0 & \text{if $r-j$ is odd,}\\
        r!\,(-1)^{(r-j)/2} & \text{if $r-j$ is even.}
          \end{cases}
\end{equation}
```

There may be more than two cases in the environment, as in the example reproduced here from page 130:

$$y = \begin{cases} -1 & : \quad x < 0 \\ 0 & : \quad x = 0 \\ +1 & : \quad x > 0 \end{cases}$$

```
\[ y = \begin{cases} -1 &:\quad x<0\\
      \hfill 0 &:\quad x=0\\ +1 &:\quad x>0
          \end{cases} \]
```

E.3.7   Equation numbering

Numbering hierarchy

With the standard LATEX classes book and report, equations are given a double number with the chapter designation and then a sequential number starting at 1 for each new chapter. For the article class, the equations are numbered sequentially throughout the work.

With the amsmath package, it is possible to alter this hierarchy. For example, if an article is to have the equations numbered within each section, with the section number, give

```
\numberwithin{equation}{section}
```

to redefine the equation numbers to include the section number and to make the equation counter reset every time the section counter is incremented. This is as though the equation counter had been created with \newcounter{equation}[section] (Section 7.1.2), something which the user cannot normally bring about. Furthermore, \theequation is redefined to be \thesection.\arabic{equation}, something that is in the user's power but which is not much use if the equation counter is never reset.

Subnumbering equations

On page 199 we give an example of how equations may be subnumbered, that is, the main equation number stays the same and a letter is appended to it, as 1.8a, 1.8b, 1.8c .... The amsmath package provides this feature with the subequations environment. Numbered equations appearing within

```
\begin{subequations} ...   \end{subequations}
```

will all have the same main number which is one more than that of the previous one, with sequential, lower case letters attached.

Within the environment, the `equation` counter refers to the subnumber, that is, to the letters, while the main number is to be found in the `parentequation` counter. To change the format of the subnumber, say to 1.8-A, 1.8-B, ..., give

```
\begin{subequations}
\renewcommand{\theequation}
            {\theparentequation-\Alph{equation}}
. . .
\end{subequations}
```

Referencing equation numbers

The LaTeX cross-reference system is described in Section 8.3.1 and works exactly the same way with $\mathcal{A}_{\mathcal{M}}S$-LaTeX: when `\label{`*marker*`}` is issued in a mathematical formula that receives an automatic equation number, that number can be printed anywhere in the text with `\ref{`*marker*`}`, where *marker* is arbitrary text to identify that equation.

The amsmath package adds a command `\eqref{`*marker*`}` to print the equation number in parentheses, as it appears beside the math formula. For example, the cases equation on page 425 is referred to as equation E.8 with `\ref`, or as equation (E.8) with `\eqref`.

If `\label` is given immediately after the start of a `subequations` environment, the corresponding `\ref` commands will print the main equation number without the extra letter. In this way one can refer to the entire group of equations. Later `\label` commands are associated with individual equation lines and reference them with the letters.

Page breaks within multiline formulas

Unlike the standard LaTeX eqnarray environment, the $\mathcal{A}_{\mathcal{M}}S$-LaTeX multiline math environments do not normally allow any page breaks to occur within them. The idea is that the author should have more control over where such breaks may occur. To allow or force a page break within a multiline equation, one gives

```
\displaybreak[num]
```

just before the line breaking command \\. Here the optional *num* has the same meaning as for the standard `\pagebreak` command (page 51): without it, a new page is forced, but it may take values of 0–4 to allow a break with increasing degree of encouragement, whereby 4 also forces the page break.

Alternatively, one can issue `\allowdisplaybreaks` in the document preamble to allow LaTeX to break pages automatically within multiline formulas as necessary. This command too takes an optional argument *num* with possible values between 0 and 4 which make it progressively easier for automatic page breaks to occur.

Once \allowdisplaybreaks has been given in the preamble, it is still possible to suppress page breaks within a formula by ending an equation line with * instead of with \\.

E.3.8   Package options for amsmath

The main amsmath package for $\mathcal{A}_{\mathcal{M}}S$-LATEX recognizes a number of options that may be given when it is loaded with

> \usepackage[*options*]{amsmath}

They are listed here as pairs with opposing effects. The member of each pair that is assumed if neither is given, the default, is indicated by underlining.

centertags | tbtags The equation number for a split environment (page 421) is centered vertically by default. With tbtags, it is placed either to the left of the first line or to the right of the last line, depending on the side on which numbers are to appear.

sumlimits | nosumlimits In displayed formulas, initial and final limits appear below and above the $\sum$ sign with the sumlimits option. With nosumlimits, they are placed beside the sign, raised and lowered with the usual ^ and _ characters.

Other symbols that are affected by these options are $\sum \prod \coprod \bigcup \biguplus$ $\bigcap \bigsqcup \bigvee \bigwedge \bigodot \bigotimes$ and $\bigoplus$. On the other hand, integral signs are not influenced by them.

sumlimits

$$\sum_{n=0}^{\infty} \frac{1}{2^n} = 2; \quad \prod_{i=0}^{m-1} n - i = \frac{n!}{(n-m)!}$$

nosumlimits

$$\sum\nolimits_{n=0}^{\infty} \frac{1}{2^n} = 2; \quad \prod\nolimits_{i=0}^{m-1} n - i = \frac{n!}{(n-m)!}$$

The input text is the same for both of the above cases.

intlimits | nointlimits Integral signs normally have their limits at the side; these options allow them to be placed above and below as for summations.

intlimits

$$\int_0^a \sqrt{a^2 - x^2}\, dx = \int_0^1 a^2 \sqrt{1 - \sin^2 t}\, d\sin t = a^2 \int_0^{\pi/2} \cos^2 t\, dt = \frac{\pi a^2}{4}$$

`nointlimits`

$$\int_0^a \sqrt{a^2 - x^2}\, \mathrm{d}x = \int_0^1 a^2\sqrt{1 - \sin^2 t}\, \mathrm{d}\sin t = a^2 \int_0^{\pi/2} \cos^2 t\, \mathrm{d}t = \frac{\pi a^2}{4}$$

where again the input text is the same for both cases. As a reminder, the standard LATEX treatment of limits on integral signs is the same as for `nointlimits`.

`namelimits` | `nonamelimits` The function names \det, \gcd, \inf, \lim, \liminf, \limsup \max, \min, \Pr, and \sup frequently take lower limits which normally appear below the name in displayed formulas. With `nonamelimits`, they are placed at the lower right.

namelimits   nonamelimits

$$\lim_{x \to \infty} \left(1 + \frac{1}{x}\right)^x = e = 2.7182\ldots \quad \lim_{x \to \infty} \left(1 + \frac{1}{x}\right)^x = e = 2.7182\ldots$$

The above options determine the standard placement of limits for the entire document. It is still possible to change the behavior in any particular case with the \limits and \nolimits commands, as in normal LATEX.

The remaining package options select the side for equation numbers and the horizontal positioning of equations.

`leqno` | `reqno` The standard location for equation numbers is on the right side, at the margin; with `leqno`, they are placed on the left side of the equation.

`fleqn` With this option, all displayed equations are printed flush left, set off from the left margin by an amount \mathindent. Without this option, equations are centered. (This is like the `fleqn` class option for the standard classes, page 27.)

E.4   Further $\mathcal{A}_{\mathcal{M}}S$-LATEX packages

The $\mathcal{A}_{\mathcal{M}}S$-LATEX packages described in this section must be loaded explicitly if their features are to be exploited. Unlike the amsbsy and amstext packages, they are not loaded automatically with amsmath. They may, however, be used on their own, independently of the main package.

E.4.1   Extended theorem declarations

The amsthm package offers many additional possibilities for generating theorem-like declarations described for standard LATEX in Section 4.5. It is loaded with

```
\usepackage{amsthm}
```

like any other package.

As for standard LATEX, a new theorem declaration is created with a statement like

```
\newtheorem{com}{Comment}
```

where the first argument is the name of the theorem type (here com) and the second is the title that is printed when the theorem declaration is invoked. For example,

```
\begin{com}
    Theorem declarations can have any name.
\end{com}
```

produces the declaratory text

>   **Comment 1.** *Theorem declarations can have any name.*

In addition to the two mandatory arguments, the \newtheorem command may have one of two optional ones. The complete syntax is

```
\newtheorem{type}[num_like]{title}
     or
\newtheorem{type}{title}[in_counter]
```

where *num_like* is the name of an existing theorem-like declaration which is to be numbered in the same sequence as *type*, and *in_counter* is a counter name like chapter or section to reset the numbers of the *type* declarations.

All this is standard LATEX so far. The amsthm package adds the following features.

- A \newtheorem* is provided that defines an unnumbered theorem structure.

- Three predefined theorem styles are available:

  plain in which the title and number are in bold face and the text italic;

  definition with title and number in bold face and the text in normal font;

  remark for title and number in italic and the text normal.

  The desired style is activated by first issuing \theoremstyle{*style*}; all subsequent \newtheorem statements will have this style until a new one is activated.

- A \swapnumbers can be issued to cause all following new theorem types to have the numbers appear before the title, as **1 Comment.**

- New theorem styles may be defined by the user by means of a \newtheoremstyle command, or additional predefined styles may be loaded with package options. Since this is fairly specialized and complex, it is best to examine the example file thmtest.tex or read the documentation in amsthm.dtx.

- A proof environment is available for presenting short proofs. It is an unnumbered structure with the title *Proof*. The text is terminated automatically with the Q.E.D. symbol □. This symbol may be altered by redefining the command \qedsymbol; it may be printed at any time by issuing \qed.

The amsthm package has much in common with Frank Mittelbach's theorem package in the tools collection of Section D.3.3.

E.4.2 Commutative diagrams

The extra $\mathcal{A}_{\mathcal{M}}S$-LᴬTᴇX package amscd makes it easier to generate commutative diagrams like the one here at the right. The package is loaded as usual with

\usepackage{amscd}

$$
\begin{CD}
S^{\mathcal{W}_\Lambda} \otimes T @>j>> T \\
@VVV @VV{\text{End } P}V \\
(S \otimes T)/I @= (Z \otimes T)/J
\end{CD}
$$

These diagrams are created within the CD environment using some additional arrow commands. These bear the rather unusual names @>>> @<<< @AAA and @VVV for arrows pointing right, left, upwards, and downwards, respectively. The command @= draws a horizontal double rule, a lengthened equals sign.

Any text or symbols between the first and second > or < characters will appear above the horizontal arrow in \scriptstyle font. Similarly, any text or symbols between the second and third characters will be printed below the arrow.

For vertical arrows, text or symbols between the first and second A or V are placed to the left; those between the second and third to the right, again in \scriptstyle.

The above example diagram, which is taken from the $\mathcal{A}_{\mathcal{M}}S$-LᴬTᴇX manual amsldoc.tex, was produced with

```
\[ \begin{CD}
    S^{\mathcal{W}_\Lambda}\otimes T @>j>>   T\\
    @VVV                          @VV{\End P}V\\
    (S\otimes T)/I                @= (Z\otimes T)/J
    \end{CD}   \]
```

The command \End to print the function name 'End' is not standard. It must be previously defined with \DeclareMathOperator{\End}{End} (see page 416).

E.4.3   References with upref package

Normally the numbers printed with the `\ref` and `\pageref` commands are in the current font, whether that be bold, italic, or upright. In order to ensure that the numbers are always upright, load the extra $\mathcal{A}_{\mathcal{M}}$S-LᴬTEX package `upref` with the `\usepackage` command.

E.5   The $\mathcal{A}_{\mathcal{M}}$S fonts

The $\mathcal{A}_{\mathcal{M}}$S makes a number of fonts available to complement the regular Computer Modern fonts provided with the standard TeX/LaTeX installation. They include extra math alphabets, supplemental CM bold math italic and symbol fonts in smaller sizes than 10 pt, Cyrillic fonts, and additional symbol fonts.

In the next sections we describe these various fonts and how to take advantage of them.

E.5.1   Extra CM math fonts

Standard TeX installations of Computer Modern fonts provide bold math italic `cmmib10`, the bold symbols `cmbsy10`, and math extensions `cmex10` fonts only in 10 pt size, as indicated by the suffix 10 to their names. The $\mathcal{A}_{\mathcal{M}}$S has supplemented these with versions in sizes 5–9 pt.

cmmib5	cmmib6	cmmib7	cmmib8	cmmib9
cmbsy5	cmbsy6	cmbsy7	cmbsy8	cmbsy9
		cmex7	cmex8	cmex9

The small caps font `cmcsc10` is also given companions `cmcsc8` and `cmcsc9`.

The normal LaTeX 2_ε installation automatically assumes that these fonts are on the system and incorporates them into the necessary NFSS font definition files. Substitutions will be made if they are missing.

E.5.2   Cyrillic fonts

The $\mathcal{A}_{\mathcal{M}}$S Cyrillic fonts were originally used in reviews of books published in Russian and other Slavic languages in which the titles were to be rendered in the original language. In 1988, the Humanities and Arts Computing Center of the University of Washington redesigned them for general purpose Slavic studies, adding the pre-Revolutionary and accented letters. The overall appearance was also greatly enhanced (see Layout 8 on page 445).

These fonts all bear the prefix wncy, followed by the style designation r (upright), b (bold), i (italic), sc (small caps), or ss (upright sans serif), and the design size in points.

The best way to enable the Cyrillic fonts is simply to select font encoding OT2 under NFSS, as illustrated in Section 8.5.2 on page 237. The .fd (font definition) files for the Cyrillic fonts have been so set up that the other font attributes fully parallel those of the Latin fonts. This means wncyr10 has all the same attributes as cmr10, except for the encoding: family cmr, shape n, series m. (Of course, if the CM fonts are not the current standard ones, it will require more than just selecting the OT2 encoding to activate these Cyrillic fonts.)

The layout of the Cyrillic fonts has been chosen in such a way that the input text may be entered following the regular English transliteration scheme. Thus Cyrillic C is in position 83 where Latin S is normally situated. Typing S when a Cyrillic font is active outputs the correct equivalent C. Numerals and punctuations are to be found in the standard locations and so may be typed in as usual. 'Санкт-Петербург 10?' is thus generated by {\cyr Sankt-Peterburg 10?}.

Since the Cyrillic alphabet possesses more letters than the Latin, many of them must be transliterated with multi-letter combinations. These are automatically programmed into the fonts using TEX's ligature system. For example, Ch is treated as a ligature for symbol 81 Ч just as fi is for fi in a Latin font. This means that multi-letter transliterations are simply typed in. The input for 'Хрущев' is {\cyr Khrushchev}, where Kh → X and shch → щ. The transliteration scheme is that for English; other languages have their own systems to reproduce the original pronunciation. For example, 'Горбачёв' is *Gorbatschow* in German, *Gorbaciov* in Italian, and *Gorbachev* in English. These other schemes do *not* work with these fonts.

Not all letters can be produced so automatically (for example, the ё in Горбачёв) and for this reason the $\mathcal{A}_{\mathcal{M}}\mathcal{S}$ provides a file cyracc.def containing macro definitions for accented letters and other special features, such as the hard and soft signs. When these macros are given in a Latin font, additional transliteration symbols appear.

Unfortunately, some of the accent commands in cyracc.def will not work with LaTeX 2_ε. What is needed is an encoding definition file named ot2enc.def, similar to ot1enc.def and t1enc.def, to define those commands that are encoding-specific with the help of the special definition commands in Section C.5.7.

E.5.3   Euler fonts

The fonts known as the *Euler* collection, named after the eighteenth-century mathematician Leonhard Euler, have been collected from various sources by the $\mathcal{A}_{\mathcal{M}}\mathcal{S}$. They include a 'blackboard' font, representing a professor's handwriting on a blackboard, a Fraktur or Gothic font, and a script font. Their main purpose in mathematics is to be a substitute for the CM calligraphic math alphabets.

To use the blackboard characters, one must load the package `amsfonts` (which is included automatically with the `amssymb` package) and then employ the math alphabet command \mathbb. Thus $\mathbb{A B C ..}$ produces

$$\mathbb{A B C D E F G H I J K L N M O P Q R S T U V W X Y Z}$$

The package `eucal` redefines the \mathcal command to use the Euler script characters in place of the CM calligraphic letters. If this package is loaded with the option `mathscr`, the command \mathcal is left unchanged and instead \mathscr is defined to invoke these letters. In this case, $\mathscr{A B C ..}$ produces

$$\mathscr{A B C D E F G H I J K L N M O P Q R S T U V W X Y Z}$$

Finally, the package `eufrak` defines the math alphabet command \mathfrak, with which $\mathfrak{A B C ..}$ yields

$$\mathfrak{A B C D E F G H I J K L N M O P Q R S T U V W X Y Z}$$

This math alphabet is also enabled with the `amsfonts` package.

E.5.4   Extra math symbols

The set of symbols in the CM math symbol fonts by no means exhausts the fantasies of active mathematicians. To overcome this deficiency, the $\mathcal{A}_{\mathcal{M}}S$ has produced two fonts, `msam10` and `msbm10`, containing only symbols, arrows, and the blackboard characters. They are also available in sizes 5–9 pt. Since these fonts originated in the days when TEX could only handle 128 characters in any font, that is exactly how many they contain. Today, they could be combined into one font of 256 characters.

Two packages permit access to this treasure trove of hieroglyphs:

amsfonts enables the \mathbb math alphabet command for the blackboard characters, and defines those symbol names which are otherwise only provided in the `latexsym` package (Section 5.3.3).

amssymb is the more convenient package, which loads `amsfonts` and then defines names for all the symbols in the two fonts.

For example,

```
\[ \circlearrowright \Cup \lessapprox \lll \varpropto \because
   \circeq \vDash \blacktriangle \sphericalangle \]
```

$$\circlearrowright \quad \Cup \quad \lessapprox \quad \lll \quad \varpropto \quad \because \quad \circeq \quad \vDash \quad \blacktriangle \quad \sphericalangle$$

All the symbols and their associated names from the `amssymb` package are to be found in Tables G.20–G.26 on pages 552–554.

TₑX Fonts

Computers work exclusively with numbers, or more precisely, only with bits which may be interpreted as numbers. They do not know the difference between the letter A and an apple, or that either even exists. For text processing, all symbols, both input and output, need to be represented as numbers somehow. The association between number and symbol is called the *encoding* or *layout*. The latter term derives from the common method of illustrating the encoding in the form of a table.

Encoding tables are by no means standard. The standard ASCII scheme is just one of several, and it is limited to 7 bits, or 128 characters. There are 8-bit (256 characters) versions as well, in fact, a large number exist, tailored to different computer systems and languages. The question of coding the input for LaTeX documents with more than 7 bits has been addressed in Section 8.6.

In this appendix, we look at how the underlying TeX program deals with output fonts, their nomenclature, and their encoding tables.

F.1 Font metrics and bitmaps

When TeX decides to output character *nn* in a particular font, all it needs to know is how much room to leave for it. TeX does not care what the character looks like, for that is the task of the DVI driver afterwards.

Information about the characters in each font are stored in various files, all bearing the root name of the font but with different extensions. This information is divided into files that contain only the sizes of each symbol, and those with the actual drawing or image data.

.tfm TeX font metric files are the only font files read in by TeX itself. They contain the sizes of the characters, such as width, height, and depth. For slanted fonts, they also possess the 'italic correction' (Section 3.5.1) for each letter. Furthermore, they specify for which letter combinations a different spacing is required, such as AV instead of AV, or for which a ligature is available. Finally, the .tfm files provide information about the slope of the characters (zero for an unslanted font), the standard word spacing and its stretch and shrinkage (Section 9.5.4), the width of the em and quad spacings, and the spacings at the end of a sentence. Mathematical and symbol fonts require even more information which is also included in their .tfm files.

.pk Compressed pixel files contain bitmaps (images) of the symbols in each font, in one size and one resolution. The DVI driver program uses these to send the output to a printer or previewer. These files are normally generated automatically by the METAFONT program (Section F.3) as needed.

Since these files might exist several times for a given font, in difference sizes or resolutions, many installations distinguish them by adding the resolution to the extension, as cmr10.300pk or cmr10.600pk for 300 and 600 dpi versions, respectively. On systems that cannot support extensions with more than three characters, both files are named cmr10.pk but are stored in directories dpi300 and dpi600.

Font magnification is achieved by selecting an appropriate resolution. If cmr10 is to be printed on a 300 dpi printer at double its normal size, the 600 dpi version is used instead.

.pxl Obsolete uncompressed pixel files have been replaced by the compressed .pk versions.

.mf METAFONT source files contain the drawing instructions for each of the fonts. These are not read by the DVI driver but rather by the METAFONT program which converts them into .pk bitmap files at the required size and resolution.

.vf Virtual font files are an alternative to .pk files. Instead of having bitmaps for each character, they contain instructions that refer to characters in different, real fonts, or that tell the driver to draw a black box and issue a warning message (that the character does not exist, for example).

Virtual fonts are normally associated with PostScript fonts and are therefore described in Section F.5. However, there is no reason for them not to be used with bitmap fonts as well.

F.2   Computer Modern fonts

When Donald E. Knuth invented the TₑX program, he also provided it with an extensive set of character fonts, which he named *Computer Modern,* rather than relying on the fonts available on any given printer. At that time, the printer fonts were not so good, and certainly were not uniform. With the supplied fonts, TEX could produce identical, high-quality results on all printers.

LATEX, of course, has inherited these fonts, so that they have almost become a trademark for documents produced by TEX or LATEX. This is not really necessary, for LATEX need not be married to any particular set of fonts, especially with the New Font Selection Scheme (Section 8.5) which simplifies font installation enormously. The main fonts used in this book, for example, are Lucida Bright, Lucida Sans, and Lucida Sans Typewriter, designed by Bigelow & Holmes and distributed by Y&Y Inc.

F.2.1   Font families

Typography is the study and classification of typefaces, something that goes back to Gutenberg's invention of *movable type* (not of the printing press, which was invented by the Chinese) five and a half centuries ago. Since that time, many families of fonts have been created, bearing classical names like Baskerville, Garamond, Univers, etc. Each member of such a family has the same overall design, or basic look, but vary by being slanted, italic, bold, or thin; and of course, they come in different sizes.

Font families are classified according to certain criteria that often determine to what use they will be put.

Serif fonts: are those that have little horizontal lines, or *serifs*, at their edges, to guide the eye better. Experience has shown these to be best for general reading, and so they are regularly used for the main body of text. The NFSS terminology refers to these as *Roman* fonts.

Sans serif fonts: are those that are lacking any serifs. Such fonts with their starker appearance are often employed for titling or headlining. Compare sans serif with regular text.

Fixed fonts: are those with a uniform letter width, something that has evolved from the typewriter and has been carried over to computer listings. Classically, such fonts have no business in book printing, where *proportional* fonts (the letter *i* is narrower than *m*) have always dominated.

Decorative fonts: are ones that stand out because of some unusual characteristic. They are intended to catch attention and to attract the eye. Such families are not complete, with a full range of shapes and widths, and are employed mostly by advertising.

Mathematical fonts: are collections of special symbols needed for mathematical works. Their further classification cannot be compared to that of text fonts at all.

A book designer must decide what kinds of typeface families are needed. He or she might decide on a Roman (serifed) font for the main body, a sans serif one for headlines, and then, if it is a book containing computer code, select a fixed font for setting those parts. Finally, symbol fonts will be needed if the work contains mathematical sections.

The Computer Modern set of fonts provides all these classes of families. However, since they were produced before the NFSS attribute system of classification was established, the CM font nomenclature does not conform perfectly with this scheme. The NFSS system frees the user from having to think about the CM font names, for specifying the attributes is sufficient. Of course, the NFSS installation must provide the font definition .fd files which translate any set of attributes into a real font name, or into some acceptable substitute.

F.2.2   Classification of CM fonts

All the TEX font names begin with the letters cm, which stands for 'Computer Modern', followed by one to four letters describing the style of font, and finally one or two digits specifying the design size in points. This is the root name of the font which must be given in the \newfont command (Section 4.1.6) if it is to be activated directly.

The CM fonts can be classified as text, math, decorative, or other special symbol fonts, as described in the rest of this section. Each type of font has an encoding scheme shown in the accompanying layout tables.

Text fonts

The Computer Modern fonts appropriate for straightforward text can be classified into the three families Roman, sans serif, and typewriter. Within each of these families there are upright, slanted, italic, small caps, and bold variants.

Table F.1 presents the names of these fonts together with the NFSS family/series/shape assignments from Section 8.5. They all have the same coding (OT1). The * in each name is the design size specification which takes on values of 5, 6, 7, 8, 9, 10, 12, 17. Only cmr* is available in all these sizes, while some are only to be found in size 10. (This table contains much the same information as in Table 8.3 on page 235 but in a different form.)

One sees several apparent inconsistencies in these font names, such as cmti* for the italic upright font. Why is this not cmri*? The answer is that this is a *text italic* font, as opposed to the *math italic* font cmmi* described below. These inconsistences have arisen because the CM fonts were not created with the NFSS classification system in mind, since that was established a decade later.

Table F.1: Computer Modern text fonts

Style	Series/shape	Roman cmr	Sans serif cmss	Typewriter cmtt
	Family=			
Upright	m/n	cmr*	cmss*	cmtt*
" slanted	m/sl	cmsl*	cmssi*	cmsltt*
" italic	m/it	cmti*	—	cmitt*
" small caps	m/sc	cmcsc*	—	cmtcsc*
Bold	b/m	cmb*	—	—
Bold extended	bx/m	cmbx*	cmssbx*	—
" " slanted	bx/sl	cmbxsl*	—	—
" " italic	bx/it	cmbxti*	—	—

Further inconsistencies exist in the encoding schemes for the CM text fonts. They are all nominally OT1, but there are slight deviations among them. The proper OT1 encoding is displayed in Layout 1 on the next page for font cmr10. The text italic fonts are identical to the upright ones, except that the dollar sign in position 36 is replaced by the pound sign (Layout 3 for cmti10). The slanted fonts, on the other hand, are exactly the same as the upright ones. Other deviations are to be found in the small caps (Layout 2) and all the typewriter fonts (Layout 4).

Math fonts

The Computer Modern fonts provide three types of mathematical fonts, each with its own encoding scheme.

The *mathematical italic* fonts cmmi* contain Latin and Greek letters, in upper and lower case, plus a number of extra symbols. Since variable names in formulas are set in *italics*, these are basically italic fonts. The encoding scheme is designated OML, for *old math letters*, and is displayed in Layout 5. It is available in design sizes 5–12 pt in normal weight, but only in 10 pt size in bold face, cmmib10. These fonts are used for the letters math alphabet of Section C.5.4.

Symbol fonts cmsy* provide the rest of the math symbols for formulas, except for those that appear in variable sizes. The encoding scheme is named OMS for *old math symbols* and is shown in Layout 6. It comes in sizes 5–10 pt in normal weight, and in 10 pt size in bold face, cmbsy10. The symbols math alphabet uses these fonts.

Variable sized symbols are to be found in the font cmex10, with encoding OMX, for *old math extension*. Layout 7 illustrates this set. There is no bold version of this font. It belongs to the largesymbols math alphabet of Section C.5.4.

	0	1	2	3	4	5	6	7
'00x	Γ 0	Δ 1	Θ 2	Λ 3	Ξ 4	Π 5	Σ 6	Υ 7
'01x	Φ 8	Ψ 9	Ω 10	ff 11	fi 12	fl 13	ffi 14	ffl 15
'02x	ı 16	J 17	` 18	´ 19	ˇ 20	˘ 21	¯ 22	° 23
'03x	¸ 24	ß 25	æ 26	œ 27	ø 28	Æ 29	Œ 30	Ø 31
'04x	´ 32	! 33	'' 34	# 35	$ 36	% 37	& 38	' 39
'05x	(40	) 41	* 42	+ 43	, 44	- 45	. 46	/ 47
'06x	0 48	1 49	2 50	3 51	4 52	5 53	6 54	7 55
'07x	8 56	9 57	: 58	; 59	¡ 60	= 61	¿ 62	? 63
'10x	@ 64	A 65	B 66	C 67	D 68	E 69	F 70	G 71
'11x	H 72	I 73	J 74	K 75	L 76	M 77	N 78	O 79
'12x	P 80	Q 81	R 82	S 83	T 84	U 85	V 86	W 87
'13x	X 88	Y 89	Z 90	[91	" 92	] 93	^ 94	. 95
'14x	' 96	a 97	b 98	c 99	d 100	e 101	f 102	g 103
'15x	h 104	i 105	j 106	k 107	l 108	m 109	n 110	o 111
'16x	p 112	q 113	r 114	s 115	t 116	u 117	v 118	w 119
'17x	x 120	y 121	z 122	– 123	— 124	˝ 125	~ 126	¨ 127

Font Layout 1: The character font `cmr10`. This is the standard character assignment for the `OT1` encoding scheme.

	0	1	2	3	4	5	6	7
'00x	Γ 0	Δ 1	Θ 2	Λ 3	Ξ 4	Π 5	Σ 6	Υ 7
'01x	Φ 8	Ψ 9	Ω 10	↑ 11	↓ 12	' 13	¡ 14	¿ 15
'02x	I 16	J 17	` 18	´ 19	ˇ 20	˘ 21	¯ 22	° 23
'03x	¸ 24	SS 25	Æ 26	Œ 27	ø 28	Æ 29	Œ 30	Ø 31
'04x	´ 32	! 33	'' 34	# 35	$ 36	% 37	& 38	' 39
'05x	(40	) 41	* 42	+ 43	, 44	- 45	. 46	/ 47
'06x	0 48	1 49	2 50	3 51	4 52	5 53	6 54	7 55
'07x	8 56	9 57	: 58	; 59	< 60	= 61	> 62	? 63
'10x	@ 64	A 65	B 66	C 67	D 68	E 69	F 70	G 71
'11x	H 72	I 73	J 74	K 75	L 76	M 77	N 78	O 79
'12x	P 80	Q 81	R 82	S 83	T 84	U 85	V 86	W 87
'13x	X 88	Y 89	Z 90	[91	" 92	] 93	^ 94	. 95
'14x	' 96	A 97	B 98	C 99	D 100	E 101	F 102	G 103
'15x	H 104	I 105	J 106	K 107	L 108	M 109	N 110	O 111
'16x	P 112	Q 113	R 114	S 115	T 116	U 117	V 118	W 119
'17x	X 120	Y 121	Z 122	– 123	— 124	˝ 125	~ 126	¨ 127

Font Layout 2: The character font `cmcsc10`. The differences from Layout 1 are symbols 11–15, 25, 60, and 62. The ligatures that normally appear in 11–15 have been replaced by some extra symbols.

	0	1	2	3	4	5	6	7
'00x	Γ 0	Δ 1	Θ 2	Λ 3	Ξ 4	Π 5	Σ 6	Υ 7
'01x	Φ 8	Ψ 9	Ω 10	ﬀ 11	ﬁ 12	ﬂ 13	ﬃ 14	ﬄ 15
'02x	ı 16	ȷ 17	` 18	´ 19	ˇ 20	˘ 21	¯ 22	˚ 23
'03x	¸ 24	ß 25	æ 26	œ 27	ø 28	Æ 29	Œ 30	Ø 31
'04x	ˍ 32	! 33	" 34	# 35	£ 36	% 37	& 38	' 39
'05x	(40	) 41	* 42	+ 43	, 44	- 45	. 46	/ 47
'06x	0 48	1 49	2 50	3 51	4 52	5 53	6 54	7 55
'07x	8 56	9 57	: 58	; 59	¡ 60	= 61	¿ 62	? 63
'10x	@ 64	A 65	B 66	C 67	D 68	E 69	F 70	G 71
'11x	H 72	I 73	J 74	K 75	L 76	M 77	N 78	O 79
'12x	P 80	Q 81	R 82	S 83	T 84	U 85	V 86	W 87
'13x	X 88	Y 89	Z 90	[91	" 92	] 93	^ 94	· 95
'14x	' 96	a 97	b 98	c 99	d 100	e 101	f 102	g 103
'15x	h 104	i 105	j 106	k 107	l 108	m 109	n 110	o 111
'16x	p 112	q 113	r 114	s 115	t 116	u 117	v 118	w 119
'17x	x 120	y 121	z 122	- 123	— 124	˝ 125	~ 126	¨ 127

Font Layout 3: The character font `cmti10`. The only difference from Layout 1 is symbol 36 (£ instead of $). All other *text italic* fonts have this same pattern.

	0	1	2	3	4	5	6	7
'00x	Γ 0	Δ 1	Θ 2	Λ 3	Ξ 4	Π 5	Σ 6	Υ 7
'01x	Φ 8	Ψ 9	Ω 10	↑ 11	↓ 12	' 13	ı 14	¿ 15
'02x	ı 16	ȷ 17	` 18	´ 19	ˇ 20	˘ 21	¯ 22	˚ 23
'03x	¸ 24	ß 25	æ 26	œ 27	ø 28	Æ 29	Œ 30	Ø 31
'04x	␣ 32	! 33	" 34	# 35	$ 36	% 37	& 38	' 39
'05x	(40	) 41	* 42	+ 43	, 44	- 45	. 46	/ 47
'06x	0 48	1 49	2 50	3 51	4 52	5 53	6 54	7 55
'07x	8 56	9 57	: 58	; 59	< 60	= 61	> 62	? 63
'10x	@ 64	A 65	B 66	C 67	D 68	E 69	F 70	G 71
'11x	H 72	I 73	J 74	K 75	L 76	M 77	N 78	O 79
'12x	P 80	Q 81	R 82	S 83	T 84	U 85	V 86	W 87
'13x	X 88	Y 89	Z 90	[91	\ 92	] 93	^ 94	_ 95
'14x	' 96	a 97	b 98	c 99	d 100	e 101	f 102	g 103
'15x	h 104	i 105	j 106	k 107	l 108	m 109	n 110	o 111
'16x	p 112	q 113	r 114	s 115	t 116	u 117	v 118	w 119
'17x	x 120	y 121	z 122	{ 123		124	} 125	~ 126	¨ 127

Font Layout 4: The character font `cmtt10`. All `tt` fonts are set up with the same pattern. The differences from Layout 1 lie in symbols 11–15, 60, 62, 92, 123, 124, and 125. Furthermore, the italic typewriter fonts also have £ in place of $.

The term *old* in these encoding names refers to the original encoding as introduced by Donald Knuth, in the same way that the text encoding OT1 indicated the old, or original, text encoding with 128 characters per font. The new encoding schemes contain 256 characters. For text, the encoding is named T1 (Section F.4.3); such extended math fonts do not yet exist.

The extra $\mathcal{A}_{\mathcal{M}}\mathcal{S}$ fonts in Section E.5.1 add additional bold face sizes for cmmib* and cmbsy* as well as some more sizes for cmex*.

Decorative fonts

Three families of decorative or special fonts are available in a single design size and limited attributes. They all exhibit OT1 encoding.

cmfr *Funny Roman* family, consists of two fonts, cmff10 leaning to the left and cmfi10 *which leans to the right*. Both are series m with shapes n and it respectively.

cmfib *Fibonacci* family, contains one font cmfib8 derived from the Fibonacci series of numbers. It has series m and shape n.

cmdunh *Dunhill* family, with one font cmdunh10 with series m and shape n. This font is demonstrated in the sample letterhead on page 310.

Logo fonts

The fonts logo8, logo9, logo10, logosl10, and logobf10 contain only the nine letters A, E, F, M, N, O, P, S, and T, for generating the logos

METAFONT   *METAFONT*   **METAFONT**
METAPOST   *METAPOST*   **METAPOST**

The L^AT_EX lasy fonts

As an extension of the cmsy* fonts for mathematical symbols, L^AT_EX provides some additional symbols with the font lasy* in design sizes 5–10 pt. It contains the 15 symbols: ‹› ⌄ ⊲ ⊴ ▷ ⊵ ℧ ⋈ □ ◇ ⤳ ⊏ ⊐.

In L^AT_EX 2$_\varepsilon$, these symbols are not defined unless one of the packages latexsym or amsfonts has been loaded.

Fonts for making pictures

The special picture elements for use in the L^AT_EX picture environment are stored in fonts named line10, lcircle10, linew10, and lcirclew10. The first two are used for lines, ovals, and circles when \thinlines is in effect. The sloping lines and arrow heads are to be found in line10

	0	1	2	3	4	5	6	7
'00x	Γ 0	Δ 1	Θ 2	Λ 3	Ξ 4	Π 5	Σ 6	Υ 7
'01x	Φ 8	Ψ 9	Ω 10	α 11	β 12	γ 13	δ 14	ϵ 15
'02x	ζ 16	η 17	θ 18	ι 19	κ 20	λ 21	μ 22	ν 23
'03x	ξ 24	π 25	ρ 26	σ 27	τ 28	υ 29	ϕ 30	χ 31
'04x	ψ 32	ω 33	ε 34	ϑ 35	ϖ 36	ϱ 37	ς 38	φ 39
'05x	$\leftharpoonup$ 40	$\leftharpoondown$ 41	$\rightharpoonup$ 42	$\rightharpoondown$ 43	$^\backprime$ 44	$^\prime$ 45	$\triangleright$ 46	$\triangleleft$ 47
'06x	0 48	1 49	2 50	3 51	4 52	5 53	6 54	7 55
'07x	8 56	9 57	. 58	, 59	$<$ 60	$/$ 61	$>$ 62	$\star$ 63
'10x	∂ 64	A 65	B 66	C 67	D 68	E 69	F 70	G 71
'11x	H 72	I 73	J 74	K 75	L 76	M 77	N 78	O 79
'12x	P 80	Q 81	R 82	S 83	T 84	U 85	V 86	W 87
'13x	X 88	Y 89	Z 90	$\flat$ 91	$\natural$ 92	$\sharp$ 93	$\smile$ 94	$\frown$ 95
'14x	ℓ 96	a 97	b 98	c 99	d 100	e 101	f 102	g 103
'15x	h 104	i 105	j 106	k 107	l 108	m 109	n 110	o 111
'16x	p 112	q 113	r 114	s 115	t 116	u 117	v 118	w 119
'17x	x 120	y 121	z 122	$\imath$ 123	$\jmath$ 124	$\wp$ 125	$\vec{}$ 126	$\frown$ 127

Font Layout 5: The font cmmi10. This corresponds to the OML encoding scheme, as does cmmib10. It contains lower case Greek letters in positions 11–39 and math symbols in 40–47, 60–64, and 123–127.

	0	1	2	3	4	5	6	7
'00x	$-$ 0	$\cdot$ 1	$\times$ 2	$*$ 3	$\div$ 4	$\diamond$ 5	$\pm$ 6	$\mp$ 7
'01x	$\oplus$ 8	$\ominus$ 9	$\otimes$ 10	$\oslash$ 11	$\odot$ 12	$\bigcirc$ 13	$\circ$ 14	$\bullet$ 15
'02x	$\asymp$ 16	$\equiv$ 17	$\subseteq$ 18	$\supseteq$ 19	$\leq$ 20	$\geq$ 21	$\preceq$ 22	$\succeq$ 23
'03x	$\sim$ 24	$\approx$ 25	$\subset$ 26	$\supset$ 27	$\ll$ 28	$\gg$ 29	$\prec$ 30	$\succ$ 31
'04x	$\leftarrow$ 32	$\rightarrow$ 33	$\uparrow$ 34	$\downarrow$ 35	$\leftrightarrow$ 36	$\nearrow$ 37	$\searrow$ 38	$\simeq$ 39
'05x	$\Leftarrow$ 40	$\Rightarrow$ 41	$\Uparrow$ 42	$\Downarrow$ 43	$\Leftrightarrow$ 44	$\nwarrow$ 45	$\swarrow$ 46	$\propto$ 47
'06x	$\prime$ 48	∞ 49	$\in$ 50	$\ni$ 51	$\triangle$ 52	∇ 53	$/$ 54	$\mid$ 55
'07x	$\forall$ 56	$\exists$ 57	$\neg$ 58	$\emptyset$ 59	$\Re$ 60	$\Im$ 61	$\top$ 62	$\perp$ 63
'10x	$\aleph$ 64	$\mathcal{A}$ 65	$\mathcal{B}$ 66	$\mathcal{C}$ 67	$\mathcal{D}$ 68	$\mathcal{E}$ 69	$\mathcal{F}$ 70	$\mathcal{G}$ 71
'11x	$\mathcal{H}$ 72	$\mathcal{I}$ 73	$\mathcal{J}$ 74	$\mathcal{K}$ 75	$\mathcal{L}$ 76	$\mathcal{M}$ 77	$\mathcal{N}$ 78	$\mathcal{O}$ 79
'12x	$\mathcal{P}$ 80	$\mathcal{Q}$ 81	$\mathcal{R}$ 82	$\mathcal{S}$ 83	$\mathcal{T}$ 84	$\mathcal{U}$ 85	$\mathcal{V}$ 86	$\mathcal{W}$ 87
'13x	$\mathcal{X}$ 88	$\mathcal{Y}$ 89	$\mathcal{Z}$ 90	$\cup$ 91	$\cap$ 92	$\uplus$ 93	$\wedge$ 94	$\vee$ 95
'14x	$\vdash$ 96	$\dashv$ 97	$\lfloor$ 98	$\rfloor$ 99	$\lceil$ 100	$\rceil$ 101	$\{$ 102	$\}$ 103
'15x	$\langle$ 104	$\rangle$ 105	$\mid$ 106	$\|$ 107	$\updownarrow$ 108	$\Updownarrow$ 109	$\backslash$ 110	$\wr$ 111
'16x	$\surd$ 112	$\amalg$ 113	∇ 114	$\int$ 115	$\sqcup$ 116	$\sqcap$ 117	$\sqsubseteq$ 118	$\sqsupseteq$ 119
'17x	$\S$ 120	$\dagger$ 121	$\ddagger$ 122	$\P$ 123	$\clubsuit$ 124	$\diamondsuit$ 125	$\heartsuit$ 126	$\spadesuit$ 127

Font Layout 6: The font cmsy10. The cmsy fonts, adhering to the OMS encoding scheme, contain many math symbols as well as the calligraphic letters $\mathcal{A}\ldots\mathcal{Z}$ in positions 65–90. The bold face version cmbsy10 is set up on the same pattern.

	0	1	2	3	4	5	6	7
'00x	(0	) 1	⌈ 2	⌉ 3	⌊ 4	⌋ 5	⌈ 6	⌉ 7
'01x	{ 8	} 9	⟨ 10	⟩ 11	| 12	‖ 13	/ 14	\ 15
'02x	(16	) 17	(18	) 19	⌈ 20	⌉ 21	| 22	| 23
'03x	⌈ 24	⌉ 25	{ 26	} 27	⟨ 28	⟩ 29	/ 30	\ 31
'04x	(32	) 33	⌈ 34	⌉ 35	| 36	| 37	⌈ 38	⌉ 39
'05x	{ 40	} 41	⟨ 42	⟩ 43	/ 44	\ 45	/ 46	\ 47
'06x	(48	\ 49	⌈ 50	⌉ 51	⌊ 52	| 53	| 54	| 55
'07x	⎰ 56	⎱ 57	⎩ 58	⎭ 59	} 60	} 61	. 62	| 63
'10x	⎝ 64	⎠ 65	| 66	| 67	⟨ 68	⟩ 69	⊔ 70	⊔ 71
'11x	∮ 72	∮ 73	⊙ 74	⊙ 75	⊕ 76	⊕ 77	⊗ 78	⊗ 79
'12x	Σ 80	Π 81	∫ 82	∪ 83	∩ 84	⊎ 85	∧ 86	∨ 87
'13x	Σ 88	Π 89	∫ 90	∪ 91	∩ 92	⊎ 93	∧ 94	∨ 95
'14x	⨆ 96	⨆ 97	⌢ 98	⌢ 99	⌢ 100	~ 101	~ 102	~ 103
'15x	⌈ 104	⌉ 105	⌊ 106	⌋ 107	⌈ 108	⌉ 109	{ 110	} 111
'16x	√ 112	√ 113	√ 114	√ 115	√ 116	| 117	⌐ 118	‖ 119
'17x	↑ 120	↓ 121	⌒ 122	⌍ 123	⌌ 124	⌎ 125	⇑ 126	⇓ 127

Font Layout 7: The font cmex10, containing the mathematical symbols that appear in varying sizes. This is the OMX encoding scheme.

	0	1	2	3	4	5	6	7
'00x	Њ 0	Љ 1	Џ 2	Э 3	I 4	Є 5	Ђ 6	Ћ 7
'01x	њ 8	љ 9	џ 10	э 11	i 12	є 13	ђ 14	ћ 15
'02x	Ю 16	Ж 17	Й 18	Ё 19	V 20	Ѳ 21	S 22	Я 23
'03x	ю 24	ж 25	й 26	ё 27	v 28	ѳ 29	s 30	я 31
'04x	¨ 32	! 33	" 34	Ѣ 35	˘ 36	% 37	' 38	' 39
'05x	(40	) 41	* 42	ѣ 43	, 44	- 45	. 46	/ 47
'06x	0 48	1 49	2 50	3 51	4 52	5 53	6 54	7 55
'07x	8 56	9 57	: 58	; 59	« 60	ı 61	» 62	? 63
'10x	˘ 64	А 65	Б 66	Ц 67	Д 68	Е 69	Ф 70	Г 71
'11x	Х 72	И 73	J 74	К 75	Л 76	М 77	Н 78	О 79
'12x	П 80	Ч 81	Р 82	С 83	Т 84	У 85	В 86	Ш 87
'13x	Ш 88	Ы 89	З 90	[91	" 92	] 93	Ь 94	Ъ 95
'14x	' 96	а 97	б 98	ц 99	д 100	е 101	ф 102	г 103
'15x	х 104	и 105	j 106	к 107	л 108	м 109	н 110	о 111
'16x	п 112	ч 113	р 114	с 115	т 116	у 117	в 118	щ 119
'17x	ш 120	ы 121	з 122	– 123	— 124	№ 125	ь 126	ъ 127

Font Layout 8: The character font wncyr10, one of the Cyrillic fonts from the University of Washington. Its coding scheme is designated OT2.

while the circles and oval segments (Section 6.1.4) are in lcircle10. The second pair of fonts with the added w contain thicker lines, for use with \thicklines.

F.2.3 $\mathcal{A}_{\mathcal{M}}\mathcal{S}$ Cyrillic fonts

The Cyrillic fonts that are part of the amsfonts collection are described in Section E.5.2. They conform to the font encoding scheme OT2, displayed in Layout 8. They are available as upright (wncyr*), bold (wncyb*), small caps (wncysc*), and upright sans serif (wncyss*) fonts. As for the CM fonts, * represents the design size in points.

The NFSS system assigns these fonts to families cmr and cmss; that is, they are treated as members of the Computer Modern set. This is not as absurd as it sounds since they are intended to be used with the CM fonts.

F.3   An overview of METAFONT

METAFONT is a program for designing and developing character fonts, written by the inventor of the TEX program, Donald E. Knuth (Knuth, 1986b, 1986c, 1986d). With METAFONT, the existing fonts may be regenerated with any desired magnification, larger or smaller than the design size.

The creation of a totally new set of fonts, such as for Arabic or Hebrew, is feasible, although something that should be left to experts, requiring detailed knowledge about the internal METAFONT language. This is not a book on METAFONT, so no such information will be presented here. We limit ourselves to describing how to create new pixel files for the existing fonts.

Most of today's DVI drivers and previewers are so constructed that they call METAFONT themselves whenever the required pixel files are missing. However, there might be times when the user must invoke it manually.

The METAFONT program is usually called with the command mf, but this may vary with installation. Once started, it prints the version number and the names of any extra files that are automatically loaded. Then it waits for a response from the user, prompting with the symbol **. The response must be of the form

```
\mode=localfont; mag=nn; input file
```

followed by ⟨return⟩. The value nn of mag= is a decimal number specifying the desired magnification; if it is omitted or set to 1, the design size is taken. The file entry to input is the root name of the font to be processed. There must be a file available with that root name plus the extension .mf. There should be such files for all the CM fonts in Section F.2.2 as part of the basic METAFONT implementation. They contain the character definitions in the form of 62 different parameters, which will not be elaborated here. Other .mf files, say for the amsfonts of Section E.5, may also be available.

METAFONT now proceeds to generate the pixel patterns for the selected font in the chosen magnification, printing the font name and the code numbers for the symbols to the screen. Finally, the program prints * and waits for a new user entry. The response end ⟨return⟩ terminates the program and restores control to the operating system.

The selected mode not only sets the base resolution, it also makes fine adjustments to optimize the output for that printer type. The file modes.mf defines a large number of modes for various printers. For 300 dpi, the usual mode is cx or one of its synonyms hplaser, laserwriter, etc. For 600 dpi, one often chooses mode ljfour. The last line of modes.mf equates localfont to the default mode at the local installation. The user or system manager must see that this is correctly set.

The results of the METAFONT run are two new files, one being the .tfm file for the font, the other bearing the root name and the extension .xxxgf, where gf stands for *generic font*. The xxx reflects the chosen magnification: if this is 1, xxx is the printer resolution in dots per inch (dpi). METAFONT obtains the information about the basic printer and its resolution from the selected mode.

The one .tfm file may be used for all magnifications of the same root font. In some places, localfont is modified so that the .tfm file is generated only for the design size, for example, when mag=1 is selected.

The generic font coding must now be converted into the .pk form that may be read by a driver. This is carried out with the subsidiary program gftopk. Suppose we have called METAFONT to generate cmr10 in magnification 1.44, for a 300 dpi printer; we now have the file cmr10.432gf. The call

    gftopk cmr10.432gf

creates the file cmr10.432pk. On some systems this is named cmr10.pk and stored in a directory named 432dpi or dpi432.

The usual TEX magnification steps are $\sqrt{1.2}$ and integral powers of 1.2. These may be entered into the command line mag= either as a decimal number or as magstep n to indicate 1.2^n. Here n must be a whole number. The half step $\sqrt{1.2}$ may be entered as magstep0.5.

In accordance with the development of TEX 3.0 for multilingual applications and 256 characters per font, Donald E. Knuth has extended METAFONT to a version 2.0. He now wishes to withdraw from any more new improvements of either of these programs, being prepared merely to correct any definite errors they may contain. To emphasize this decision, he will from now on give version numbers to TEX that converge to π (3.14159 ...) and to METAFONT that approach the value of e (2.71828 ...). At present, TEX is at version 3.14159 and METAFONT at 2.718. As a consequence of this decision, any further major developments to these two programs, such as undertaken by user groups, will be under new names, since Donald E. Knuth has kept the copyright to the existing names.

F.4   Extended character sets for TEX 3.0

Proper exploitation of the enhanced capabilities of TEX 3.0 requires extended fonts with 256 characters rather than 128. The additional characters should include accented letters as well as other accents missing from the original fonts.

F.4.1   Limitations and deficiencies of the CM fonts

Most of the diacritic marks (accents) used in European languages written with the Latin alphabet are contained in, or may be generated by, TEX's Computer Modern (CM) fonts. A basic set of naked accents is available for combination with other letters, such as the acute accent ´ with the letter e to make é. Other combinations may be constructed for diacritic marks that are not predefined in TEX or LATEX.

Fashioning diacritic characters as a combination of letters and special symbols has one great disadvantage for the TEX processing: words containing such characters cannot take part in the automatic word division since the hyphenation patterns include only pure letters. The accented letters, such as those in German and French and most other languages, must be treated as *single* characters in the hyphenation patterns, and must appear as single letters in the character set.

In addition to diacritic characters, a number of special letters are employed in some European languages, such as ß, Æ, æ, Œ, œ, Ø, and ø, which are provided in standard T_EX with the CM fonts (Section 2.5.6). However, other special letters, such as Ɖ, ŋ, Þ, þ, and ð, are missing completely and cannot be easily constructed from existing ones.

F.4.2   The Cork proposal

At the 1990 International T_EX Conference in Cork, Ireland, an extension of the Latin alphabet and its assignments within the 256 character positions was proposed and accepted. This extension includes the majority of special and diacritic letters as single characters for many languages written with the Latin alphabet. Hyphenation patterns for such languages may include the special and diacritic letters as single letters for optimal word division by T_EX and L^AT_EX.

Character fonts conforming to the Cork scheme are to bear the identifying letters ec in their names for 'Extended Computer' in place of the cm for 'Computer Modern'.

F.4.3   The Extended Computer fonts

The Cork proposal for extending the T_EX fonts to 256 characters was first implemented by Norbert Schwarz, who produced an initial set of META-FONT source files. He also selected the designation dc to emphasize that this was a preliminary realization of the EC fonts. Some work was still needed to fine-tune the design of several symbols.

After issuing versions 1.2 and 1.3 of the DC fonts in 1995 and 1996, Jörg Knappen released the first set of true EC fonts in January, 1997. Font Layout 9 presents his font ecrm1000, the extended version of cmr10. The EC fonts are now considered to be stable in that neither their encoding nor their metrics (the .tfm files) will be changed in future. Thus their behavior as far as T_EX and L^AT_EX are concerned is finalized. The actual printed characters might be modified slightly in later updates.

The EC font names are of the form ec*xxnnnn*, where *xx* represents two letters specifying the font characteristics, and *nnnn* the design size in hundredths of points. Thus ecrm1000 is the upright Roman font in size 10 points.

The METAFONT source files are available for the following extended fonts (without the size specification):

	0	1	2	3	4	5	6	7
'00x	` 0	´ 1	^ 2	~ 3	¨ 4	˝ 5	° 6	ˇ 7
'01x	˘ 8	¯ 9	˙ 10	˛ 11	¸ 12	‚ 13	‹ 14	› 15
'02x	" 16	" 17	„ 18	« 19	» 20	– 21	— 22	23
'03x	0 24	1 25	J 26	ff 27	fi 28	fl 29	ffi 30	ffl 31
'04x	␣ 32	! 33	" 34	# 35	$ 36	% 37	& 38	' 39
'05x	(40	) 41	* 42	+ 43	, 44	- 45	. 46	/ 47
'06x	0 48	1 49	2 50	3 51	4 52	5 53	6 54	7 55
'07x	8 56	9 57	: 58	; 59	< 60	= 61	> 62	? 63
'10x	@ 64	A 65	B 66	C 67	D 68	E 69	F 70	G 71
'11x	H 72	I 73	J 74	K 75	L 76	M 77	N 78	O 79
'12x	P 80	Q 81	R 82	S 83	T 84	U 85	V 86	W 87
'13x	X 88	Y 89	Z 90	[91	\ 92	] 93	^ 94	_ 95
'14x	' 96	a 97	b 98	c 99	d 100	e 101	f 102	g 103
'15x	h 104	i 105	j 106	k 107	l 108	m 109	n 110	o 111
'16x	p 112	q 113	r 114	s 115	t 116	u 117	v 118	w 119
'17x	x 120	y 121	z 122	{ 123	| 124	} 125	~ 126	- 127
'20x	Ă 128	Ą 129	Ć 130	Č 131	Ď 132	Ě 133	Ę 134	Ğ 135
'21x	Ĺ 136	Ľ 137	Ł 138	Ń 139	Ň 140	Ŋ 141	Ő 142	Ŕ 143
'22x	Ř 144	Ś 145	Š 146	Ş 147	Ť 148	Ţ 149	Ű 150	Ů 151
'23x	Ÿ 152	Ź 153	Ž 154	Ż 155	IJ 156	İ 157	đ 158	§ 159
'24x	ă 160	ą 161	ć 162	č 163	ď 164	ě 165	ę 166	ğ 167
'25x	ĺ 168	ľ 169	ł 170	ń 171	ň 172	ŋ 173	ő 174	ŕ 175
'26x	ř 176	ś 177	š 178	ş 179	ť 180	ţ 181	ű 182	ů 183
'27x	ÿ 184	ź 185	ž 186	ż 187	ij 188	ı 189	¿ 190	£ 191
'30x	À 192	Á 193	Â 194	Ã 195	Ä 196	Å 197	Æ 198	Ç 199
'31x	È 200	É 201	Ê 202	Ë 203	Ì 204	Í 205	Î 206	Ï 207
'32x	Đ 208	Ñ 209	Ò 210	Ó 211	Ô 212	Õ 213	Ö 214	Œ 215
'33x	Ø 216	Ù 217	Ú 218	Û 219	Ü 220	Ý 221	Þ 222	SS 223
'34x	à 224	á 225	â 226	ã 227	ä 228	å 229	æ 230	ç 231
'35x	è 232	é 233	ê 234	ë 235	ì 236	í 237	î 238	ï 239
'36x	ð 240	ñ 241	ò 242	ó 243	ô 244	õ 245	ö 246	œ 247
'37x	ø 248	ù 249	ú 250	û 251	ü 252	ý 253	þ 254	ß 255

Font Layout 9: The extended font `ecrm1000` with T1 encoding.

ecrm	ecrb	eccc	ecci	ecvi	ecss	ecdh
ecbx	ecti	ecxc	ectt	ecvi	ecsi	
ecbl	ecui	ecsc	ecit	ectc	ecsx	
ecrb	ecbi	ecoc	ecvt	ecst	ecso	

Comparing these with the root names of the CM fonts from Table F.1 on page 439, one may easily recognize the correspondence. For example, ecbx* is the extended bold font corresponding to the CM font cmbx*.

It is intended that the EC fonts should exist in most design sizes. The present distribution contains almost all the fonts in sizes from 5 to 35.83 pt, that is, with size specifications:

0500	0600	0700	0800	0900	1000	1095
1200	1440	1728	2074	2488	2986	3583

In contrast to the CM text fonts which exhibit differences in the symbol assignments among them, as illustrated in Layouts 1–4, the EC fonts *all* have exactly the same encoding scheme, as presented in Layout 9.

The METAFONT files for the EC fonts may be obtained from the sources mentioned in Section D.6. What one receives is the key file ecstdedt.tex, which when processed under TEX produces the necessary .mf files. It also generates a batch job file to run METAFONT on these sources. Since such a batch file is system dependent, one is allowed to customize ecstdedt.tex for one's own operating system.

A documentation file dcdoc.tex is also provided, presenting the EC character assignments and listing all the languages that are supported.

A parallel set of fonts called *text companion*, or TC, fonts is also provided, the sources of which are extracted by processing tcstdedt.tex. These fonts contain special symbols for text that are normally found in the CM math fonts, if at all, such as currency symbols and degree signs. These fonts are still somewhat experimental so that the symbol assignments are not yet stable. The current contents are shown in Layout 10.

F.4.4   Invoking the EC and TC fonts

The EC fonts correspond to the NFSS encoding scheme T1 (Section 8.5.1). The simplest way to activate them in LATEX 2$_\varepsilon$ is to place

```
\usepackage[T1]{fontenc}
```

in the preamble of the document. All this really does is to make T1 the standard encoding by redefining \encodingdefault to be T1, and it loads the file t1enc.def which redefines the accent and special letter commands by means of the encoding commands of Section C.5.7. Which set of T1 fonts is actually implemented (EC, old or new DC) depends on the choice of .fd files taken during the LATEX 2$_\varepsilon$ installation.

To obtain access to the symbols in the TC fonts, one can load

	0	1	2	3	4	5	6	7
'00x	` 0	´ 1	^ 2	~ 3	¨ 4	˜ 5	° 6	˘ 7
'01x	˘ 8	¯ 9	˙ 10	، 11	¸ 12	؛ 13	14	15
'02x	16	17	‖ 18	19	20	– 21	— 22	23
'03x	← 24	→ 25	⌢ 26	⌢ 27	⌢ 28	⌢ 29	30	31
'04x	ƀ 32	33	34	35	$ 36	37	38	' 39
'05x	40	41	* 42	43	, 44	= 45	. 46	/ 47
'06x	0 48	1 49	2 50	3 51	4 52	5 53	6 54	7 55
'07x	8 56	9 57	58	59	⟨ 60	— 61	⟩ 62	63
'10x	64	65	66	67	68	69	70	71
'11x	72	73	74	75	76	℧ 77	78	◯ 79
'12x	80	81	82	83	84	85	86	Ω 87
'13x	88	89	90	⟦ 91	92	⟧ 93	↑ 94	↓ 95
'14x	` 96	97	★ 98	⁰∕₀ 99	† 100	101	102	103
'15x	104	105	106	107	☙ 108	∞ 109	♪ 110	111
'16x	112	113	114	115	116	117	118	119
'17x	120	121	122	123	124	125	~ 126	꞊ 127
'20x	˘ 128	˘ 129	″ 130	‶ 131	† 132	‡ 133	‖ 134	‰ 135
'21x	• 136	°C 137	$ 138	¢ 139	f 140	ℭ 141	W 142	N 143
'22x	G 144	P 145	£ 146	R 147	? 148	¿ 149	đ 150	™ 151
'23x	‰ 152	¶ 153	B 154	№ 155	℀ 156	e 157	o 158	℠ 159
'24x	{ 160	} 161	¢ 162	£ 163	¤ 164	¥ 165	¦ 166	§ 167
'25x	¨ 168	© 169	ª 170	↻ 171	¬ 172	Ⓟ 173	® 174	¯ 175
'26x	° 176	± 177	² 178	³ 179	´ 180	µ 181	¶ 182	· 183
'27x	※ 184	¹ 185	º 186	√ 187	¼ 188	½ 189	¾ 190	€ 191
'30x	192	193	194	195	196	197	198	199
'31x	200	201	202	203	204	205	206	207
'32x	208	209	210	211	212	213	× 214	215
'33x	216	217	218	219	220	221	222	223
'34x	224	225	226	227	228	229	230	231
'35x	232	233	234	235	236	237	238	239
'36x	240	241	242	243	244	245	÷ 246	247
'37x	248	249	250	251	252	253	254	255

Font Layout 10: The text companion font `tcrm1000` with the TS1 encoding scheme.

```
\usepackage{textcomp}
```

This not only redefines several existing symbol commands, it also adds many new ones. For example, `\copyright`, which is normally defined to be `\textcircle{c}`, is changed to print character 169 from an appropriate TC font. Character 191 is the symbol for the euro, the European currency unit; however, the official symbol should be printed in a sans serif font. This is best achieved by defining

```
\newcommand{\euro}{\textsf{\texteuro}}
```

so that `\euro20` produces €20.

F.4.5   Special character commands

Inspecting Font Layout 9, one notices that the EC fonts contain not only many single characters that are formed out of two CM symbols (like Ä = A + ¨) but also several characters that have no correspondence in the CM font layout at all. The first type is accommodated by internally redefining the action of accent and special character commands. The second set requires new commands that are recognized only when the T1 encoding is active. These are

the ogonek accent `\k{o}`: ǫ

special letters   `\DH` = Ð   `\DJ` = Đ   `\NG` = Ŋ   `\TH` = Þ
                   `\dh` = ð   `\dj` = đ   `\ng` = ŋ   `\th` = þ

special symbols   `\guillemotleft` = «   `\guillemotright` = »
                   `\guilsingleft` = ‹   `\guilsinglright` = ›
                   `\quotedblbase` = „   `\quotesinglbase` = ‚
                   `\textquotedbl` = "

When issued in OT1 encoding, these commands print an error message.

Note: the `\guillemotleft` and `\guillemotright` are not misprints even though the proper word for the French quotations marks is *guillemet*. The PostScript fonts contain these erroneous names for these symbols and this mistake has propagated to such an extent that it can never be removed from all the software that includes it. A *guillemot* is in fact an Arctic bird, not a French quotation mark.

F.5   PostScript fonts

PostScript fonts, or more correctly Type 1 fonts, are treated exactly the same way as the METAFONT fonts as far as LaTeX is concerned: during the processing, a `.tfm` file is read in for each font specifying the character sizes and other properties. This is all that LaTeX needs to know, for it is the task of the driver program to print the actual character that fills the reserved space.

The driver makes use of the *virtual font* mechanism, which means that what LATEX sees are actually artificial fonts that do not really exist on their own. There are .tfm files for these fonts, so the LATEX processing proceeds as normal. What the driver then does is to read a .vf file instead of the pixel .pk files. The instructions in the virtual font file tell the driver how to create each character: they may be drawn, taken from other fonts, or distorted. This is how PostScript slanted and small caps fonts are emulated, for such fonts do not exist in the 'raw' form.

Even the font layout can be redesigned with virtual fonts. The raw PostScript fonts have an encoding scheme that conforms to neither OT1 nor T1, but which is used as a pool of symbols for constructing virtual fonts that do conform to these schemes. The raw encoding scheme is presented in Layout 11 on page 455. In particular, the Computer Modern fonts contain upper case Greek letters in the first 11 slots (Layout 1 on page 440) which are to be found in the PostScript symbol font only. The virtual font ptmr7t conforms to this by taking its characters from both the raw Times-Roman and Symbol fonts.

F.5.1 Naming scheme for PostScript fonts

In order to be compatible with all possible operating systems, it is necessary to reduce the names of the PostScript font files to a maximum of eight characters. This makes for an extremely abbreviated and cryptic nomenclature.

The most commonly used scheme is that of Karl Berry. Here, the first letter of the name specifies the supplier of the font, for example p for Adobe (stands for PostScript), or h for Bigelow & Holmes, who designed the Lucida fonts used in this book, or m for Monotype, or l for Linotype, and so on.

A two-letter typeface code follows the supplier letter, such as tm for Times-Roman. Next come various letters to specify weight, for example r for regular (upright Roman) or b for bold, and variant, like i for italic, o for oblique (slanted). Next come a number plus letter to indicate the encoding scheme, followed by possible width code letters. The Berry names for the 35 standard PostScript fonts that should be loaded in every printer are listed in Table F.2 on the next page, without the encoding suffixes.

The most important encoding suffixes for our purposes are listed in Table F.3. The 7-bit encodings have already been explained in Section F.2.2 and shown in Layouts 1, 5–7, while the 8t or T1 scheme is to be found in Layout 9 in Section F.4.3.

The other 8-bit schemes listed in Table F.3 are the raw encoding, 8r, shown in Table 11 and the Adobe standard encoding, 8a, which is the default for most Type 1 fonts if no re-encoding is specified. This scheme

Table F.2: Root names of the 35 standard PostScript fonts

pagd	AvantGarde-Demi	phvrrn	Helvetica-Narrow
pagdo	AvantGarde-DemiOblique	phvron	Helvetica-Narrow-Oblique
pagk	AvantGarde-Book	pncb	NewCenturySchlbk-Bold
pagko	AvantGarde-BookOblique	pncbi	NewCenturySchlbk-BoldItalic
pbkd	Bookman-Demi	pncr	NewCenturySchlbk-Roman
pbkdi	Bookman-DemiItalic	pncri	NewCenturySchlbk-Italic
pbkl	Bookman-Light	pplb	Palatino-Bold
pbkli	Bookman-LightItalic	pplbi	Palatino-BoldItalic
pcrb	Courier-Bold	pplr	Palatino-Roman
pcrbo	Courier-BoldOblique	pplri	Palatino-Italic
pcrr	Courier	psyr	Symbol
pcrro	Courier-Oblique	ptmb	Times-Bold
phvb	Helvetica-Bold	ptmbi	Times-BoldItalic
phvbo	Helvetica-BoldOblique	ptmr	Times-Roman
phvbrn	Helvetica-Narrow-Bold	ptmri	Times-Italic
phvbon	Helvetica-Narrow-BoldOblique	pzcmi	ZapfChancery-MediumItalic
phvr	Helvetica	pzdr	ZapfDingbats
phvro	Helvetica-Oblique		

Table F.3: A selection of the encoding suffixes in the Berry nomenclature

Suffix	Encoding	NFSS designation	Page
7t	7-bit TₑX text encoding	OT1	440
7m	TₑX math italics encoding	OML	443
7v	TₑX math extension encoding	OMX	444
7y	TₑX math symbols encoding	OMS	443
8t	8-bit Cork encoding	T1	449
8a	Adobe standard encoding		456
8r	TeXBase1Encoding		455

is illustrated in Layout 12 on page 456 although it plays absolutely no role in the NFSS installation for PostScript fonts.

Virtual fonts `ptmr7t` and `ptmr8t` are both constructed from the same raw font `ptmr8r`, differing only in the character assignments. There are also raw fonts that are modifications of the basic fonts. For example, `ptmro8t` is an oblique Times-Roman virtual font, based on `ptmro8r`, the raw version. However, there is no such font in the repertoire of Table F.2. This pseudo font is generated by applying a PostScript slanting operation to Times-Roman, as illustrated in Section F.5.4.

F.5.2 Installing PostScript fonts

The standard 35 PostScript fonts are preloaded into every PostScript printer but to address them properly with LATₑX and the NFSS system, one

	0	1	2	3	4	5	6	7
'00x	0	· 1	fi 2	fl 3	/ 4	″ 5	Ł 6	ł 7
'01x	˛ 8	° 9	10	˘ 11	‒ 12	13	Ž 14	ž 15
'02x	˘ 16	ı 17	J 18	ff 19	ffi 20	ffl 21	22	23
'03x	24	25	26	27	28	29	` 30	' 31
'04x	32	! 33	" 34	# 35	$ 36	% 37	& 38	' 39
'05x	(40	) 41	* 42	+ 43	, 44	- 45	. 46	/ 47
'06x	0 48	1 49	2 50	3 51	4 52	5 53	6 54	7 55
'07x	8 56	9 57	: 58	; 59	< 60	= 61	> 62	? 63
'10x	@ 64	A 65	B 66	C 67	D 68	E 69	F 70	G 71
'11x	H 72	I 73	J 74	K 75	L 76	M 77	N 78	O 79
'12x	P 80	Q 81	R 82	S 83	T 84	U 85	V 86	W 87
'13x	X 88	Y 89	Z 90	[91	\ 92	] 93	^ 94	_ 95
'14x	' 96	a 97	b 98	c 99	d 100	e 101	f 102	g 103
'15x	h 104	i 105	j 106	k 107	l 108	m 109	n 110	o 111
'16x	p 112	q 113	r 114	s 115	t 116	u 117	v 118	w 119
'17x	x 120	y 121	z 122	{ 123	| 124	} 125	~ 126	127
'20x	128	129	‚ 130	f 131	„ 132	… 133	† 134	‡ 135
'21x	ˆ 136	‰ 137	Š 138	‹ 139	Œ 140	141	142	143
'22x	144	145	146	" 147	" 148	· 149	– 150	— 151
'23x	˜ 152	™ 153	š 154	› 155	œ 156	157	158	Ÿ 159
'24x	160	¡ 161	¢ 162	£ 163	¤ 164	¥ 165	¦ 166	§ 167
'25x	¨ 168	© 169	ª 170	« 171	¬ 172	173	® 174	¯ 175
'26x	° 176	± 177	² 178	³ 179	´ 180	µ 181	¶ 182	· 183
'27x	¸ 184	¹ 185	º 186	» 187	¼ 188	½ 189	¾ 190	¿ 191
'30x	À 192	Á 193	Â 194	Ã 195	Ä 196	Å 197	Æ 198	Ç 199
'31x	È 200	É 201	Ê 202	Ë 203	Ì 204	Í 205	Î 206	Ï 207
'32x	Ð 208	Ñ 209	Ò 210	Ó 211	Ô 212	Õ 213	Ö 214	× 215
'33x	Ø 216	Ù 217	Ú 218	Û 219	Ü 220	Ý 221	Þ 222	ß 223
'34x	à 224	á 225	â 226	ã 227	ä 228	å 229	æ 230	ç 231
'35x	è 232	é 233	ê 234	ë 235	ì 236	í 237	î 238	ï 239
'36x	ð 240	ñ 241	ò 242	ó 243	ô 244	õ 245	ö 246	÷ 247
'37x	ø 248	ù 249	ú 250	û 251	ü 252	ý 253	þ 254	ÿ 255

Font Layout 11: The 'raw' encoding *TeXBase1Encoding*, or 8r for PostScript fonts, the basis for all other encodings.

	0	1	2	3	4	5	6	7
'00x	0	1	2	3	4	5	6	7
'01x	8	9	10	11	12	13	14	15
'02x	16	17	18	19	20	21	22	23
'03x	24	25	26	27	28	29	30	31
'04x	32	! 33	" 34	# 35	$ 36	% 37	& 38	' 39
'05x	(40	) 41	* 42	+ 43	, 44	- 45	. 46	/ 47
'06x	0 48	1 49	2 50	3 51	4 52	5 53	6 54	7 55
'07x	8 56	9 57	: 58	; 59	< 60	= 61	> 62	? 63
'10x	@ 64	A 65	B 66	C 67	D 68	E 69	F 70	G 71
'11x	H 72	I 73	J 74	K 75	L 76	M 77	N 78	O 79
'12x	P 80	Q 81	R 82	S 83	T 84	U 85	V 86	W 87
'13x	X 88	Y 89	Z 90	[91	\ 92	] 93	^ 94	_ 95
'14x	` 96	a 97	b 98	c 99	d 100	e 101	f 102	g 103
'15x	h 104	i 105	j 106	k 107	l 108	m 109	n 110	o 111
'16x	p 112	q 113	r 114	s 115	t 116	u 117	v 118	w 119
'17x	x 120	y 121	z 122	{ 123	| 124	} 125	~ 126	127
'20x	128	129	130	131	132	133	134	135
'21x	136	137	138	139	140	141	142	143
'22x	144	145	146	147	148	149	150	151
'23x	152	153	154	155	156	157	158	159
'24x	160	¡ 161	¢ 162	£ 163	/ 164	¥ 165	ƒ 166	§ 167
'25x	¤ 168	' 169	" 170	« 171	‹ 172	› 173	fi 174	fl 175
'26x	176	– 177	† 178	‡ 179	· 180	181	¶ 182	· 183
'27x	, 184	„ 185	" 186	» 187	… 188	‰ 189	190	¿ 191
'30x	192	` 193	´ 194	^ 195	~ 196	¯ 197	˘ 198	· 199
'31x	¨ 200	201	° 202	¸ 203	204	˝ 205	˛ 206	ˇ 207
'32x	— 208	209	210	211	212	213	214	215
'33x	216	217	218	219	220	221	222	223
'34x	224	Æ 225	226	ª 227	228	229	230	231
'35x	Ł 232	Ø 233	Œ 234	º 235	236	237	238	239
'36x	240	æ 241	242	243	244	ı 245	246	247
'37x	ł 248	ø 249	œ 250	ß 251	252	253	254	255

Font Layout 1,2: The Adobe *StandardEncoding* or 8a for PostScript fonts, the usual default coding without re-encoding.

needs the following files:

.tfm the font metric files for both the virtual and raw fonts;

.fd the NFSS font definition files associating the font attributes to precise virtual font names; these and the .tfm files are the only ones read by LATEX itself;

.vf virtual font files that instruct the DVI driver how to produce the characters; usually this refers to characters in a real font with different encoding;

.map mapping files that tell the DVI driver the true internal names of the real fonts, plus any re-encodings or distortions that must be undertaken; thus ptmr8r is translated to Times-Roman with 8r encoding;

config.* optional configuration files to ensure that the mapping files are read by the driver. .

These files must be placed in the proper locations where LATEX and the driver program can find them. Furthermore, the names of the .map files must be added to the config.ps file that configures dvips. For example, to include the Times fonts, add

        p +ptm.map

to this file. Alternatively, invoke config.ptm which contains just this line, by calling dvips with the option -Pptm. The -P option may be issued several times in one command line.

All these support files can be obtained presorted into appropriate directories with the archive file lw35nfss.zip from CTAN (Figure D.4 on page 401) in directory fonts/psfonts/.

For fonts other than the 35 standard ones, the actual character files containing the drawing instructions will also be needed. These usually have the extensions .pfa or .pfb for ASCII or binary form. The mapping files inform the driver if such files need to be loaded.

F.5.3 Computer Modern as PostScript fonts

The Computer Modern fonts have also been converted to Type 1 fonts, as a joint project of Y&Y Inc. and Blue Sky Research. Originally part of their commercial TEX installations, these fonts are now available free of charge and can be found on CTAN under fonts/cm/ps-type1/bluesky. An alternative, earlier set by Basil Malyshev is located in fonts/cm/ps-type1/bakoma.

These fonts are especially recommended for use with pdfTEX (page 395). In fact, for PDF output one should not even consider applying the bitmap fonts for they produce terrible results, whether generated with pdfTEX or with the Distiller program.

F.5.4   Long road from input code to output character

Let us illustrate with an example how LATEX translates an input character with given font attributes into a particular symbol in a specific font. Suppose we have the following attributes (Section 8.5.1):

Encoding: OT1   Family: ptm   Series: b   Shape: sl

The encoding and family indicate the font definition file that associates these attributes with real font names, in this case ot1ptm.fd. In that file, one finds this set of attributes to be assigned to font ptmbo7t; this is the Berry name for *Times-BoldOblique, 7-bit-text.*

Our sample input code is \AE; according to file ot1enc.def, this corresponds to character 29 in the OT1 encoding. LATEX thus writes instructions to the .dvi file to output the symbol in position 29 of font ptmbo7t. And with that, LATEX is finished with its part of the operation.

The next step is to process the .dvi file with a DVI driver like dvips. When this program searches for a font, it first looks in certain font mapping files, then it seeks the .vf virtual font files, and finally it tries to find it as a .pk pixel file. In our case, it discovers the virtual font ptmbo7t.vf, which tells the driver that character 29 is to be symbol 198 of font ptmbo8r. This is the same basic font but with 8-bit-raw encoding. Again dvips searches for this next font, and finds it in the font mapping file ptm.map, which contains the line

```
ptmbo8r Times-Bold ".167 SlantFont TeXBase1Encoding
                                    ReEncodeFont" <8r.enc
```

This tells the driver that it is to use the font with the internal name *Times-Bold.* Furthermore, the text in quotes is to be inserted into the output file; this instructs the printer to slant the font by a factor 0.167 and to apply the TeXBase1Encoding encoding vector. To this end, the file 8r.enc, containing the definition of this vector, must also be copied into the output.

Finally, the output .ps file is sent to a PostScript printer or other PostScript interpreter. The specified encoding vector indicates that character 198 is the symbol with the internal name AE, which is then printed from the Times-Bold font, distorted as required. The long road has come full circle.

Note that Type 1 fonts are essentially a collection of symbols identified by name, whereas application programs like LATEX address characters by positional number. The virtual font mechanism may reorder these assignments, but it is ultimately the encoding vectors that relate each number to a definite symbol.

Command Summary

This appendix contains a brief description of all the LaTeX commands, in alphabetical (ASCII) order, neglecting the leading backslash \, along with some TeX commands that have been explained in this book. In the following section, the commands are presented in their logical grouping in a number of tables and figures. At the end, those TeX commands that may not be used during LaTeX processing are listed.

G.1 Brief description of the LaTeX commands

For each command in the following summary, the section and page number is given where it is introduced and described in detail. The numbers are shown in the form: '(*Section*) – *Page*': for example (2.5.1) – 13 means 'Subsection 2.5.1, page 13'. If these numbers are missing, then the command has not been presented in the book but is only mentioned here.

The following notations may be added to the commands:

[m] those permitted in math mode only;

[a] those belonging to $\mathcal{A}_{\mathcal{M}}$S-LaTeX;

[p] those allowed only in the preamble.

Any commands that were added after the first official release of LaTeX 2_ε in June 1994 have the date of their first appearance indicated, as for example *(1994/12/01)*. Files containing them should include this or a later date as an optional argument to \NeedsTeXFormat{LaTeX2e}[1994/12/01].

\␣ · (2.1), (3.5.1) – 16, 45

Normal space between words after a command without arguments or after a period that is not the end of a sentence.

! . . . . (8.4) – 230

Field separation character within the \index command. For example:
with \index{command!fragile} one produces an index sub-entry
'fragile' under the main entry 'command'.

!'   produces ¡ . . . . (2.5.6) – 21

\!   [m] . . . . (5.5.1) – 143

In math mode, a negative space of $-1/6$ quad: $xx\backslash!x = xxx$.

"   . . . . (2.5.2), (8.4), (B.2) – 20, 231, 313

1. In normal text, this produces the double closing quote ".

2. Literal sign for MakeIndex, in order to print one of the special
   characters !, @, |, or ". Example: \index{"!} to enter character
   ! without it being interpreted as a separation character.

3. Delimiter for a text field in BibTeX. Example:
   AUTHOR = "Donald E. Knuth".

\"   . . . . (2.5.7) – 21

Produces an umlaut accent: \"{a} = ä.

. . . . (7.3.2), (7.4.2) – 196, 206

Argument replacement character in a user-defined command or environment.

. . . . (7.5.7) – 212

Replacement character for an *internal* argument within a *nested* user-
defined command or environment.

\#   . . . . (2.5.4) – 21

Command to produce a hash symbol: \# = #.

$ . . . . (5.1) – 117

Toggle character for switching between text and in-line math modes.
On the first appearance (text to math) it behaves the same as \(
or \begin{math}, while the second call (math to text) is as \) or
\end{math}.

\$ . . . . (2.5.4) – 21

Command to produce a dollar sign: \$ = $.

%   . . . . (4.11) – 115

Comment character. The rest of the line of text following % is ignored
by the TeX processing.

\%   . . . . (2.5.4) – 21

Command to produce a per cent sign: \% = %.

& . . . . (4.8.1) – 95

Indicates a new column in `array` and `tabular` environments.

\& . . . . (2.5.4) – 21

Command to produce an ampersand symbol: \& = &.

\' . . . . (2.5.7), (4.6.4) – 21, 82

1. Command to produce an acute accent: \'a = á.

2. Within the `tabbing` environment, a command to jump to the end of the current column.

( ) . . . . (6.1.2), (B.2) – 153, 314

For a picture command in `picture` environment, specifies a coordinate pair. In BIBTEX, an alternative form for the outermost grouping of the entry type.

\(. . . . (5.1) – 117

Switches from text to in-line math mode to produce formulas within a line of text. It functions the same as \begin{math} and as a $ sign in text mode.

\) . . . . (5.1) – 117

Switches back from in-line math mode to text mode. It functions the same as \end{math} and as a $ sign in math mode.

\+ . . . . (4.6.3) – 81

Within the `tabbing` environment, increments the left margin by one tab stop (moves it to the right).

\, ... (3.5.1), (5.5.1) – 46, 143

Small space, the size of 1/6 quad, for use in text and math mode: xx\,x = $xx\,x$.

- . . . . (2.5.3) – 21

As -, produces the hyphen - for compound words and word division, as --, the en dash –, and as ---, the em dash —.

\- . . . . (3.6.1), (4.6.3) – 54, 81

1. Denotes possible word division. If a word contains at least one \- the normal word division rules are suspended for that word and division may occur *only* at those locations.

2. Within the `tabbing` environment, decrements the left margin by one tap stop (moves it to the left).

\. . . . . \'a . . (2.5.7) – 21

Command to produce a dot accent: \.o = ȯ.

\/ . . . . (3.5.1), (3.5.1) – 46, 46
Italic correction, the extra space at the end of slanted fonts, or the command to break up ligatures.

\: [m] . . . . (5.5.1) – 143
In math mode, a medium space, the size of 2/9 quad: $xx\backslash:x = xx\,x$.

\; [m] . . . . (5.5.1) – 143
In math mode, a large space, the size of 5/18 quad: $xx\backslash;x = xx\ x$.

\< . . . . (4.6.3) – 81
Within the `tabbing` environment, moves to the left by one tab stop.

\= . . . . (2.5.7), (4.6.1) – 21, 80
1. Command to produce a macron accent: \=o = ō.
2. Within the `tabbing` environment, sets a tab stop at the current position within the line.

\> . . . . (4.6.1) – 80
Within the `tabbing` environment, moves right to the next tab stop.

?' produces ¿ . . . . (2.5.6) – 21
@ . . . . (8.4), (B.2) – 230, 313
1. In MakeIndex, separates an entry in an \index command into a lexical (for alphabetization) and printing part. Example: \index{sum@$\sum$} means that the entry appears in the index at the location of the word 'sum' but what is printed is the summation sign $\sum$.
2. In BibTEX, denotes the *entry type*. Example: @BOOK indicates that the following literature entries correspond to those of a book.

\@ . . . . (3.5.1) – 45
Extra space at the end of a sentence ending with a capital letter.

[ ] . . . . (2.1) – 16
After commands or environment calls, specifies an optional argument.

\[. . . . (5.1) – 118
Switches from text mode to displayed math mode for putting a formula on a line by itself.   Has the same effect as \begin{displaymath}.

\\[*space*] . . . . (3.5.2) – 48
Ends the current line (without right justifying it). The optional argument [*len*] inserts additional vertical spacing of length *len* before the next line.

`\\*`[*space*] . . . . (3.5.2) – 48

The same as `\\` but prevents a page break from occurring between the current and next line.

`\]` . . . . (5.1) – 118

Switches back from displayed math mode to text mode. Has the same effect as `\end{displaymath}`.

`^` [m] . . . . (5.2.2) – 119

Exponents and superscripts in equations: `x^2` $= x^2$, `x^{-2n}` $= x^{-2n}$.

`\^` . . . . (2.5.7) – 21

Command to produce a circumflex accent: `\^o` = ô.

`_` [m] . . . . (5.2.2) – 119

Subscripts in equations: `a_n` $= a_n$, `a_{i,j,k}` $= a_{i,j,k}$.

`\_` . . . . (2.5.4) – 21

Command to produce the underbar sign: `t\_v` = t.v.

`\`` . . . . (2.5.7), (4.6.4) – 21, 82

1. Command to produce a grave accent: `` \`o `` = ò.

2. Within the `tabbing` environment, pushes the following text up against the right margin of the line.

`{ }` . . . . (2.1), (2.2), (B.2) – 16, 17, 313

1. After a command or environment call, specifies a mandatory argument.

2. Grouping a section of text to create a nameless environment.

3. In BibTeX, delimiting the name of an entry type, as well as an alternative delimiter for the text field.

`\{` . . . . (2.5.4) – 21

Command to produce a left curly brace: `\{` = {.

`|` [m] produces | . . . . (5.3.4), (5.4.1) – 124, 130
`|` . . . . (8.4) – 230

In MakeIndex, the command character within a `\index` command.

1. After the command `\newcommand{\ii}[1]{\textit{#1}}` has been defined, `\index{entry|ii}` produces the page number for 'entry' in the index in italic type.

2. The cross-reference command `\see` from `makeidx.sty` can be invoked with `\index{bison|see{buffalo}}` to produce cross-references within the index.

\| [m] produces ‖ . . . . (5.3.6) – 125

\} . . . . (2.5.4) – 21

> Command to produce a right curly brace: \} = }.

~ . . . . (3.5.1) – 45

> A normal space between words, but without the possibility that the line will be broken there. Example: Prof.~Jones ensures that 'Prof.' and 'Jones' both remain on the same line.

\~ . . . . (2.5.7) – 21

> Command to produce a tilde accent: \~n = ñ.

\a= . . . . (4.6.4) – 82

> Produces a macron accent within tabbing environment: \a=o = ō.

\a' . . . . (4.6.4) – 82

> Produces an acute accent within tabbing environment: \a'o = ó.

\a' . . . . (4.6.4) – 82

> Produces a grave accent within tabbing environment: \a'o = ò.

\AA produces Å . . . . (2.5.6) – 21

\aa produces å . . . . (2.5.6) – 21

\abovedisplayskip [m] . . . . (5.5.4) – 147

> Vertical space between a *long* displayed equation and the preceding line of text. A new value may be assigned with the \setlength command:
>
>   \setlength{\abovedisplayskip}{10pt plus2pt minus5pt}

\abovedisplayshortskip [m] . . . . (5.5.4) – 147

> Vertical space between a *short* displayed equation and the preceding line of text. A new value may be assigned with the \setlength command as in the above example.

\abstractname . . . . (C.4.1) – 351

> Command containing the heading for the abstract. In English, this is 'Abstract' but may be altered for adaptation to other languages.

\acute{x} [m] . . . . (5.3.9) – 127

> Acute accent over math variable x: \acute{a} = $á$

\Acute{x} [m][a] . . . . (E.3.2) – 412

> With the amsmath package, can be used like \acute, but with multiple $\mathcal{A}_{\mathcal{M}}S$-LaTeX math accents the positioning will be correct.

`\addcontentsline{`*file*`}{`*format*`}{`*entry*`}` . . .   (3.4.3), (3.4.4) – 43, 44

> Manual addition of *entry* into the list file `.toc`, `.lof`, or `.lot`, according
> to the value of *file*, to be formatted as the heading of a sectioning
> command, as given by *format*, for example
>
>   \addcontentsline{toc}{section}{References}

`\address{`*sender*`}`   . . . .   (A.1) – 298

> In the `letter` document class, enters the sender's address. Multiple
> lines in *sender* are separated by \\.

2ε `\addtime{`*secs*`}` . . . .   (8.10.3) – 252

> In the `slides` class, if the option `clock` has been selected, a time
> marker, in minutes, appears at the bottom of the notes. This com-
> mand adds the specified number of seconds to the marker. See also
> `\settime`.

`\addtocontents{`*file*`}{`*entry*`}` . . . .   (3.4.3), (3.4.4) – 43, 44

> Manual addition of *entry* into the list file `.toc`, `.lof`, or `.lot`, according
> to the value of *file*. Example:
>
>   \addtocontents{lof}{\protect\newpage}

`\addtocounter{`*counter*`}{`*number*`}` . . . .   (7.1.3) – 190

> Adds *number* to the current value of the number stored in *counter*.

`\addtolength{\`*length_name*`}{`*length*`}`   . . . .   (7.2) – 192

> Adds the quantity *length* to the current value of the length command
> `\`*length_name*.

`\addvspace{`*length*`}`   . . . .   (7.2) – 193

> Inserts vertical spacing of amount *length* between paragraphs at the
> point where the command is given. If other vertical spacing exists,
> the total will not exceed *length*.

`\AE` produces Æ . . . .   (2.5.6) – 21
`\ae` produces æ . . . .   (2.5.6) – 21
`\aleph` [m] produces ℵ . . . .   (5.3.6) – 125
`\allowdisplaybreaks[`*num*`]` [p][a] . . . .   (E.3.7) – 427

> With the `amsmath` package, allows automatic page breaks to occur
> within multiline math formulas. If *num* is present, it takes a value
> of 0-4 to increase the ease with which page breaks occur. Without
> this command, a manual page break can be made at the end of any
> formula line with `\displaybreak`.

`\Alph{`*counter*`}`   . . . .   (7.1.4) – 191

> Prints the current value of *counter* as a capital letter.

`\alph{`*counter*`}`   . . . .   (7.1.4) – 191

> Prints the current value of *counter* as a lower case letter.

`\alpha` [m] produces α . . . . (5.3.1) – 123
`\alsoname`   . . . . (C.4.1) – 352

> Command for use in modified `makeidx` package. It prints the text for a command `\seealso`. In English, this is 'see also' but may be altered for adaptation to other languages. (This command is not standardized!)

`\amalg` [m] produces $\amalg$ . . . . (5.3.3) – 124
`\and` . . . . (3.3.1) – 37

> Used to separate author names within the `\author` command for generating a title page with `\maketitle`.

`\angle` [m] produces $\angle$ . . . . (5.3.6) – 125
`\appendixname`   . . . . (C.4.1) – 351

> Command containing the heading for the appendix. In English, this is 'Appendix' but may be altered for adaptation to other languages.

`\approx` [m] produces $\approx$ . . . . (5.3.4) – 124
`\arabic{`*counter*`}` . . . . (7.1.4) – 191

> Prints the current value of *counter* as an Arabic number.

`\arccos` [m] . . . . (5.3.8) – 126

> Command to produce the function name 'arccos' in equations.

`\arcsin` [m] . . . . (5.3.8) – 126

> Command to produce the function name 'arcsin' in equations.

`\arctan` [m] . . . . (5.3.8) – 126

> Command to produce the function name 'arctan' in equations.

`\arg` [m] . . . . (5.3.8) – 126

> Command to produce the function name 'arg' in equations.

`\arraycolsep` . . . . (4.8.2) – 96

> Half the width of the intercolumn spacing in the `array` environment. Value is assigned with the LaTeX command `\setlength`:
>
>   \setlength{\arraycolsep}{3mm}

`\arrayrulewidth` . . . . (4.8.2) – 96

> The thickness of vertical and horizontal lines in the `array` and `tabular` environments. Its value is assigned to a length with `\setlength`:
>
>   \setlength{\arrayrulewidth}{0.5mm}

`\arraystretch` . . . . (4.8.2) – 96

> Factor to change the spacing between lines in a table, normal value being 1. Spacing is multiplied by this factor, which is set to a new value with `\renewcommand{\arraystretch}{`*factor*`}`.

`\ast` [m] produces $*$ . . . . (5.3.3) – 124

`\asymp` [m] produces $\asymp$ . . . . (5.3.4) – 124

$\boxed{2_\varepsilon}$ `\AtBeginDocument{`*code*`}` [p] . . . . (C.2.4) – 335

> Stores the *code* to be inserted into the processing stream when `\begin{document}` is executed. Commands that are only allowed in the preamble may be part of *code*. A package might include coding in this way to ensure that it is not overwritten by another package.

$\boxed{2_\varepsilon}$ `\AtEndDocument{`*code*`}` [p] . . . . (C.2.4) – 335

> Stores the *code* to be inserted into the processing stream when `\end{document}` is executed. A package might include coding in this way to have additional features printed automatically at the end of the document.

$\boxed{2_\varepsilon}$ `\AtEndOfClass{`*code*`}` [p] . . . . (C.2.4) – 335

> Stores the *code* to be inserted into the processing stream when the current class file has finished being read. May only be given in a class file, or in another file that is read by a class file. May be used by a local configuration file to overwrite defaults in the class file itself.

$\boxed{2_\varepsilon}$ `\AtEndOfPackage{`*code*`}` [p] . . . . (C.2.4) – 335

> Stores the *code* to be inserted into the processing stream when the current package file has finished being read. May only be given in a package file, or in another file that is read by a package file. May be used by a local configuration file to overwrite defaults in the package file itself.

`\author{`*name*`}` . . . . (3.3.1) – 37

> Enters the author name(s) for a title page produced by the `\maketitle` command.

`\b{`*x*`}` . . . . (2.5.7) – 21

> Command to produce an underbar accent: `\b{o}` = o̠.

$\boxed{2_\varepsilon}$ `\backmatter` . . . . (3.3.5) – 42

> In the book class, introduces the material that comes at the end (bibliography, index) by turning off the chapter numbering of the `\chapter` command.

`\backslash` [m] produces $\backslash$ . . . . (5.3.6) – 125

`\bar{`*x*`}` [m] . . . . (5.3.9) – 127

> Macron accent over the math variable x: `\bar{a}` = $\bar{a}$.

`\Bar{`*x*`}` [m][a] . . . . (E.3.2) – 412

> With the amsmath package, can be used like `\bar`, but with multiple $\mathcal{A}_{\mathcal{M}}S$-LATEX math accents the positioning will be correct.

`\baselineskip` . . . . (3.2.3) – 33

Interline spacing within a paragraph. Every font has its own internal line spacing. A new value (a rubber length) may be assigned with `\setlength`:

`\setlength{\baselineskip}{12pt plus2pt minus1pt}`

`\baselinestretch` . . . . (3.2.3), (4.1.2) – 33, 59

A factor with the normal value of 1 by which the internal length `\baselineskip` is multiplied to produce the actual interline spacing. May be changed with:

`\renewcommand{\baselinestretch}{`*factor*`}`

The new value takes effect after the next change in font size!

`\begin{`*envrnmnt*`}` . . . . (2.2) – 17

Start of an environment with the name *envrnmnt*. This command must be coupled with `\end{`*envrnmt*`}` to terminate the environment. The environment name in both these commands must be identical.

`\begin{abstract}` . . . . (3.3.2) – 39

Start of the environment `abstract` to produce an abstract. With document class `article`, font size `\small` and the `quotation` environment are selected. With `report`, the abstract appears on a separate page with normal font size and line width. In both cases, the heading **Abstract** is centered above the text.

`\begin{align}` [a] . . . . (E.3.6) – 422

With the amsmath package, switches to displayed math mode to produce a set of aligned equations. Line are terminated by `\\` commands. The lines are split into columns aligned on the first, third, fifth … `&` characters. Each line receives an equation number unless the *-form of the environment has been selected.

`\begin{alignat}{`*num*`}` [a] . . . . (E.3.6) – 424

Is the same as the `align` environment except that no spacing is inserted automatically between the column pairs. The argument *num* is the number of column pairs $= (1 + n_\&)/2$ where $n_\&$ is the number of `&` signs in one row. Explicit spacing may be placed between column pairs, especially if the left part of that pair is otherwise empty.

`\begin{aligned}` [*pos*] [m][a] . . . . (E.3.6) – 425

With the amsmath package, is like the `align` environment but is used as an element within math mode. The optional argument *pos* determines the vertical positioning relative to neighboring elements: t or b for top or bottom, no argument for centering.

`\begin{appendix}` . . . . (3.3.4) – 41

Start of the environment `appendix` to produce an appendix. The main section counter is reset to zero and its numbering appears as capital letters.

\begin{array}[*pos*]{*col*} [m] . . . . (4.8.1), (5.4.3) – 93, 132

Start of the environment array to produce matrices and arrays in math mode. The column definition *col* contains a formatting character for each column. Thus \begin{array}{lcr} produces an array with three columns: one *left justified*, one *centered*, and one *right justified*. The optional parameter *pos* determines how the array is aligned vertically with text outside it on the same line: t with the top line, b with the bottom line, while the default is with the center. See also \begin{tabular}

\begin{center} . . . . (4.2.1) – 63

Start of the environment center. Each line of text terminated by \\ appears centered. See also \centering.

\begin{*command_name*} . . . . (2.2) – 17

Most declaration commands, such as the font styles and sizes, can be used as environment names. For example, \begin{small} switches to font size \small until the countercommand \end{small} is given.

\begin{alltt} . . . . (4.9.1) – 109

When the alltt package has been loaded, this environment prints out original text in typewriter typeface, maintaining line breaks, special symbols, and so on, except for \ { } which function as usual. This allows commands to be executed within the typewriter text.

\begin{bmatrix} [m][a] . . . . (E.3.4) – 415

Like the matrix environment, but enclosed in square brackets [].

\begin{Bmatrix} [m][a] . . . . (E.3.4) – 415

Like the matrix environment, but enclosed in curly braces { }.

\begin{cases} [m][a] . . . . (E.3.6) – 426

With the amsmath package, writes math expressions on several lines, terminated by \\, in left-justified columns, separated by &, with a curly brace enclosing all lines at the left, the whole being centered vertically.

\begin{description} . . . . (4.3.3) – 67

Start of the environment description to produce an indented list with labels. The label text is the argument *label* in the command \item[*label*].

\begin{displaymath} . . . . (5.1) – 118

Switches from text to displayed math mode for producing a formula on a line by itself. Functions the same as \[.

`\begin{document}` . . . . (1.5.2) – 12

Start of the outermost environment of a text document. This command terminates the preamble. It is obligatory for every LaTeX document, as is its counterpart `\end{document}` for ending the document.

`\begin{enumerate}` . . . . (4.3.2) – 66

Start of the environment enumerate to produce a numbered, indented listing. The style of numbering depends on the depth of nesting; at the first level, it consists of a running Arabic number that is incremented with each call to `\item`.

`\begin{eqnarray}` . . . . (5.4.7) – 136

Switches from text to displayed math mode to produce a set of equations or a multiline formula in the form of a three-column table {rcl}. The individual lines of the formula are ended with the command \\; the fields within a line are separated by & characters. Each line is given a sequential equation number unless the command `\nonumber` appears within it.

`\begin{eqnarray*}` . . . . (5.4.7) – 136

Is the same as the eqnarray environment except that no equation numbers are printed.

`\begin{equation}` . . . . (5.1) – 118

Switches from text to displayed math mode to produce a formula on a line by itself, including an automatic sequential equation number.

`\begin{falign}` [a] . . . . (E.3.6) – 424

With the amsmath package, is the same as the align environment except that spacing is inserted between the column pairs to fill up the entire line.

`\begin{figure}[loc]` . . . . (6.4.1) – 179

Float environment for entering text for a figure. The optional argument *loc* can be any combination of the letters h, t, b, and p to determine the various positioning possibilities. Default is tbp. Under LaTeX 2_ε, the character ! may additionally be given to ignore all float spacing and number restrictions set by the float style parameters.

`\begin{figure*}[loc]` . . . . (6.4.1) – 179

The same as the figure environment except that the figure is inserted over the width of two columns when the option twocolumn or the command \twocolumn has been selected. The standard form figure will only fill the width of one column.

2ε `\begin{filecontents}{`*file_name*`}` [p] . . . . (C.2.9) – 339

> An environment that may be given only before `\documentclass`, it writes its lines literally to a file of the specified name, if that file does not already exist. It also adds comment lines stating its source. In this way, non-standard files may be included in the main document file for shipment to other installations. If a file with the stated name already exists, it is not overwritten, but a warning message is issued.

2ε `\begin{filecontents*}{`*file_name*`}` [p] . . . . (C.2.9) – 339

> Is the same as the `filecontents` environment, except that no comment lines are written to the file. The file will contain exactly the contents of the environment, and nothing more.

`\begin{flushleft}` . . . . (4.2.2) – 63

> Start of the `flushleft` environment in which each line of text is left justified, that is, it begins flush with the left margin but is not expanded to match the right edge. The equivalent declaration is `\raggedright`.

`\begin{flushright}` . . . . (4.2.2) – 63

> Start of the `flushright` environment in which each line of text is right justified, that is, the right-hand side is flush with the right margin, but the line is not expanded to start exactly at the left edge. The equivalent declaration is `\raggedleft`.

`\begin{gather}` [a] . . . . (E.3.6) – 422

> With the `amsmath` package, switches to displayed math mode to produce several lines of equations, all centered with no alignment. Lines are terminated by `\\` commands. Each line receives an equation number, unless the *-form of the environment has been selected.

`\begin{gathered}` [*pos*] [m][a] . . . . (E.3.6) – 425

> With the `amsmath` package, is like the `gather` environment but is used as an element within math mode. The optional argument *pos* determines the vertical positioning relative to neighboring elements: t or b for top or bottom, no argument for centering.

`\begin{itemize}` . . . . (4.3.1) – 66

> Start of the `itemize` environment for producing labeled, indented listings. The type of label depends on the depth of nesting; at the first level it is a • generated by each `\item` command.

`\begin{letter}{`*recipient*`}` . . . . (A.1) – 298

> Start of a letter with the document class `letter`. Name and address of the *recipient* are given within the second pair of brackets; lines of text within this argument are ended with the command `\\`.

\begin{list}{*standard_label*}{*list_decl*} . . . . (4.4) – 73

Start of a generalized list environment. The label is defined by *standard_label*, which is generated by each \item command. The desired list declarations are contained in *list_decl* (see page 74).

2ε \begin{lrbox}{*boxname*} . . . . (4.7.1) – 86

Functions in a way similar to the command \sbox, except that it is the text of the environment that is stored in the LR box named *boxname*, which has previously been created with \newsavebox{*boxname*}. The contents of the box may be printed as often as desired with \usebox{*boxname*}.

\begin{math} . . . . (5.1) – 117

Switches from text to in-line math mode to produce formulas within a line of text. This environment has the same effect as \(or $ in text mode.

\begin{matrix} [m][a] . . . . (E.3.4) – 415

With the amsmath package, is the same as the array environment except that the column specifier argument may be omitted, without which up to 10 centered columns may be entered. This maximum may be changed with the counter MaxMatrixCols. The environments pmatrix, bmatrix, Bmatrix, vmatrix, and Vmatrix function the same as matrix but are enclosed in braces (), [], {}, | | and ‖ ‖, respectively.

\begin{minipage}[*pos*][*height*][*inner_pos*]{*width*}

. . . . (4.7.3), (4.7.5) – 87, 89

Environment to format text within a 'minipage' of width *width*. Its vertical positioning with respect to the surrounding text is determined by the optional argument *pos*: t for alignment with its top line, b with its bottom line, and centered with no argument. The other two optional arguments (not available with LATEX 2.09) are: *height* to give the total height, and *inner_pos* to specify how the text is to be positioned inside it. Possible values are t for top, b for bottom, c for centered, and s to be stretched out to fill the whole vertical space. The default is the value of the external positioning *pos* option. The *height* argument may contain the parameters \height, \depth, \width, and \totalheight.

\begin{multicols}{*num_cols*}[*header*][*pre_space*] . . (D.3.3) – 387

This environment is provided by the multicol package in the tools collection (Section D.3.3). It switches to printing the text in *num_cols* columns, with *header* printed in one column across the top. A new page is inserted only if the remaining space on the current page is less than \premulticols or the optional argument *pre_space*. A new page is inserted at the end if the remaining space is less than \postmulticols. The columns on the last page are balanced. Column separation and rule are set by the lengths \columnsep and \columnseprule.

With the amsmath package, switches to displayed math mode to produce a single equation over several lines. Line breaks are forced by \\ commands. The first line is to the far left, the last to the right, all others centered. With \shoveleft{*form*} and \shoveright{*form*}, single lines consisting of *form* may be pushed to the left or right. The single equation number appears at the right of the last line, or at the left of the first line, depending on class options reqno (default) and leqno, respectively. With the *-form of the environment, the equation number is suppressed.

In slides class, the environment for producing a note for the current slide. For LaTeX 2.09, this is in black and white only. Notes are numbered with the current slide number followed by a hyphen and running number, for example 8-1, 8-2, etc.

In slides class, the environment for producing an overlay for the current slide. Overlays are numbered with the current slide number followed by a lower case letter, for example 3-a, 3-b, etc. See also \begin{slide}.

Environment to generate a picture with the width *x_dimen* and height *y_dimen*, where the unit of length has previously been specified by the declaration \unitlength.

Most general form of the call to the picture environment. The picture is displaced to the left by *x_offset* and downwards by *y_offset*.

Like the matrix environment, but enclosed in round parentheses ().

Start of the quotation environment in which text is indented on both sides relative to the normal page margins. Paragraphs within the environment are marked with an additional indentation of the first line.

The same as the quotation environment except that the first line of a paragraph is not indented but instead additional line spacing comes between paragraphs.

In slides class, the main environment for producing a slide.

\begin{sloppypar} . . . . (3.6.3) – 54

Inside this environment word spacings are allowed to stretch more generously than usual so that paragraphs are broken up into lines with fewer word divisions. See also \sloppy. The countercommand is \fussy.

\begin{split} [m][a] . . . . (E.3.6) – 421

With the amsmath package, is used within a math environment such as equation to write a formula over several lines. Line breaks are forced with \\ commands and the lines are horizontally aligned on the & alignment marker. Any equation number is generated by the outer environment. It is either centered with the class option centertags (default) or with tbtags it appears at the right of the last line, or at the left of the first line, depending on class options reqno (default) and leqno, respectively.

\begin{subarray}{*pos*}{*1st line*\\..*last line*} [ma][a]   (E.3.2) – 410

With the amsmath package, sets multiline text for superscripts and subscripts, like \substack, but the parameter *pos* may take on values c or l for centered or left-justified lines.

\begin{subequations} [a] . . . . (E.3.7) – 426

With the amsmath package, numbers equations within it with a fixed main number and sequence of lower case letters attached, as 7a, 7b, 7c, ...

\begin{tabbing} . . . . (4.6.1) – 80

Start of the tabbing environment in which special tabbing commands become operational: \= sets a tab stop, \> jumps to the next stop, \< goes back a stop, \\ terminates and starts a new line, \+ sets the left margin one tab stop further, \- moves the left margin back one stop.

\begin{table}[*loc*] . . . . (4.8.5), (6.4.1) – 107, 179

Float environment for entering text for a table. The optional argument *loc* can be any combination of the letters h, t, b, and p to determine the various positioning possibilities. Default is tbp. Under LaTeX 2_ε, the character ! may additionally be given to ignore all float spacing and number restrictions set by the float style parameters.

\begin{table*}[*loc*] . . . . (6.4.1) – 179

The same as the table environment except that the table is inserted over the width of two columns when the option twocolumn or the command \twocolumn has been selected. The standard form table will only fill the width of one column.

`\begin{tabular}[`*pos*`]{`*cols*`}` . . . . (4.8.1) – 93

Start of the `tabular` environment for producing tables. The argument *cols* contains a formatting character for each column in the table: c for centered text, l for left, r for right justification, or p{*wd*} for a column of width *wd* in which the text may extend over several lines.

When the entry @{*text*} appears between any two of the above column formatting characters, *text* is inserted in every row between those two columns. Where the character | appears, a vertical line is drawn in every row.

The optional argument *pos* specifies how the table is to be vertically aligned with the surrounding text: with no argument, it is centered, otherwise with t the top line, with b the bottom line is aligned with the external baseline.

The text entries of the individual columns are separated by & and the rows are terminated by \\.

`\begin{tabular*}{`*width*`}[`*pos*`]{`*cols*`}` . . . . (4.8.1) – 93

The same as `\begin{tabular}` except that the total width of the table is given by the argument *width*. This may only be achieved successfully if there is rubber spacing between the columns. This may be added with @{\extracolsep\fill} somewhere within the *cols* format definition.

`\begin{thebibliography}{`*sample_label*`}` . . (4.3.6), (8.3.3) – 70, 226

Environment to generate a list of literature references. The *sample_label* is the longest reference marker that will appear. Each entry in the bibliography starts with the command \bibitem which prints the marker for that entry; lines after the first are indented by an amount equal to the width of *sample_label*.

`\begin{theindex}` . . . . (8.3.4) – 227

Environment to produce a keyword index in two-column format. Entries are made with the \item, \subitem, \subsubitem, or \indexspace commands.

`\begin{`*theorem_type*`}[`*extra_title*`]` . . . . (4.5) – 79

Environment to invoke a theorem-like structure that has previously been defined by the user with the \newtheorem command. The name of the environment, *theorem_type*, something like theorem or axiom, is the first argument of the \newtheorem command. The *extra_title* is text that added after the name and number of the structure in () parentheses.

`\begin{titlepage}` . . . . (3.3.1) – 36

Environment to produce a title page without a page number. The user has total control over the composition of this page.

\begin{trivlist} . . . . (4.4.5) – 78

Environment to generate a trivial list *without* a sample label and list declarations. The parameters \leftmargin, \labelwidth, and \itemsep are all set to 0 pt while \listparindent = \parindent and \parsep = \parskip.

\begin{verbatim} . . . . (4.9) – 108

Environment to print out original text, that is, as from a typewriter. Blank lines, line breaking, and commands are all output literally without any interpretation or formatting.

\begin{verbatim*} . . . . (4.9) – 108

The same as the verbatim environment except that blanks are printed as ␣ to make them visible.

\begin{verse} . . . . (4.2.4) – 64

Environment for setting rhymes, poems, verses, etc. Stanzas are separated by blank lines, individual lines by the \\ command.

\begin{vmatrix} [m][a] . . . . (E.3.4) – 415

Like the matrix environment, but enclosed in vertical lines | |.

\begin{Vmatrix} [m][a] . . . . (E.3.4) – 415

Like the matrix environment, but enclosed in double vertical lines ‖ ‖.

\belowdisplayskip [m] . . . . (5.5.4) – 147

Vertical spacing between a *long* displayed formula and the following text. A new value may be assigned with the \setlength command:

    \setlength{\belowdisplayskip}{\abovedisplayskip}

sets \belowdisplayskip to the same value as \abovedisplayskip. See further examples under \abovedisplayskip.

\belowdisplayshortskip [m] . . . . (5.5.4) – 147

Vertical spacing between a *short* displayed formula and the following text. Value is set with the \setlength command as in the above example.

\beta [m] produces β . . . . (5.3.1) – 123

2.09 **\bezier{*num*}**$(x_1, y_1)(x_2, y_2)(x_3, y_3)$ . . . . (6.1.4) – 163

This command can be given within the picture environment to draw a quadratic Bézier curve from point (x_1, y_1) to (x_3, y_3), using (x_2, y_2) as the extra Bézier point. The curve is drawn as $num + 1$ dots. This is the same as \qbezier except that *num* is mandatory.

In LaTeX 2.09, the option bezier must be included in the \documentstyle line to enable this command.

2.09 \bf . . . . (4.1.5) – 62

Switches to the **Roman, upright, bold** typeface.

2ε \bfdefault . . . . (C.5.1) – 353

This command defines the series attribute that is selected with the
\bfseries command. It may be redefined with \renewcommand:

\renewcommand{\bfdefault}{b}

2ε \bfseries . . . . (4.1.3), (8.5.2) – 60, 236

This declaration switches to a font in the current family and shape,
but with the **bold** series attribute.

\bibitem[*label*]{*key*} *entry_text* . . . . (4.3.6), (8.3.3) – 71, 226

Command to enter the text for a literature reference in the
thebibliography environment. The reference word *key* is used
in the main body of the text with the \cite command to refer to this
entry. The \cite command is replaced either by the optional *label*
text or by a standard label, such as a sequential number in square
brackets.

\bibliography{*file*} . . . . (8.3.3), (B.1) – 227, 311

For producing a bibliography with the aid of the BibTeX program;
file is the root name of one or more files containing the literature
databases to be searched.

\bibliographystyle{*style*} . . . . (B.1) – 312

In conjunction with the BibTeX program, this command selects the
style in which the bibliography entries are to be written. Choices
for *style* are plain, unsrt, alpha, and abbrv, where the first is the
default. Other non-standard styles may also exist.

\bibname . . . . (C.4.1) – 351

Command containing the heading for the bibliography in book and
report document classes. In English, this is 'Bibliography' but may
be altered for adaptation to other languages.

\big*br_symbol* [m] . . . . (5.5.3) – 146

A bracket symbol larger than normal, but smaller than \Big. Example: \big(.

\Big*br_symbol* [m] . . . . (5.5.3) – 146

A bracket symbol larger than \big, but smaller than \bigg. Example:
\Big[.

\bigcap [m] produces ∩ . . . . (5.3.7) – 126
\bigcirc [m] produces ○ . . . . (5.3.3) – 124
\bigcup [m] produces ∪ . . . . (5.3.7) – 126
\bigg*br_symbol* [m] . . . . (5.5.3) – 146

A bracket symbol larger than \Big, but smaller than \Bigg. Example:
\bigg|.

\Bigg*br_symbol* [m] . . . . (5.5.3) – 146
> The largest bracket symbol. Example: \Bigg\langle.

\bigg*lbr_symbol* [m] . . . . (5.5.3) – 146
> The same as \bigg but is also a logical left (opening) bracket.

\Bigg*lbr_symbol* [m] . . . . (5.5.3) – 146
> The same as \Bigg but is also a logical left (opening) bracket.

\bigg*mbr_symbol* [m] . . . . (5.5.3) – 146
> The same as \bigg but with larger horizontal spacings on either side
> (relation operator).

\Bigg*mbr_symbol* [m] . . . . (5.5.3) – 146
> The same as \Bigg but with larger horizontal spacings on either side
> (relation operator).

\bigg*rbr_symbol* [m] . . . . (5.5.3) – 146
> The same as \bigg but is also a logical right (closing) bracket.

\Bigg*rbr_symbol* [m] . . . . (5.5.3) – 146
> The same as \Bigg but is also a logical right (closing) bracket.

\big*lbr_symbol* [m] . . . . (5.5.3) – 146
> The same as \big but is also a logical left (opening) bracket.

\Big*lbr_symbol* [m] . . . . (5.5.3) – 146
> The same as \Big but is also a logical left (opening) bracket.

\big*mbr_symbol* [m] . . . . (5.5.3) – 146
> The same as \big but with larger horizontal spacings on either side
> (relation operator).

\Big*mbr_symbol* [m] . . . . (5.5.3) – 146
> The same as \Big but with larger horizontal spacings on either side
> (relation operator).

\bigodot [m] produces $\odot$ . . . . (5.3.7) – 126
\bigoplus [m] produces $\oplus$ . . . . (5.3.7) – 126
\bigotimes [m] produces $\otimes$ . . . . (5.3.7) – 126
\big*rbr_symbol* [m] . . . . (5.5.3) – 146
> The same as \big but is also a logical right (closing) bracket.

\Big*rbr_symbol* [m] . . . . (5.5.3) – 146
> The same as \Big but is also a logical right (closing) bracket.

> Inserts large vertical spacing of the amount `\bigskipamount`. See also `\medskip` and `\smallskip`.

`\bigskipamount`

> Standard value for the amount of vertical spacing that is inserted with the command `\bigskip`. May be changed with the `\setlength` command:
>
> `\setlength{\bigskipamount}{5ex plus1.5ex minus2ex}`

> With the amsmath package, prints a binomial expression:
>
> `\[\binom{n}{k}\]` yields $\binom{n}{k}$.

> Command to produce the function name 'mod' in the form
>
> `a\bmod b` = $a \bmod b$

> Switches to bold face for math mode. This command must be given in text mode, however, before going into math mode. To set only part of a formula in bold face, use `\mbox{\boldmath$...$}` to return temporarily to text mode.

> When one of the packages amsmath or amsbsy has been loaded, this command prints *symbol* in bold face. Unlike `\mathbf`, it also affects math symbols and lower case Greek letters.

> A command that is executed before a float at the bottom of a page. It is normally defined to do nothing, but may be redefined to add a rule between the float and the main text. It must not add any net vertical spacing.
>
> `\renewcommand{\botfigrule}{\vspace*{-.4pt}`
> `\rule{\columnwidth}{.4pt}}`

`\bottomfraction` . . . . (6.4.3) – 182

Maximum fraction of a page that may be taken up by floats at the bottom. May be set to a new value with:

`\renewcommand{\bottomfraction}{`*decimal_frac*`}`.

`bottomnumber` . . . . (6.4.3) – 182

Maximum number of floats that may appear at the bottom of a page. Set to a new number with `\setcounter{bottomnumber}{`*num*`}`.

`\bowtie` [m] produces ⋈ . . . . (5.3.4) – 124
`\Box` [m] produces □ . . . . (5.3.3) – 124
`\boxed{`*formula*`}` [m][a] . . . . (E.3.5) – 419

With the `amsmath` package, sets the mathematical *formula* in a box.

`\breve{`*x*`}` [m] . . . . (5.3.9) – 127

Breve accent over the math variable x: `\breve{a}` = $\breve{a}$.

`\Breve{`*x*`}` [m][a] . . . . (E.3.2) – 412

With the `amsmath` package, can be used like `\breve`, but with multiple $\mathcal{A}_{\mathcal{M}}$S-LaTeX math accents the positioning will be correct.

`\bullet` [m] produces • . . . . (5.3.3) – 124

`\c{`*x*`}` . . . . (2.5.7) – 21

Produces a cedilla under x: `\c{C}` = Ç.

2.09 `\cal` [m] . . . . (5.3.2) – 123

A declaration in LaTeX 2.09 to select *calligraphic* letters in math mode. It has been replaced by the command `\mathcal` in LaTeX 2ε.

`\cap` [m] produces ∩ . . . . (5.3.3) – 124
`\caption[`*short_form*`]{`*caption_text*`}` . . . . (6.4.4) – 183

Produces a numbered title or caption with the text *caption_text* within the float environments `figure` or `table`. The *short_form* is the abbreviated text appearing in the list of figures or tables, which is the same as the *caption_text* if it is omitted.

`\captions`*language* . . . . (C.4.1), (D.1.3) – 351, 373

A command used in several language adaptations to redefine the headings of special sections such as 'Chapter' and 'Contents'. It occurs in packages `esperant` and `german` as well as in the `babel` system. This command is normally part of the definition of the `\selectlanguage` command.

`\cc{`*list*`}` . . . . (A.1) – 299

Command within document class `letter` to generate 'cc:', copies, followed by a list of names *list* at the end of the letter.

Command in the `letter` document class containing the word to be printed by the \cc command. In English, this is 'cc' but may be altered for adaptation to other languages.

Declaration to switch to centered lines of text, each input line being terminated by \\. See also \begin{center}.

An additional TEX command that sets *text* centered on a horizontal line by itself.

With the `amsmath` package, produces a continued fraction. The optional argument *pos* may be l or r to have the numerator left or right justified on the horizontal rule, otherwise it is centered.

Starts a new chapter on a new page, with an automatic sequential chapter number and *title* as header. If the optional *short title* is given, it appears in place of *title* in the table of contents and in the running head at the top of the pages.

Starts a new chapter on a new page, with *title* as header, but without a chapter number. The entry does not appear in the table of contents.

Command containing the chapter heading. In English, this is 'Chapter' but may be altered for adaptation to other languages.

Háček accent over the math variable x: \check{a} = ǎ.

With the `amsmath` package, can be used like \check, but with multiple $\mathcal{A}_{\mathcal{M}}\mathcal{S}$-LATEX math accents the positioning will be correct.

Tests that the current definition of *com_name* is as expected. If not, an error message is issued. This is used to ensure that important commands have not been altered by other packages.

(1994/12/01)

\chi [m] produces χ . . . . (5.3.1) – 123
\circ [m] produces ∘ . . . . (5.3.3) – 124
\circle{*diameter*} . . . . (6.1.4) – 159

> Picture element command to produce a circle with diameter *diameter* in the picture environment. To be used as an argument in a \put or \multiput command.

\circle*{*diameter*} . . . . (6.1.4) – 159

> Like \circle but produces a solid circle, filled in black.

\cite[*note*]{*key*} . . . . (4.3.6), (8.3.3), (B.1) – 71, 226, 312

> Literature citation using the keyword *key* to produce a reference label in the text. The optional *note* text is included with the label.

$\boxed{2\varepsilon}$ \ClassError{*class_name*}{*error_text*}{*help*} [p] . . . . . (C.2.7) – 337

> Writes an error message *error_text* to the monitor and transcript file, labeled with the class name, and halts processing, waiting for a user response as for a LATEX error. If H⟨*return*⟩ is typed, the *help* text is printed. Both *error_text* and *help* may contain \MessageBreak for a new line, \space for a forced space, and \protect before commands that are to have their names printed literally and not interpreted.

$\boxed{2\varepsilon}$ \ClassInfo{*class_name*}{*info_text*} [p] . . . . (C.2.7) – 338

> Is like \ClassWarningNoLine except that the text *info_text* is only written to the transcript file, and not to the monitor.

$\boxed{2\varepsilon}$ \ClassWarning{*class_name*}{*warning_text*} [p] . . . . . (C.2.7) – 337

> Writes *warning_text* to the monitor and transcript file, labeled with the class name and the current line number of the input file. Processing continues. The *warning_text* is formatted in the same way as that for \ClassError.

$\boxed{2\varepsilon}$ \ClassWarningNoLine{*class_name*}{*warning_text*} [p] . . (C.2.7) – 337

> Is like \ClassWarning except that the current line number of the input file is not printed.

\cleardoublepage . . . . (3.5.5) – 52

> Ends the current page and outputs all unprocessed floats on to one or more float pages. The next page will be a *right-hand* one, with an odd page number.

\clearpage . . . . (3.5.5) – 51

> Ends the current page and outputs all unprocessed floats on to one or more float pages.

\cline{*n – m*} . . . . (4.8.1) – 95

> In tabular environment, produces a horizontal rule from the beginning of column *n* to the end of column *m*. Example: \cline{2-5}.

\closing{*regards*} . . . . (A.1) – 299

End of the text in the letter environment; *regards* stands for the desired terminating text.

\clubsuit [m] produces ♣ . . . . (5.3.6) – 125

[2ε] \color *col_spec* . . . . (6.3) – 177

A command made available with the color package. It is a declaration that switches the color in which the text is printed to that specified. It remains in effect until the end of the current environment or until countermanded by another \color command. The *col_spec* is either the name of a color defined (or predefined) by \definecolor or of the form [*model*]{*specs*}, where the arguments have the same meaning as they do for \definecolor. Examples:

\color[rgb]{0.5,0.5,0} \color{magenta}

[2ε] \colorbox *col_spec*{*text*} . . . . (6.3) – 177

A command made available with the color package. The *text* is set in an LR box with the specified color as the background color. The *col_spec* is the same as for \color.

\columnsep . . . . (3.1.1) – 28

Declaration for the amount of intercolumn spacing in two-column page formatting. May be changed with the \setlength command:

\setlength{\columnsep}{1pt}

\columnseprule . . . . (3.1.1) – 28

Declaration for the thickness of the vertical rule separating the columns in two-column page formatting. Value is set with the \setlength command:

\setlength{\columnseprule}{1pt}

\cong [m] produces ≅ . . . . (5.3.4) – 124

\contentsline{*sec_type*}{\numberline{*sec_num*}*title_text*}{*page*}

Command that appears in the .toc file for every entry in the table of contents, which is read when the \tableofcontents command is given. Such commands may be altered or added to the .toc file by means of the text editor. The entry *sec_type* stands for the sectioning level, such as section, while *sec_num* is its number (for example, 2.3) and *page* is the page number where the entry appears.

\contentsname . . . . (C.4.1) – 351

Command containing the heading for the table of contents. In English, this is 'Contents' but may be altered for adaptation to other languages.

\coprod [m] produces ∐ . . . . (5.3.7) – 126

\copyright produces © . . . . (2.5.5) – 21

\cos [m] . . . . (5.3.8) – 126

Command to produce the function name 'cos' in formulas.

`\cosh` [m] . . . . (5.3.8) – 126

> Command to produce the function name 'cosh' in formulas.

`\cot` [m] . . . . (5.3.8) – 126

> Command to produce the function name 'cot' in formulas.

`\coth` [m] . . . . (5.3.8) – 126

> Command to produce the function name 'coth' in formulas.

`\csc` [m] . . . . (5.3.8) – 126

> Command to produce the function name 'csc' in formulas.

`\cup` [m] produces ∪ . . . . (5.3.3) – 124

`2ε` `\CurrentOption` [p] . . . . (C.2.3) – 334

> A command that may only be used in the definition of options, especially for default options. It contains the name of the option being processed.

`\d{x}` . . . . (2.5.7) – 21

> Produces a 'dot under' accent: `\d{o}` = ọ.

`\dag` produces † . . . . (2.5.5) – 21

`\dagger` [m] produces † . . . . (5.3.3) – 124

`\dashbox{dash}(x_dimen,y_dimen)[pos]{text}` . . . . . (6.1.4) – 155

> Picture element command to produce a dashed frame with width *x_dimen* and height *y_dimen*, using a dash length of *dash* in the `picture` environment. Without the optional *pos*, the contents *text* are centered within the frame, otherwise they are positioned at the left (1), right (r), top (t), or bottom (b), or a combination thereof, such as 1t. This command is used as an argument in a `\put` or `\multiput` command.

`\dashv` [m] produces ⊣ . . . . (5.3.6) – 125

`\date{date_text}` . . . . (3.3.1), (A.1) – 37, 299

> 1. The command `\maketitle` normally prints the current date on the title page. The declaration `\date` will replace the date with whatever text is given in *date_text*.
>
> 2. Prints the text *date_text* instead of the automatic current date in a letter.

`\date`*language* . . . . (D.1.1), (D.1.3) – 366, 373

> A command used in several language adaptations to redefine the `\today` command according to the requirements of *language*. It occurs in packages `esperanto` and `german` as well as in the `babel` system. It may also be used for 'dialects', such a `\dateUSenglish` and `\dateenglish`. This command is normally part of the definition of the `\selectlanguage` command.

2ε \dblfigrule . . . . (6.4.3) – 183

A command that is executed after a two-column float at the top of a page. It is normally defined to do nothing, but may be redefined to add a rule between the float and the main text. It must not add any net vertical spacing.

```
\renewcommand{\dblfigrule}{\vspace*{-.4pt}
   \rule{\textwidth}{.4pt}}
```

\dblfloatpagefraction . . . . (6.4.3) – 182

For two-column page formatting, the fraction of a float page that must be filled with floats before a new page is called. A new value is assigned with

```
\renewcommand{\dblfloatpagefraction}{decimal_frac}
```

\dbinom{*over*}{*under*} [m][a] . . . . (E.3.3) – 413

With the amsmath package, produces a binomial as \binom does, but in \displaystyle size.

\dblfloatsep . . . . (6.4.3) – 182

For two-column page formatting, the vertical spacing between floats that extend over both columns. A new value is set with the \setlength command:

```
\setlength{\dblfloatsep}{12pt plus 2pt minus 4pt}
```

\dbltextfloatsep . . . . (6.4.3) – 183

For two-column page formatting, the vertical spacing between floats extending over both columns at the top of the page and the following text. A new value is set with the \setlength command.

\dbltopfraction . . . . (6.4.3) – 182

For two-column page formatting, the maximum fraction of a page that may be occupied at the top by floats extending over both columns. A new value is assigned with

```
\renewcommand{\dbltopfraction}{decimal_frac}.
```

dbltopnumber . . . . (6.4.3) – 182

For two-column page formatting, the maximum number of floats that may appear at the top of a page extending over both columns. A new value is assigned with

```
\setcounter{dbltopnumber}{num}.
```

\ddag produces ‡ . . . . (2.5.5) – 21
\ddagger [m] produces ‡ . . . . (5.3.3) – 124
\dddot{*x*} [m][a] . . . . (E.3.2) – 412

With the amsmath package, a triple dot accent in math formulas:
\dddot{a} = $\dddot{a}$.

\ddddot{*x*} [m][a]   . . . . (E.3.2) – 412

> With the `amsmath` package, a four-dot accent in math formulas:
> \ddddot{a} = $\ddddot{a}$.

\ddot{*x*} [m]   . . . . (5.3.9) – 127

> A double dot accent in mathematical formulas: \ddot{a} = $\ddot{a}$.

\Ddot{*x*} [m][a] . . . . (E.3.2) – 412

> With the `amsmath` package, can be used like \ddot, but with multiple
> $\mathcal{A}_{\mathcal{M}}\mathcal{S}$-LaTeX math accents the positioning will be correct.

\ddots [m] produces $\ddots$ . . . . (5.2.6) – 121

⌐2ε⌐ \DeclareErrorFont{*code*}{*fam*}{*ser*}{*shp*}{*sz*} [p]   . . (C.5.6) – 357

> If no valid font can be found, even after substituting the attributes
> as given by \DeclareFontSubstitution, the font declared with this
> command is selected as a last resort.

⌐2ε⌐ \DeclareFixedFont{*cmd*}{*code*}{*fam*}{*ser*}{*shp*}{*sz*} [p]
. . . . (C.5.2) – 354

> Defines *cmd* to be a font *declaration* that selects a font with the
> fixed attributes given in the definition.

⌐2ε⌐ \DeclareFontEncoding{*code*}{*text_set*}{*math_set*} [p]   . (C.5.6) – 357

> Declares *code* to be the name of a new encoding scheme; *text_set* is a
> set of commands that is to be executed when switching to text mode,
> *math_set* a set for math mode.

⌐2ε⌐ \DeclareFontEncodingDefaults{*text_set*}{*math_set*} [p] (C.5.6) – 357

> Declares the sets of commands to be executed by all encodings when
> switching to text and math modes; the additional commands specific
> to each encoding are executed afterwards.

⌐2ε⌐ \DeclareFontFamily{*code*}{*fam*}{*opt*} [p] . . . . (C.5.6) – 357

> Declares *fam* to be a new font family with the encoding *code*; the
> encoding must previously have been declared. The commands in *opt*
> are executed every time this family is selected.

⌐2ε⌐ \DeclareFontShape{*code*}{*fam*}{*ser*}{*shp*}{*def*}{*opt*} [p]
. . . . (C.5.6) – 358

> Associates external font names with the given font attribute values;
> the actual definition *def* relates font sizes to font names, as explained
> on page 358. The commands in *opt* are executed every time this shape
> is selected.

[2ε] `\DeclareFontSubstitution{`*code*`}{`*fam*`}{`*ser*`}{`*shp*`}` [p]   (C.5.6) – 357

Declares the font attributes that are to be substituted in case there is no valid font corresponding to the set of attributes selected. The order of substitution is *shape*, *series*, and *family*; the encoding is never substituted.

[2ε] `\DeclareGraphicsExtensions{`*ext_list*`}` [p] . . . . (6.2.4) – 175

Establishes the list of default extensions for graphics files that can be imported with the `\includegraphics` command and the `graphics` or `graphicx` packages; *ext_list* is a comma-separated list of file extensions, such as `.eps,.ps`.

[2ε] `\DeclareGraphicsRule{`*ext*`}{`*type*`}{`*bb*`}{`*cmd*`}` [p] . . . (6.2.4) – 176

Associates a graphics file extension *ext* with a graphics file type and a file extension (*bb*) where the bounding box information is to be read, and an operating command (*cmd*) that is to be executed on the file to make it available for importation. For example

```
\DeclareGraphicsRule{.eps.gz}{eps}
    {.eps.bb}{`gunzip -c #1}
```
Here the command must be prefixed with ` and #1 represents the name of the file to be processed.

[2ε] `\DeclareMathAccent{\`*cmd*`}{`*type*`}{`*sym_fnt*`}{`*pos*`}` [p] . (C.5.4) – 356

Declares `\`*cmd* to be a math accent command, printed with the character in position *pos* of the symbol font with the internal name *sym_fnt*. The *type* is either `\mathord` or `\mathalpha`; in the latter case the symbol changes with math alphabet.

[2ε] `\DeclareMathAlphabet{\`*cmd*`}{`*code*`}{`*fam*`}{`*ser*`}{`*shp*`}` [p]
. . . . (C.5.3) – 355

Defines `\`*cmd* to be a math alphabet for setting letters in math mode; the font selected *in all math versions* is that with the specified font attributes. If a different font is to be invoked for certain math versions, they are defined individually with `\SetMathAlphabet` afterwards. However, if the shape attribute is left empty, the alphabet command is indeed created, but remains undefined in *all* versions, requiring an explicit `\SetMathAlphabet` declaration for each one.

[2ε] `\DeclareMathDelimiter{\`*cmd*`}{`*type*`}{`*sym_fnt1*`}{`*pos1*`}`
`{`*sym_fnt2*`}{`*pos2*`}` [p] . . . . (C.5.4) – 356

Declares `\`*cmd* to be a math delimiter in two sizes: the smaller variant is the character in position *pos1* of the symbol font with the internal name *sym_fnt1*, while the larger is from position *pos2* of font *sym_fnt2*.

`\DeclareMathOperator{`*\cmd*`}{`*name*`}` [p][a] . . . . . . (E.3.5) – 416

With the `amsopn` or `amsmath` packages, defines *\cmd* to be a math mode command to print the text *name* as a function name in an upright font with appropriate spacing. With the *-form, the raising and lowering operators ^ and _ produce limits, above or below the name.

⎡2ε⎤ `\DeclareMathRadical{`*\cmd*`}{`*sym_fnt1*`}{`*pos1*`}`
      `{`*sym_fnt2*`}{`*pos2*`}` [p] . . . . (C.5.4) – 356

Declares *\cmd* to be a math radical symbol: the smaller variant is the character in position *pos1* of the symbol font with the internal name *sym_fnt1*, while the larger is from position *pos2* of font *sym_fnt2*.

⎡2ε⎤ `\DeclareMathSizes{`*text*`}{`*math_text*`}{`*script*`}{`*sscript*`}` [p] (C.5.4) – 356

Sets the point sizes for the three math styles `\textstyle`, `\scriptstyle`, and `\scriptscriptstyle` when the text is being printed in point size *text*. The unit pt is not included.

⎡2ε⎤ `\DeclareMathSymbol{`*\symbol*`}{`*type*`}{`*sym_fnt*`}{`*pos*`}` [p] (C.5.4) – 356

Declares *\symbol* to be a command that prints the character in position *pos* with the symbol font with the internal name *sym_fnt*. This same symbol is printed in all math alphabets, but may be different for other math versions if the symbol font has been `\Set...` to have different attibutes in that math version.

⎡2ε⎤ `\DeclareMathVersion{`*ver*`}` [p] . . . . (C.5.3) – 355

Declares *ver* to be a new math version for math alphabets and symbol fonts. Initially, the version will use those fonts defined by `\Declare...` commands, which may be redefined for this math version with appropriate `\Set...` commands.

⎡2ε⎤ `\DeclareOldFontCommand{`*\cmd*`}{`*text_specs*`}{`*math_specs*`}` [p]
. . . . (C.5.2) – 354

Defines *\cmd* to be a font *declaration* that invokes *text_specs* in text mode, and *math_specs* in math mode. The new command behaves in math mode as a 2.09 font declaration, although *math_specs* must be a math font command without its argument. It is meant to define commands to be compatible with LaTeX 2.09 and should generally be avoided.

⎡2ε⎤ `\DeclareOption{`*option*`}{`*code*`}` [p] . . . . (C.2.3) – 334

In a class or package file, this command defines the set of commands (*code*) that is to be associated with the given *option*. These commands are executed when `\ExecuteOptions` or `\ProcessOptions` is called. After the latter, all definitions are erased, to save memory. The *code* is internally stored in a command named `\ds@`*option*.

[2ε] `\DeclareOption*{`*code*`}` [p] . . . . (C.2.3) – 334

In a class or package file, this command defines the default set of commands that is associated with every undefined option. Special commands that may be used within *code* are `\CurrentOption` (the name of the option) and `\OptionNotUsed`. Example:

```
\DeclareOption*{\InputIfFileExists
  {\CurrentOption.sty}{}{\OptionNotUsed}}
```

[2ε] `\DeclareRobustCommand{`*\cmd*`}[`*narg*`][`*opt*`]{`*def*`}` . . . (C.2.5) – 336

Defines or redefines the command *\cmd* in the same way as `\newcommand` except that the result is robust: it may be used in the argument of another command without a `\protect` command before it.

[2ε] `\DeclareRobustCommand*{`*\cmd*`}[`*narg*`][`*opt*`]{`*def*`}` . . (C.2.6) – 336

The same as `\DeclareRobustCommand` except that the arguments to *\cmd* must be 'short', not containing any new paragraphs. *(1994/12/01)*

[2ε] `\DeclareSymbolFont{`*sym_fnt*`}{`*code*`}{`*fam*`}{`*ser*`}{`*shp*`}` [p]
. . . . (C.5.4) – 355

Declares the math font with the given attributes to be a symbol font that may be addressed by other commands with the name *sym_fnt*. The symbol font applies to all math versions unless redefined with a `\SetSymbolFont` command.

[2ε] `\DeclareSymbolFontAlphabet{`*\cmd*`}{`*sym_fnt*`}` [p] . . (C.5.4) – 356

Defines *\cmd* to be a math alphabet that uses the font declared with a `\DeclareSymbolFont` command to have the internal name *sym_fnt*. This is preferred over `\DeclareMathAlphabet` if there is a defined symbol font with the necessary attributes for the math alphabet.

[2ε] `\DeclareTextAccent{`*\cmd*`}{`*code*`}{`*pos*`}` [p] . . . . . . (C.5.7) – 360

Defines *\cmd* to be an accent command when encoding *code* is active; it uses the symbol in font position *pos* as the accent.

[2ε] `\DeclareTextAccentDefault{`*\cmd*`}{`*code*`}` [p] . . . . . (C.5.7) – 361

Declares the default encoding that is to be taken if the accent command *\cmd* is called in an encoding for which it is not explicitly defined. *(1994/12/01)*

[2ε] `\DeclareTextCommand{`*\cmd*`}{`*code*`}[`*narg*`][`*opt*`]{`*def*`}` [p]
. . . . (C.5.7) – 360

Defines *\cmd* in the same way as `\newcommand` except the definition is only valid when encoding *code* is active.

`2ε` \DeclareTextCommandDefault{*cmd*}[*narg*][*opt*]{*def*} [p]

. . . . (C.5.7) – 361

Creates a default definition for the command *cmd* for all encodings for which it is not explicitly defined. *(1994/12/01)*

`2ε` \DeclareTextComposite{*cmd*}{*code*}{*let*}{*pos*} [p] . . (C.5.7) – 360

Defines *cmd* followed by the letter *let* to be the single character in font position *pos* when encoding *code* is active. It thus defines the action of accent commands in T1 encoding when the accented letter exists as a separate symbol. For example:

    \DeclareTextComposite{\'}{T1}{e}{233}

The command must already have been defined for the encoding with either \DeclareTextAccent or \DeclareTextCommand (with one argument).

`2ε` \DeclareTextCompositeCommand{*cmd*}{*code*}{*let*}{*def*} [p]

. . . . (C.5.7) – 360

Is the same as \DeclareTextComposite except that any definition may be assigned to the command/letter combination, and not just a single symbol. *(1994/12/01)*

`2ε` \DeclareTextFontCommand{*cmd*}{*font_specs*} [p] . . . (C.5.2) – 354

Defines *cmd* to be a text font command that sets its argument with the *font_spec* specifications. Example:

    \DeclareTextFontCommand{\textbf}{\bfseries}

`2ε` \DeclareTextSymbol{*cmd*}{*code*}{*pos*} [p] . . . . . (C.5.7) – 360

Defines *cmd* to print the symbol in font position *pos* when encoding *code* is active.

`2ε` \DeclareTextSymbolDefault{*cmd*}{*code*} [p] . . . . . (C.5.7) – 361

Declares the default encoding that is to be taken if the symbol command *cmd* is called in an encoding for which it is not explicitly defined. *(1994/12/01)*

`2ε` \definecolor{*name*}{*model*}{*specs*} . . . . (6.3) – 177

A command made available with the color package. It associates the name of a color (*name*) with the specifications *specs* according to a certain *model*. Possible values for *model* are rgb (red, green, blue), cmyk (cyan, magenta, yellow, black), gray, and named. In each case, *specs* is a comma-separated list of numbers between 0 and 1 specifying the strength of the relevant component. In the case of the named model, *specs* is an internal name for the color that is known by the driver program. Examples:

    \definecolor{litegrn}{cmyk}{0.25,0,0.75,0}
    \definecolor{brown}{named}{RawSienna}

A number of colors are predefined in all color drivers: red, green, blue, yellow, cyan, magenta, black, and white.

\deg [m] . . . . (5.3.8) – 126
> Command to produce the function name 'deg' in formulas.

[2ε] \DeleteShortVerb{\c} . . . . (4.9.1, 8.8.3) – 109, 242
> When the standard package shortvrb has been loaded, this command counteracts the effects of a previous \MakeShortVerb{\c}, allowing the character c to have its original meaning once more.

\Delta [m] produces Δ . . . . (5.3.1) – 123
\delta [m] produces δ . . . . (5.3.1) – 123

[2ε] \depth . . . . (4.7.1) – 85
> A length parameter equal to the depth of a box (baseline to bottom); it may only be used in the *width* specification of \makebox, \framebox, or \savebox, or in the *height* specification of a \parbox or minipage environment.
>
> > \framebox[20\depth]{text}

\det [m] . . . . (5.3.8) – 126
> Command to produce the function name 'det' in formulas. Can be combined with a lower limit by means of the subscript command.

\dfrac{*numerator*}{*denominator*} [m][a] . . . . (E.3.3) – 413
> With the amsmath package, produces a fraction as \frac does, but in \displaystyle size.

[2ε] \DH   . . . . (F.4.5) – 452
> When T1 encoding is active, prints the character Đ. *(1994/12/01)*

[2ε] \dh   . . . . (F.4.5) – 452
> When T1 encoding is active, prints the character ð. *(1994/12/01)*

\Diamond [m] produces ◊   . . . . (5.3.3) – 124
\diamond [m] produces ◊ . . . . (5.3.3) – 124
\diamondsuit [m] produces ◆ . . . . (5.3.6) – 125
\dim [m] . . . . (5.3.8) – 126
> Command to produce the function name 'dim' in formulas.

\discretionary{*before*}{*after*}{*without*}   . . . . (3.6.1) – 54
> Hyphenation suggestion within a word. The word may be divided such that *before* is at the end of one line, and *after* at the start of the next line. If no division occurs, *without* is printed.

\displaybreak[*num*] [m][a] . . . . (E.3.7) – 427
> With the amsmath package, allows a manual page break in a multiline math formula when given just before \\; if *num* is present, it takes on values of 0-4 with increasing encouragement for a break, where 4 is the same as no value, a forced page break. Automatic page breaks are impossible in multiline formulas unless \allowdisplaybreaks has been issued in the preamble.

`\displaystyle` [m] . . . . (5.5.2) – 144

Switches to font size `\displaystyle` as the active font within a math formula.

`\div` [m] produces ÷ . . . . (5.3.3) – 124

$\boxed{2\varepsilon}$ `\DJ` . . . . (F.4.5) – 452

When T1 encoding is active, prints the character Đ. *(1994/12/01)*

$\boxed{2\varepsilon}$ `\dj` . . . . (F.4.5) – 452

When T1 encoding is active, prints the character đ. *(1994/12/01)*

$\boxed{2\varepsilon}$ `\documentclass`[*options*]{*class*}[*version*] [p] . . . . (3.1) – 25

Normally the first command in a LATEX document, determining the overall characteristics. Standard values for *class* are:

`article, report, book, letter,` and `slides`

of which only one may be selected. In addition, various options may be chosen, their names separated by commas. Possibilities are:

`10pt, 11pt, 12pt,`
`letterpaper, legalpaper, executivepaper,`
`a4paper, a5paper, b5paper, landscape,`
`onecolumn, twocolumn,`
`oneside, twoside,`
`notitlepage, titlepage,`
`leqno, fleqn, openbib,`
`draft, final`

These and any additional options are all global, meaning that they also apply to any packages specified with a following `\usepackage` command.

The optional *version* is a date, given in the form yyyy/mm/dd, as for example 1994/08/01. If the date of the class file read in is earlier than this, a warning message is printed.

$\boxed{2.09}$ `\documentstyle`[*options*]{*style*} [p] . . . . (3.1) – 25

In LATEX 2.09, normally the first command in a document, determining the overall characteristics. This command has been replaced in LATEX 2ε by `\documentclass`, but it may still be issued to initiate the compatibility mode for processing older documents.

`\dot{`*x*`}` [m] . . . . (5.3.9) – 127

A dot accent in mathematical formulas: `\dot{a}` = $\dot{a}$.

`\Dot{`*x*`}` [m][a] . . . . (E.3.2) – 412

With the amsmath package, can be used like `\dot`, but with multiple $\mathcal{A}_{\mathcal{M}}\mathcal{S}$-LATEX math accents the positioning will be correct.

`\doteq` [m] produces $\doteq$ . . . . (5.3.4) – 124
`\dotfill` . . . . (3.5.1) – 48

    Fills up the space in a line with a dotted leader: . . . . = `\dotfill`.

`\dots` produces . . . . (5.2.6) – 121
`\dots` [m][a] . . . . (E.3.2) – 412

    With the `amsmath` package, places continuation dots in math mode automatically at a height determined by the following symbol.

`\dotsb` [m][a] produces dots for binary operator: $\cdots$ . . (E.3.2) – 412
`\dotsc` [m][a] produces dots for commas: . . . . (E.3.2) – 412
`\dotsi` [m][a] produces dots for integral signs: $\cdots$ . . . (E.3.2) – 412
`\dotsm` [m][a] produces dots for multiplication: $\cdots$ . . . (E.3.2) – 412
`\doublerulesep` . . . . (4.8.2) – 96

    The distance between double rules inside the `tabular` and `array` environments. New value assigned with `\setlength` outside of the environment.

`\Downarrow` [m] produces $\Downarrow$ . . . . (5.3.5) – 125
`\downarrow` [m] produces $\downarrow$ . . . . (5.3.5) – 125

`\ell` [m] produces ℓ . . . . (5.3.6) – 125
`\em` . . . . (4.1.1) – 57

    This declaration switches to an emphatic font, one that has the current family and series, but with a different shape attribute. It normally toggles between an upright and an italic shape.

$\boxed{2\varepsilon}$ `\emph{`*text*`}` . . . . (4.1.1) – 57

    This command sets its argument in an emphatic font, one that has the current family and series, but with a different shape attribute. It normally toggles between an upright and an italic shape. It is equivalent to `{\em `*text*`\/}`, that is, it includes the italic correction.

`\emptyset` [m] produces $\emptyset$ . . . . (5.3.6) – 125
`\encl{`*enclosures*`}` . . . . (A.1) – 299

    Command in the document class `letter` to add the word 'encl:' with the list *enclosures* at the end of a letter. For International LaTeX, the actual word printed is contained in the command `\enclname`.

`\enclname` . . . . (C.4.1) – 352

    Command in the `letter` document class containing the word to be printed by the `\encl` command. In English, this is 'encl' but may be altered for adaptation to other languages.

`\end{`*environment*`}` . . . . (2.2) – 17

    Command to terminate an environment started with a command `\begin{`*environment*`}`.

[2ε] \enlargethispage{*size*} . . . . (3.5.5) – 52

The \textheight parameter is temporarily increased by the length *size*, in order to improve a bad page break. On the following pages, \textheight will have its normal value once more.

[2ε] \enlargethispage*{*size*} . . . . (3.5.5) – 52

Is the same as \enlargethispage except that any additional spacing between the lines is removed as necessary to maximize the amount of text on the page.

[2ε] \ensuremath{*math cmds*} . . . . (7.3.1) – 194

Sets *math cmds* in math mode; may be called in both text and math modes. Its main use is to define new commands that require math mode but can be called from either mode.

\epsilon [m] produces ϵ . . . . (5.3.1) – 123
\eqref{*marker*} [a] . . . . (E.3.7) – 427

With the amsmath package, is a variation on the \ref command, and prints the equation number defined with \label{*marker*} in parentheses, as (5.6).

\equiv [m] produces $\equiv$ . . . . (5.3.4) – 124
\eta [m] produces η . . . . (5.3.1) – 123
\evensidemargin . . . . (3.2.4) – 35

Sets the left margin for the even-numbered pages. It is effective in the document class book and, when the option twoside has been selected, in the other classes. A new value is assigned with the \setlength command:

        \setlength{\evensidemargin}{2.5cm}

[2ε] \ExecuteOptions{*option_list*} [p] . . . . (C.2.3) – 334

In a class or package file, this command executes all the option definitions in *option_list*. This is normally invoked just prior to \ProcessOptions to establish certain options as default.

\exists [m] produces $\exists$ . . . . (5.3.6) – 125
\exp [m] . . . . (5.3.8) – 126

Command to produce the function name 'exp' in formulas.

\extracolsep{*extra_width*} . . . . (4.8.1) – 94

Tabular command for setting extra spacing between all the following columns in a table. This command is inserted as an @-expression into the column definition field of the tabular environment:

        \begin{tabular}{lr@{\extracolsep{2.5mm}}lcr}

\fbox{*text*} produces a frame around $\boxed{\textit{text}}$ . . . . (4.7.1) – 84
\fboxrule . . . . (4.7.8) – 92

> The line thickness for the frames drawn by \fbox and \framebox commands. A new value is assigned with the \setlength command:
>
>   \setlength{\fboxrule}{1pt}

\fboxsep . . . . (4.7.8) – 92

> The distance between the frame and text in the \fbox and \framebox commands. A new value is assigned with the \setlength command:
>
>   \setlength{\fboxsep}{1mm}

$\boxed{2\varepsilon}$ \fcolorbox *col_spec1 col_spec2*{*text*} . . . . (6.3) – 177

> A command made available with the color package. Like \colorbox, the *text* is set in an LR box with the *col_spec2* as the background color, but with a frame of color *col_spec1* around it. The *col_spec*s are either both defined names or employ the same model. Examples:
>
>   \fcolorbox[rgb]{1,0,0}{0,1,0}{Text}
>   \fcolorbox{red}{green}{Text}

\figurename   . . . . (C.4.1) – 351

> Command containing the name for a figure caption. In English, this is 'Figure' but may be altered for adaptation to other languages.

\fill   . . . . (2.4.2) – 19

> A rubber length with a natural size of zero that can stretch to any size necessary to fill up the horizontal or vertical space available.

\flat [m] produces ♭ . . . . (5.3.6) – 125
\floatpagefraction . . . . (6.4.3) – 182

> The fraction of a float page that must be filled with floats before a new page is called. A new value is assigned with
>
>   \renewcommand{\floatpagefraction}{*decimal_frac*}

\floatsep . . . . (6.4.3) – 182

> The vertical spacing between floats that appear at the top or bottom of a page. A new value is set with the \setlength command:
>
>   \setlength{\floatsep}{12pt plus 2pt minus 4pt}

\flushbottom . . . . (3.2.4) – 34

> A declaration that puts vertical spacing between paragraphs so that the last line on every page is at the same position. Standard for the book document class and for the twoside option.

\fnsymbol{*counter*} . . . . (7.1.4) – 191

> Prints the current value of the given *counter* as a 'footnote symbol':
> * † ‡ § ¶ ‖ ** †† ‡‡

[2ε] \fontencoding{*enc*} . . . . (8.5.1) – 233

This command selects the font encoding scheme. Possible values of *enc* are OT1 for the standard and T1 for the Cork encodings. Other values are also possible.

[2ε] \fontfamily{*fam*} . . . . (8.5.1) – 233

This command selects the 'family' of fonts. Possible values of *fam* for standard LaTeX with the Computer Modern fonts are cmr, cmss, cmtt, and cmfi.

[2ε] \fontseries{*ser*} . . . . (8.5.1) – 233

This command selects the 'series' of fonts within a 'family'. Possible values of *ser* for standard LaTeX are m (medium) and bx (bold extended).

[2ε] \fontshape{*form*} . . . . (8.5.1) – 233

This command selects the 'shape' of fonts. Possible values of *form* are n (normal), it (italic), sl (slanted), sc (small caps), and u ('unslanted' italic).

[2ε] \fontsize{*sz*}{*line_sp*} . . . . (8.5.1) – 233

This command selects the font size. The argument *sz* specifies the size of the characters in points (without the dimension pt) and *line_sp* determines the value of the interline spacing (\baselineskip), with an explicit dimension. Example: \fontsize{12}{14pt}.

\footnote[*num*]{*footnote_text*} . . . . . (4.10.1), (4.10.2) – 110, 111

Produces a footnote containing the text *footnote_tex*. The optional argument *num* will be used as the footnote number in place of the next number in the automatic sequence.

\footnotemark[*num*] . . . . (4.10.3) – 111

Produces a footnote marker in the current text. The optional argument *num* will be used as the footnote number in place of the next number in the automatic sequence. May be used within structures where \footnote is not normally permitted, such as LR boxes, tables, math formulas.

\footnoterule . . . . (4.10.6) – 115

This is an internal command to produce the horizontal rule between the regular text on a page and the footnote text at the bottom. May be changed with

    \renewcommand{\footnoterule}
    {\rule{*wdth*}{*hght*}\vspace{-*hght*}}

\footnotesep . . . . (4.10.6) – 115

The vertical spacing between two footnotes. A new value is assigned with the \setlength command:

    \setlength{\footnotesep}{6.5pt}

Switches to the font size \footnotesize, which is smaller than \small but larger than \scriptsize.

Produces a footnote with the text *footnote_text* but without generating a marker in the current text. The marker that is used for the footnote itself at the bottom of the page derives from the current value of the counter footnote, which remains unchanged, or from the value of the optional argument *num*. This command may be used together with the \footnotemark command to insert footnotes into structures where they are otherwise not allowed, such as LR boxes, tables, and math formulas. The \footnotetext command must be given *outside* of that structure.

The distance from the bottom edge of the text body to the lower edge of the footline. A value is assigned with the \setlength command:

    \setlength{\footskip}{25pt}

In the babel system, sets a short *text* in the selected *language*.

Math command for generating fractions.

Produces a frame without any intervening spacing around text. Mainly used as a picture element in a \put or \multiput command within the picture environment.

Produces a frame of width *width* around *text*. By default, the text is centered within the frame, but may be left or right justified by giving the optional argument *pos* as l or r. In LATEX 2ε, it may also have the value s, to stretch the text to the given width.

Picture element command to produce a frame of width *x_dimen* and height *y_dimen* within the picture environment. Without the optional argument *pos*, the text is centered vertically and horizontally. The text may be left or right justified, and/or aligned at the top or bottom, by setting *pos* to a combination of the letters l, r, t, and b, such as tr for top, right. In LATEX 2ε, *pos* may also contain s to stretch the text to the full width. The command is to be used as the argument of a \put or \multiput command.

\frenchspacing . . . . (3.5.1) – 45

After this command has been given, no additional horizontal spacing is inserted at the end of a sentence. The countercommand is \nonfrenchspacing.

2ε \frontmatter . . . . (3.3.5) – 42

In the book class, introduces the material that comes at the beginning (preface, table of contents) by turning off the chapter numbering of the \chapter command and switching to Roman numbers for the pagination.

\frown [m] produces ⌢ . . . . (5.3.4) – 124

\fussy . . . . (3.6.3) – 55

Countercommand of \sloppy that allows larger interword spacings than normal. After \fussy has been given, the normal spacings apply once more.

\Gamma [m] produces Γ . . . . (5.3.1) – 123

\gamma [m] produces y . . . . (5.3.1) – 123

\gcd [m] . . . . (5.3.8) – 126

Command to produce the function name 'gcd' in formulas. A lower limit may be given as a subscript.

\ge [m] produces ≥ . . . . (5.3.4) – 124

\genfrac{*left*}{*right*}{*thkns*}{*mathsz*}{*over*}{*under*} [m][a]
. . . . (E.3.3) – 413

With the amsmath package, produces a generalized fraction with delimiters *left* and *right*, line thickness *thkns*, math font size *mathsz* 0-3, and with *over* on top of *under*. If *mathsz* is empty, the sizing is automatic.

\geq [m] produces ≥ . . . . (5.3.4) – 124

\gets [m] produces ← . . . . (5.3.5) – 125

\gg [m] produces ≫ . . . . (5.3.4) – 124

\glossary{*glossary_entry*} . . . . (8.3.5) – 229

Write a \glossaryentry command to the .glo file if \makeglossary has been issued in the preamble; else it does nothing.

\glossaryentry{*glossary_entry*}{*page_number*} . . . . (8.3.5) – 229

The form in which the entry is written to the .glo file by the \glossary command.

2ε \graphpaper[*num*](*x*,*y*)(*lx*,*ly*) . . . . (6.1.6) – 165

A command added with the graphpap package for use in the picture environment. It plots a labeled grid system with the lower left corner at (x, y), *lx* wide and *ly* high. Grid lines are drawn every *num* units, with the fifth ones thicker. If *num* is not specified, it is assumed to be 10. All arguments must be integers, not decimal fractions.

\grave{*x*} [m] . . . . (5.3.9) – 127

  A grave accent over the math variable *x*: \grave{a} = *à*.

\Grave{*x*} [m][a] . . . . (E.3.2) – 412

  With the amsmath package, can be used like \grave, but with multiple
  $\mathcal{A}_{\mathcal{M}}\mathcal{S}$-LaTeX math accents the positioning will be correct.

[2ε] \guillemotleft . . . . (F.4.5) – 452

  When T1 encoding is active, prints the symbol «. *(1994/12/01)*

[2ε] \guillemotright . . . . (F.4.5) – 452

  When T1 encoding is active, prints the symbol ». *(1994/12/01)*

[2ε] \guilsinglleft . . . . (F.4.5) – 452

  When T1 encoding is active, prints the symbol ‹. *(1994/12/01)*

[2ε] \guilsinglright . . . . (F.4.5) – 452

  When T1 encoding is active, prints the symbol ›. *(1994/12/01)*

\H{*x*} . . . . (2.5.7) – 21

  Hungarian double acute accent: \H{o} = ő.

\hat{*x*} [m] . . . . (5.3.9) – 127

  Circumflex over the math variable *x*: \hat{a} = *â*.

\Hat{*x*} [m][a] . . . . (E.3.2) – 412

  With the amsmath package, can be used like \hat, but with multiple
  $\mathcal{A}_{\mathcal{M}}\mathcal{S}$-LaTeX math accents the positioning will be correct.

\hbar [m] produces $\hbar$ . . . . (5.3.6) – 125

\headheight . . . . (3.2.4) – 35

  The height of the head at the top of each page. A new value is
  assigned with the \setlength command:

      \setlength{\headheight}{25pt}

\headsep . . . . (3.2.4) – 35

  Vertical spacing between the lower edge of the page head and the
  top of the main text. A new value is assigned with the \setlength
  command:

      \setlength{\headsep}{0.25in}

\headtoname . . . . (C.4.1) – 352

  Command in the letter document class containing the text that
  precedes the recipient's name in the headline after the first page.
  In English, this is 'To' but may be altered for adaptation to other
  languages.

\heartsuit [m] produces ♥ . . . . (5.3.6) – 125

2ε \height . . . . (4.7.1) – 85

A length parameter equal to the natural height of a box (distance from the baseline to the top); it may only be used in the *width* specification of \makebox, \framebox, or \savebox, or in the *height* specification of a \parbox or a minipage environment.

> \framebox[6\height]{text}

\hfill . . . . (3.5.1) – 47

A horizontal rubber spacing with a natural length of zero that can be stretched to any value. Used to fill up a horizontal line with blank spacing. This command is an abbreviation for \hspace{\fill}.

\hline . . . . (4.8.1) – 95

Produces a horizontal line in the array and tabular environments over the width of the entire table.

\hoffset 555, 556

Horizontal offset of the output page from the printer border set by the printer driver. This printer border is normally 1 inch from the left edge of the paper. The standard value of \hoffset is 0 pt so that the left reference margin of the page is identical with the printer margin. A new value is assigned with the \setlength command:

> \setlength{\hoffset}{-1in}

\hom [m] . . . . (5.3.8) – 126

Command to produce the function name 'hom' in formulas.

\hookleftarrow [m] produces ↩ . . . . (5.3.5) – 125
\hookrightarrow [m] produces ↪ . . . . (5.3.5) – 125
\hrulefill . . . . (3.5.1) – 48

Fills up the space in a line with a rule: _____= \hrulefill.

\hspace{*width*} . . . . (3.5.1) – 47

Produces horizontal spacing of length *width*. It is ignored if it occurs at the beginning or end of a line.

\hspace*{*width*} . . . . (3.5.1) – 47

Produces horizontal spacing of length *width* even at the beginning or end of a line.

\huge . . . . (4.1.2) – 58

Switches to the font size \huge, which is smaller than \Huge but larger than \LARGE.

\Huge . . . . (4.1.2) – 58

Switches to the largest font size available \Huge, which is larger than \huge.

\hyphenation{*hyphenation_list*} [p] . . . . (3.6.2) – 54

> Sets up a list of hyphenation exceptions. The *hyphenation_list* consists of a collection of words containing hyphens at the places where word division may occur: hy-phen-a-tion per-mit-ted.

\i produces ı . . . . (2.5.7) – 22

\idotsint [m][a] produces ∫ ··· ∫ . . . . (E.3.2) – 409

\iff [m] produces ⟺ . . . . (5.3.5) – 125

2ε \IfFileExists{*file_name*}{*true*}{*false*} . . . . (C.2.8) – 338

> Tests if the file *file_name* can be found in the places where LATEX looks for files; if so, the code *true* is executed; otherwise, *false*. This is like \InputIfFileExists except the file is not input.

\iflanguage{*language*}{*yes_text*}{*no_text*} . . . . (D.1.3) – 372

> In the multilingual babel system, tests if *language* is the currently selected language and, if so, executes *yes_text*, otherwise *no_text*.

\ifthenelse{*test*}{*then_text*}{*else_text*} . . . . (7.3.5) – 201

> A conditional command available when the standard package ifthen has been loaded. If the logical statement *test* evaluates to ⟨*true*⟩ then *then_text* is inserted, otherwise *else_text*. The logical statement may be relational (two numbers with one of < = > between them), an even–odd test (\isodd{*number*}), a comparison of two texts (\equal{*text1*}{*text2*}), a comparison of two lengths (\lengthtest{*length1* op *length2*}, *op* is one of < = >), or a test of a boolean switch (\boolean{*switch*}). Switches are created with \newboolean{*switch*} and set with \setboolean{*switch*}{*value*}, where *value* is true or false. Logical statements may be combined with logical operators \and, \or, and \not, and grouped with \(and \).
>
> The tests \isodd, \lengthtest, and \boolean are new to LATEX 2ε.

\Im [m] produces ℑ . . . . (5.3.6) – 125

\imath [m] produces ı . . . . (5.3.6) – 125

\in [m] produces ∈ . . . . (5.3.4) – 124

\include{*file*} . . . . (8.1.2) – 219

> Inserts the contents of the file with the root name *file* and extension .tex into the current text at the point where the command appears. A new page is always started! Together with \includeonly, this command allows portions of the document to be processed as though the rest of the text were present.

2ε \includegraphics[*llx,lly*][*urx,ury*]{*file_name*} . . . . . (6.2.1) – 168

> A command made available with the graphics package that imports external graphics stored in the file *file_name*. The coordinates of the bounding box are given by *llx, lly* (lower left corner) and by *urx, ury*

(upper right corner). It is the contents of this bounding box that are used for further manipulation: scaling, rotating. It is also the (manipulated) bounding box that is used to reserve space in the text; any graphics that extend beyond the limits of this box will also be printed, but overlapping any other material that may be beside it.

The bounding box coordinates may have units attached to them; the default units are big points bp (72 per inch).

If the bounding box coordinates are omitted, the information is obtained in some other manner, depending on the type of graphics file. For an encapsulated PostScript file, this information is taken from the graphics file itself.

If *llx* and *lly* are not specified, they are assumed to be 0. That is, if only one set of optional coordinates are given, they refer to the upper right corner.

[2ε] \includegraphics*[*llx,lly*] [*urx,ury*] {*file_name*} . . . . (6.2.1) – 168

The same as \includegraphics except that any graphics extending beyond the bounding box are *not* included: the figure is clipped.

[2ε] \includegraphics[*key=value*, ...]{*file_name*} . . . . (6.2.1) – 170

With the graphicx package, this command has a different syntax in which the scaling, rotating, clipping are effected through *key=value* pairs, like width=7cm, angle=90, scale=.5.

\includeonly{*file_list*} [p] . . . . (8.1.2) – 219

Only those files whose names are in *file_list*, separated by commas, will be read in by the \include commands. The \include commands for other file names are ignored. Nevertheless, all the auxiliary files are read in so that the page and section numbers will be correct, as are the cross-reference markers.

\indent . . . . (3.5.4) – 50

The first line of the *next* paragraph is to be indented.

\index{*index_entry*} . . . . (8.3.4), (8.4) – 228, 230

Writes a \indexentry command to the .idx file if the \makeindex command has been issued in the preamble; otherwise it does nothing. The MakeIndex program (Section 8.4) can process this file if the entries are in the forms

\index{*main_entry*}
\index{*main_entry*!*sub_entry*}
\index{*main_entry*!*sub_entry*!*sub_sub_entry*}

making up a theindex environment with the entries alphabetically ordered and organized with \item, \subitem, and \subsubitem commands.

\indexentry{*index_entry*}{*page_number*} . . . . (8.3.4) – 228

> The form in which the entry is written to the .idx file by the \index command.

\indexname . . . . (C.4.1) – 351

> Command containing the heading for the index. In English, this is 'Index' but may be altered for adaptation to other languages.

\indexspace . . . . (8.3.4) – 228

> Command within theindex environment to produce a blank line.

\inf [m] . . . . (5.3.8) – 126

> Command to produce the function name 'inf' in formulas. A lower limit may be set as a subscript.

\infty [m] produces ∞ . . . . (5.3.6) – 125
\intertext{*insert_text*} [m][a] . . . . (E.3.1) – 408

> When the package amsmath is loaded, this command inserts text as a left-justified line between lines of an equation, without affecting their horizontal alignment.

\input{*file*} . . . . (8.1.1) – 217

> Inserts the contents of the file with the root name *file* and extension .tex into the current text at the point where the command appears. The file that is read in may also contain further \input commands.

$\boxed{2\varepsilon}$ \InputIfFileExists{*file_name*}{*true*}{*false*} . . . . . (C.2.8) – 338

> Tests if the file *file_name* can be found in the places where LaTeX looks for files; if so, the code *true* is executed and the file is input; otherwise, *false* is executed.

\int [m] produces $\int$ . . . . (5.2.5) – 121
\iint [m][a] produces $\iint$ . . . . (E.3.2) – 409
\iiint [m][a] produces $\iiint$ . . . . (E.3.2) – 409
\iiiint [m][a] produces $\iiiint$ . . . . (E.3.2) – 409
\intextsep . . . . (6.4.3) – 182

> The vertical spacing between floats in the middle of a page and the surrounding text. A new value is assigned with the \setlength command:

> > \setlength{\intextsep}{10pt plus2pt minus3pt}

\invisible . . . . (8.10.2) – 251

> In slides class, a declaration that makes the following text be printed in 'invisible ink', that is, it takes up as much space as though it were there. It remains in effect until the end of the environment, or end of the curly braces, in which it was issued, or until \visible is given. It is used for making overlays.

\iota [m] produces ι . . . . (5.3.1) – 123

2.09 \it . . . . (4.1.5) – 62

Switches to the *Roman, italic, medium* typeface.

2ε \itdefault   . . . . (C.5.1) – 353

This command defines the shape attribute that is selected with the
\itshape command. It may be redefined with \renewcommand:

    \renewcommand{\itdefault}{it}

\item[*label*] . . . . (4.3), (4.4.1) – 66, 73

Produces a label and the start of an item text in a list environment.
Without the optional argument, the label is generated according to
the type of environment, for example numbers for the enumerate
environment. The optional argument inserts *label* in place of this
standard item label.

\item{*entry*} . . . . (8.3.4) – 227

Produces a main entry in the index environment.

\itemindent . . . . (4.4.2) – 76

The amount by which the label and the text of the first line after each
\item is indented in a list environment. The standard value is 0 pt,
but a new value may be assigned with the \setlength command:

    \setlength{\itemindent}{1em}

\itemsep . . . . (4.4.2) – 74

The amount of vertical spacing in addition to \parsep that is inserted
between the \item texts in a list environment. A new value may be
assigned with the \setlength command:

    \setlength{\itemsep}{2pt plus1pt minus1pt}

2ε \itshape . . . . (4.1.3), (8.5.2) – 60, 237

This declaration switches to a font in the current family and series,
but with the *italic* shape attribute.

\j produces $\jmath$ . . . . (2.5.7) – 22

\jmath [m] produces $\jmath$ . . . . (5.3.6) – 125

\Join [m] produces $\bowtie$ . . . . (5.3.6) – 125

\jot . . . . (5.5.4) – 147

The amount of vertical spacing between the formula lines of an
eqnarray or eqnarray* environment. Standard value is 3 pt. A new
value may be assigned with the \setlength command:

    \setlength{\jot}{4.5pt}

2ε \k{*x*} . . . . (F.4.5) – 452

>When T1 encoding is active, prints the ogonek accent \k{A} = Ą.
>*(1994/12/01)*

\kappa [m] produces κ . . . . (5.3.1) – 123
\ker [m] . . . . (5.3.8) – 126

>Command to produce the function name 'ker' in formulas.

\kill . . . . (4.6.2) – 81

>Removes the preceding sample line in a tabbing environment that
>was given only to set the tabs and not to be printed at this point.

\L produces Ł . . . . (2.5.6) – 21
\l produces ł . . . . (2.5.6) – 21
\label{*marker*} . . . . (8.3.1) – 224

>Sets a marker in the text at this position with the name *marker*.
>It may be referred to either earlier or later in the document with
>the command \ref{*marker*} to output the counter that was then
>current, such as the section, equation, or figure number, or with the
>command \pageref{*marker*} to print the page number where the
>marker was set.

\labelenum*n* . . . . (4.3.5) – 69

>A set of commands to produce the standard labels for the nesting
>levels of the enumerate environments, where *n* is one of i, ii, iii,
>or iv. For example,
>
>>\renewcommand{\labelenumii}{\arabic{enumii}.)}
>
>changes the standard labels of the second-level enumerate environ-
>ment to be 1.), 2.), etc.

\labelitem*n* . . . . (4.3.5) – 69

>A set of commands to produce the standard labels for the nesting
>levels of the itemize environments, where *n* is one of i, ii, iii, or
>iv. For example,
>
>>\renewcommand{\labelitemi}{$\Rightarrow$}
>
>changes the standard labels of the outermost itemize environment
>to ⇒.

\labelsep . . . . (4.4.2) – 75

>In a list environment, the distance between the label box and the
>list text. A new value is assigned with the \setlength command:
>
>>\setlength{\labelsep}{5pt}

\labelwidth . . . . (4.4.2) – 75

>In a list environment, the width of the box reserved for the label. A
>new value is assigned with the \setlength command:
>
>>\setlength{labelwidth}{2.2cm}

In TeX versions 3.0 and later, the set of hyphenation patterns number *num* is made active. The patterns must be previously loaded into the format file by an `initex` run in which \language{*num*} was given before those patterns were read in.

Switches to the font size \large, which is smaller than \Large but larger than \normalsize.

Switches to the font size \Large, which is smaller than \LARGE but larger than \large.

Switches to the font size \LARGE, which is smaller than \huge but larger than \Large.

Adjusts the size of the bracket symbol *lbrack* to fit the height of the formula between the \left ... \right pair. For example, \left[. If there is to be no matching bracket, the \left and \right commands must still be given to specify the part of the formula to be sized, but the missing bracket is given as a period (for example, \right.).

A command inside the `eqnarray` environment that outputs its argument as though it had zero width, thus having no effect on the column widths. It is used mainly for the first row of a multi-row formula.

`\leftmargin` . . . . (4.4.2) – 74

In a `list` environment, the amount by which the left edge of the text is indented relative to the surrounding text. A new value is assigned with the `\setlength` command. For nested `list` environments, different values for the indentation can be specified by adding i ... vi to the declaration name, such as

        `\setlength{\leftmarginiii}{0.5cm}`

See also   . . . . (4.3.4) – 67.

`\Leftrightarrow` [m] produces ⇔ . . . . (5.3.5) – 125
`\leftrightarrow` [m] produces ↔ . . . . (5.3.5) – 125
`\leftroot`{*shift*} [m][a] . . . . (E.3.5) – 417

With the `amsmath` package, used in the index to a `\sqrt` command to shift it slightly to the left. The *shift* is a number specifying how many units to move it. Example:

        `\sqrt[\leftroot{-1}\uproot{3}\beta]{k}`

`\leq` [m] produces ≤ . . . . (5.3.4) – 124
`\lfloor` [m] produces ⌊ . . . . (5.4.1) – 130
`\lg` [m] . . . . (5.3.8) – 126

Command to produce the function name 'lg' in formulas.

`\lhd` [m] produces ◁ . . . . (5.3.3) – 124
`\lim` [m] . . . . (5.3.8) – 126

Command to produce the function name 'lim' in formulas. A lower limit may be set as a subscript.

`\liminf` [m] . . . . (5.3.8) – 126

Command to produce the function name 'lim inf' in formulas. A lower limit may be set as a subscript.

`\limits` [m] . . . . (5.2.5), (5.3.7) – 121, 126

Places the upper and lower limits above and below the appropriate symbols where these would normally go just after them.

`\limsup` [m] . . . . (5.3.8) – 126

Command to produce the function name 'lim sup' in formulas. A lower limit may be set as a subscript.

`\line`($\Delta x, \Delta y$){*length*}   . . . . (6.1.4) – 157

A picture element command within a `picture` environment for drawing horizontal and vertical lines of any length as well as slanted lines at a limited number of angles. For horizontal and vertical lines, the *length* argument is the actual length in units of `\unitlength`. For slanted lines, *length* is the length of the projection on to the x-axis (horizontal displacement). The slope is determined by the $(\Delta x, \Delta y)$ arguments, which take on integral values such that $-6 \leq \Delta x \leq 6$ and $-6 \leq \Delta y \leq 6$. This command is the argument of a `\put` or `\multiput` command.

`\linebreak[n]` . . . . (3.5.2) – 49

A recommendation to break the line of text at this point such that
it fills the horizontal space available (left and right justified). The
urgency of the recommendation is given by the integral number
n between 0 and 4, with the higher numbers meaning a stronger
recommendation. A value of 4 is the same as the command without
the optional argument and means an obligatory line break.

`\linethickness{`*thickness*`}` . . . . (6.1.4) – 164

Sets the thickness of the horizontal and vertical lines in the `picture`
environment. The argument *thickness* is a length specification with
units, for example, `1.2mm`.

`\listfigurename` . . . . (C.4.1) – 351

Command containing the heading for the list of figures. In English,
this is 'List of Figures' but may be altered for adaptation to other
languages.

[2ε] `\listfiles [p]` . . . . (C.2.9) – 339

When given in the preamble, causes a list of all files read in during
the processing to be printed to the monitor and to the transcript file
at the end of the run. The list includes version number, date, and
any additional information entered with one of the `\Provides...`
commands.

`\listoffigures` . . . . (3.4.4) – 44

Produces a list of figures with the entries from all the `\caption`
commands in `figure` environments.

`\listoftables` . . . . (3.4.4) – 44

Produces a list of tables with the entries from all the `\caption`
commands in `table` environments.

`\listparindent` . . . . (4.4.2) – 75

Depth of indentation for the first line of a paragraph inside a `list`
environment. A new value may be assigned with the `\setlength`
command:

    \setlength{\listparindent}{1em}

`\listtablename` . . . . (C.4.1) – 351

Command containing the heading for the list of tables. In English,
this is 'List of Tables' but may be altered for adaptation to other
languages.

`\ll [m]` produces ≪ . . . . (5.3.4) – 124
`\ln [m]`. . . . . (5.3.8) – 126

Command to produce the function name 'ln' in formulas.

$\boxed{2_\varepsilon}$ \LoadClass[*options*]{*class*}[*version*] [p] . . . . (C.2.2) – 333

> This command may only be invoked within a class file to load another class file. It may only be called once within any class file. The file loaded must have the extension .cls. Any options specified in the \documentclass command are *not* passed over as global options.
>
> The optional *version* is a date, given in the form yyyy/mm/dd, as for example 1994/08/01. If the date of the class file is earlier than this, a warning message is printed.

$\boxed{2_\varepsilon}$ \LoadClassWithOptions{*class*}[*version*] [p] . . . . . . (C.2.2) – 333

> Like \LoadClass except all the currently specified options are automatically passed to *class*. *(1995/12/01)*

\location{*number*} . . . . (A.1) – 298

> In the letter document class, enters the sender's room number. In the standard LaTeX letter class, *number* is only output if \address has not been called. It is intended to be used in company letterheads.

\log [m] . . . . (5.3.8) – 126

> Command to produce the function name 'log' in formulas.

\Longleftarrow [m] produces $\Leftarrow$ . . . . (5.3.5) – 125
\longleftarrow [m] produces $\leftarrow$ . . . . (5.3.5) – 125
\Longleftrightarrow [m] produces $\Leftrightarrow$ . . . . (5.3.5) – 125
\longleftrightarrow [m] produces $\longleftrightarrow$ . . . . (5.3.5) – 125
\longmapsto[m] produces $\longmapsto$ . . . . (5.3.5) – 125
\Longrightarrow [m] produces $\Rightarrow$ . . . . (5.3.5) – 125
\longrightarrow [m] produces $\rightarrow$ . . . . (5.3.5) – 125
\lq produces ', identical to the ' symbol.
\lvert [m][a] produces | (left delimiter) . . . . (E.3.5) – 418
\lVert [m][a] produces ‖ (left delimiter) . . . . (E.3.5) – 418

$\boxed{2_\varepsilon}$ \mainmatter . . . . (3.3.5) – 42

> In the book class, introduces the main body of text after the front matter, by resetting the page numbering to 1 with Arabic numbers, and by reactivating the chapter numbering with the \chapter command. It undoes the effects of \frontmatter.

\makebox[*width*][*pos*]{*text*} . . . . (4.7.1) – 84

> Produces a box of width *width* containing *text* centered horizontally, unless *pos* is given to specify that it is to be left (l) or right (r) justified. In LaTeX 2_ε, it may also have the value s, to stretch the text to *width*.

\makebox(*x_dimen,y_dimen*)[*pos*]{*text*} . . . . (6.1.4) – 155

Picture element command to produce a box of width *x_dimen* and height *y_dimen* within the picture environment. Without the optional argument *pos*, the text is centered vertically and horizontally. The text may be left or right justified, and/or aligned at the top or bottom, by setting *pos* to a combination of the letters l, r, t, and b, such as tr for top, right. In LaTeX 2_ε, *pos* may also contain s to stretch the text to the full width. The command is used as the argument of a \put or \multiput command.

\makeglossary [p] . . . . (8.3.5) – 229

Command to activate the \glossary commands in the text.

\makeindex [p] . . . . (8.3.4) – 228

Command to activate the \index commands in the text.

\makelabel{*text*} . . . . (4.4.1), (7.5.9) – 73, 213

An internal command that is called by the \item command to produce the actual label *text* within a list environment.

\makelabels . . . . (A.1) – 301

Produces address labels in the letter document class using the entries from the \begin{letter} environment.

[2ε] \MakeLowercase{*text_cmd*} . . . . (C.3.2) – 346

Converts *text_cmd* (text and commands) to lower case. *(1995/06/01)*

[2ε] \MakeShortVerb{\c} . . . . (4.9.1, 8.8.3) – 109, 242

When the standard package shortvrb has been loaded, this command makes the character *c* a shorthand form for \verbc: everything that appears between two occurrences of *c* is printed literally, in typewriter type. With \DeleteShortVerb{\c}, the character is restored to its normal meaning. Example: \MakeShortVerb{\|}

\maketitle . . . . (3.3.1) – 38

Produces a title page using entries in the \author and \title commands, and optionally those in the \date and \thanks commands.

[2ε] \MakeUppercase{*text_cmd*} . . . . (C.3.2) – 346

Converts *text_cmd* (text and commands) to upper case. *(1995/06/01)*

\mapsto [m] produces ↦ . . . . (5.3.5) – 125
\marginpar[*left_text*]{*right_text*} . . . . (4.10.5) – 114

Produces a marginal note at the right of the text containing *right_text*. With two-sided formatting, the marginal note goes into the left margin on the even pages, in which case the optional *left_text* will be written instead. For two-column text, the marginal notes always go into the 'outer' margin, and again *left_text* will be used for the left margin.

\marginparpush . . . . (4.10.6) – 115

> The minimum vertical separation between two marginal notes. A new value may be assigned with the \setlength command.

\marginparsep . . . . (4.10.6) – 115

> The spacing between the edge of the text and a marginal note. A new value may be assigned with the \setlength command.

\marginparwidth . . . . (4.10.6) – 115

> The width of the box reserved for marginal notes. A new value may be assigned with the \setlength command.

\markboth{*left_head*}{*right_head*} . . . . (3.2.1) – 31

> Sets the text entries for the left and right page headlines in two-sided formatting when the page style myheadings has been selected, or when the automatic entries of page style headings are to be changed.

\markright{*head*} . . . . (3.2.1) – 31

> Sets the text entry for the page headline when the page style myheadings has been selected, or when the automatic entry of page style headings is to be manually changed. In two-sided formatting, only the right headline is set with this command.

[2ε] \mathbf{*text*} [m] . . . . (5.4.2), (C.5.3) – 131, 355

> This command sets *text* in a bold font (\bfseries) within math mode. Spaces are ignored as usual.

[2ε] \mathcal{*text*} [m] . . . . (5.3.2), (5.4.2), (C.5.3) – 123, 131, 355

> This command sets *text* in calligraphic letters within math mode. It replaces the \cal declaration in LATEX 2.09.

\mathindent . . . . (3.1.1) – 28

> The indentation of displayed formulas from the left margin when the option fleqn has been selected. A new value may be assigned with the \setlength command:
>
>   \setlength{\mathindent}{25pt}

[2ε] \mathit{*text*} [m] . . . . (5.4.2), (C.5.3) – 131, 355

> This command sets *text* in a text italic font (\itshape) within math mode. It differs from \mathnormal in that the spacing between the letters is as in regular text. Compare *mathit* and *mathnormal*.

[2ε] \mathnormal{*text*} [m] . . . . . (5.3.1), (5.4.2), (C.5.3) – 123, 131, 355

> This command sets *text* in the normal (italic) math font within math mode. It replaces the \mit declaration in LATEX 2.09. In this font, capital Greek letters are also set in italics: $\Gamma\mathnormal{\Gamma}$ = $\Gamma\Gamma$.

$\boxed{2_\varepsilon}$ \mathring{x} [m] . . . . (5.3.9) – 127

A ring accent in mathematical formulas: \mathring{a} = å.
(1998/06/01)

$\boxed{2_\varepsilon}$ \mathrm{*text*} [m] . . . . (5.4.2), (C.5.3) – 131, 355

This command sets *text* in a Roman font (\rmfamily) within math mode. Spaces are ignored as usual.

$\boxed{2_\varepsilon}$ \mathsf{*text*} [m] . . . . (5.4.2), (C.5.3) – 131, 355

This command sets *text* in a sans serif font (\sffamily) within math mode. Spaces are ignored as usual.

$\boxed{2_\varepsilon}$ \mathtt{*text*} [m] . . . . (5.4.2), (C.5.3) – 131, 355

This command sets *text* in a typewriter font (\ttfamily) within math mode. Spaces are ignored as usual.

$\boxed{2_\varepsilon}$ \mathversion{*ver*} . . . . (C.5.3) – 354

Selects the current math version. Possible values are bold and normal; new values may be created with the \DeclareMathVersion command.

\max [m] . . . . (5.3.8) – 126

Command to produce the function name 'max' in formulas. A lower limit may be set as a subscript.

\mbox{*text*} produces an LR box around *text* . . . . (4.7.1) – 84

$\boxed{2_\varepsilon}$ \mddefault . . . . (C.5.1) – 353

This command defines the series attribute that is selected with the \mdseries command. It may be redefined with \renewcommand:

        \renewcommand{\mddefault}{m}

$\boxed{2_\varepsilon}$ \mdseries . . . . (4.1.3), (8.5.2) – 60, 236

This declaration switches to a font in the current family and shape, **but with the** medium **series attribute.**

\medskip . . . . (3.5.3) – 50

Inserts large vertical spacing of the amount \medskipamount. See also \bigskip and \smallskip.

\medskipamount

Standard value for the amount of vertical spacing that is inserted with the command \medskip. May be changed with the \setlength command:

        \setlength{\medskipamount}{3ex plus1ex minus1ex}

\medspace [m][a] . . . . (E.3.5) – 418

> With the amsmath package, this is an alias for \:, a medium space in a math formula.

[2ε] \MessageBreak . . . . (C.2.7) – 337

> Forces a new line in the texts of error, warning, and information messages. These are the only places where it may be invoked, otherwise it does nothing.

\mho [m] produces ℧ . . . . (5.3.6) – 125
\mid [m] produces | . . . . (5.3.4) – 124
\min [m] . . . . (5.3.8) – 126

> Command to produce the function name 'min' in formulas. A lower limit may be set as a subscript.

[2.09] \mit [m] . . . . (5.3.1) – 123

> A declaration in LaTeX 2.09 that switches to the font style 'math italic', the standard in math mode. In LaTeX 2_ε, it has been replaced by the command \mathnormal.

\mod{*arg*} [m][a] . . . . (E.3.5) – 417

> With the amsopn or amsmath packages, command to produce the function name 'mod' in formulas in the form:
>
> y\mod{a+b} = $y \mod a + b$

\models [m] produces ⊨ . . . . (5.3.4) – 124
\mp [m] produces ∓ . . . . (5.3.3) – 124
\mspace{*mu*} [m][a] . . . . (E.3.5) – 418

> With the amsmath package, inserts spacing in math formulas; *mu* is a math space in units of mu (=1/18 em): \mspace{-4mu}.

\mu [m] produces μ . . . . (5.3.1) – 123
\multicolumn{*n*}{*col*}{*text*} . . . . (4.8.1) – 95

> Merges the next *n* columns in the array and tabular environments, formatting the text entry *text* according to the single column definition *col*, which may be l, c, r as well as |.

\multiput(*x*, *y*)(Δ*x*, Δ*y*){*n*}{*pic_elem*} . . . . (6.1.3) – 153

> Multiple positioning command in the picture environment. The object *pic_elem* is placed *n* times, at (x, y), $(x + \Delta x, y + \Delta y)$, ... $(x + (n - 1)\Delta x, y + (n - 1)\Delta y)$.

\multlinegap [m][a] . . . . (E.3.6) – 420

> A length that determines the left and right margins of formulas produced with the $\mathcal{A}_{\mathcal{M}}S$-LaTeX multline environment; initial value is 10 pt but may be reset by the user.

\nabla [m] produces ∇ . . . . (5.3.6) – 125
\name{*sender*} . . . . (A.2) – 302

> In the letter document class, enters the sender's name.

\natural [m] produces $\natural$ . . . . (5.3.6) – 125
\nearrow [m] produces $\nearrow$ . . . . (5.3.5) – 125
$\boxed{2\varepsilon}$ \NeedsTeXFormat{*format*}[*version*] [p] . . . . (C.2.1) – 332

> Declares the TeX format that is necessary for processing the file. This should be the first statement in the file. At the moment, the only legitimate value for *format* is LaTeX2e. The *version*, if included, must be given as a date in the form yyyy/mm/dd, specifying the earliest possible release date of the format that is consistent with all the features employed in the file. Example:
>
> \NeedsTeXFormat{LaTeX2e}[1994/06/01]

\neg [m] produces $\neg$ . . . . (5.3.6) – 125
\negmedspace [m][a] . . . . (E.3.5) – 418

> With the amsmath package, this inserts a negative medium space in a math formula.

\negthickspace [m][a] . . . . (E.3.5) – 418

> With the amsmath package, this inserts a negative thick space in a math formula.

\negthinspace [m][a] . . . . (E.3.5) – 418

> With the amsmath package, this is an alias for \!, a negative thin space in a math formula.

\neq [m] produces $\neq$ . . . . (5.3.4) – 125
$\boxed{2\varepsilon}$ \newboolean{*switch*} . . . . (7.3.5) – 203

> Creates a new boolean switch. The value of the switch is set with \setboolean{*switch*}{*value*}, where *value* is true or false. Its value is tested with \boolean{*switch*}, which may be used as a logical statement in the *test* part of \ifthenelse and \whiledo. Requires the standard LaTeX package ifthen.

\newcommand{*com_name*}[*narg*][*opt*]{*def*} . . . . (7.3) – 193

> Defines a user command with the name *com_name* to be *def*. The first optional argument *narg* $\leq$ 9 specifies how many variable arguments the command is to have, which appear in the *def* as the replacement characters #1 to #*narg*. If the second optional argument is present (not available with LaTeX 2.09), the first argument of the new command is optional, and takes on the value *opt* if it is not explicitly given.

$\boxed{2\varepsilon}$ \newcommand*{*com_name*}[*narg*][*opt*]{*def*} . . . . . (C.2.6) – 336

> The same as \newcommand except that the arguments to *com_name* must be 'short', not containing any new paragraphs. *(1994/12/01)*

\newcounter{*counter_name*}[*in_counter*] . . . . (7.1.2) – 190

>   Establishes a new counter with the name *counter_name*. The optional argument *in_counter* is the name of an existing counter which, when incremented, resets the new counter to zero; that is, the new counter is a sub-counter of *in_counter*.

\newenvironment{*env_name*}[*narg*][*opt*]{*beg_def*}{*end_def*}
. . . . (7.4) – 204

>   Defines a user environment with the name *env_name* which has the \begin definition *beg_def* and the \end definition *end_def*. The optional argument *narg* ≤ 9 specifies how many variable arguments the environment is to have, which appear in the *beg_def* as the replacement characters #1 to #*narg*. If the second optional argument is present (not available with LATEX 2.09), the first argument of the \begin command is optional, and takes on the value *opt* if it is not explicitly given.

2ε \newenvironment*{*env_name*}[*narg*][*opt*]{*beg_def*}{*end_def*}
. . . . (C.2.6) – 336

>   The same as \newenvironment except that the arguments to \begin{*env_name*} must be 'short', not containing any new paragraphs. *(1994/12/01)*

\newfont{*font_cmd*}{*font_name* scaled *size*} . . . . . (4.1.6) – 62

>   Establishes the relation between the font file name *file_name* magnified by the scaling factor *size* and a font selection command *font_cmd*. After *font_cmd* has been called, \baselineskip, the interline spacing, still has its previous value.

\newlength{*length_cmd*} . . . . (7.2) – 193

>   Creates a new length command with the name *length_cmd* and initializes it to 0 pt. New values may be assigned as for all length commands with the \setlength command:
>   \setlength{*length_cmd*}{*length*}
>   The quantity *length* must have units (cm, pt, etc.) and may be a rubber length.

\newline . . . . (3.5.2) – 48

>   Terminates and starts a line of text *without* right justifying it.

\newpage . . . . (3.5.5) – 51

>   Terminates and starts a new page, leaving the rest of the page blank.

\newsavebox{*boxname*} . . . . (4.7.1) – 86

>   Creates a storage box with the name *boxname* in which LR boxes may be saved with the \savebox command.

\newtheorem{*type*} [*num_like*] {*title*}
\newtheorem{*type*} {*title*} [*in_ctr*]   . . . .   (4.5) – 79

Defines a new theorem-like environment named *type* which when called prints a theorem declaration with the name *title* in **bold face**, followed by an automatic sequential number, and the actual text of the environment in *italic*. The optional argument *num_like* is the name of another theorem structure which is to share the same numbering counter. The other optional argument *in_ctr* is the name of a sectioning counter, such as chapter, which is to reset the theorem counter every time it is incremented. That is, the theorem counter is a sub-counter of *in_ctr*. Only one of the optional arguments may be given.

[2ε] \NG   . . . .   (F.4.5) – 452

When T1 encoding is active, prints the character Ŋ. *(1994/12/01)*

[2ε] \ng   . . . .   (F.4.5) – 452

When T1 encoding is active, prints the character ŋ. *(1994/12/01)*

\ni [m] produces ∋ . . . .   (5.3.4) – 124
\nocite{*key*}   . . . .   (8.3.3), (B.1) – 227, 312

The entry in the literature database with the keyword *key* will be included in the bibliography without any citation (reference) in the text.

[2ε] \nocorr   . . . .   (4.1.4) – 61

To be used with font commands \emph, \textsl, \textit to suppress automatic italic correction.

    \nocorr\emph{emphatic text}   or
    \emph{emphatic text\nocorr}

\nofiles [p] . . . .   (8.9) – 244

Issued in the preamble, this command suppresses the output of the auxiliary files .aux, .glo, .idx, .lof, .lot, and .toc.

\noindent   . . . .   (3.5.4) – 50

The first line of the *next* paragraph will *not* be indented.

\nolimits [m] . . . .   (5.3.7) – 126

Places the upper and lower limits after the appropriate symbols where these would normally go just above or below them.

\nolinebreak[*n*]   . . . .   (3.5.2) – 49

A recommendation *not* to break the line of text at this point. The urgency of the recommendation is given by the integral number *n* between 0 and 4, with the higher numbers meaning a stronger recommendation. A value of 4 is the same as the command without the optional argument and means absolutely no line break here.

`\nonfrenchspacing` . . . . (3.5.1) – 45

> Countermands `\frenchspacing`, switching back to the standard formatting in which extra word spacing is inserted at the end of a sentence.

`\nonumber` [m] . . . . (5.4.7) – 137

> The formula line in an `eqnarray` environment in which this command appears will *not* contain an equation number.

`\nopagebreak[n]` . . . . (3.5.5) – 51

> A recommendation *not* to break the page at this point. The urgency of the recommendation is given by the integral number n between 0 and 4, with the higher numbers meaning a stronger recommendation. A value of 4 is the same as the command without the optional argument and means absolutely no page break here.

$\boxed{2\varepsilon}$ `\normalcolor` . . . . (6.3) – 177

> A command that normally does nothing. However, if the `color` package has been loaded, it resets the color for text to be the color that was in effect at the end of the preamble, normally black. A `\color` command in the preamble can alter this 'standard' color.

> This command is called by many internal LaTeX macros, to reset the text color when printing headlines and headings. Other packages should also use it so that they are consistent with the `color` package.

$\boxed{2\varepsilon}$ `\normalfont` . . . . (4.1.3), (8.5.2) – 61, 237

> This declaration switches to the font with the default family, shape, and series attributes.

`\normalmarginpar` . . . . (4.10.5) – 114

> Countermands `\reversemarginpar`, switching back to the standard placement of marginal notes in the 'outer' margin.

`\normalsize` . . . . (4.1.2) – 58

> Switches to the font size `\normalsize`, the size selected by the option in the `\documentclass` or `\documentstyle` commands. It is smaller than `\large` but larger than `\small`.

`\not` [m] . . . . (5.3.4) – 125

> Changes the following comparison symbol into its negative counterpart:
>
> $\qquad$ `\not\cong` $= \ncong$

`\notag{mark}` [m][a] . . . . (E.3.6) – 420

> Within one of the $\mathcal{AMS}$-LaTeX alignment environments, suppresses the automatic equation number.

`\notesname` . . . . (C.4.1) – 352

> Command for some extensions to LaTeX containing the text for a command `\notes`. In English, this is 'notes' but may be altered for adaptation to other languages. (This command is not standardized!)

`\notin` [m] produces ∉ . . . . (5.3.4) – 125
`\nu` [m] produces ν . . . . (5.3.1) – 123
`\numberwithin{`*ctr*`}{`*in_ctr*`}` [a] . . . . (E.3.7) – 426

> With the `amsmath` package, redefines the counter *ctr* to be a sub-counter of *in_ctr*, meaning it is reset every time *in_ctr* is incremented. The value of *in_ctr* is printed with that of *ctr*. This is normally used to make equations in an article to be numbered within sections:
>
> `\numberwithin{equation}{section}`

`\nwarrow` [m] produces ↖ . . . . (5.3.5) – 125

`\O`   produces Ø . . . . (2.5.6) – 21
`\o`   produces ø . . . . (2.5.6) – 21
`\oddsidemargin` . . . . (3.2.4) – 35

> Sets the left margin for the odd-numbered pages in document class book or when the option `twoside` has been selected for other classes. In all other cases, it sets the left margin for *all* pages. A new value is assigned with the `\setlength` command:
>
> `\setlength{\evensidemargin}{1.5cm}`

`\odot` [m] produces ⊙ . . . . (5.3.3) – 124
`\OE`   produces Œ . . . . (2.5.6) – 21
`\oe`   produces œ . . . . (2.5.6) – 21
`\oint` [m] produces ∮ . . . . (5.3.7) – 126
`\Omega` [m] produces Ω . . . . (5.3.1) – 123
`\omega` [m] produces ω . . . . (5.3.1) – 123
`\ominus` [m] produces ⊖ . . . . (5.3.3) – 124
`\onecolumn` . . . . (3.2.5) – 36

> Starts a new page and switches from two-column to one-column page formatting.

`\onlynotes{`*note_nums*`}` . . . . (8.10.3) – 252

> In `slides` class, a command to be issued in the preamble to generate only those notes whose numbers appear in *note_nums*. The command behaves the same as `\onlyslides` for slides.

`\onlyslides{`*slide_nums*`}` . . . . (8.10.3) – 252

> In `slides` class, a command to be issued in the preamble to generate only those slides whose numbers appear in *slide_nums*. The numbers are separated by commas, and may include a range with a hyphen:
> `\onlyslides{4,10-13,23}`.

\opening{*dear*} . . . . (A.1) – 299

> In the letter environment of the letter class, this sets the form of the salutation at the start of the letter text; for example, \opening{Dear George,}.

\oplus [m] produces $\oplus$ . . . . (5.3.3) – 124

2ε \OptionNotUsed [p] . . . . (C.2.3) – 334

> A command that may only be used in the definition of options, especially default options. It declares the \CurrentOption to be unprocessed. This is used if the processing of a default option should fail, say because some file is missing. LaTeX is then informed that this requested option is still outstanding.

\oslash [m] produces $\oslash$ . . . . (5.3.3) – 124
\otimes [m] produces $\otimes$ . . . . (5.3.3) – 124
\oval (*x_dimen,y_dimen*) [*part*] . . . . (6.1.4) – 160

> Picture element command to produce an oval with width *x_dimen* and height *y_dimen* in the picture environment. The optional *part* argument may take on values of t, b, l, and r to draw only the top, bottom, left, or right halves of the oval. A combination of these values may be given to draw only a quarter of the oval, such as tl or lt for the top left part. To be used as an argument in a \put or \multiput command.

\overbrace{*sub_form*} [m] . . . . (5.4.4) – 134

> Produces a horizontal curly brace over the math formula *sub_form*. Any following superscript will be placed centered above the horizontal brace.
>
> $$\overbrace{a+b} = \overbrace{a + b}$$
>
> $$\overbrace{x+y+z}\char94\{\xi\eta\zeta\} = \overbrace{x + y + z}^{\xi\eta\zeta}$$

\overleftarrow{*expr*} [m][a] . . . . (E.3.2) – 411

> With the amsmath package, places a long leftwards pointing arrow over the mathematical expression *expr*.

\overleftrightarrow{*expr*} [m][a] . . . . (E.3.2) – 411

> With the amsmath package, places a long double arrow over the mathematical expression *expr*.

\overline{*sub_form*} [m] . . . . (5.4.4) – 134

> Produces a horizontal bar over the math formula *sub_form*:
>
> $$\overline{a-b} = \overline{a - b}$$

\overrightarrow{*expr*} [m][a] . . . . (E.3.2) – 411

> With the amsmath package, places a long rightwards pointing arrow over the mathematical expression *expr*.

\overset{*char*}{*symbol*} [m][a] . . . . (E.3.2) – 410

> With the amsmath package, places *char* over the math symbol *symbol*
> in superscript size.

\P produces ¶ . . . . (2.5.5) – 21

[2ε] \PackageError{*pkg_name*}{*error_text*}{*help*} [p] . . . . (C.2.7) – 337

> Writes an error message *error_text* to the monitor and transcript file,
> labeled with the package name, and halts processing, waiting for a
> user response as for a LaTeX error. If H⟨*return*⟩ is typed, the *help* text is
> printed. Both *error_text* and *help* may contain \MessageBreak for a
> new line, \space for a forced space, and \protect before commands
> that are to have their names printed literally and not interpreted.

[2ε] \PackageInfo{*pkg_name*}{*info_text*} [p] . . . . (C.2.7) – 338

> Is like \PackageWarningNoLine except that the text *info_text* is only
> written to the transcript file, and not to the monitor.

[2ε] \PackageWarning{*pkg_name*}{*warn_text*} [p] . . . . . . (C.2.7) – 337

> Writes *warn_text* to the monitor and transcript file, labeled with
> the package name and the current line number of the input file.
> Processing continues. The *warn_text* is formatted in the same way as
> that for \PackageError.

[2ε] \PackageWarningNoLine{*pkg_name*}{*warn_text*} [p] . . (C.2.7) – 337

> Is like \PackageWarning except that the current line number of the
> input file is not printed.

\pagebreak[*n*] . . . . (3.5.5) – 51

> A recommendation to break the page at this point. The urgency of
> the recommendation is given by the integral number *n* between 0 and
> 4, with the higher numbers meaning a stronger recommendation. A
> value of 4 is the same as the command without the optional argument
> and means an obligatory page break.

[2ε] \pagecolor *col_spec* . . . . (6.3) – 177

> A command made available with the color package. Sets the back-
> ground color starting with the current page. All following pages have
> the same background color until \pagecolor is called once more.
> The *col_spec* is the same as for \color.

\pagename . . . . (C.4.1) – 352

> Command in the letter document class containing the text for page
> numbers after the first page. In English, this is 'Page' but may be
> altered for adaptation to other languages.

\pagenumbering{*style*} . . . . (3.2.2) – 32

Determines the style of the page numbering and resets the page counter to 1. Possible values for *style* are: arabic, roman, Roman, alph, and Alph.

\pageref{*marker*} . . . . (8.3.1) – 224

Prints the number of the page where *marker* has been set by a \label{*marker*} command.

\pagestyle{*style*} [p] . . . . (3.2) – 31

Determines the page style, that is, the contents of the head and footlines on every page. Possible values for *style* are: plain, empty, headings, and myheadings.

[2ε] \paperheight . . . . (3.2.4) – 35

The total height of the page as specified by the page size option in the \documentclass command line. With the default lettersize option, this is 11 in; with a4paper, it is 29.7 cm. The additional option landscape interchanges the values of \paperwidth and \paperheight.

[2ε] \paperwidth . . . . (3.2.4) – 35

The total width of the page as specified by the page size option in the \documentclass command line. With the default lettersize option, this is 8.5 in; with a4paper, it is 21 cm. The additional option landscape interchanges the values of \paperwidth and \paperheight.

\par . . . . (3.5.3) – 50

Ends the current paragraph and begins a new one. This command is equivalent to a blank line.

\paragraph[*short title*]{*title*} . . . . (3.3.3) – 39

The second last command in the sectioning hierarchy, coming between \subsubsection and \subparagraph. It formats *title* with the current sub-subsection number and an automatic sequential paragraph number. If the optional *short title* is given, it appears in place of *title* in the table of contents.

\paragraph*{*title*} . . . . (3.3.3) – 39

The same as \paragraph but without a number or an entry in the table of contents.

\parallel [m] produces ‖ . . . . (5.3.4) – 124
\parbox[*pos*][*height*][*inner_pos*]{*width*}{*text*} (4.7.3), (4.7.5) – 87, 89

Produces a vertical box of width *width* in which *text* is set in lines that are left and right justified to this width. The vertical positioning with respect to the surrounding text is determined by the optional argument *pos*: t for alignment with its top line, b with its bottom line, and centered with no argument. The two additional optional arguments (not available with LATEX 2.09): *height* to give the total height, and *inner_pos* to specify how the text is to be positioned inside it. Possible values are t for top, b for bottom, c for centered, and s to be stretched out to fill the whole vertical space. The default is the value of the external positioning *pos* option. The *height* argument may contain the parameters \height, \depth, \width, and \totalheight.

\parindent . . . . (3.2.3) – 33

The amount of indentation for the first line of a paragraph. A new value may be assigned with the \setlength command:

    \setlength{\parindent}{1.5em}

\parsep . . . . (4.4.2) – 74

The vertical spacing between paragraphs within a list environment. A new value may be assigned with the \setlength command:

    \setlength{\parsep}{2pt plus1pt minus1pt}

\parskip . . . . (3.2.3) – 33

The vertical spacing between paragraphs. A new value may be assigned with the \setlength command:

    \setlength{\parskip}{3pt plus1pt minus2pt}

\part[*short title*]{*title*} . . . . (3.3.3) – 39

The highest command in the sectioning hierarchy. It begins a new 'Part' with an automatic sequential part number and the heading *title*. The following sectioning numbers are not influenced by the part number. If the optional *short title* is given, it appears in place of *title* in the table of contents.

\part*{*title*} . . . . (3.3.3) – 39

The same as \part but without a number or an entry in the table of contents.

\partial [m] produces ∂ . . . . (5.3.6) – 125
\partname . . . . (C.4.1) – 351

Command containing the part heading. In English, this is 'Part' but may be altered for adaptation to other languages.

\partopsep . . . . (4.4.2) – 74

The additional vertical spacing at the beginning and/or end of a listing when a blank line precedes or follows the environment commands. A new value may be assigned with the \setlength command:

    \setlength{\partopsep}{2pt plus1pt minus1pt}

[2ε] \PassOptionsToClass{*options*}{*class*} [p] . . . . (C.2.3) – 334

> Assigns the options in the list *options* to the specified class file, which is later loaded with \LoadClass. This command must be called from a class file, or from another file input by a class file. It may be used in the definition of options, or in a configuration file to activate options.

[2ε] \PassOptionsToPackage{*options*}{*package*} [p] . . . . (C.2.3) – 334

> Assigns the options in the list *options* to the specified package file, which is later loaded with \RequirePackage. This command may be called from a class or package file. It may be used in the definition of options, or in a configuration file to activate certain options.

\perp [m] produces ⊥ . . . . (5.3.4) – 124
\Phi [m] produces Φ . . . . (5.3.1) – 123
\phi [m] produces ϕ . . . . (5.3.1) – 123
\Pi [m] produces Π . . . . (5.3.1) – 123
\pi [m] produces π . . . . (5.3.1) – 123
\pm [m] produces ± . . . . (5.3.4) – 124
\pmb{*symbol*} [m][a] . . . . (E.3.1) – 407

> When one of the packages amsmath or amsbsy has been loaded, this command prints *symbol* in simulated bold face. This is done by printing it several times slightly displaced.

\pmod{*arg*} [m] . . . . (5.3.8), (E.3.5) – 127, 417

> Command to produce the function name 'mod' in formulas in the form:
>
> y\pmod{a+b} = y (mod $a + b$)

\pod{*arg*} [m][a] . . . . (E.3.5) – 417

> With the amsopn or amsmath packages, command to produce the function name 'mod' in formulas in the form:
>
> y\pod{a+b} = y $(a + b)$

\poptabs . . . . (4.6.4) – 82

> Restores the last set of tabular stops in the tabbing environment that has been saved with \pushtabs.

\pounds produces £ . . . . (2.5.5) – 21
\Pr [m] . . . . (5.3.8) – 126

> Command to produce the function name 'Pr' in formulas. A lower limit may be set as a subscript.

\prec [m] produces ≺ . . . . (5.3.4) – 124
\preceq [m] produces ≼ . . . . (5.3.4) – 124
\prefacename . . . . (C.4.1) – 352

> Command for some extensions to LaTeX containing the text for a command \preface. In English, this is 'Preface' but may be altered for adaptation to other languages. (This command is not standardized!)

\prime [m] produces ′ (identical to the ' symbol) . . . . (5.3.6) – 125
\printindex . . . . (8.4) – 230

> A command defined in the makeidx.sty file that generates the index environment after the program MakeIndex has processed the .idx file.

|2ε| \ProcessOptions [p] . . . . (C.2.3) – 334

> In a class or package file, this command processes the requested options by executing the \ds@ commands for each one *in the order in which they were defined.* The \ds@ commands are then erased.

|2ε| \ProcessOptions* [p] . . . . (C.2.3) – 334

> This is the same as \ProcessOptions except that the \ds@ commands are executed *in the order in which the options were requested.* It is equivalent to the LaTeX 2.09 command \@options, which is still maintained for compatibility.

\prod [m] produces ∏ . . . . (5.3.7) – 126
\propto [m] produces ∝ . . . . (5.3.4) – 124
\protect . . . . (2.6) – 23

> Fragile commands may be used in moving arguments when they are preceded by the \protect command. Example:
>
> \section{The \protect\pounds{} Sign}.

|2ε| \providecommand{*com_name*}[*narg*][*opt*]{*def*} . . . (7.3.1) – 195

> The same as \newcommand except that if a command with the name *com_name* already exists, the new definition is ignored.

|2ε| \providecommand*{*com_name*}[*narg*][*opt*]{*def*} . . . (C.2.6) – 336

> The same as \providecommand except that the arguments to *com_name* must be 'short', not containing any new paragraphs. *(1994/12/01)*

|2ε| \ProvidesClass{*class*}[*version*] [p] . . . . (C.2.1) – 332

> At the beginning of a class file, this statement declares the name of the class and its version, to be checked against the name and version in the \documentclass or \LoadClass command that input it. The *version* specification, if present, consists of three parts: date, version number, and additional information. Example:
>
> \ProvidesClass{thesis}[1995/01/25 v3.8 U of Saigon]

|2ε| \ProvidesFile{*file_name*}[*version*] . . . . (C.2.1) – 333

> At the beginning of a general file, this statement declares its name and version. No checking is done when the file is read in with \input, but the information is printed out if \listfiles has been activated. This command is not limited to the preamble as the other \Provides.. commands are.

$\boxed{2_\varepsilon}$ `\ProvidesPackage{`*class*`}[`*version*`] [p]` . . . . (C.2.1) – 332

At the beginning of a package file, this statement declares the name of the package and its version, to be checked against the name and version in the `\usepackage` or `\RequirePackage` command that input it. The *version* specification, if present, consists of three parts: date, version number, and additional information. Example:

`\ProvidesPackage{notes}[1995/02/13 1.2 G. Smith]`

$\boxed{2_\varepsilon}$ `\ProvideTextCommand{\`*cmd*`}{`*code*`}[`*narg*`][`*opt*`]{`*def*`} [p]`
. . . . (C.5.7) – 360

Defines `\`*cmd* in the same way as `\providecommand` except the definition is only valid when encoding *code* is active. If the command `\`*cmd* is already defined for that encoding, it is not redefined. *(1994/12/01)*

$\boxed{2_\varepsilon}$ `\ProvideTextCommandDefault{\`*cmd*`}{`*code*`}[`*narg*`][`*opt*`]{`*def*`} [p]`
. . . . (C.5.7) – 361

Is the same as `\DeclareTextCommandDefault` except that if a default encoding definition already exists for `\`*cmd*, then no redefinition occurs and the previous definition is retained. *(1994/12/01)*

`\ps` *text* . . . . (A.1) – 299

Adds a postscript to a letter in the `letter` document class.

`\Psi [m]` produces Ψ . . . . (5.3.1) – 123
`\psi [m]` produces ψ . . . . (5.3.1) – 123
`\pushtabs` . . . . (4.6.4) – 82

Saves the current set of tabulator stops in the `tabbing` environment. It may be recalled with the `\poptabs` command.

`\put(`x,y`){`*pic_elem*`}` . . . . (6.1.3) – 153

The positioning command within a `picture` environment. The picture element *pic_elem* is placed with its reference point at the location (x,y).

$\boxed{2_\varepsilon}$ `\qbezier[`*num*`](`x_1,y_1`)(`x_2,y_2`)(`x_3,y_3`)` . . . . . . (6.1.4) – 163

This command can be given within the `picture` environment to draw a quadratic Bézier curve from point (x_1,y_1) to (x_3,y_3), using (x_2,y_2) as the extra Bézier point. The curve is drawn as $num + 1$ dots if the optional argument *num* is given, otherwise *num* is calculated automatically to produce a solid line. It is the same as `\bezier` except that *num* is optional.

`\quad` . . . . (3.5.1) – 47

Inserts horizontal spacing of size 1 em.

`\qquad`   . . . . (3.5.1) – 47

> Inserts horizontal spacing of size 2 em.

[2ε] `\quotedblbase`   . . . . (F.4.5) – 452

> When T1 encoding is active, prints the symbol „. *(1994/12/01)*

[2ε] `\quotesinglbase`   . . . . (F.4.5) – 452

> When T1 encoding is active, prints the symbol ‚. *(1994/12/01)*

[2ε] `\r{x}` . . . . (2.5.7) – 21

> Produces a circle accent: `\r{o}` = o̊.

`\raggedbottom` . . . . (3.2.4) – 34

> The standard page formatting for `article`, `report`, and `letter` document classes when the `twoside` option has *not* been selected. The spacing between paragraphs is fixed so that the last line will vary from page to page. The opposite command is `\flushbottom`.

`\raggedleft` . . . . (4.2.2) – 64

> After this declaration, the lines of text will only be right justified and the left margin will be uneven. The individual lines are terminated by `\\`. See also `\begin{flushright}`.

`\raggedright` . . . . (4.2.2) – 64

> After this declaration, the lines of text will only be left justified and the right margin will be uneven. The individual lines are terminated by `\\`. See also `\begin{flushleft}`.

`\raisebox{`*lift*`}[`*height*`][`*depth*`]{`*text*`}`   . . . . (4.7.2) – 86

> An LR box containing *text* is raised an amount *lift* above the current baseline. If *lift* is negative, the box is lowered. The optional arguments state that it is to be treated as though it extended by *height* above and by *depth* below the baseline regardless of its true extents.

`\raisetag{`*len*`}` [m][a]   . . . . (E.3.6) – 420

> Within one of the $\mathcal{A}_{\mathcal{M}}S$-LaTeX alignment environments, raises the equation number or marker by *len* above its normal position.

`\rangle` [m] produces ⟩ . . . . (5.4.1) – 130
`\rceil` [m] produces ⌉   . . . . (5.4.1) – 130
`\Re` [m] produces ℜ   . . . . (5.3.6) – 125
`\ref{`*marker*`}`   . . . . (8.3.1) – 224

> Prints the number of the section, equation, figure, or table where *marker* has been set by a `\label{`*marker*`}` command.

⌐2ε⌐ `\reflectbox{`*text*`}` . . . . (6.2.1) – 169

A command made available with the `graphics` package that reflects the contents *text* as an LR box such that left and right are reversed.

`\refname`   . . . . (C.4.1) – 351

Command containing the heading for the bibliography in `article` document class. In English, this is 'References' but may be altered for adaptation to other languages.

`\refstepcounter{`*counter*`}` . . . . (7.1.3) – 190

Increases the value of the number stored in *counter* by one, the same as `\stepcounter`, but also makes the specified counter the relevent one for the `\label`-`\ref` cross-referencing commands.

`\renewcommand{`*com_name*`}[`*narg*`][`*opt*`]{`*def*`}` . . . . . . (7.3) – 193

The same as `\newcommand` except that the command `\`*com_name* must already exist, otherwise an error message is printed.

⌐2ε⌐ `\renewcommand*{`*com_name*`}[`*narg*`][`*opt*`]{`*def*`}` . . . . (C.2.6) – 336

The same as `\renewcommand` except that the arguments to `\`*com_name* must be 'short', not containing any new paragraphs. *(1994/12/01)*

`\renewenvironment{`*env*`}[`*narg*`][`*opt*`]{`*beg*`}{`*end*`}` . . . . (7.4) – 204

The same as `\newenvironment` except that the environment *env* must already exist, otherwise an error message is printed.

⌐2ε⌐ `\renewenvironment*{`*env*`}[`*narg*`][`*opt*`]{`*beg*`}{`*end*`}` . . (C.2.6) – 336

The same as `\renewenvironment` except that the arguments to `\begin{`*env*`}` must be 'short', not containing any new paragraphs. *(1994/12/01)*

⌐2ε⌐ `\RequirePackage[`*options*`]{`*packages*`}[`*version*`] [p]` . . . (C.2.2) – 333

This command is the equivalent of `\usepackage` within a class or package file. It loads one or more package files with the extension `.sty`. More than one package may be specified in *packages*, the names being separated by commas. Any *options* listed will be applied to all packages. Furthermore, any options listed in the `\documentclass` command will also be applied to the package files.

The optional *version* is a date, given in the form yyyy/mm/dd, as for example 1994/08/01. If the date of the package file is earlier than this, a warning message is printed.

⌐2ε⌐ `\RequirePackageWithOptions{`*package*`}[`*version*`] [p]` . (C.2.2) – 333

Like `\RequirePackage` except all the currently specified options are automatically passed to *package*. *(1995/12/01)*

[2ε] \resizebox{*h_length*}{*v_length*}{*text*} . . . . (6.2.1) – 169

> A command made available with the `graphics` package that scales the contents *text* as an LR box such that the horizontal size becomes *h_length* and the vertical size *v_length*. If either size is given as !, the one scale factor is applied to both dimensions.

[2ε] \resizebox*{*h_length*}{*v_length*}{*text*} . . . . (6.2.1) – 169

> The same as \resizebox except that the vertical size *v_length* refers to the total height plus depth of the LR box.

\reversemarginpar . . . . (4.10.5) – 114

> Changes the placement of marginal notes from the standard (right or 'outer' margin) to the opposite side. Can be countermanded with \normalmarginpar.

\rfloor [m] produces ⌋ . . . . (5.4.1) – 130
\rhd [m] produces ▷ . . . . (5.3.3) – 124
\rho [m] produces ρ . . . . (5.3.1) – 123
\right*rbrack* [m] . . . . (5.4.1) – 129

> Adjusts the size of the bracket symbol *rbrack* to fit the height of the formula between the \left ... \right pair. For example, \right]. If there is to be no matching bracket, the \left and \right commands must still be given to specify the part of the formula to be sized, but the missing bracket is given as a period (for example, \left.).

\Rightarrow [m] produces ⇒ . . . . (5.3.5) – 125
\rightarrow [m] produces → . . . . (5.3.5) – 125
\rightharpoondown [m] produces ⇁ . . . . (5.3.5) – 125
\rightharpoonup [m] produces ⇀ . . . . (5.3.5) – 125
\rightleftharpoons [m] produces ⇌ . . . . (5.3.5) – 125
\rightmargin . . . . (4.4.2) – 75

> In a `list` environment, the amount by which the right edge of the text is indented relative to the right side of the surrounding text. Standard value is 0 pt. A new value is assigned with the \setlength command:
>
>   \setlength{\rightmargin}{0.5cm}

[2.09] \rm . . . . (4.1.5) – 62

> Switches to the Roman, upright, medium typeface (the default).

[2ε] \rmdefault . . . . (C.5.1) – 353

> This command defines the family attribute that is selected with the \rmfamily command. It may be redefined with \renewcommand:
>
>   \renewcommand{\rmdefault}{ptm}

[2ε] \rmfamily . . . . (4.1.3), (8.5.2) – 60, 236
>   This declaration switches to a font in the current series and shape, but with the Roman family attribute.

\Roman{*counter*} . . . . (7.1.4) – 191
>   Prints the current value of the *counter* as an upper case Roman numeral.

\roman{*counter*} . . . . (7.1.4) – 191
>   Prints the current value of the *counter* as a lower case Roman numeral.

[2ε] \rotatebox{*angle*}{*text*} . . . . (6.2.1) – 169
>   A command made available with the graphics package that rotates the contents *text* as an LR box through the *angle* expressed in degrees. The rotation is counterclockwise about the left-hand end of the baseline of the box.

\rq   produces ', identical to the ' symbol.
\rule[*lift*]{*width*}{*height*} . . . . (4.7.6) – 90
>   Produces a black rectangle of width *width* and height *height*, raised above the baseline by an amount *lift*, if this optional argument is given. A value of '0 pt' for either the *width* or *height* creates an invisible horizontal or vertical *strut* that may be used to make spacing.

\rvert [m][a] produces | (right delimiter)   . . . . (E.3.5) – 418
\rVert [m][a] produces ‖ (right delimiter)   . . . . (E.3.5) – 418

\S produces §   . . . . (2.5.5) – 21
[2ε] \SS produces SS, the upper case version of \ss, ß   . . . . . (2.5.6) – 21
\savebox{*boxname*}[*width*][*pos*]{*text*} . . . . (4.7.1) – 86
>   Functions the same as the \makebox command except that the box contents are not output but saved under the name *boxname*, which has been previously defined with \newsavebox. The box may be set any place in the text as often as desired with the command \usebox{*boxname*}.

\savebox{*sub_pic_name*}(*x_dim,y_dim*)[*pos*]{*sub_pic*} . (6.1.4) – 164
>   In the picture environment, a sub-picture *sub_pic* may be stored as a box of width *x_dim* and height *y_dim* under the name *boxname*, which has been previously defined with \newsavebox. The *pos* argument functions as it does for \makebox (picture). The box may be set any place in the picture environment with the command \usebox{*boxname*}.

\sb   produces a subscript, identical to the _ symbol.
\sbox{*boxname*}{*text*}   . . . . (4.7.1) – 86
>   Stores *text* in an LR box named *boxname* that has previously been created with \newsavebox{*boxname*}. The contents of the box may be printed as often as desired with \usebox{*boxname*}.

$\boxed{2.09}$ \sc . . . . (4.1.5) – 62

Switches to the ROMAN, CAPS AND SMALL CAPS, MEDIUM typeface.

$\boxed{2\varepsilon}$ \scalebox{*h_scale*}[*v_scale*]{*text*} . . . . (6.2.1) – 169

A command made available with the graphics package that scales the contents *text* as an LR box with the horizontal factor *h_scale* and optionally with the (different) vertical factor *v_scale*. If *v_scale* is missing, it is the same as *h_scale*.

$\boxed{2\varepsilon}$ \scdefault . . . . (C.5.1) – 353

This command defines the shape attribute that is selected with the \scshape command. It may be redefined with \renewcommand:

   \renewcommand{\scdefault}{sc}

\scriptscriptstyle [m] . . . . (5.5.2) – 144

Switches to font size \scriptscriptstyle as the active font inside a math formula.

\scriptsize . . . . (4.1.2) – 58

Switches to the font size \scriptsize, which is smaller than \footnotesize but larger than \tiny.

\scriptstyle [m] . . . . (5.5.2) – 144

Switches to font size \scriptstyle as the active font inside a math formula.

$\boxed{2\varepsilon}$ \scshape . . . . (4.1.3), (8.5.2) – 60, 237

This declaration switches to a font in the current family and series, but with the CAPS AND SMALL CAPS shape attribute.

\searrow [m] produces ↘ . . . . (5.3.5) – 125

\sec [m] . . . . (5.3.8) – 126

Command to produce the function name 'sec' in formulas.

\section[*short title*]{*title*} . . . . (3.3.3) – 39

Begins a new section, formatting *title* with the current chapter number (book and report classes only) and an automatic sequential section number. If the optional *short title* is given, it appears in place of *title* in the table of contents and the running head at the top of the pages.

\section*{*title*} . . . . (3.3.3) – 39

The same as \section but without a number or an entry in the table of contents.

\see . . . . (8.4) – 230

A command defined in the file makeidx.sty for use with the MakeIndex program. It is called within an \index command to refer to
another entry in the keyword index, in the form:

\index{*entry*|see{*reference*}}

Note: the above text is correct with | in place of \ for |see.

\seename . . . . (C.4.1) – 352

A command defined in the file makeidx.sty containing the text for
the command \see. In English, this is 'see' but may be altered for
adaptation to other languages.

2ε \selectfont . . . . (8.5.1) – 235

Activates the font with the current set of attributes, making it the
current font in which text is set. It should normally follow an attribute
selection command. Example:

\fontshape{sl}\selectfont

\selectlanguage{*language*} . . . . (D.1.1), (D.1.3) – 366, 373

Command in multi-language packages such as esperant, german
and in the babel system for changing the language. The names
of titles, the form of the date command \today, special language-
specific commands, and the hyphenation patterns (TeX version 3) are
all changed. Example:

\selectlanguage{english}

2ε \setboolean{*switch*}{*value*} . . . . (7.3.5) – 203

Sets the value of a boolean switch to ⟨*true*⟩ or ⟨*false*⟩, depending
on *value*, which must be true or false. The switch must have
been created with \newboolean{*switch*}. Its value is tested with
\boolean{*switch*}, which may be used as a logical statement in the
test part of \ifthenelse and \whiledo. Requires the standard LaTeX
package ifthen.

\setcounter{*counter*}{*value*} . . . . (7.1.3) – 190

Assigns the integral number *value* to the counter *counter*.

\setlength{*length_cmd*}{*length_spec*} . . . . (7.2) – 192

The length command with the name *length_cmd* is assigned the
length value *length_spec*, which may be a fixed or rubber length.
See also . . . . (2.4.1), (2.4.2) – 19, 19

2ε \SetMathAlphabet{*cmd*}{*ver*}{*code*}{*fam*}{*ser*}{*shp*} [p]
. . . . (C.5.3) – 355

Defines the math alphabet that has been declared with
\DeclareMathAlphabet to take the specified font attibutes in the
math version *ver*.

`\setminus` [m] produces \ . . . . (5.3.3) – 124

`[2ε]` `\SetSymbolFont{`*sym_fnt*`}{`*ver*`}{`*code*`}{`*fam*`}{`*ser*`}{`*shp*`}` [p]
. . . . (C.5.4) – 355

Defines the symbol font that has been declared with `\DeclareSymbolFont` to take the specified font attibutes in the math version *ver*.

`[2ε]` `\settime{`*secs*`}` . . . . (8.10.3) – 252

In the `slides` class, if the option `clock` has been selected, a time marker, in minutes, appears at the bottom of the notes. This command sets the internal timer to the specified number of seconds. See also `\addtime`.

`[2ε]` `\settodepth{\`*length_cmd*`}{`*text*`}` . . . . (7.2) – 193

The length command with the name `\`*length_cmd* is assigned a value equal to the depth of *text* below the baseline.

`[2ε]` `\settoheight{\`*length_cmd*`}{`*text*`}` . . . . (7.2) – 193

The length command with the name `\`*length_cmd* is assigned a value equal to the height of *text* above the baseline.

`\settowidth{\`*length_cmd*`}{`*text*`}` . . . . (7.2) – 193

The length command with the name `\`*length_cmd* is assigned a value equal to the length of *text* as it would be set in an LR box.

`[2.09]` `\sf` . . . . (4.1.5) – 62

Switches to the sans serif, upright, medium typeface.

`[2ε]` `\sfdefault` . . . . (C.5.1) – 353

This command defines the family attribute that is selected with the `\sffamily` command. It may be redefined with `\renewcommand`:

  `\renewcommand{\sfdefault}{phv}`

`[2ε]` `\sffamily` . . . . (4.1.3), (8.5.2) – 60, 236

This declaration switches to a font in the current series and shape, but with the sans serif family attribute.

`\sharp` [m] produces ♯ . . . . (5.3.6) – 125

`\shortstack[`*pos*`]{`*text*`}` . . . . (6.1.4) – 161

Formats the *text* into a single column, where the individual rows are terminated by \\. The optional positioning argument *pos* takes on values of l or r to set the text left or right justified, otherwise it is centered. Example:

aa
bbb
cc
x
yy
zzz

  `\shortstack{aa\\bbb\\cc\\x\\yy\\zzz}`

\showhyphens{*word_list*} . . . . (3.6.5) – 56

Displays on the monitor screen the possible divisions for the words in *word_list*.

\sideset{*pre*}{*post*}*symbol* [m][a] . . . . (E.3.2) – 410

With the amsmath package, places superscripts and subscripts snuggly before (*pre*) and after (*post*) the math symbol *symbol*. Example: $\sideset{_\dag^*}{_\dag^*}\prod$ yields $\sideset{_\dag^*}{_\dag^*}\prod$

\Sigma [m] produces Σ . . . . (5.3.1) – 123
\sigma [m] produces σ . . . . (5.3.1) – 123
\signature{*name*} . . . . (A.1) – 298

In the letter document class, supplies the name of the writer that should go below the signature if this is different from the entry in \name.

\sim [m] produces $\sim$ . . . . (5.3.4) – 124
\simeq [m] produces $\simeq$ . . . . (5.3.4) – 124
\sin [m] . . . . (5.3.8) – 126

Command to produce the function name 'sin' in formulas.

\sinh [m] . . . . (5.3.8) – 126

Command to produce the function name 'sinh' in formulas.

[2.09] \sl . . . . (4.1.5) – 62

Switches to the *Roman, slanted, medium* typeface.

[2ε] \sldefault . . . . (C.5.1) – 353

This command defines the shape attribute that is selected with the \slshape command. It may be redefined with \renewcommand:
    \renewcommand{\sldefault}{sl}

\sloppy . . . . (3.6.3) – 55

After this command has been given, word spacings are allowed to stretch more generously than usual so that paragraphs are broken up into lines with fewer word divisions. It is countermanded by \fussy. See also \begin{sloppypar}.

[2ε] \slshape . . . . (4.1.3), (8.5.2) – 60, 237

This declaration switches to a font in the current family and series, but with the *slanted* shape attribute.

\small . . . . (4.1.2) – 58

Switches to the font size \small, which is smaller than \normalsize but larger than \footnotesize.

`\smallskip` . . . . (3.5.3) – 50

> Inserts large vertical spacing of amount `\smallskipamount`. See also `\medskip` and `\bigskip`.

`\smallskipamount`

> Standard value for the amount of vertical spacing that is inserted with the command `\smallskip`. May be changed with the `\setlength` command:
>
> `\setlength{\smallskipamount}{1ex plus0.5ex minus0.3ex}`

`\smash[`*pos*`]{`*text*`}` [m][a] . . . . (E.3.5) – 418

> With the `amsmath` package, the TeX command acquires an optional argument *pos* that may be b or t, to effectively zero the depth or height of the *text*. With no *pos*, both height and depth are zeroed.

`\smile` [m] produces ⌣ . . . . (5.3.4) – 124
`\sp` produces a superscript, identical to the ˆ symbol.
`\spadesuit` [m] produces ♠ . . . . (5.3.6) – 125
`\sqcap` [m] produces ⊓ . . . . (5.3.3) – 124
`\sqcup` [m] produces ⊔ . . . . (5.3.3) – 124
`\sqrt[`*n*`]{`*arg*`}` [m] . . . . (5.2.4) – 120

> Basic math command to produce a root sign. The height and length of the sign are made to fit the contents *arg*. The optional argument *n* is the degree of the root: `\sqrt[3]{2}` = $\sqrt[3]{2}$, `\sqrt{2}` = $\sqrt{2}$.

`\sqsubset` [m] produces ⊏ . . . . (5.3.4) – 124
`\sqsubseteq` [m] produces ⊑ . . . . (5.3.4) – 124
`\sqsupset` [m] produces ⊐ . . . . (5.3.4) – 124
`\sqsupseteq` [m] produces ⊒ . . . . (5.3.4) – 124
`\ss` produces ß . . . . (2.5.6) – 21
`\stackrel{`*upper*`}{`*lower*`}` [m] . . . . (5.4.5) – 135

> Places one mathematical symbol *upper* on top of another *lower*, such that the upper one appears in a smaller typeface:
>
> `\stackrel{\alpha}{\longrightarrow}` = $\stackrel{\alpha}{\longrightarrow}$

`\star` [m] produces ⋆ . . . . (5.3.3) – 124
`\stepcounter{`*counter*`}` . . . . (7.1.3) – 190

> Increases the value of the number stored in *counter* by one.

`\stretch{`*decimal_num*`}` . . . . (7.2) – 193

> A rubber length with a natural value of 0 pt but with a stretchability that is *decimal_num* times that of `\fill`.

`\subitem{`*sub_entry*`}` . . . . (8.3.4) – 227

> In `theindex` environment, a command to produce a second-level entry after an `\item` command.

\subjectname . . . . (C.4.1) – 352

Command for use in modified letter document classes. It prints the text of the \subject command. In English, this is 'Subject' but may be altered for adaptation to other languages. (This command is not standardized!)

\subparagraph[*short title*]{*title*} . . . . (3.3.3) – 39

The last command in the sectioning hierarchy, coming after \paragraph. It formats *title* with the current paragraph number and an automatic sequential sub-paragraph number. If the optional *short title* is given, it appears in place of *title* in the table of contents.

\subparagraph*{*title*} . . . . (3.3.3) – 39

The same as \subparagraph but without a number or an entry in the table of contents.

\subsection[*short title*]{*title*} . . . . (3.3.3) – 39

The command in the sectioning hierarchy that comes between the \section and \subsubsection. It formats *title* with the current section number and an automatic sequential subsection number. If the optional *short title* is given, it appears in place of *title* in the table of contents.

\subsection*{*title*} . . . . (3.3.3) – 39

The same as \subsection but without a number or an entry in the table of contents.

\substack{*1st line*\\..*last line*} [m][a] . . . . (E.3.2) – 409

With the amsmath package, produces centered multiline indices or limits; it must immediately follow ^ or _ and be enclosed in { }.

\subsubitem{*sub_sub_entry*} . . . . (8.3.4) – 227

In the index environment, a command to produce a third-level entry under a \subitem command.

\subsubsection[*short form*]{*title*} . . . . (3.3.3) – 39

The command in the sectioning hierarchy coming between \subsection and \paragraph. It formats *title* with the current sub-section number and an automatic sequential sub-subsection number. If the optional *short title* is given, it appears in place of *title* in the table of contents.

\subsubsection*{*title*} . . . . (3.3.3) – 39

The same as \subsubsection but without a number or an entry in the table of contents.

\subset [m] produces ⊂ . . . . (5.3.4) – 124
\subseteq [m] produces ⊆ . . . . (5.3.4) – 124
\succ [m] produces ≻ . . . . (5.3.4) – 124
\succeq [m] produces ≽ . . . . (5.3.4) – 124
\sum [m] produces ∑ . . . . (5.2.5) – 121
\sup [m] . . . . (5.3.8) – 126

>   Command to produce the function name 'sup' in formulas. A lower
>   limit may be set as a subscript.

2ε \suppressfloats[*loc*] . . . . (6.4.2) – 181

>   Any floats given between this command and the end of the current
>   page will be suspended at least until the next page. If the optional *loc*
>   is given as one of t or b (not both), only floats with that placement
>   parameter are suspended.

\supset [m] produces ⊃ . . . . (5.3.4) – 124
\supseteq [m] produces ⊇ . . . . (5.3.4) – 124
\surd [m] produces √ . . . . (5.3.6) – 125
\swarrow [m] produces ↙ . . . . (5.3.5) – 125
\symbol{*n*} . . . . (4.1.7) – 63

>   Produces the symbol in the current character font that is stored
>   under the internal number *n*.

\t{*xy*} . . . . (2.5.7) – 21

>   Produces a 'tie-after' accent over two letters: \t{oo} = o͡o.

\tabbingsep . . . . (4.6.4) – 82

>   Determines the spacing between the text *ltext* and the current tabular
>   stop when *ltext*\' is given in a tabbing environment. A new value
>   may be assigned with the \setlength command.

\tabcolsep . . . . (4.8.2) – 96

>   Determines the half-column spacing in the tabular environment. A
>   new value may be assigned with the \setlength command:
>
>   \setlength{\tabcolsep}{3mm}

\tableofcontents . . . . (3.4.2) – 43

>   Prints the table of contents from information in the sectioning com-
>   mands and additional entries.

\tablename . . . . (C.4.1) – 351

>   Command containing the name for a table caption. In English, this
>   is 'Table' but may be altered for adaptation to other languages.

$\boxed{2\varepsilon}$ \tabularnewline[*len*] . . . . (4.8.1) – 96

Terminates a row in the `tabular` or `array` environments, adding vertical spacing *len* if it is specified. This is equivalent to \\[*len*] except that there is no ambiguity as to whether it is terminating a row in the table or a line of text within a column entry. If something like \raggedright is given in the last column, then this command *must* be used in place of \\. *(1994/12/01)*

\tag{*mark*} [m][a] . . . . (E.3.6) – 420

Within one of the $\mathcal{A}_{\mathcal{M}}\mathcal{S}$-LATEX alignment environments, prints *mark* in place of the equation number, in parentheses. The *-form prints it without parentheses.

\tan [m] . . . . (5.3.8) – 126

Command to produce the function name 'tan' in formulas.

\tanh [m] . . . . (5.3.8) – 126

Command to produce the function name 'tanh' in formulas.

\tau [m] produces τ . . . . (5.3.1) – 123

\tbinom{*over*}{*under*} [m][a] . . . . (E.3.3) – 413

With the `amsmath` package, produces a binomial as \binom does, but in \textstyle size.

\telephone{*number*} . . . . (A.1), (A.2) – 298, 302

In the `letter` document class, enters the sender's telephone number. In the standard LATEX `letter.sty`, *number* is only output if \address has not been called. It is intended to be used in company letter styles such as `mpletter`.

\TeX produces TEX . . . . (2.1) – 17

\text{*short_text*} [m][a] . . . . (E.3.1) – 407

When one of the packages `amsmath` or `amstext` has been loaded, this command prints *short_text* as normal text within a math formula. If used in subscripts or superscripts, automatic sizing takes place.

$\boxed{2\varepsilon}$ \text*sym_name* . . . . (8.7) – 238

An alternative means to produce certain special symbols that otherwise are only available in math mode or through ligature combinations:
\textbullet (•); \textemdash (—); \textendash (–); \textexclamdown (¡); \textperiodcentered (·); \textquestiondown (¿); \textquotedblleft ("); \textquotedblright ("); \textquoteleft ('); \textquoteright ('); \textvisiblespace (␣) *(1994/12/01)* \textasciicircum (^); \textasciitilde (˜); \textbackslash (\); \textbar (|); \textgreater (>); \textless (<) *(1995/12/01)*

2ε `\textbf{`*text*`}`   . . . . (4.1.4) – 61

> This command sets its argument in a font in the current family
> and shape, but with the **bold** series attribute. It is equivalent to
> `{\bfseries` *text*`}`.

2ε `\textcircled{`*char*`}`   . . . . (8.7) – 238

> Produces the specified character in a circle: `\textcircled{s}` = Ⓢ.
> *(1994/12/01)*

2ε `\textcolor` *col_spec*`{`*text*`}`   . . . . (6.3) – 177

> A command made available with the `color` package. The *text* is set
> in the specified color. The *col_spec* is the same as for `\color`.

2ε `\textcompwordmark`   . . . . (8.7) – 238

> Prints an invisible character that may be used to break ligatures:
> `f\textcompwordmark i` = fi
> *(1994/12/01)*

`\textfloatsep` . . . . (6.4.3) – 182

> The vertical spacing between floats at the top of the page and the
> following text or between text and floats at the bottom of the page.
> A new value is set with the `\setlength` command.
>
> > `\setlength{\textfloatsep}{20pt plus 2pt minus 4pt}`

`\textfraction` . . . . (6.4.3) – 182

> The minimum fraction of a page containing text and floats that must
> be filled with text. A new value is set with
>
> > `\renewcommand{\textfraction}{`*decimal_frac*`}`

`\textheight` . . . . (3.2.4) – 35

> The total height reserved for the text on each page, excluding head
> and footlines. A new value may be assigned with the `\setlength`
> command:
>
> > `\setlength{\textheight}{45\baselineskip}`

2ε `\textit{`*text*`}`   . . . . (4.1.4) – 61

> This command sets its argument in a font in the current family
> and series, but with the *italic* shape attribute. It is equivalent to
> `{\itshape` *text*`}`.

2ε `\textmd{`*text*`}`   . . . . (4.1.4) – 61

> This command sets its argument in a font in the current family and
> shape, **but with the** medium **series attribute.** It is equivalent to
> `{\mdseries` *text*`}`.

[2ε] \textnormal{*text*} . . . . (4.1.3) – 61

> This command sets its argument in the font with the default family, series, and shape attributes. It is equivalent to {\normalfont *text*}.

[2ε] \textquotedbl . . . . (F.4.5) – 452

> When T1 encoding is active, prints the symbol ". *(1994/12/01)*

[2ε] \textregistered produces ® *(1995/12/01)* . . . . . . (8.7) – 238

[2ε] \textrm{*text*} . . . . (4.1.4) – 61

> Sets *text* in a font in the current series and shape, but with the Roman family attribute. It is equivalent to {\rmfamily *text*}.

[2ε] \textsc{*text*} . . . . (4.1.4) – 61

> Sets *text* in a font in the current family and series, but with the CAPS AND SMALL CAPS shape attribute. It is equivalent to {\scshape *text*}.

[2ε] \textsf{*text*} . . . . (4.1.4) – 61

> Sets *text* in a font in the current series and shape, but with the sans serif family attribute. It is equivalent to {\sffamily *text*}.

[2ε] \textsl{*text*} . . . . (4.1.4) – 61

> Sets *text* in a font in the current family and series, but with the *slanted* shape attribute. It is equivalent to {\slshape *text*}.

\textstyle [m] . . . . (5.5.2) – 144

> Switches to font size \textstyle inside a math formula.

[2ε] ^{*char*} . . . . (8.7) – 238

> Produces a superscript in the current text, rather than math, font: ^{12} = ^{12}. *(1995/06/01)*

[2ε] \texttrademark produces ™ *(1995/12/01)* . . . . (8.7) – 238

[2ε] \texttt{*text*} . . . . (4.1.4) – 61

> Sets *text* in a font in the current series and shape, but with the typewriter family attribute. It is equivalent to {\ttfamily *text*}.

[2ε] \textup{*text*} . . . . (4.1.4) – 61

> Sets *text* in a font in the current family and series, *but with the* upright *shape attribute*. It is equivalent to {\upshape *text*}.

\textwidth . . . . (3.2.4) – 35

> The total width reserved for the text on a page. For two-column formatting, this is the width of both columns plus the gap between them. A new value may be assigned with the \setlength command.

\tfrac{*numerator*}{*denominator*} [m][a] . . . . (E.3.3) – 413

> With the amsmath package, produces a fraction as \frac does, but in \textstyle size.

2ε \TH . . . . (F.4.5) – 452

> When T1 encoding is active, prints the character Þ.

2ε \th . . . . (F.4.5) – 452

> When T1 encoding is active, prints the character þ.

\thanks{*footnote_text*} . . . . (3.3.1) – 38

> Produces a footnote to an author's name on the title page when \maketitle is called.

\the*counter* . . . . (7.1.4) – 191

> Internal commands for formatting and printing counter values, making possible use of other counters. For example, \thesubsection might be defined to be \thesection.\roman{subsection}. A new definition may be made with \renewcommand{\the*counter*}{*def*}.

\Theta [m] produces Θ . . . . (5.3.1) – 123
\theta [m] produces θ . . . . (5.3.1) – 123
\thicklines . . . . (6.1.4) – 164

> In the picture environment, this command sets all the sloping lines and arrows, circles, and ovals to be drawn with thicker than normal lines.

\thickspace [m][a] . . . . (E.3.5) – 418

> With the amsmath package, this is an alias for \;, a thick space in a math formula.

\thinlines . . . . (6.1.4) – 164

> In the picture environment, resets the line thickness for sloping lines and arrows, circles, and ovals back to the standard value after \thicklines has been given.

\thinspace [m][a] . . . . (E.3.5) – 418

> With the amsmath package, this is an alias for \,, a thin space in a math formula.

\thispagestyle{*style*} . . . . (3.2) – 31

> Changes the page style for the current page only. Possible values for *style* are: plain, empty, headings, and myheadings.

\tilde{*x*} [m] . . . . (5.3.9) – 127

> Produces a tilde (squiggle) over the math variable x: \tilde{a} = $\tilde{a}$.

\Tilde{*x*} [m][a] . . . . (E.3.2) – 412
> With the amsmath package, can be used like \tilde, but with multiple
> *AMS*-LaTeX math accents the positioning will be correct.

\tiny . . . . (4.1.2) – 58
> Switches to the smallest font size available \tiny, smaller than
> \scriptsize.

\times [m] produces × . . . . (5.3.3) – 124
\title{*text*} . . . . (3.3.1) – 37
> Enters the *text* for the title page that is produced by \maketitle.

\to [m] produces → . . . . (5.3.5) – 125
\today . . . . (2.5.9), (C.4.2) – 22, 352
> Prints the current date in the American fashion. This form may be
> changed to British or to that of other languages by redefining the
> command with the help of the internal TeX commands \day, \month,
> and \year.

\top [m] produces ⊤ . . . . (5.3.6) – 125

⟦2ε⟧ \topfigrule . . . . (6.4.3) – 183
> A command that is executed after a float at the top of a page. It is
> normally defined to do nothing, but may be redefined to add a rule
> between the float and the main text. It must not add any net vertical
> spacing.
>
>   \renewcommand{\topfigrule}{\vspace*{-.4pt}
>   \rule{\columnwidth}{.4pt}}

\topfraction . . . . (6.4.3) – 182
> The maximum fraction of a page that may be occupied at the top by
> floats at the top of the page. A new value is assigned with
>
>   \renewcommand{\topfraction}{*decimal_frac*}

\topmargin . . . . (3.2.4) – 35
> The size of the margin from the top of the page to the page head. A
> new value may be assigned by the \setlength command:
>
>   \setlength{topmargin}{0.5in}

topnumber . . . . (6.4.3) – 181
> The maximum number of floats that may appear at the top of a page.
> A new value is assigned with:
>
>   \setcounter{topnumber}{*num*}

\topsep . . . . (4.4.2), (5.5.4) – 74, 147
> The extra vertical spacing, in addition to \parskip, inserted at the
> beginning and end of a listing environment. When document class
> option fleqn has been chosen, it is also inserted at the beginning
> and end of displayed math formulas. A new value may be assigned
> with the \setlength command:
>
>   \setlength{\topsep}{4pt plus2pt minus2pt}

`\topskip` . . . . (3.2.4) – 35

The vertical distance from the top of the page body to the baseline of the first line of text. A new value may be assigned with the `\setlength` command:

`\setlength{\topskip}{12pt}`

`2ε` `\totalheight` . . . . (4.7.1) – 85

A length parameter equal to the total natural height of a box (height plus depth); it may only be used in the *width* specification of `\makebox`, `\framebox`, or `\savebox`, or in the *height* specification of a `\parbox` or a `minipage` environment.

`\framebox[6\totalheight]{text}`

`totalnumber` . . . . (6.4.3) – 182

The total number of floats that may appear on a page regardless of their positions. A new value is assigned with:

`\setcounter{totalnumber}{num}`

`\triangle` [m] produces △ . . . . (5.3.6) – 125
`\triangleleft` [m] produces ◁ . . . . (5.3.3) – 124
`\triangleright` [m] produces ▷ . . . . (5.3.3) – 124

`2.09` `\tt` . . . . (4.1.5) – 62

Switches to the `typewriter, upright, medium` typeface.

`2ε` `\ttdefault` . . . . (C.5.1) – 353

This command defines the family attribute that is selected with the `\ttfamily` command. It may be redefined with `\renewcommand`:

`\renewcommand{\ttdefault}{pcr}`

`2ε` `\ttfamily` . . . . (4.1.3), 8.5.2 – 60, 236

This declaration switches to a font in the current series and shape, but with the `typewriter` family attribute.

`\twocolumn[text]` . . . . (3.2.5) – 36

Begins a new page and switches to two-column page format. The optional *text* is set in one column extending over the two columns.

`\typein[\cmd]{message}` . . . . (8.1.3) – 221

Prints the *message* to the monitor screen and stops the program, waiting for a reply from the user. The text of the response is assigned to the LaTeX command named `\@typein`, or to `\cmd` if the optional argument has been given. After the return key is pressed, the processing continues. The typed-in text is inserted in place of `\@typein` if the optional argument was not given, otherwise it may be inserted as one pleases with the `\cmd` command.

`\typeout{`*message*`}` . . . . (8.1.3) – 220

> Prints the *message* to the monitor screen and continues the processing. The *message* is also written to the `.log` file.

`\u{`*x*`}` . . . . (2.5.7) – 21

> Produces a breve accent: `\u{o}` = ŏ.

`\unboldmath` . . . . (5.4.9) – 141

> Countermands the `\boldmath` command. It must be given outside of the math mode. Afterwards, formulas are set in standard '*math italics*' once more.

`\underbrace{`*sub_form*`}` [m] . . . . (5.4.4) – 134

> Produces a horizontal curly brace beneath the math formula *sub_form*. Any following subscript will be placed centered below the horizontal brace.
>
> $\underbrace{a+b}$: $\underbrace{a + b}$
>
> $\underbrace{x+y+z}_{\xi\eta\zeta}$: $\underbrace{x + y + z}_{\xi\eta\zeta}$

`\underleftarrow{`*expr*`}` [m][a] . . . . (E.3.2) – 411

> With the `amsmath` package, places a long leftwards pointing arrow beneath the mathematical expression *expr*.

`\underleftrightarrow{`*expr*`}` [m][a] . . . . (E.3.2) – 411

> With the `amsmath` package, places a long double arrow beneath the mathematical expression *expr*.

`\underline{`*text*`}` . . . . (5.4.4) – 134

> Underlines the *text* in both math and normal text modes: `\underline{Text}` = T̲e̲x̲t̲.

`\underrightarrow{`*expr*`}` [m][a] . . . . (E.3.2) – 411

> With the `amsmath` package, places a long rightwards pointing arrow beneath the mathematical expression *expr*.

`\underset{`*char*`}{\`*symbol*`}` [m][a] . . . . (E.3.2) – 410

> With the `amsmath` package, places *char* below the math symbol `\`*symbol* in subscript size.

`\unitlength` . . . . (6.1.1) – 152

> Defines the unit of length for the following `picture` environments. A value is assigned with the `\setlength` command:
>
> `\setlength{\unitlength}{1.2cm}`

This command defines the shape attribute that is selected with the \upshape command. It may be redefined with \renewcommand:

```
\renewcommand{\updefault}{n}
```

This declaration switches to a font in the current family and series, *but with the* upright *shape attribute.*

With the amsmath package, used in the index to a \sqrt command to shift it slightly upwards. The *shift* is a number specifying how many units to move it. Example:

```
\sqrt[\leftroot{-1}\uproot{3}\beta]{k}
```

Inserts into the text the contents of the box that was saved with the \sbox or \savebox command under the name *boxname*, which has been previously created with the \newsavebox command.

Command in the list environment that specifies which counter is to be employed in the standard labels with the \item commands. This counter is incremented by one with each \item call.

Activates the font with the given set of attributes in the current size. Is equivalent to selecting the given font attributes and then calling \selectfont.

Loads one or more package files containing additional LaTeX or TeX definitions. The files have the extension .sty. More than one package may be specified in *packages*, the names being separated by commas. Any *options* listed will be applied to all packages. Furthermore, any options listed in the \documentclass command will also be applied to the package files.

The optional *version* is a date, given in the form yyyy/mm/dd, as for example 1994/08/01. If the date of the package file is earlier than this, a warning message is printed.

Example:

```
\usepackage{bezier,ifthen}[1994/06/01]
```

\v{*x*} . . . . (2.5.7) – 21

Produces háček accent: \v{o} = ǒ.

\value{*counter*} . . . . (7.1.3) – 191

The current value of the number stored in *counter* for use with commands that require a number. It does *not* output this number. For example, \setcounter{*counter1*}{\value{*counter2*}} sets *counter1* to the same value as that of *counter2*.

\varepsilon [m] produces ε . . . . (5.3.1) – 123
\varinjlim [m][a] produces '$\varinjlim$' in formulas . . . . . (E.3.5) – 416
\varliminf [m][a] produces '$\varliminf$' in formulas . . . . . (E.3.5) – 416
\varlimsup [m][a] produces '$\varlimsup$' in formulas . . . . . (E.3.5) – 416
\varphi [m] produces φ . . . . (5.3.1) – 123
\varpi [m] produces ϖ . . . . (5.3.1) – 123
\varprojlim [m][a] produces '$\varprojlim$' in formulas . . . . . (E.3.5) – 416
\varrho [m] produces ϱ . . . . (5.3.1) – 123
\varsigma [m] produces ς . . . . (5.3.1) – 123
\vartheta [m] produces ϑ . . . . (5.3.1) – 123
\vdash [m] produces $\vdash$ . . . . (5.3.6) – 125

\vdots [m] produces $\vdots$ . . . . (5.2.6) – 121
\vec{*x*} [m] . . . . (5.3.9) – 127

A vector symbol over the variable x: \vec{a} = $\vec{a}$.

\Vec{*x*} [m][a] . . . . (E.3.2) – 412

With the amsmath package, can be used like \vec, but with multiple $\mathcal{A}_{\mathcal{M}}\mathcal{S}$-LaTeX math accents the positioning will be correct.

\vector($\Delta x, \Delta y$){*length*} . . . . (6.1.4) – 159

A picture element command within a picture environment for drawing horizontal and vertical arrows of any length as well as slanted arrows at a limited number of angles. For horizontal and vertical arrows, the *length* argument is the actual length in units of \unitlength. For slanted arrows, *length* is the length of the projection on to the x-axis (horizontal displacement). The slope is determined by the ($\Delta x, \Delta y$) arguments, which take on integral values such that $-4 \leq \Delta x \leq 4$ and $-4 \leq \Delta y \leq 4$. This command is the argument of a \put or \multiput command.

\vee [m] produces ∨ . . . . (5.3.3) – 124

\verb|*source_text*| . . . . (4.9) – 108

Everything that comes between the |...| symbols is output in the typewriter font exactly as is with no interpretation of special symbols or commands. Any symbol other than * may be used as the switch character, illustrated here as |, as long as it does not appear in *source_text*.

\verb*|*source_text*| . . . . (4.9) – 108

The same as \verb except that blanks are made visible with the symbol ␣.

\vfill . . . . (3.5.3) – 50

A vertical rubber spacing with a natural length of zero that can be stretched to any value. Used to fill up parts of a page with blank spacing. This command is an abbreviation for \vspace{\fill}.

$\boxed{2\varepsilon}$ \visible . . . . (8.10.2) – 251

In slides class, a declaration that countermands a previous \invisible command, making text printed again. It remains in effect until the end of the environment, or end of the curly braces, in which it was issued, or until \invisible is given. It is used for making overlays.

\vline . . . . (4.8.1) – 95

Prints a vertical rule within the column entry of a table in the tabular environment.

\voffset 555, 556

Vertical offset of the output page from the printer border set by the printer driver. This printer border is normally 1 inch from the top edge of the paper. The standard value of \voffset is 0 pt so that the top reference margin of the page is identical with the printer margin. A new value is assigned with the \setlength command:

\setlength{\voffset}{-1in}

\vspace{*height*} . . . . (3.5.3) – 49

Produces vertical spacing of length *height*. It is ignored if it occurs at the beginning or end of a page.

\vspace*{*height*} . . . . (3.5.3) – 49

Produces vertical spacing of length *height* even at the beginning or end of a page.

\wedge [m] produces ∧ . . . . (5.3.3) – 124

\whiledo{*test*}{*do_text*} . . . . (7.3.5) – 201

A conditional command available when the standard package ifthen has been loaded. The *do_text* is inserted repeatedly as long as the logical statement *test* evaluates to ⟨*true*⟩. The logical statement may be relational (two numbers with one of < = > between them), an even–odd test (\isodd{*number*}), a comparison of two texts (\equal{*text1*}{*text2*}), a comparison of two lengths (\lengthtest{*length1* op *length2*}, *op* is one of < = >), or a test of a boolean switch (\boolean{*switch*}). Switches are created with \newboolean{*switch*} and set with \setboolean{*switch*}{*value*}, where *value* is true or false. Logical statements may be combined with logical operators \and, \or, and \not, and grouped with \(and \).

The tests \isodd, \lengthtest, and \boolean are new to LaTeX 2ε.

\widehat{*xyz*} [m] . . . . (5.3.9) – 127

Produces a wide \hat symbol over several characters: \widehat{xyz} = $\widehat{xyz}$.

\widetilde{*xyz*} [m] . . . . (5.3.9) – 127

Produces a wide \tilde symbol over several characters: \widetilde{xyz} = $\widetilde{xyz}$.

$\boxed{\text{2ε}}$ \width . . . . (4.7.1) – 85

A length parameter equal to the natural width of a box; it may only be used in the *width* specification of \makebox, \framebox, or \savebox, or in the *height* specification of a \parbox or a minipage environment.

    \framebox[2\width]{text}

\wp [m] produces $\wp$ . . . . (5.3.6) – 125
\wr [m] produces $\wr$ . . . . (5.3.3) – 124

\Xi [m] produces Ξ . . . . (5.3.1) – 123
\xi [m] produces ξ . . . . (5.3.1) – 123
\xleftarrow[*below*]{*above*} [m][a] . . . . (E.3.2) – 411

With the amsmath package, draws leftward pointing arrow with *above* printed over it in superscript size, and optionally *below* beneath it in subscript size.

\xrightarrow[*below*]{*above*} [m][a] . . . . (E.3.2) – 411

With the amsmath package, draws rightward pointing arrow with *above* printed over it in superscript size, and optionally *below* beneath it in subscript size.

\zeta [m] produces ζ . . . . (5.3.1) – 123

G.2   Summary tables and figures

Table G.1   Font attribute commands   (4.1.3) – p. 60.

\rmfamily	\textrm{*text*}	Roman
\sffamily	\textsf{*text*}	sans serif
\ttfamily	\texttt{*text*}	typewriter
\upshape	\textup{*text*}	upright
\itshape	\textit{*text*}	*italic*
\slshape	\textsl{*text*}	*slanted*
\scshape	\textsc{*text*}	SMALL CAPS
\mdseries	\textmd{*text*}	medium
\bfseries	\textbf{*text*}	**bold face**

Table G.2   Math alphabet commands   (5.4.2) – p. 131.

\mathrm{*text*}	Roman
\mathsf{*text*}	sansserif
\mathnormal{*text*}	*normal*
\mathtt{*text*}	typewriter
\mathit{*text*}	*italic*
\mathbf{*text*}	**boldface**
\mathcal{*text*}	$\mathcal{CAL}$

Table G.3   Font sizes   (4.1.2) – p. 58.

\tiny	smallest
\scriptsize	very small
\footnotesize	smaller
\small	small
\normalsize	normal
\large	large
\Large	larger
\LARGE	even larger
\huge	still larger
\Huge	largest

Table G.4   LaTeX 2.09 font declarations   (4.1.5) – p. 62.

\rm	Roman	\it	*Italic*	\sc	SMALL CAPS
\bf	**Bold Face**	\sl	*Slanted*	\sf	Sans Serif
\tt	Typewriter	\mit	$\Gamma\Pi\Phi$	\cal	$\mathcal{CAL}$

Table G.5 Dimensions (2.4.1) – p. 19.

mm	millimeter	bp	big point (1 in = 72 bp)
cm	centimeter	dd	(1157 dd = 1238 pt)
in	inch (1 in = 2.54 cm)	cc	cicero (1 cc = 12 dd)
pt	point (1 in = 72.27 pt)	sp	(1 pt = 65536 sp)
pc	pica (1 pc = 12 pt)		

em	The current width of a capital M
ex	The current height of the letter x

Table G.6 Accents (2.5.7) – p. 21.

ò =\'{o}	ó=\'{o}	ô=\^{o}	ö=\"{o}	õ=\~{o}
ō =\={o}	ȯ=\.{o}	ŏ=\u{o}	ǒ=\v{o}	ő=\H{o}
o͡o=\t{oo}	o̧=\c{o}	o̠=\d{o}	o̲=\b{o}	o̊=\r{o}

Table G.7 Special letters from other languages (2.5.6) – p. 21.

œ={\oe}	Œ={\OE}	æ={\ae}	Æ={\AE}	å={\aa}	Å ={\AA}	¡=!'
ø ={\o}	Ø ={\O}	ł ={\l}	Ł ={\L}	ß={\ss}	SS={\SS}	¿=?'

(The commands \r and \SS are only valid for LaTeX 2_ε.)

Table G.8 Special symbols (2.5.5) – 21.

† \dag § \S © \copyright ‡ \ddag ¶ \P £ \pounds

Table G.9 Command symbols (2.5.4) – 21.

$ \$ % \% { \{ _ _ & \& # \# } \}

Table G.10 Greek letters (5.3.1) – p. 123.

Lower case letters

α	\alpha	θ	\theta	o	o	τ	\tau
β	\beta	ϑ	\vartheta	π	\pi	υ	\upsilon
γ	\gamma	ι	\iota	ϖ	\varpi	ϕ	\phi
δ	\delta	κ	\kappa	ρ	\rho	φ	\varphi
ϵ	\epsilon	λ	\lambda	ϱ	\varrho	χ	\chi
ε	\varepsilon	μ	\mu	σ	\sigma	ψ	\psi
ζ	\zeta	ν	\nu	ς	\varsigma	ω	\omega
η	\eta	ξ	\xi				

Upper case letters

Γ	\Gamma	Λ	\Lambda	Σ	\Sigma	Ψ	\Psi
Δ	\Delta	Ξ	\Xi	Υ	\Upsilon	Ω	\Omega
Θ	\Theta	Π	\Pi	Φ	\Phi		

Table G.11   Binary operation symbols   (5.3.3) – p. 124.

±	\pm	∩	\cap	∘	\circ	○	\bigcirc
∓	\mp	∪	\cup	•	\bullet	□	\Box
×	\times	⊎	\uplus	◇	\diamond	◇	\Diamond
÷	\div	⊓	\sqcap	◁	\lhd	△	\bigtriangleup
·	\cdot	⊔	\sqcup	▷	\rhd	▽	\bigtriangledown
∗	\ast	∨	\vee	⊴	\unlhd	◁	\triangleleft
⋆	\star	∧	\wedge	⊵	\unrhd	▷	\triangleright
†	\dagger	⊕	\oplus	⊘	\oslash	\	\setminus
‡	\ddagger	⊖	\ominus	⊙	\odot	≀	\wr
⨿	\amalg	⊗	\otimes				

Table G.12   Relational symbols   (5.3.4) – p. 124.

≤	\le	\leq	≥	\ge	\geq	≠	\neq	~	\sim
≪	\ll		≫	\gg		≐	\doteq	≃	\simeq
⊂	\subset		⊃	\supset		≈	\approx	≍	\asymp
⊆	\subseteq		⊇	\supseteq		≅	\cong	⌣	\smile
⊏	\sqsubset		⊐	\sqsupset		≡	\equiv	⌢	\frown
⊑	\sqsubseteq		⊒	\sqsupseteq		∝	\propto	⋈	\bowtie
∈	\in		∋	\ni		≺	\prec	≻	\succ
⊢	\vdash		⊣	\dashv		⪯	\preceq	⪰	\succeq
⊨	\models		⊥	\perp		∥	\parallel |	|	\mid	

Table G.13   Negated relational symbols   (5.3.4) – p. 125.

≮	\not<	≯	\not>	≠	\not=
≰	\not\le	≱	\not\ge	≢	\not\equiv
⊀	\not\prec	⊁	\not\succ	≁	\not\sim
⋠	\not\preceq	⋡	\not\succeq	≄	\not\simeq
⊄	\not\subset	⊅	\not\supset	≉	\not\approx
⊈	\not\subseteq	⊉	\not\supseteq	≇	\not\cong
⋢	\not\sqsubseteq	⋣	\not\sqsupseteq	≭	\not\asymp
∉	\not\in	∉	\notin		

Table G.14   Brackets   (5.4.1) – p. 130.

(	(	)	)	⌊	\lfloor	⌋	\rfloor
[	[	]	]	⌈	\lceil	⌉	\rceil
{	\{	}	\}	⟨	\langle	⟩	\rangle
|	|	‖	|	↑	\uparrow	⇑	\Uparrow
/	/	\	\backslash	↓	\downarrow	⇓	\Downarrow
				↕	\updownarrow	⇕	\Updownarrow

Table G.15 Arrows (5.3.5) – p. 125.

← `\leftarrow` `\gets`	⟵ `\longleftarrow`	↑ `\uparrow`
⇐ `\Leftarrow`	⟸ `\Longleftarrow`	⇑ `\Uparrow`
→ `\rightarrow` `\to`	⟶ `\longrightarrow`	↓ `\downarrow`
⇒ `\Rightarrow`	⟹ `\Longrightarrow`	⇓ `\Downarrow`
↔ `\leftrightarrow`	⟷ `\longleftrightarrow`	↕ `\updownarrow`
⇔ `\Leftrightarrow`	⟺ `\Longleftrightarrow`	⇕ `\Updownarrow`
↦ `\mapsto`	⟼ `\longmapsto`	↗ `\nearrow`
↩ `\hookleftarrow`	↪ `\hookrightarrow`	↘ `\searrow`
↼ `\leftharpoonup`	⇀ `\rightharpoonup`	↙ `\swarrow`
↽ `\leftharpoondown`	⇁ `\rightharpoondown`	↖ `\nwarrow`
⇌ `\rightleftharpoons`	⤳ <u>`\leadsto`</u>	

Table G.16 Miscellaneous symbols (5.3.6) – p. 125.

ℵ `\aleph`	′ `\prime`	∀ `\forall`	□ <u>`\Box`</u>
ℏ `\hbar`	∅ `\emptyset`	∃ `\exists`	◇ <u>`\Diamond`</u>
ı `\imath`	∇ `\nabla`	¬ `\neg`	△ `\triangle`
ȷ `\jmath`	√ `\surd`	♭ `\flat`	♣ `\clubsuit`
ℓ `\ell`	∂ `\partial`	♮ `\natural`	♦ `\diamondsuit`
℘ `\wp`	⊤ `\top`	♯ `\sharp`	♥ `\heartsuit`
ℜ `\Re`	⊥ `\bot`	‖ `|`	♠ `\spadesuit`
ℑ `\Im`	⊢ `\vdash`	∠ `\angle`	⋈ <u>`\Join`</u>
℧ <u>`\mho`</u>	⊣ `\dashv`	\ `\backslash`	∞ `\infty`

The underlined commands in Tables G.11, G.12, G.15, and G.16 can only be used in LATEX 2ε if one of the packages latexsym or amsfonts has been loaded.

Table G.17 Mathematical symbols in two sizes (5.3.7) – p. 126.

Σ ∑	`\sum`	∩ ⋂	`\bigcap`	⊙ ⨀	`\bigodot`
∫ ∫	`\int`	∪ ⋃	`\bigcup`	⊗ ⨂	`\bigotimes`
∮ ∮	`\oint`	⊔ ⨆	`\bigsqcup`	⊕ ⨁	`\bigoplus`
∏ ∏	`\prod`	∨ ⋁	`\bigvee`	⊎ ⨄	`\biguplus`
∐ ∐	`\coprod`	∧ ⋀	`\bigwedge`		

Table G.18 Function names (5.3.8) – p. 126.

`\arccos`	`\cosh`	`\det`	`\inf`	`\limsup`	`\Pr`	`\tan`
`\arcsin`	`\cot`	`\dim`	`\ker`	`\ln`	`\sec`	`\tanh`
`\arctan`	`\coth`	`\exp`	`\lg`	`\log`	`\sin`	
`\arg`	`\csc`	`\gcd`	`\lim`	`\max`	`\sinh`	
`\cos`	`\deg`	`\hom`	`\liminf`	`\min`	`\sup`	

Table G.19   Math accents   (5.3.9) – p. 127.

$\hat{a}$ \hat{a}	$\breve{a}$ \breve{a}	$\grave{a}$ \grave{a}	$\bar{a}$ \bar{a}
$\check{a}$ \check{a}	$\acute{a}$ \acute{a}	$\tilde{a}$ \tilde{a}	$\vec{a}$ \vec{a}
$\dot{a}$ \dot{a}	$\ddot{a}$ \ddot{a}	$\mathring{a}$ \mathring{a}	

The following symbols are made available with the package amssymb.

Table G.20 $\mathcal{AMS}$ arrows

⇢	\dashrightarrow	⇠	\dashleftarrow
⇇	\leftleftarrows	⇆	\leftrightarrows
⇚	\Lleftarrow	↞	\twoheadleftarrow
↢	\leftarrowtail	↬	\looparrowleft
⇋	\leftrightharpoons	↶	\curvearrowleft
↺	\circlearrowleft	↰	\Lsh
⇈	\upuparrows	↿	\upharpoonleft
↓	\downharpoonleft	⊸	\multimap
↭	\leftrightsquigarrow	⇉	\rightrightarrows
⇄	\rightleftarrows	⇉	\rightrightarrows
⇄	\rightleftarrows	↠	\twoheadrightarrow
↣	\rightarrowtail	↪	\looparrowright
⇌	\rightleftharpoons	↷	\curvearrowright
↻	\circlearrowright	↱	\Rsh
⇊	\downdownarrows	↾	\upharpoonright
↓	\downharpoonright	⇝	\rightsquigarrow

Negated arrows

↚	\nleftarrow	↛	\nrightarrow
⇍	\nLeftarrow	⇏	\nRightarrow
↮	\nleftrightarrow	⇎	\nLeftrightarrow

Table G.21 $\mathcal{AMS}$ binary operation symbols

∔	\dotplus	∖	\smallsetminus
⋒	\Cap	⋓	\Cup
⊼	\barwedge	⊻	\veebar
⩞	\doublebarwedge	⊟	\boxminus
⊠	\boxtimes	⊡	\boxdot
⊞	\boxplus	⋇	\divideontimes
⋉	\ltimes	⋊	\rtimes
⋋	\leftthreetimes	⋌	\rightthreetimes
⋏	\curlywedge	⋎	\curlyvee
⊖	\circleddash	⊛	\circledast
⊚	\circledcirc	·	\centerdot
⊺	\intercal		

Table G.22 $\mathcal{A}_{\mathcal{M}}\mathcal{S}$ Greek and Hebrew letters

F	\digamma	$\varkappa$	\varkappa
ℶ	\beth	ℸ	\daleth ℷ \gimel

Table G.23 $\mathcal{A}_{\mathcal{M}}\mathcal{S}$ delimiters

⌜	\ulcorner	⌝	\urcorner	⌞	\llcorner	⌟	\lrcorner

Table G.24 $\mathcal{A}_{\mathcal{M}}\mathcal{S}$ relational symbols

≦	\leqq	⩽	\leqslant
⋜	\eqslantless	≲	\lesssim
⪅	\lessapprox	≊	\approxeq
⋖	\lessdot	⋘	\lll
≶	\lessgtr	⪋	\lesseqgtr
⪚	\lesseqqgtr	≑	\doteqdot
≓	\risingdotseq	≒	\fallingdotseq
∽	\backsim	⋍	\backsimeq
⊆	\subseteqq	⋐	\Subset
⊏	\sqsubset	≼	\preccurlyeq
⋞	\curlyeqprec	≾	\precsim
⪷	\precapprox	◁	\vartriangleleft
⊴	\trianglelefteq	⊨	\vDash
⊪	\Vvdash	⌣	\smallsmile
⌢	\smallfrown	≏	\bumpeq
≎	\Bumpeq	≧	\geqq
⩾	\geqslant	⋝	\eqslantgtr
≳	\gtrsim	⪆	\gtrapprox
⋗	\gtrdot	⋙	\ggg
≷	\gtrless	⪌	\gtreqless
⪛	\gtreqqless	≖	\eqcirc
≗	\circeq	≜	\triangleq
∼	\thicksim	≈	\thickapprox
⊇	\supseteqq	⋑	\Supset
⊐	\sqsupset	≽	\succcurlyeq
⋟	\curlyeqsucc	≿	\succsim
⪸	\succapprox	▷	\vartriangleright
⊵	\trianglerighteq	⊩	\Vdash
∣	\shortmid	∥	\shortparallel
≬	\between	⋔	\pitchfork
∝	\varpropto	◀	\blacktriangleleft
∴	\therefore	϶	\backepsilon
▶	\blacktriangleright	∵	\because

Table G.25 $\mathcal{A}_{\mathcal{M}}\mathcal{S}$ negated relational symbols

≮	\nless	≰	\nleq
⪇	\nleqslant	≦̸	\nleqq
⪇	\lneq	≨	\lneqq
⪇	\lvertneqq	⪉	\lnsim
⪹	\lnapprox	⊀	\nprec
⋠	\npreceq	⋨	\precnsim
⪹	\precnapprox	≁	\nsim
∤	\nshortmid	∤	\nmid
⊬	\nvdash	⊭	\nvDash
⋪	\ntriangleleft	⋬	\ntrianglelefteq
⊈	\nsubseteq	⊊	\subsetneq
⊊	\varsubsetneq	⊊	\subsetneqq
⊊	\varsubsetneqq	≯	\ngtr
≱	\ngeq	≱	\ngeqslant
≱	\ngeqq	⪈	\gneq
≩	\gneqq	⪈	\gvertneqq
⪊	\gnsim	⪺	\gnapprox
⊁	\nsucc	⋡	\nsucceq
⋡	\nsucceq	⋩	\succnsim
⪺	\succnapprox	≇	\ncong
∦	\nshortparallel	∦	\nparallel
⊭	\nvDash	⊯	\nVDash
⋫	\ntriangleright	⋭	\ntrianglerighteq
⊉	\nsupseteq	⊉	\nsupseteqq
⊋	\supsetneq	⊋	\varsupsetneq
⊋	\supsetneqq	⊋	\varsupsetneqq

Table G.26 Miscellaneous $\mathcal{A}_{\mathcal{M}}\mathcal{S}$ symbols

ℏ	\hbar	ℏ	\hslash
△	\vartriangle	▽	\triangledown
□	\square	◇	\lozenge
Ⓢ	\circledS	∠	\angle
∡	\measuredangle	∄	\nexists
℧	\mho	⅁	\Finv
⅁	\Game	k	\Bbbk
‵	\backprime	∅	\varnothing
▲	\blacktriangle	▼	\blacktriangledown
■	\blacksquare	◆	\blacklozenge
★	\bigstar	∢	\sphericalangle
∁	\complement	∂	\eth
╱	\diagup	╲	\diagdown

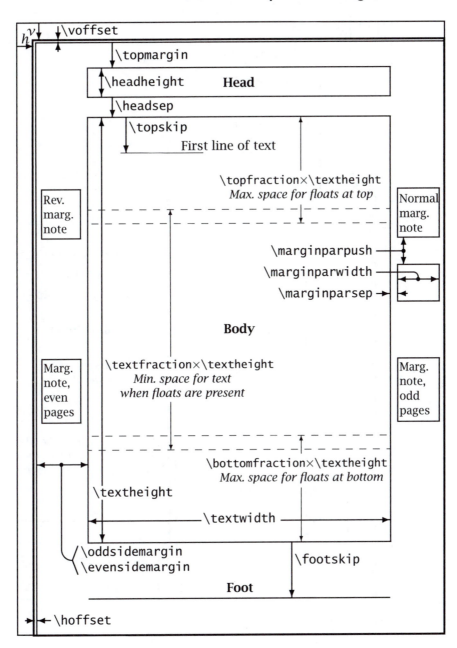

Figure G.1 Single column page format
(3.2.4), p. 35 – (4.10.6), p. 115 – (6.4.3), p. 182.

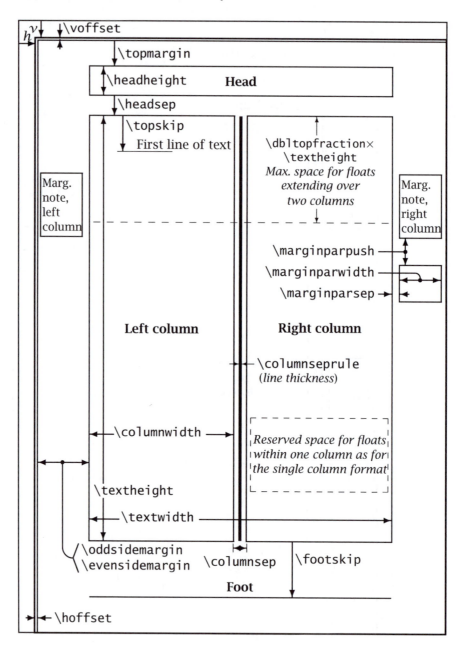

Figure G.2   Double column page format
(3.2.4), p. 35 – (3.1.1), p. 28 – (4.10.6), p. 115 – (6.4.3), p. 182.

Remarks on the page format figures

The reference margins in the LATEX processing are shifted from the *logical* margins by the amounts \hoffset and \voffset. These in turn are displaced from the *physical* margins by *h* and *v* in the DVI driver. The default values for \hoffset and \voffset are 0 pt, so that the reference margins are equal to the *logical* ones. The usual values for *h* and *v* are 1 inch. Thus the logical page margins on the left and at the top are shifted 1 inch from the physical edge of the paper. The user may alter this by changing the values of \hoffset and \voffset.

LATEX 2ε recognizes the parameters \paperwidth and \paperheight which contain the full dimensions of the paper, including the 1 inch margins. These are set by the paper size option to the class specification.

The parameter \footheight was specified but never used in LATEX 2.09; it has been dropped from LATEX 2ε.

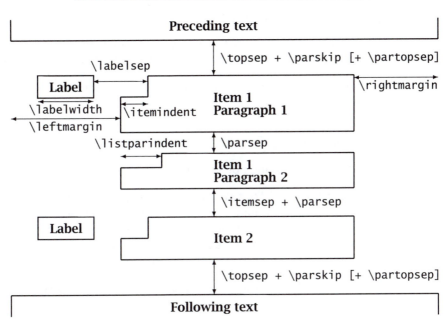

Figure G.3   Format of the list environment   (4.4.2) – p. 74.

Note 1: The default values for the three parameters \itemindent, \listparindent, \rightmargin are 0 pt.

Note 2: The default values in the trivlist environment are 0 pt for \itemindent, \leftmargin, \rightmargin, and \labelwidth; on the other hand, \parsep and \listparindent are assigned the respective values of \parskip and \parindent.

G.3   Invalid Plain TEX commands

As mentioned in several places in this book, LATEX is a set of macros, definitions, procedures, auxiliary files, etc., that have been built on top of the more basic Plain TEX format. This means that LATEX is an extension of Plain TEX, so that all its commands are also available under LATEX. At least, that is what one might expect.

LATEX users who are also familiar with the complexity of the TEX system might find it at times useful to resort to that set of more basic commands. This is not recommended for 'normal' users. We have illustrated a few of these commands within this book and they should be sufficient. For the advanced user it is necessary to point out that a number of Plain TEX commands will not function at all under LATEX, since they have been altered or undefined by the higher-level system. We list these invalid TEX commands according to their logical functionality.

G.3.1   TEX tabulator commands

The following TEX tabulator commands will not work under LATEX. They have been replaced by the tabbing environment and its commands:

```
\tabs       \tabset     \tabsdone      \cleartabs
\settabs    \tabalign   \+
```

G.3.2   Page formatting, footnotes, floats

The Plain TEX commands below have been deleted:

```
\advancepageno   \footstrut     \nopagenumbers  \pageno
\dosupereject    \headline      \normalbottom   \plainoutput
\endinsert       \makefootline  \pagebody       \topins
\folio           \makeheadline  \pagecontents   \topinsert
\footline        \midinsert     \pageinsert     \vfootnote
```

Instead one achieves the same results by means of \pagestyle, LATEX footnote commands, and the figure and table environments.

G.3.3   TEX font commands

The Plain TEX font commands

```
\fivei      \fiverm     \fivesy     \fivebf
\seveni     \sevenbf    \sevensy    \teni        \oldstyle
```

are not defined in LATEX. Similar results may be produced with the typeface and font size commands.

G.3.4   Math commands

All the T_EX math commands may be applied in L^AT_EX with the exceptions of

    `\eqalign`   `\eqalignno`   `\leqalingnno`

which are replaced by the `eqnarray` and `eqnarray*` environments.

G.3.5   Miscellaneous

The `\beginsection` command in Plain T_EX has been taken over by L^AT_EX's sectioning commands. The text is not terminated with `\end` or with `\bye` but rather with the command `\end{document}`. The `\centering` command in T_EX has been replaced by a L^AT_EX command with the same name. In T_EX, there is a command `\line` (for writing a line of text) which conflicts with L^AT_EX's command of the same name in the `picture` environment (to draw a line). Most applications of T_EX's `\line` are accomplished with `\center`, `\flushleft`, and `\flushright` in L^AT_EX.

The T_EX command `\magnification` has been fully removed. It is actually more reasonable and efficient to carry out the magnification as an option to the DVI driver program, provided this is available on your driver. This requires no change to the `.tex` or `.dvi` files, altering the printer output directly. Check your driver manual or computer center.

Bibliography

Abrahams P. W., with Hargreaves K. A. and Berry K. (1990). *TEX for the Impatient*. Reading MA: Addison-Wesley

Beccari, C. (1997). Typesetting mathematics for science and technology according to ISO 31/XI. *TUGboat*, **18**(1), 39–48

Botway L. and Biemesderfer C. (1985). *LATEX Command Summary*. Providence RI: TEX Users Group

Eijkhout V. (1992). *TEX by Topic, a TEXnician's Reference*. Harlow: Addison-Wesley

Flynn P. (1995). HTML & TEX: Making them sweat. *TUGboat*, **16**(2), 146–150

Haralambous Y. and Rahtz S. (1995). *LATEX*, hypertext and PDF, or the entry of TEX into the world of hypertext. *TUGboat*, **16**(2), 162–173

Goossens M. and Saarela J. (1995). From LATEX to HTML and back. *TUGboat*, **16**(2), 174–214

Goossens M., Mittelbach F. and Samarin A. (1994). *The LATEX Companion*. Reading MA: Addison-Wesley

Goossens M., Rahtz S. and Mittelbach F. (1997). *The LATEX Graphics Companion*. Reading MA: Addison-Wesley

Goossens M. and Rahtz S. (1999). *The LATEX Web Companion*. Reading MA: Addison-Wesley

Knuth D. E. (1984). *The TEXbook*, Computers and Typesetting, Vol. A. Reading MA: Addison-Wesley

Knuth D. E. (1986a). *TEX: The Program*, Computers and Typesetting, Vol. B. Reading MA: Addison-Wesley

Knuth D. E. (1986b). *The METAFONT book*, Computers and Typesetting, Vol. C. Reading MA: Addison-Wesley

Knuth D. E. (1986c). *METAFONT: The Program*, Computers and Typesetting, Vol. D. Reading MA: Addison-Wesley

Knuth D. E. (1986d). *Computer Modern Typefaces*, Computers and Typesetting, Vol. E. Reading MA: Addison-Wesley

Kopka H. (1994). *LATEX, Band 1, eine Einführung*. Bonn: Addison-Wesley

Kopka H. (1995). *LATEX, Band 2, Ergänzungen, mit einer Einführung in METAFONT*. Bonn: Addison-Wesley

Kopka H. (1997). *LATEX, Band 3, Erweiterungen*. Bonn: Addison-Wesley

Lamport L. (1985). *LATEX—A Document Preparation System*. Reading MA: Addison-Wesley

Lamport L. (1994). *LATEX—A Document Preparation System*, 2nd edn. for LATEX 2_ε. Reading MA: Addison-Wesley

Merz T. (1997a). *PostScript & Acrobat/PDF*. Berlin: Springer-Verlag

Merz T. (1997b). *Web Publishing with Acrobat/PDF*. Berlin: Springer-Verlag

Rokicki T. (1985). Packed (PK) font file format. *TUGboat*, **6**(3), 115–20

Reckdahl, K. (1996a). Using EPS graphics in LATEX 2_ε documents, part 1: The graphics and graphicx packages. *TUGboat*, **17**(1), 43–53

Reckdahl K. (1996b). Using EPS graphics in LATEX 2_ε documents, part 2: Floating figures, boxed figures, captions, and math in figures. *TUGboat*, **17**(3), 288–310

Samuel A. L. (1985). *First Grade TEX: A Beginner's TEX Manual*. Providence, RI: TEX Users Group

Schwarz N. (1990). *Introduction to TEX*. Reading MA: Addison-Wesley

Snow W. (1992). *TEX for the Beginner*. Reading MA: Addison-Wesley

Sojka P., Thành H. T. and Zlatuška J. The joy of TEX2PDF—Acrobatics with an alternative to DVI format. *TUGboat*, **17**(3), 244–251

Spivak M. (1986). *The Joy of TEX*. Providence RI: American Mathematical Society

Taylor P. Computer typesetting or electronic publishing? New trends in scientific publication. *TUGboat*, **17**(4), 367–381

Urban M. (1986). *An Introduction to LATEX*. Providence RI: TEX Users Group

Index

For purposes of alphabetization, the backslash character \ is ignored at the start of an entry. Otherwise, the ordering is by the ASCII sequence.

Bold page numbers indicate the place where the command or concept is introduced, explained, or defined. Slanted page numbers refer to the Command Summary, Appendix G.

The keyword index is set up with main and two sub-entries. If a keyword cannot be found as a main entry, one should try to find it as a sub-entry to some more general term. Such major topics are

$\mathcal{A}_{\mathcal{M}}$S-LʌTEX, bibliographic database, box, command, command (user-defined), cross-reference, environment, environment (user-defined), error messages, file, float, fonts, footnote, formula, hyphenation, LʌTEX, letter, line breaking, lists, page breaking, picture, programming, sectioning, spacing, style parameter, slides, symbols, tabbing, table, table examples, TEX, text.

! (MakeIndex), **230**, *459*
\!, **143**, 418, *460*
! ', **21**, *460*
", **20**, *460*
" (BɪʙTEX), 313, *460*
" (MakeIndex), **231**, *460*
" (German commands), 364
\" (¨ accent), **21**, 364, *460*
#, 15, **196**, 211, 283, 287, *460*
##, **212**, *460*
\#, **21**, 283, *460*
$, 15, **118**, 270, 283, 284, *460*
$$, 118
\$, 15, **21**, 23, 270, *460*
%, 15, 20, 44, **116**, *460*
\%, **21**, 23, *460*
&, 15, **95**, 137, 283, 284, 421, 423, *461*
\&, **21**, *461*
', 20
'', 20
\' (´ accent), **21**, *461*
\' (tabbing), **82**, *461*

(, 130, 153, 314, *461*
\(, **117**, 273, 283, *461*
), 130, 153, 314, *461*
\), **117**, 273, *461*
\+ (tabbing), **81**, 82, 278, *461*
\,, **46**, 121, 142, **143**, 418, *461*
-, **21**, *461*
\-, 286
\- (hyphenation), 28, **53**, 293, *461*
--, **21**, *461*
---, **21**, *461*
\- (tabbing), **81**, 82, 278, *461*
\. (˙ accent), **21**, *461*
/, **130**
\/, **46**, *462*
\:, 132, **143**, 418, *462*
\;, 132, **143**, 418, *462*
\< (tabbing), **81**, 273, 278, *462*
\= (¯ accent), **21**, *462*
\= (tabbing), **80**, 81, 82, 277, *462*
\> (tabbing), **80**, 81–3, 278, *462*
? ', **21**, *462*
@ (BɪʙTEX), 314, 316, *462*

563